HISTOLOGY
AND HUMAN MICROANATOMY

MICHAEL HANS ELIAS, Ph. D
Senior Author, Editions 1-4
1907 - 1985

ELIAS-PAULY'S

HISTOLOGY
AND HUMAN MICROANATOMY

Peter S. Amenta

*Professor and Chairman
Department of Anatomy
Hahnemann University
Philadelphia, Pennsylvania*

Fifth Edition

A Wiley Medical Publication
JOHN WILEY & SONS,
New York - Toronto

U.S.A. and Canada distributed exclusively by
John Wiley and Sons, Inc.
605 Third Avenue, New York, N.Y. 10016, U.S.A.

Library of Congress Cataloging in Publication Data

Elias, Hans, 1907-1985
Elias-Pauly's Histology and Human Microanatomy.

(Wiley Medical Publication)
Includes index.
1. Histology.
I. Pauly, John E. II. Amenta, Peter Sebastian, 1927- III. Title. IV. Series.
QM551.E4 1986   611'.018   86-5559
ISBN 0-471-83646-X

All rights reserved.
No part of this book may be reprinted or reproduced
in any way without the publisher's written permission

ISBN 88-299-0598-4

Printed in Italy

© 1987 by Piccin Nuova Libraria, S.p.A., Padova

Contents

PREFACE TO THE FIFTH EDITION	IX

INTRODUCTION	1
Units of Measurement	1
Fixation and Staining	7

Capter 1
THE CELL	11
Unit Membrane	15
Nucleus	15
Cytoplasm	20
Organelles	20
Cell Division	29
Chromatin	31
Autoradiography	34
DNA Replication	36
Mitosis	38
Karyotypes	39
Protein Synthesis	40
Cell Size and Shape	41
Connections between Cells	46
Activities of Cells	49
Transformation of Cells	50

Chapter 2
EPITHELIUM	53
Definition and Classification	53
Surface Specializations	60
Epithelial like Tissues	64
Metaplasia	64

Chapter 3
CONNECTIVE TISSUE	65
Intercellular Material	66
Cells of C.T.	72
Types of C.T.	78

Chapter 4
SKELETAL TISSUE	81
Cartilage and Bone as Building Materials	81
Formation and Kinds of Cartilage	86
Structure of Bone	90
Classification of Bone	93
Joints	100

Chapter 5
BLOOD	103
Erythrocytes	104
Leukocytes	104
Hemocytopoesis	107
Bone marrow structural components	107

Chapter 6
MUSCLE	117
Non-striated muscle tissue	118
Skeletal Muscle Tissue	123
Cardiac striated Muscle Tissue	128

Chapter 7
NERVOUS TISSUE	135
Nerve cell or Neuron	137

Types of Nerve Cells	140
Receptors	142
Motor End Plate	145
Synapses	145
Neuroglia	149
Ependyma	149
Choroid Plexus	149
Regions of the CNS	151
Meninges	157
Ganglia	160
Peripheral Nerves	161

Chapter 8
ORGANIZATION OF THE BODY ... 167
Inner and Outer Worlds	167
Limiting Membranes	167
Surface Increase	172
Glands	174
Structural Forms	175

Chapter 9
CARDIOVASCULAR SYSTEM ... 185
Capillaries	185
Arteries	193
Veins	200
AVAs	205
Glomera	205
Plexuses and End-Arteries	207
Heart	208

Chapter 10
DEFENSE SYSTEM ... 215
Reticuloendothelial System	215
Lymphatic and Lymphoid System	217
Lymph Vessels	218
Lymphoid Tissue	220
Lymph Nodules	220
Lymph Nodes	222
Lymphoendothelial Organs: Tonsils, Ileum, Thymus, Bursa	226
Spleen	229

Chapter 11
INTEGUMENT ... 237
Skin	239
Epidermis	239
Dermis	246
Tela subcutanea	249
Appendages	251
Sudoriferous Glands	251
Sebaceous Glands	255
Hair	256
Nails	264

Chapter 12
DIGESTIVE APPARATUS ... 265
Mouth	268
Pharynx	281
Esophagus	282
Enteric Plexuses	282
Esophageal - Cardiac Junction	284
Stomach	284
Small Intestine	299
Large Intestine	314
Liver	319
Gallbladder	336
Pancreas	342

Chapter 13
RESPIRATORY APPARATUS ... 347
Nose	349
Larynx	351
Trachea	354
Bronchial Passages	357
Alveoli	365
Respiratory Mechanics	369
Pleura	369
Evolution of the Lung	370

Chapter 14
URINARY ORGANS ... 373
Kidney	373
Juxtaglomerular Apparatus	380
Glomerulus	383
Tubules	394
Urinary passages	401
Pelvis and Ureter	401
Bladder	408

Chapter 15
ENDOCRINE SYSTEM ... 409
Hypophysis cerebri	410
Thyroid Gland	414
Parathyroid Glands	417
Suprarenal Glands	418
Epiphysis Cerebri	427

Chapter 16
THE EYE ... 431
Coats of the Eyeball	431
Dioptric Apparatus	433
Retina	437
Stereoscopy	441
Accessory organs	441
Neurea theory	443

Chapter 17
THE EAR 445
Outer and Middle Ear 447
The Labyrinth 447

Chapter 18
REPRODUCTIVE SYSTEM 455
Male Genital Organs 457
Testes 457
Spermatogenesis 457
Spermiogenesis 461
Conducting Passages 465
Accessory Glands 468
Urethra 469
Penis 473
Scrotum 479
Female Genital Organs 479
Ovaries 479
Maturation of the Egg 482
Fertilization 483
Uterine Tube 491
Uterus 493
Menstrual Cycle 494
Cervix Uteri 498
Vagina 501
Female External Genital Organs 502
Breast 503

Appendix I
MICROSCOPES 511
Magnification 511
Simple lenses 512
Compound Microscope 513
Aberrations 515
Resolving Power and Resolving Limit 516
The Condenser 517
Depth of focus 518
Various problems in microscopy 519
Ultramicroscope-Darkfield-Illumination ... 520
Polarization 520
Phase Microscopy 521
Fluorescence 522
Camera Lucida 522
Electron Microscopy 522

Appendix II
QUANTITATIVE MICROSCOPY 525
Basic Definitions 526
Counting Patterns 531
Volume 532
Surface Area per Volume 535
Length 537
Size Distribution 538
Shape 543
Curvature 551
Holmes Effect 551
Direct measurements 553
Infinite vs. Finite Objects 553
Reconstruction 553
Basic Stereological Formulae 554

Index 555

Preface to the Fifth Edition

This book has been written especially to meet the needs of students of microscopic anatomy of which histology, cytology and cell biology are subdivisions. Microscopic anatomy is an exciting field, which offers visual enjoyment and stimulates students to think creatively, challenging them to interpret flat images in terms of three dimensions and even four dimensions. Every histological image at the light and electron microscopic level is comparable to a photograph, the result of past development and a prediction of future function. As a guard in the Monteverdi Opera, "L'Incoronazione di Poppea", says: "Impariamo dagli occhi" (we learn through our eyes). The eye is the portal of entry for information on the activity of cells and organs. We are dealing with shapes and their arrangement in space and time.

For many this may be the first experience with organized, purposeful visual learning. Memorization helps little. Nevertheless, many technical terms and new terminology must be assimilated to facilitate accurate communication with others. Throughout the text, the 1983 NOMINA ANATOMICA, 5th Edition, is adhered to as closely as possible. Eponyms are carefully avoided in the text, but for reference only, some are placed in parentheses after the preferred term. It is worthy of note that only one adjectival eponym is retained in the current NOMINA HISTOLOGICA: golgiensis (note spelling in the lower case). Every effort should be made to use terminology, as approved at the Eleventh International Congress of Anatomists in Mexico (1980) and published by Williams & Wilkins in 1983.

An attempt is made to present complicated matters in as simple a manner as possible. The basic aim of the text was and is to facilitate entrance to medical training in a fashion consistent with sound educational principles. Controversial opinions are limited in acknowledgment of the professor's perogative to selects and discuss these with students. Information offered serves to prepare students for advanced courses and continuing education. A veritable treasure of illustrations is offered which will be invaluable reference sources in pathology. This book is not an encyclopedia, but an important primer of microscopic anatomy.

As Senior Editor of this fifth edition, I am pleased to acknowledge the immense amount of work by my predecessors, the late Hans Elias and John E. Pauly, and the valuable contributions of their many colleagues acknowledged in previous editions. The text format and method of presentation is retained, but I assume full responsibility for editing the current edition.

Editorial assistance for this fifth edition was provided by Dennis M. DePace, Ph.D. (Nervous Tissue), Suzanne C. Zarro, M.D. (Muscle Tissue), Harris Clearfield, M.D. (Liver), P.S. Amenta, M.D., Ph.D. (Basement Membranes and Digestive System), and R. Peter Meyer, Ph.D. (Eye, Ear and Appendices II, III). Special thanks are conveyed to a special group of MAMP graduate students for their insolicited help.

To the late Professor Hans Elias, senior author of the first four editions, I am indebted for his confidence in entrusting his classic text to my stewardship. Every effort has been made to continue his original goals and objectives. For his valued substantive contributions, during the writing of the text, I am most grateful.

To my secretarial staff sincere gratitude for preparing the manuscript for the publisher. To Lynn Reynolds for artistic contributions, I am deeply grateful. I offer my appreciation to Alfredo DiLelio III, VRDF of Rome, Italy for his generous support and assistance.

I wish to acknowledge Bertram S. Brown, M.D., President, and Israel Zwerling, M.D., Ph. D. Senior Vice-President for Academic Affairs and Dean of the School of Medicine of Hahnemann University, for creating an atmosphere conducive to academic endeavors, and for encouraging me to assume this pleasant responsibility.

This work could be accomplished only with the loving support of my wife Rose and daughters who provided a peaceful, comfortable home where most of the research and editing was conducted.

PETER S. AMENTA, PH.D.

Introduction

Human anatomy is the study of the structure of the human body. The word anatomy derives from two Greek words: *ana-* (up) + *temnein* (to cut); literally to "cut-up". In gross anatomy the scalpel, probe, fingers and eyes are the instruments used. In the dissecting room, students study whole organs, while in microscopic anatomy, the intricate composition of tissues is studied. Anatomy is concerned with the *morphology* (structure) of tissues which combine to form organs, systems and the body as a whole.

Histology (*histos-* in Greek means web or tissue) is that branch of anatomy that reveals the microscopic and *submicroscopic* (electron microscopic) structure of tissues and organs. Tissues are formed of varying proportions of cells, fibers and intercellular substances. Students should learn to recognize these components, for in pathology, disease processes may involve one or more of these components. The study of cells themselves is a specialty of anatomy called *cytology* or *cell biology*.

Microscopic anatomy deals with objects too small and too pale to be visible with the unaided eye. These are measured in *micrometers* (μm) and *nanometers* (nm). The SI system of measurements (Systeme Internationale d'Unities), considers the

Fig. I-1. The histologist observes extremely thin slices, erroneously called "sections". In practice, they are not cut with a kitchen knife, but with a very sharp razor or with a precision instrument called a microtome.

The unimaginative scientist sees only the flat image, not noticing the rich, three dimensional reality behind it.

I-2. Confusing a section with reality.

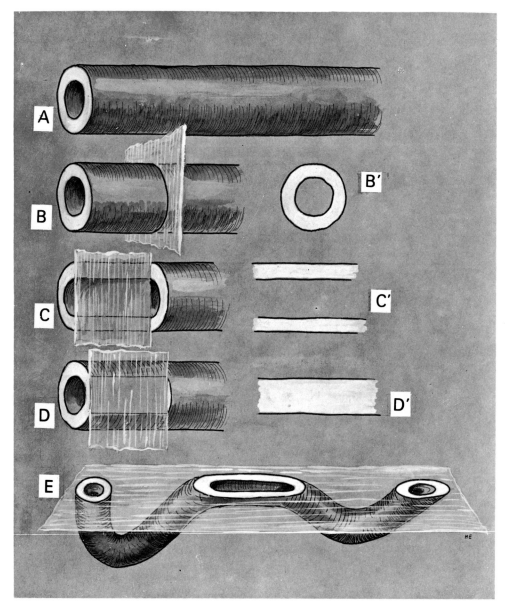

Fig. I-3. Various directions of cutting tubes. B, Transverse (cross) section; C, longitudinal section; D, tangential section; E, a curved tube cut.

older unit, the Angstrom (Å) obsolete, and hence will be avoided in this edition. The basic unit of measurement to which all others are referred to is the *meter*, abbreviated by an *m* not to be followed by a period except at the end of a sentence. It was originally defined as 1/10,000,000 of the distance from the equator to the North Pole. The standard meter bar stored in Paris deviates slightly from this value. In anatomy, the meter is the largest unit of length used, since an average adult man is about 1.7 m tall. Units of measurement frequently used in microanatomy are:

Centimeter (1 cm) = 1/100 m = 10^{-2}m
Millimeter (1 mm) = 1/1000 m = 10^{-3}m
Micrometer (1 µm) (formerly = micron = = 1 µ) = 1/1000 mm = 10^{-6}m
Nanometer (1 nm) = 1/1000 µm = 10^{-9}m

The study of microanatomy is carried

out largely by observing extremely thin slices thinner than paper (Fig. I-1). These slices, usually 5-10 μm thick for light microscopy, are placed flat on glass slides and stained to render visible the usually colorless tissues.

For electron microscopy, tissue slices averaging 50 nm in thickness are laid on tiny copper grids. Although very thin, the slices are of finite thickness, but, erroneously, are referred to as "ultrathin sections".[1] It is important to realize that the thickness of a true section equals zero. Metalurgists deal with true sections because their materials are opaque. In microanatomy, slices are translucent and therefore the thickness, however minute, must often be taken into consideration. Habit causes one to equate the two dimensional image with three dimensional reality (Fig. I-2). Students must guard against falling into this habit.

Generally, components of a tissue or an organ are oriented randomly in space in relation to the cutting plane. To achieve exact cross sections or longitudinal sections of a tube is an extremely difficult technical achievement. Imagine a tube (Fig. I-3, A). This can be cut transversely (Fig. I-3, B) (cross section). The result will be a ring-shaped image (Fig. I-3, B'). If the knife passes along the axis of the tube, it produces a longitudinal, axial section (Fig. I-3, C), which appears as a pair of parallel stripes (Fig. I-3, C'). If the cut is made parallel to the axis of the tube, but within its wall, the result is a tangential section (Fig. I-3, D). It appears as a single, wide stripe (Fig. I-3, D'). Only if the tube is visible to the naked eye will it be possible, with much patience and great skill, to mount it in the *microtome* (a cutting machine) so that longitudinal or tangential sections will result. If the tube is curved in space (Fig. I-3, E) it will be impossible to

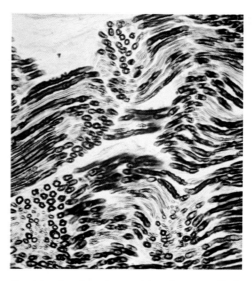

Fig. I-4. Section of a peripheral nerve. Each myelinated nerve fiber is a cylinder. The section was intended for longitudinal cutting, but the specimen curled during preparation. Therefore, this particular image shows some almost longitudinal, some almost transverse and many oblique sections through fibers.

section it longitudinally or tangentially, except for short stretches. Random sections will be oblique, with elliptical shapes (short ellipses predominate). To conclude: if oblong objects are randomly arranged in space, most of their sections will appear as short ovals. If cylinders are crooked, images as shown in Figure I-3 can be duplicated. Figure I-4, for example, shows a peripheral nerve which was intended to be cut longitudinally, but after a piece of the nerve had been severed, some of the nerve fibers curled a little. The result is a mixture of sections, transverse, oblique and longitudinal.

---

[1] Before ultrathin sections are cut with glass or diamond knives from a tiny block of plastic (epoxy resin) embedded tissue, a "thick" section, about one μm is produced. This is stained and observed with the light microscope to locate the most interesting area, before the block is trimmed again.

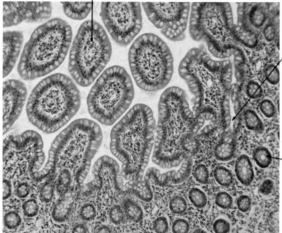

Fig. I-5. Section almost tangential through the wall of the small intestine. Villi and glands are cut nearly transversely. H&E.

Further Examples

Examples of cutting cylindrical structures are shown in Figures I-5 and I-6. These "sections" pass through intestinal villi and glands. In Figure I-5 both villi (solid cylinders surrounded by empty space) and glands (tubules with a central cavity, called a lumen) are cut transversely, hence their sections appear round. The

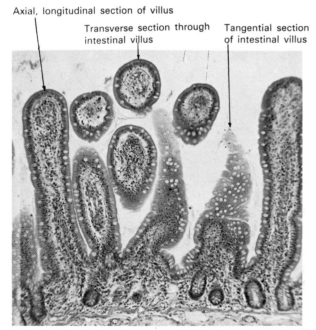

Fig. I-6. Section almost perpendicular to the wall of the small intestine. H&E.

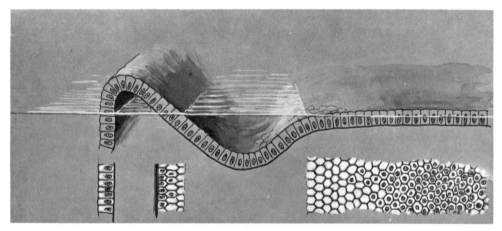

Fig. I-7. Cutting a sheet, thicker than the slice. The particular sheet shown here is a simple, columnar epithelium. One can observe that the number of layers appears to change with the direction of cutting. At the right, one notices that it will depend on the level of cutting whether cells appear to have a nucleus or not.

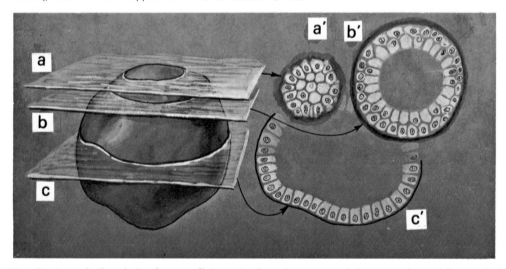

Fig. I-8. Cutting a hollow ball whose wall is a simple columnar epithelium, at three different levels.

next Figure (I-6) shows villi and glands cut longitudinally. The plane of sectioning passing near their midline produces axial or paraxial sections; or through their walls, yields tangential sections.

A sheet (Fig. I-7), when cut, will yield a stripe as a section. If the knife passes through it perpendicularly, the resulting "normal" section will be a stripe exactly as wide as the sheet is thick, and its outlines will usually appear sharp. The more the cut slants against the sheet the wider will be the section, and its outlines will appear hazy. Only parallel sections will show the sheet in its full extent. Parallel sections of a sheet can be obtained only with great difficulty, but only if the sheet is not warped. *Please read the captions and labels of all figures; for they add information not found in the text.*

A hollow, spheroid object (Fig. I-8) cut near its pole (Fig. I-8, a) will yield a tangential section passing only through its wall. The section derived will form a small disk (Fig. I-8, a') which does not show the cavity of the ball. A section slightly closer to the equatorial center (Fig. I-8, b') will produce a broad ring (Fig. I-8, b), while a section through the center (Fig. I-8, c) will be a large ring

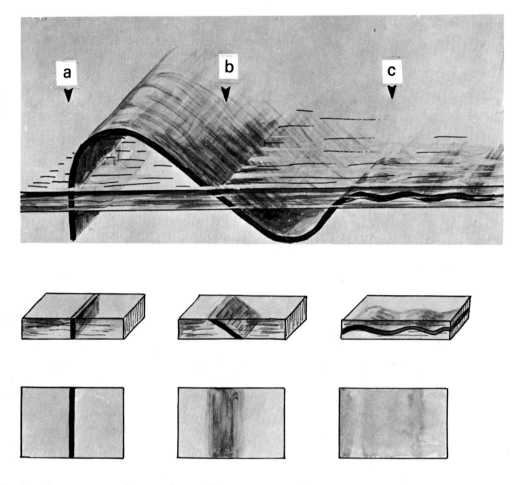

Fig. I-9. Cutting a very thin membrane. The appearance of the membrane in the microscope depends on the direction of cutting.

(Fig. I-8, c') with a rim the same width as that of the wall of the ball.

A perpendicular section through a stained membrane, which is much thinner than the slice, will be sharply outlined and appears as a sharp, dark line (Fig. I-9, a). An oblique section through the same membrane appears broad and hazy (Fig. I-9, b), while a tangential section will be very faint (Fig. I-9, c).

Figure I-10, a section through adipose tissue, shows fat globules separated by thin membranes of cytoplasm and collagen tissue. Where these partitions stand perpendicularly to the cutting plane, they appear narrow, dark and crisp. Those cut at oblique angles appear wide, delicate and hazy.

Staining Qualities of Body Components

Cells and intercellular substance consist of various materials which, with the exception of pigments, are colorless. The histologist stains them with dyes to make them more visible and to distinguish them. Specific chemical properties of tissue components give them an affinity (or -philia) for or a phobia against certain dyes.

Alkaline substances which attract acid dyes are *acidophilic*. Eosin (red), picric acid (yellow), fast green, orange G, and acid fuchsin (red) are some of the acid dyes commonly employed by the histologist.

Acid substances which attract basic dyes are *basophilic*. Hematoxylin (blue), safranin (red), carmine (red), thionin (blue), and basic fuchsin (red) are some useful basic dyes. The nuclei of cells contain *deoxyribonucleic acid* (DNA) and are therefore basophilic. Granular endoplasmic reticulum with ribonucleoprotein granules attracts basic dyes and hence is basophilic.

Cells which have an affinity for silver, are either argentaffin or argyrophil (argyrophilic). *Argentaffin* cells reduce silver nitrate directly, while *argyrophil* cells require pretreatment with a reducing agent and thus its action in reducing silver is indirect. Chromaffin substances attract chromium salts (brown). Lipids (fats) stain red with sudan and are blackened by osmium

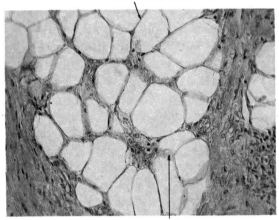

Fig. I-10. "Section", i.e. a stained slice through fat cells (adipose tissue). The fat has been dissolved, and the thin walls between fat drops are stained. The widths and darkness of these thin walls depend on the angles they form with the cutting plane. H&E, 500 ×.

tetroxide. There are many more delicate affinities than the few enumerated here.

A substance which cannot easily be stained with any dye is called chromophobic, in contrast to those easily stained, which collectively are called chromophilic.

SOME COMMON FIXATIVES AND STAINS

Fixatives

Fresh tissues are perishable and too soft to be cut into thin slices. Therefore, they usually are "fixed", i.e., immersed in or perfused with a coagulating liquid that both hardens and preserves the material. The most common "fixative" is 10% *formalin* (a 4% aqueous solution of formaldehyde). Specimens should be immersed in 10% formalin for at least 24-48 hours, but they may be stored in it indefinitely. Ten percent formalin is compatible with most stains. Some other fixatives are:

Bouin's mixtures (good for hardening soft material and particularly compatible with varieties of Masson stains):

saturated picric acid solution, 15 parts;
formalin, 5 parts;
glacial acetic acid, 1 part.

Bouin's mixture penetrates well and fixes rapidly. Although it is possible to store materials indefinitely in Bouin's mixture, it is better to fix them only 4-12 hours, then wash in 50% alcohol for 4-6 hours and store in 70% alcohol. In the past this technique was used widely. However, routine use has been greatly curtailed because of the dangers involved in storing picric acid.

Carnoy's mixture (preserves glycogen and permits good staining of mucopolysaccharides, e.g., basement membranes and brush borders):

absolute alcohol, 6 parts
chloroform, 3 parts
glacial acetic acid, 1 part

Carnoy's mixture penetrates well resulting in very rapid fixation (1-2 hours). Fixed tissues should be washed in absolute alcohol and stored in thin cedarwood oil or liquid petroleum.

Sublimate (maintains delicate pigments and increases affinity of cytoplasm for acid dyes): saturated solution of mercuric chloride.

Sublimate penetrates pooly and will shrink tissues; so usually it is used in combination with acetic acid, chromates, formalin or alcohol. Tissues will be ruined by mercuric crystals after several weeks; therefore, they should be washed and stored in 70% alcohol plus a few drops of iodine (which will remove excess mercuric crystals).

Except for those fixed in Carnoy's mixture, fixed tissues are washed in water. Alcohol is miscible with water, and subsequently tissues can be dehydrated by passing them through a series of alcohols of increasing concentration. Then the alcohol must be substituted by a fat solvent, e.g., xylol, toluene, benzol, cedarwood oil or chloroform. Except for chloroform, which has a high specific gravity and a low refractive index, most fat solvents have high refractive indices and render the tissues transparent or "clear". "Clearing" provides a visual test for good dehydration and infiltration with fat solvents.

Following replacement of alcohol by a clearing agent that is miscible with paraffin, tissues are infiltrated by immersing them in several changes of melted paraffin. Finally, they are embedded in pure paraffin in a cubical mold. After cooling, the material is hardened sufficiently to permit cutting into fairly thin slices, called "sections". These are mounted on glass slides after floating them on luke-warm water. When the sections are completely dry, they are deparaffinized in xylol and rehydrated (to water) by passing them through a series of alcohols of decreasing concentration. They now are ready to be stained.

Stains

Hematoxylin and eosin (H&E)

Hematoxylin and eosin is probably the most common combination of stains in use. The hematoxylin is considered a basic stain (cation) with an affinity for acidic substances such as deoxyribonucleic acid (DNA) and ribonucleic acid (RNA). Because nuclei, ribosomes and cartilage matrix are stained bluish violet, these structures are referred to as "basophilic". Eosin is considered an acid stain (anion) with an affinity for basic substances such as protein. It stains almost everything red except DNA and RNA. There are several methods for the preparation and use of this useful combination.

There are many other dyes, and some of these are specific for certain structures or substances:

Azan according to Heidenhain.

Azocarmine stains nuclei red. Anilin-orange-acetic acid stains muscle red, orange or brown; erythrocytes red; connective tissue intense blue; and cytoplasm pale blue.

Mallory's stain.

Though similar to Azan it is easier to use, but unfortunately the results are less permanent.

Goldner's stain.

Iron hematoxylin stains nuclei black. Acid fuchsin-phosphomolybdic acid-orange-light-green stains cytoplasm red, erythrocytes vermillion, connective tissue and mucin green, and muscle dark red.

Fat stains

Lipids can be stained with sudan or blackened with osmium tetroxide. The latter, used also as a fixative, reveals myelin.

After staining, sections are dehydrated again and sealed with a transparent, fat-soluble substance. For optical flatness, a thin glass cover slip covers and protects the preparation.

If lipids are to be demonstrated, the tissue must not be exposed to any fat solvent such as alcohol and xylol. Tissues are frozen and cut with a freezing microtome; and after slices are stained, they are covered with glycerin jelly and a cover slip or with a mixture of gum arabic and sugar.

For electron microscopy, tissues may be fixed directly with buffered osmium tetroxide or post-fixed following perfusion with glutaraldehyde and/or paraformaldehyde. Fixed tissues are embedded in epoxy resins which are sectioned best with glass or diamond knives. Contrasting ("staining") of tissue sections occurs by deposition of heavy metals (lead citrate and/or uranyl acetate).

The foregoing remarks serve to introduce the student to some of the reasons for employing different techniques in the study of histology. Actually, histochemistry and cytochemistry are well developed fields of science. Methods have been and continue to be developed to demonstrate and even quantify a host of enzymes, cytoplasmic structures, pigments, mucins, polysaccharides, connective tissue fibers and lipids. Students should consult books on histopathologic techniques, histochemistry and cytochemistry for necessary clarification and additional information.

1...
The Cell

General

Except for primitive organisms, such as viruses, bacteria and blue-green algae, all living things are composed of cells, symplasms and intercellular substances. The size of the organism does not depend on the size of the cells which compose it, but on the number of cells (e.g., salamander cells are much larger than those of an ele-

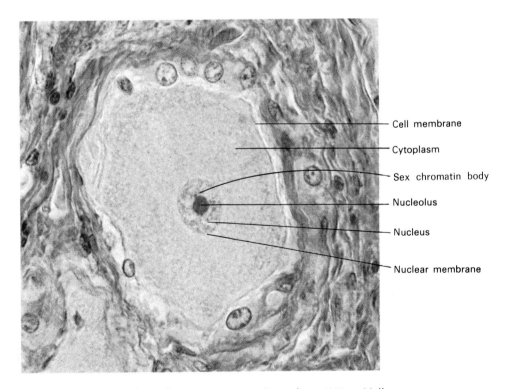

Fig. 1-1. A nerve cell from the superior cervical ganglion, 1600×, Mallory.

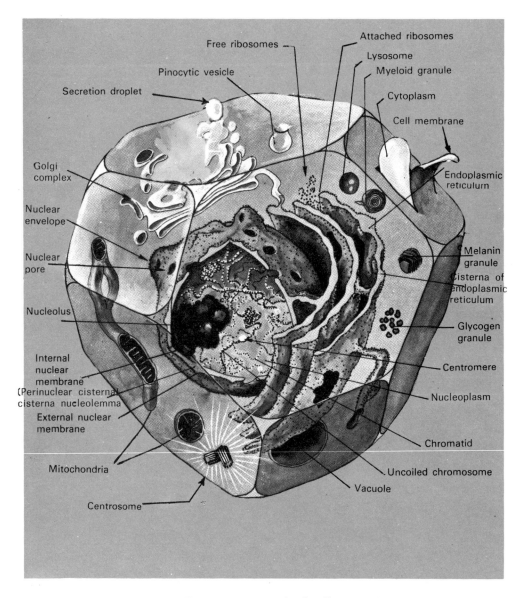

Fig. 1-2. A generalized cell.

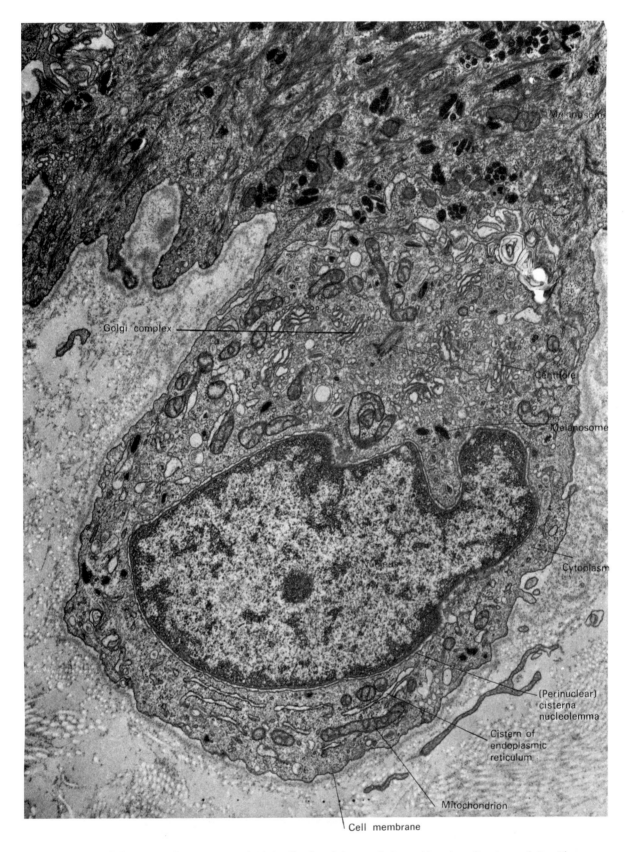

Fig. 1-3. A human melanocyte attached to the basal layer of the epidermis. Courtesy of Dr. Klaus Konrad, Vienna.

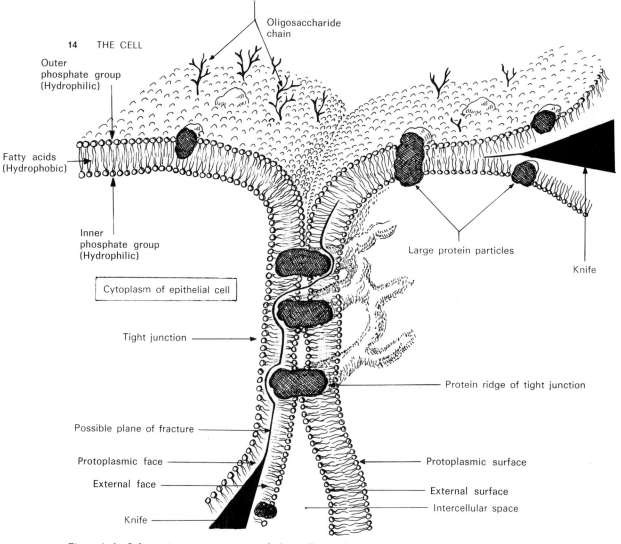

Fig. 1-4 A. Schematic representation of the cell membranes of two adjacent epithelial cells joined by a tight junction. Green: knives used for freeze fracture. Red: places where fractures most frequently occur.

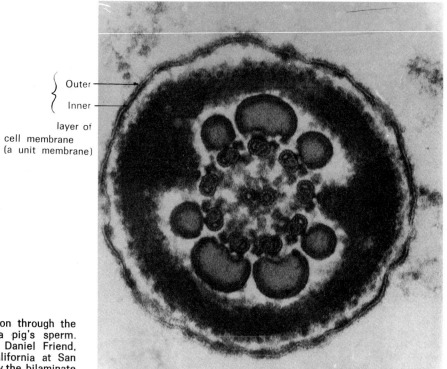

Fig. 1-4, B. Section through the tail of a Guinea pig's sperm. Courtesy of Dr. Daniel Friend, University of California at San Francisco to show the bilaminate plasma membrane.

phant). Only within narrow taxonomic groups are cell size and body size related. It has been estimated that adult human beings possess some 50 trillion cells. Most of these have a brief life span which after their death, are replaced by new cells. Others, such as nerve cells, ovocytes and perhaps striated muscle cells, are not renewed after early childhood, and consequently cannot be replaced.

Parts of the Cell

A typical cell consists of the parts shown in Figs. 1-1, 1-2 and 1-3. Membranes separate the cell from its environment and subdivide it internally. "Double" membranes (in mitochondria) and "single" membranes (around lysosomes) exist. The Latin term *membrana cellularis* (*cell membrane*) is the name applied to the membrane complex within and surrounding a cell. The term *plasmalemma* specifies only the limiting (surrounding) membrane of the cell, but it does show deep infoldings which serve to increase cell surface considerably.

Cell membrane

The cell membrane (frequently designated the *unit membrane*) can be resolved by the transmission electron microscope as two dark (electron dense) layers, each approximately 2 to 4 nm thick, sandwiching a clear (electron lucent) layer 3 to 3.5 nm thick (Fig. 1-4, B). It is generally agreed that the cell membrane consists of a bilayer of ionic and polar head groups (phospholipid molecules attached to fatty acid chains) (Fig 1-4, A). Integral protein complexes are embedded between the polar head groups, conferring a "mosaic" appearance on the molecular structure of the cell membrane. On the outer (external) surface of the membrane, and the inner (protoplasmic) surface, there are other protein molecules which give the molecular structure an asymmetrical appearance. This inherent asymmetry is amplified further by oligosaccharide chains protruding from the external surface of the plasmalemma. The lipid and integral proteins can move freely within the plasmalemma. Protein particles facilitate the passage of water soluble substances. This hypothetical structure helps to account for the known physiological property of membrane semipermeability. In Fig. 1-4B, the plasmalemma (appearing as double electron dense lines) surrounds the tail of a guinea pig sperm. This double line structure is resolvable only at very high electron microscopic magnifications. In routine electron micrographs, the cell membrane appears as a single electron dense line. When using the term "double membranes", it should be understood that two complete unit membranes are next to one another. The technique of freeze-fracture permits splitting cell membranes (albeit not in predictable planes) through the electron lucent layer. In Fig. 1-4A, possible fracture planes are demonstrated. That portion of the plasmalemma in contact with cytoplasm is designated by the letters PS (protoplasmic surface), while the outer surface is the ES (external surface). Fracture planes split the membrane so that two internal faces of the membrane may be viewed: a PF (protoplasmic face) and an EF (external face).

NUCLEUS. *The nucleus* is a mass of chromatin suspended in nucleoplasm, and confined by "double membranes". *Chromatin*, consisting primarily of the genetic material deoxyribonucleic acid (DNA), exists in two morphological forms: *euchromatin* and *heterochromatin*. Euchromatin is the metabolically active form, whereas heterochromatin is metabolically inactive.

The *nucleolus*, a small dark staining body within the nucleus, consists of ribonucleic acid (RNA) and protein. Electron

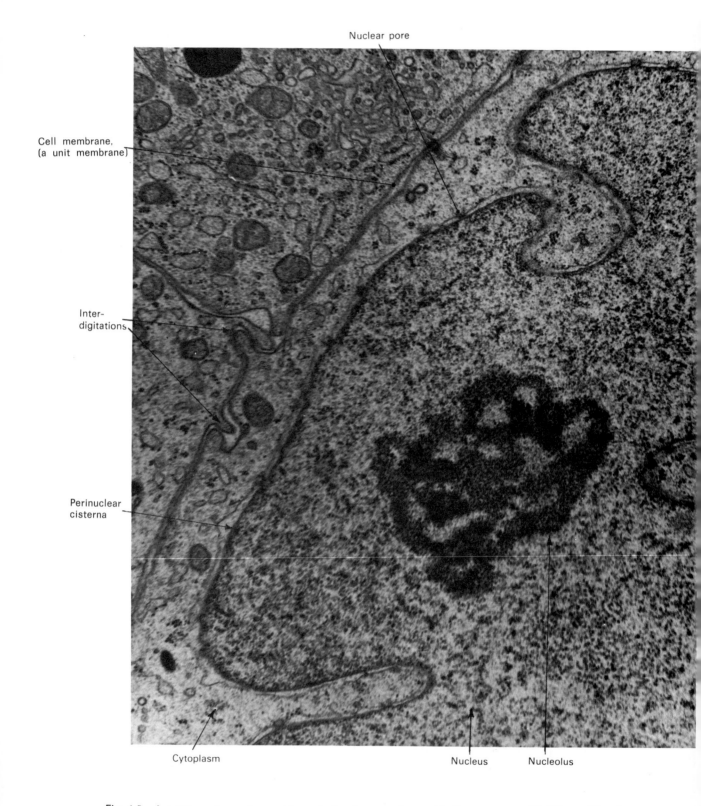

Fig. 1-5. A section of a nucleus with nucleolus from oviduct epithelium, approx. 100,000 ×. (Courtesy of Prof. Louise Odor).

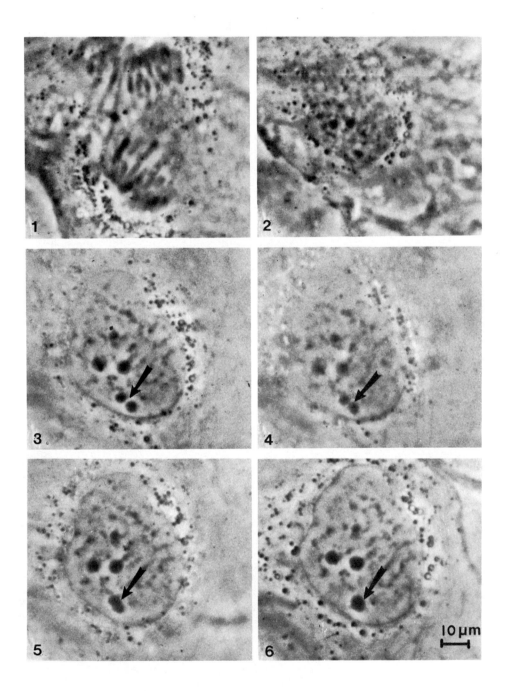

Fig. 1-5A. Selected frames from a time-lapse cinematographic record (phase contrast microscopy). Fusion of nucleoli is observed in daughter cells after anaphase (1) and reconstruction (2). Fusion is complete in frame (6). From P.S. Amenta, 1961, Fusion of nucleoli in cells cultured from the heart of *Triturus viridescens*. Anat. Rec., vol. 139, pp. 155-166.

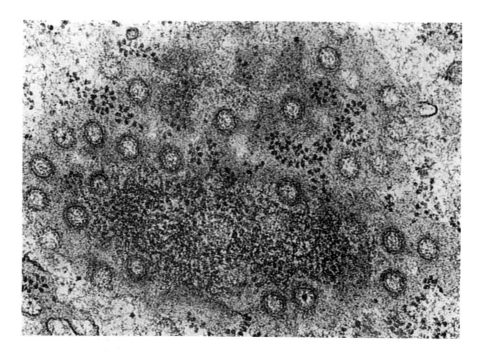

Fig. 1-6. Almost tangential section through the nuclear envelope of a liver cell showing nuclear pores. Courtesy of Dr. Claude Biava.

Fig. 1-6, A. Nucleus with pores of a pancreatic acinar cell of a rat. Freeze etching. (Courtesy of Dr. D. Friend).

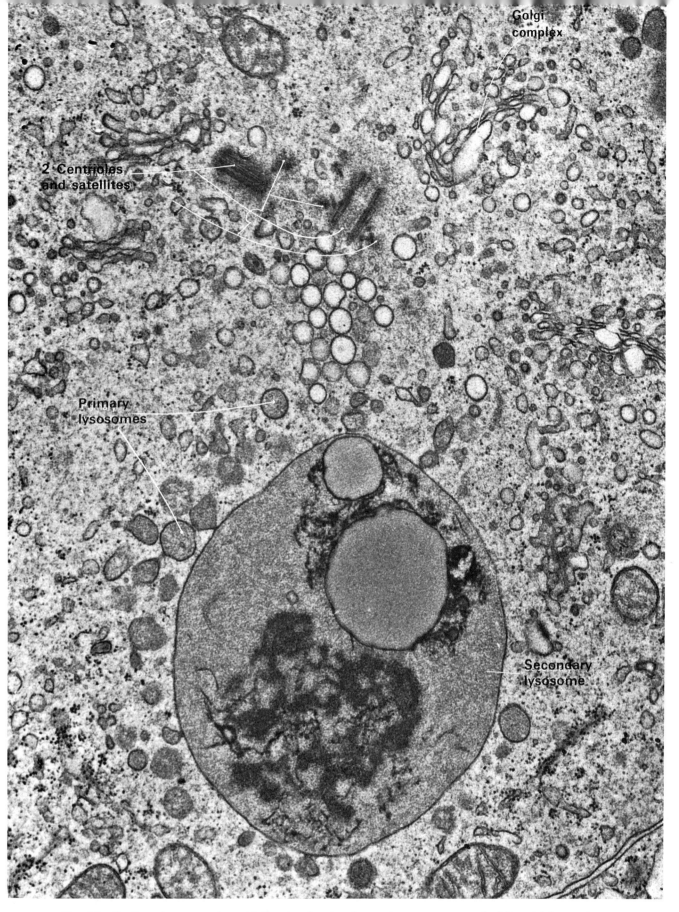

Fig. 1-7. Portion of a macrophage histiocyte, showing the centrosome composed of 2 centrioles, and a large, secondary lysosome, 50,000 ×. (Courtesy of Prof. L. Odor).

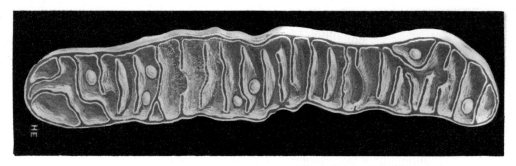

Fig. 1-8. A mitochondrion.

microscopy reveals a honeycomb structure called the *nucleolonema* consisting of two parts: a thread-like portion, the *pars filamentosa* embedded in a granular portion, the *pars granulosa*. Via light microscopy, the nucleolus appears homogeneous and sharply outlined, but it is not membrane bound. In certain metabolically active cells engaged in intense protein synthesis (e.g., embryonic cells and cancer cells), the nucleolus enlarges and may contact the nuclear membrane. Nucleoli are not static. Indeed, viewed in tissue cultures by time-lapse microcinematography they maneuver actively and may fuse with other nucleoli and again may separate (Fig. 1-5, A).

The nuclear membrane, via electron microscopy is a double membrane (Figs. 1-2, 1-3, and 1-5) with pores ranging in size from 30 to 100 nm (Fig. 1-6). Occasional bulges appearing on the membrane serve to increase the surface area (Fig. 1-5). The space between the unit membranes is termed the *cisterna nucleolemmae* (perinuclear cisterna). This space measures between 40 and 70 nm in normal subjects (Figs. 1-2 and 1-5) and much larger in pathological specimens. Fig. 1-6 clearly demonstrates the nuclear pores and the cisterna nucleolemma of a pancreatic exocrine cell (rat).

CYTOPLASM. The *cytoplasm*, a complicated network of protein molecules suspended in an aqueous medium in which organic and inorganic substances are dissolved, is mainly alkaline and therefore acidophilic (attracts acid dyes). It contains discrete lipid droplets and vacuoles filled with a variety of liquid substances, some bound by single membranes, others by double membranes.

CYTOPLASMIC ORGANELLES AND INCLUSIONS. Previously, all cellular structures were classified as organelles (living components) or inclusions (non-living, transient substances). Numerous studies have cast doubt on the efficacy of this classification, and hence per the 1983 edition of the *Nomina Histologica*, no distinction will be made between the so-called cytoplasmic organelles and inclusions. Structures are discussed with no designation as organelle or inclusion.

Lysosomes are cytoplasmic bodies inhabiting all eukaryotic cells. The total population varies between different cell types, but as an example the typical liver cell possesses approximately 300 per cell comprising 1% of the total cell volume. Lysosomes may be characterized as spherical membranous bags filled with hydrolytic enzymes intended for the controlled digestion of assimilated intracellular macromolecules. Among the many enzymes identified in liver cells are: (1) acid nucleases, (2) phosphoprotein phosphatase, (3) α-glucosidase, (4) β-N-acetyl-glucosaminadase, (5) β-glucuronidase, (6) β-galactosidase, (7) α-mannosidase and (8) aryl-sulfatase, all of which are optimally active at pH 5.

The lysosomal membrane probably pos-

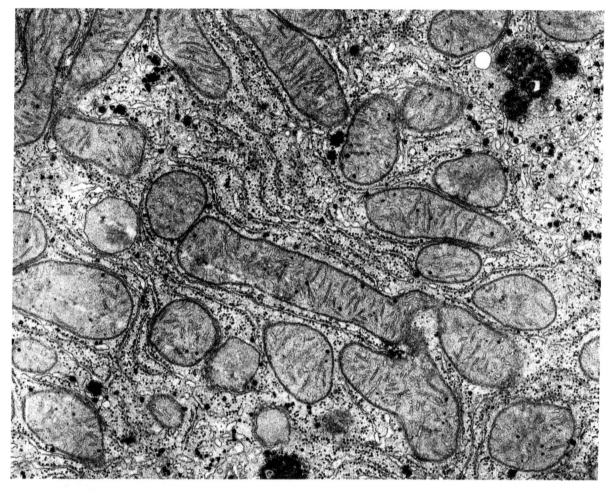

Fig. 1-9. A (above) and B (facing page). Sections through mitochondria of various shapes in liver cells. Courtesy of Dr. Albrecht Reith, Oslo.

sesses a specialized receptor which serves as a "docking marker" for *phagosomes* (transport vesicles) enclosing extracellular substances engulfed by phagocytosis or pinocytosis. Two morphological forms of lysosomes are distinguished via electron microscopy: (1) *primary lysosomes* (25 nm to 0.5 μm in diameter) which transport hydrolytic enzymes toward the phagosomes and (2) *secondary lysosomes* (0.5 μm to 1.5 μm in diameter) which represent the products of fusion of primary lysomes with phagosomes (Fig. 1-7). Waste products of intracellular digestion are enveloped in part of the 6 nm thick membrane of the secondary lysosome and discharged into the cytoplasm for *exocytosis* (external secretion) or they may remain within the cytoplasm as *residual bodies*.

Accumulating residual bodies (*lipofuscin,* or *age pigment*) serve as indicators of "wear and tear".

The remarkable lysosomal membrane prevents the confined enzymes from digesting vital proteins and nucleic acids of the host cell. Trauma and excessive variations in the milieu (osmotic pressure, temperature, pH) tend to rupture this unique membrane causing enzymes to escape and permeate the cytoplasm where *autolysis* (self digestion) will occur.

Lysosomal enzymes synthesized in the granular (rough) endoplasmic reticulum

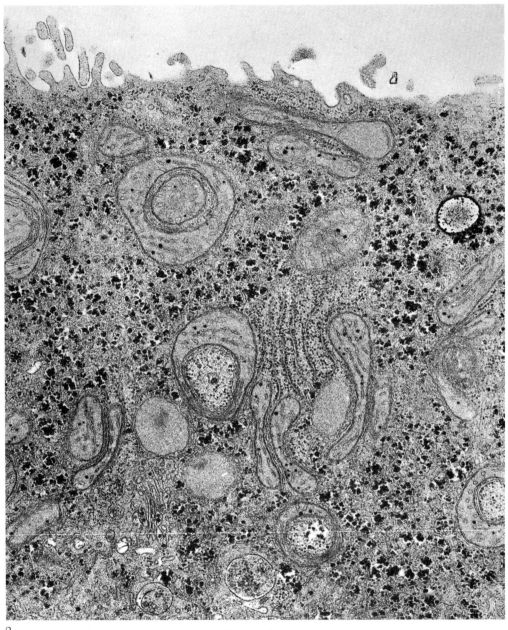

Fig. 1-9, B.

are transported in small vesicles to the *golgi complex*, where the enzymes are glycosylated and enveloped in a new membrane as *nascent granules*. These enlarge into primary lysosomes that can be transported to: (1) a free cell surface where the contents are discharged via *exocytosis*, (2) a *phagosome* or (3) an *autophagosome* containing intrinsic cellular products. Fusion of a primary lysosome with a phagosome or an autophagosome results in the formation of the secondary lysosome.

Mitochondria (Fig. 1-8) are minute bodies enclosing a matrix within double unit membranes. The inner unit membrane folds into the mitochondrial matrix to form the cristae, which exist in one form or combination of forms, depending on cell function. The lamelliform, crescentic

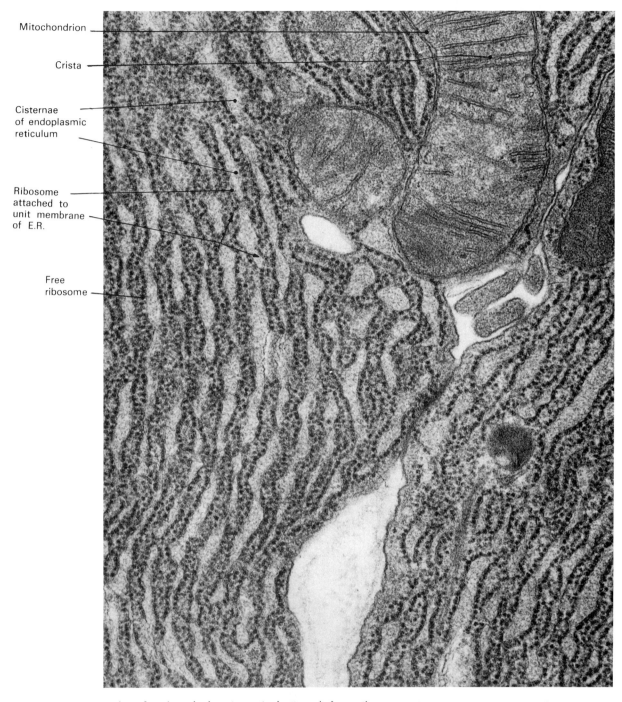

Fig. 1-10. Rough-surfaced endoplasmic reticulum and free ribosomes in mouse pancreatic cells. (Courtesy of Drs. J. K. Sherman and K. C. Liu, University of Arkansas for Medical Sciences).

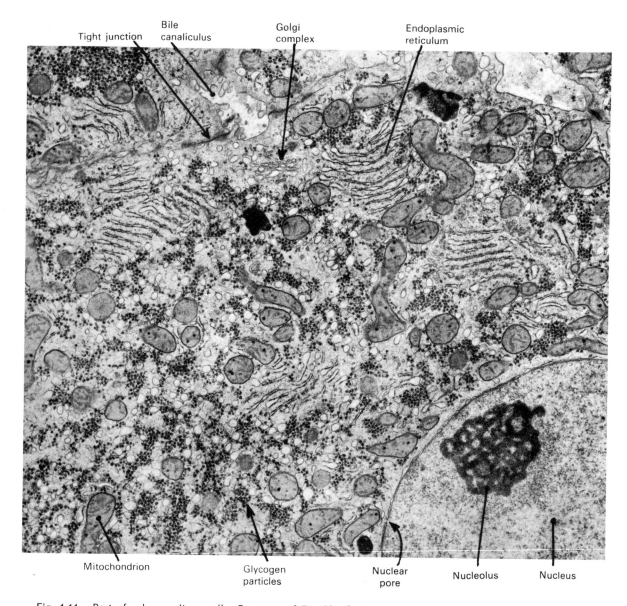

Fig. 1-11. Part of a human liver cell. Courtesy of Dr. Claude Biava.

type is most commonly encountered. Tubular and vesicular forms are characteristic of steroidogenic cells (endocrine cells of the testes, ovary and suprarenal cortex). The inner membrane, when observed in the EM via negative staining, shows numerous small spheres attached by small stalks to the inner surface. Mitochondria are pleomorphic, i.e., they present many different shapes (Fig. 1-9A, 1-9B): spheroid, disk-shaped, cup-shaped (liver cells) or rod-shaped (kidney cells). They may appear as very long threads forming rings or branches, which in vivo and in vitro are continuously changing shape or dividing and joining.

Mitochondria are instrumental in oxidative phosphorylation, the process whereby the cell obtains energy. The enzymes of the tricarboxylic acid (Krebs) cycle are associated with mitochondria, hence the popular appellation: "powerhouse of the cell". Enzymes of the respiratory chain and the mechanism for ATP synthesis are located on the inner membrane. The inner membrane contains a specific transporter

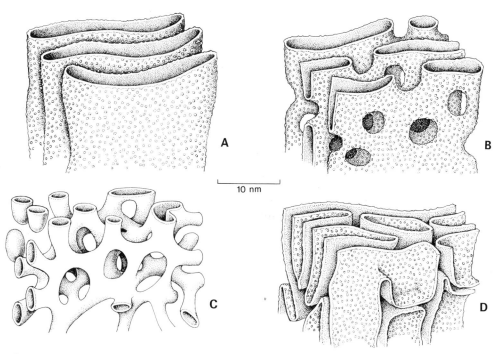

Fig. 1-12. Various forms of endoplasmic reticulum.

which moves calcium ions in and out of the mitochondrial matrix, thus maintaining the correct cytoplasmic calcium ion levels. Mitochondria possess specific forms of DNA, RNA and ribosomes resembling those of bacteria, leading to the speculation that mitochondria may possess the capacity for independent existence. Some investigators consider mitochondria to be symbionts. Like bacteria, mitochondria replicate by budding.

Ribosomes (Fig. 1-10). Ribosomes are electron dense 15 nm × 25 nm ribonucleoprotein particles which participate in the synthesis of proteins. One molecule of *messenger ribonucleic acid* (mRNA) associated with several ribosomes is required to initiate protein synthesis. Under the light microscope the degree of cytoplasmic *basophilia* (affinity for basic dyes) is directly proportional to the quantity and concentration of ribosomes. A ribosome unit consists of two subunits: (1) a small one composed of one molecule of ribonucleic acid (RNA) plus 30 associated proteins attached to (2) a larger one possessing two molecules of RNA plus 40 proteins.

Ribosomes exist freely in single units, in clusters (*polyribosomes*) or bound to endoplasmic reticulum membranes.

Endoplasmic Reticulum (cytoplasmic reticulum). Ribosomes are often associated with endoplasmic reticulum in a variety of forms: parallel, double membranes enclosing flat cisternae (Figs. 1-10, 1-12A), parallel cribriform double membranes (Figs. 1-12B), anastomosing canals (Fig. 1-12C, left); isolated vesicles (Fig. 1-12C, right); the muralium structure (Fig. 1-12D), or irregularly shaped cisterns. The cisternae always are separated completely from the cytoplasm; but may, occasionally, communicate with the cisterna nucleolemma, (Figs. 1-2, 1-13) and the exterior of the cell. It has been suggested that endoplasmic reticulum is instrumental in the active transport of substances through the cell. Membranes of endoplasmic reticulum appear in two forms, rough and smooth. If ribosomes are attached to the membrane surface facing the cytoplasm it is called

Fig. 1-13. Connection of perinuclear cistern with endoplasmatic cistern (K. Porter, from Scope, The Cell, 1962; courtesy of Upjohn Co.).

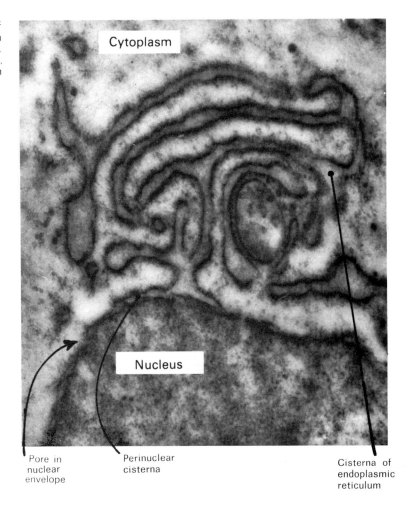

Pore in nuclear envelope

Perinuclear cisterna

Cisterna of endoplasmic reticulum

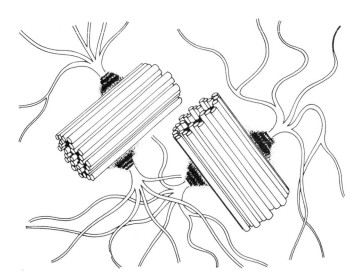

Fig. 1-14. A centrosome with satellites and microtubules. Adapted from various sources.

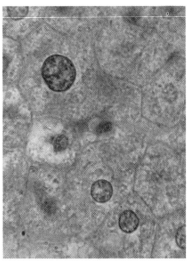

Fig. 1-15. Nuclei of liver cells of different sizes. This biopsy picture contains one dodecaploid, 2 diploid and (cut by the right margin of the picture) one tetraploid nucleus.

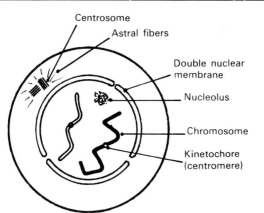

A. Early interphase (G₁)
For simplicity only one pair of homologous chromosomes is shown.

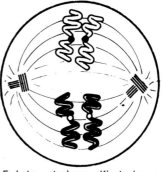

E. Late metaphase. Kinetochores «split», Chromosomes «split».

B. Late interphase (G₂)
The S phase has occurred, and each chromosome now is composed of two chromatids.

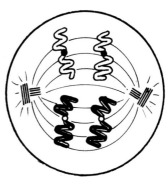

F. Early anaphase
Daughter chromosomes part.

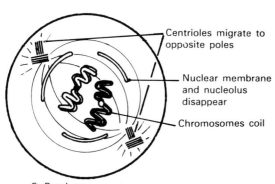

C. Prophase
The chromosomes coil.

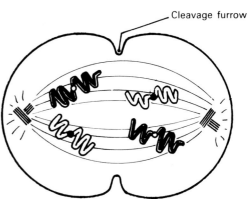

G. Late anaphase
Cleavage furrow starts to develop and daughter chromosomes collect.

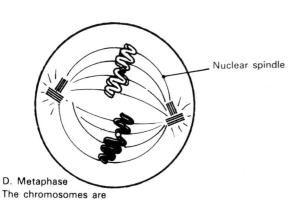

D. Metaphase
The chromosomes are arranged on the equatorial plate.

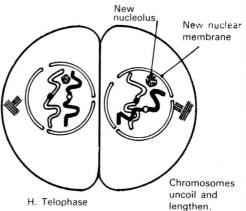

H. Telophase

Chromosomes uncoil and lengthen.

Fig. 1-16. Schematical presentation of mitosis.

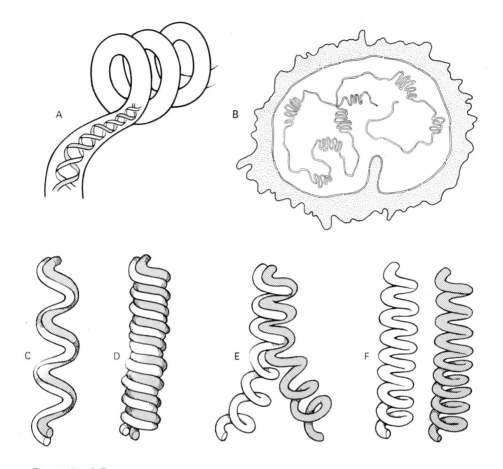

Fig. 1-17. A-F.
A. Internal structure of a chromatid during anaphase. B. Euchromatin vs. Heterochromatin.
C-D. Coiling of the chromosome. E-F. Separation of Aaughter chromosomes.

rough or *granular endoplasmic reticulum* (RER) (Fig. 1-10). Membranes relatively devoid of ribosomes form smooth or *agranular endoplasmic reticulum* (SER) associated with steroid synthesis (suprarenal, testis, ovary), production of hydrochloric acid (parietal cells of stomach) and release of calcium ions (skeletal muscle). Unattached ribosomes, frequently arranged in small rosettes in the cytoplasm, are called polyribosomes or free ribosomes (Fig. 1-10). Cells possessing a well developed RER and golgi complex may be synthesizing protein for extracellular use (secretion). Cells with an abundance of free ribosomes and minimal RER synthesize protein for intracellular use. Increasing amounts of cytoplasmic RNA, in any form, impart varying degrees of basophilia.

Golgi Complex. A complex of stacked smooth surfaced vesicles and canals (usually near the nucleus), was first described by Camillo Golgi (1898) as the *apparato reticolare interno* (Figs. 1-2, 1-3, 1-7, 1-11). The current acceptable terminology according to the International Nomina Histologica is golgi complex (note spelling in lower case). The vesicles of the golgi complex are usually large and flat; but when filled with secretion they proliferate vesicles of varying sizes as secretion granules.

The prominent *golgi complex* in glandular cells concentrates, modifies and packages substances synthesized in granular endoplasmic reticulum. The oriented, disk shaped stack of anastomosing vesicles presents two surfaces, convex (*immature face*) and concave (*mature face*). Most cell biologists consider the convex surface as the receptor site for transfer vesicles (enclosing synthesized proteins) which budded from ribosome free surfaces of granular endoplasmic reticulum. Protein processing within the golgi complex entails the transfer of glycosyl groups (*glycosylation*) to the various proteins and enzymes advancing toward concave surface. On reaching the concave surface, new vesicles filled with useful protein pinch off and migrate to the free surface of the cell, or remain within the cell for intrinsic use. Fusion of protein laden vesicles with the free surface of the cell facilitates (1) discharging vesicle contents and (2) renewal of the cell membrane and its glycoprotein coat.

Cytocentrum (centrosome). The *cytocentrum* is a specialized area of cytoplasm containing a pair of centrioles (the diplosome). Usually two centrioles exist in one cell. They appear as short cylinders made of nine bundles, each composed of three parallel tubules (Figs. 1-7 and 1-14). The centrioles contain DNA and are instrumental in cell division. Each one replicates not by horizontal or longitudinal division as once thought, but *de novo*. New centrioles develop in close proximity to pre-existing centrioles in end to end relationships. After the new centrioles appear, each pair moves to opposite ends of the cell.

Filaments are present in all cells. Those ranging in thickness from 3 to 8 nm are referred to as *microfilaments* and those from 8 to 12 nm are called *intermediate filaments* (tonofilaments). Bundles of filaments make up the *fibrils* of light microscopy. Length of microfilaments is difficult to estimate because they form complex meshworks which vary from cell to cell. The term *cytoskeleton* describes their major role in providing cellular internal support. Cells photographed via time-lapse microcinematography are seen moving rather freely. Evidence exists that most microfilaments are contractile and may be responsible for changes in the shape of cells. In muscle cells, actin filaments must act together with myosin filaments to produce muscle contraction. The larger (8 to 12 nm thick) intermediate filaments do not appear to be contractile and hence they serve primarily in support and protection. Cells exposed to trauma, such as the surface cells of the esophageal epithelium are subject to "wear and tear". Hence these intermediate filaments (tonofilaments) are suited admirably to protect epithelial cells.

Other filamentous structures, the *microtubules,* differ from filaments as follows: (1) they are twice as large with a thickness ranging from 18 to 30 nm (24 nm average), (2) may exceed several μm in length, and (3) present a circular profile in cross section, with walls composed of amino acid sequences called *tubulin* (resembling actin). Functionally, microtubules are important components of the mitotic spindle and the so-called "astral rays" of the centrosomal region. Microtubules also are major components of cilia and sperm flagellae.

Other. Other materials such as glycogen, pigments and fat also reside in the cytoplasm. Melanin (dark brown pigment) granules often are composed of parallel lamellae (Figs. 1-3 and 11-18).

Cell Division

Most cells grow and divide after reaching a maximum volume by two means: direct cell division and indirect cell division.

DIRECT CELL DIVISION (AMITOSIS). In a few tissues *cell division* is encountered, i.e., constriction and division of the nucleus (*nucleokinesis*), followed by constriction and division of the cytoplasm (*cytokinesis*). Direct cell division occurs frequently in liver cells, but is very rare in other tissues. In the liver, direct nucleokinesis may occur without cytokinesis, resulting in large multinucleated cells.

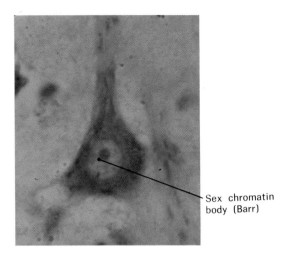

Fig. 1-18. Pyramidal cell in the cerebral cortex of a woman.

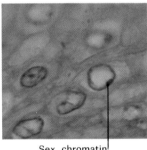

Fig. 1-19. Gingival epithelium of a woman.

In the liver, intranuclear replication of chromatin without nucleokinesis can occur. As a result, such liver cells contain diploid to dodecaploid nuclei, (i.e., giant nuclei with up to twelve sets of chromosomes) (Fig. 1-15).

INDIRECT CELL DIVISION (MITOSIS) (Fig. 1-16). While a cell performs its function and grows, its nucleus has the characteristics described above, a condition called *interphase* (Fig. 1-16, A and B). At rhythmically occurring intervals, the cell prepares for, undergoes and completes its division. The entire process of mitosis is relatively brief (35 to 120 minutes). Interphase may last many hours, days, weeks or years; or it may be perpetual as in nerve cells. Cell division, as with almost all phenomena of life, is subject to a *circadian rhythm* (circa = about. dies = day i.e. about 24 hours) during which a peak and trough in the mitotic index (number of mitotic figures per 1,000 cells) occur.

Indirect cell division or mitosis brings about the allotment of identical sets of genetic units called genes to both daughter cells. The genes are components of the chromosomes.

In every male and female sex cell, there are 23 *chromosomes* (actually *chromatids*). Twenty-two of these are called *autosomes*, and one is the *sex chromosome* (X or Y). These 23 chromosomes constitute the *haploid* number. *Conception* or *fertilization* is the union of the male and female sex cells (*gametes*) to form the *conceptus* (fertilized egg or *zygote*). At that time the 23 chromosomes of the father unite with the 23 chromosomes of the mother to establish the *diploid number*. From that moment forward, every *somatic* (body) cell of the human body contains 46 chromosomes, that is, 22 pairs of autosomes and one pair of sex chromosomes. Because each member of the 22 pairs of autosomes is morphologically similar to its partner, they are said to form *homologous pairs*. One member is of paternal origin, the other member, of maternal origin. Although the sex chromosomes are referred to as a homologous pair, they are morphologically similar only in the female where there is a pair of X chromosomes. In the male, there is one X and one Y chromosome, each delete morphologically distinct from the other.

Cell reproduction normally occurs by mitosis, a process that provides for duplication of the number of chromosomes in such a manner that each daughter cell ends up with exactly the same 23 pairs of chromosomes as the parent cell. This is not the case in the formation of sex cells. Although primordial sex cells have the usual 23 pairs of chromosomes, i.e., 46 all together, they ultimately divide so that each daughter cell receives one member of each chromosomal pair, i.e., 23 single chromosomes (haploid number). This

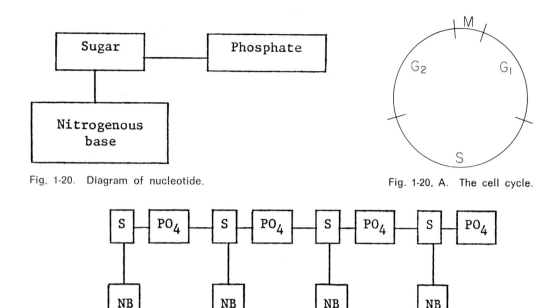

Fig. 1-20. Diagram of nucleotide.

Fig. 1-20, A. The cell cycle.

Fig. 1-21. Diagram of polynucleotide. S = Sugar, NB = Nitrogenous base, PO_4 = Phosphate group.

process, called *meiosis*, will be detailed in Chapter 18.

Chromosomes are so named because during mitosis they appear as small dark bodies after they have been stained with a basic dye. Since these small dark bodies are thread-like, indirect cell division is called *mitosis*, i.e., "presence of threads". During interphase, chromosomes are scattered throughout the nucleus, and members of a homologous pair appear to have no spatial relationship to each other. A cell, not engaged in mitosis is said to be in *interphase* (the intermitotic period). The chromosomes of such a cell are difficult to visualize via light or electron microscopy, because they are tangled as very long, thin threads. Certain segments of these threads are coiled or condensed in such a manner that they can be resolved by light microscopy (L.M.) after they have been stained. These coiled segments are referred to as *chromatin granules*. Thus, chromosomes have both extended and condensed parts. The extended parts, which cannot be resolved via L.M., constitute the *euchromatin*; while the condensed visible parts constitute *heterochromatin* (Fig. 1-17, B).

Euchromatin is the metabolically active component of the chromosome (used as a template for RNA synthesis), and heterochromatin is metabolically inactive while condensed.

The chromosomes contain the *genes* which are the biologic units of heredity. All of the genes in the 46 chromosomes constitute an individual's total genetic make-up or the *genome*. The environmental condition of the cytoplasm influences the chromosomes of the nucleus by activating some genes and repressing others. The set of molecules (basic proteins called histones and possibly RNA molecules or chromosomal RNA) responsible for specific repression of parts of the genome is termed the *epigenome*. Some workers would include in a definition of epigenome, anything that influences or affects genetic expression. Thus, in any given chromosome, only some of the genes are active at any one time. Different cells have different genes at work. However, all genes are present in all cells. Genes responsible for synthesizing pancreatic amylase are inactive in every body cell except in certain cells in the exocrine

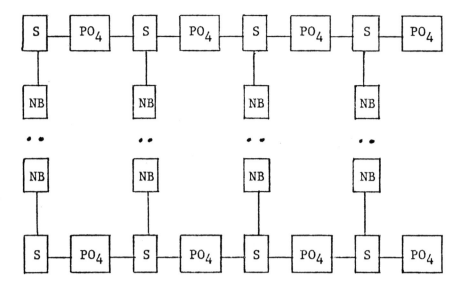

Fig. 1-22. Diagram of DNA.

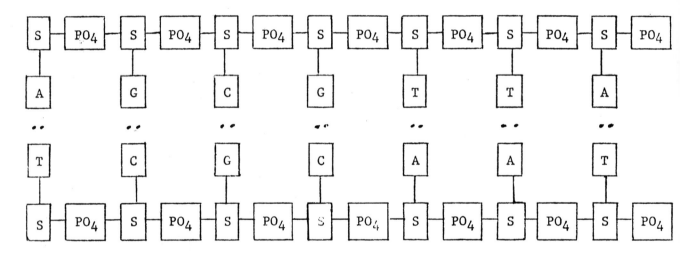

Fig. 1-23. More specific diagram of DNA.

pancreas. These genes control the cell by directing the synthesis of protein. The epigenome in these pancreatic cells are different from those in a liver cell. When a whole chromosome becomes inactive, as is normally the case of one of the X chromosomes in a female (who has two X chromosomes), it condenses in such a manner that it is resolvable by light microscopy even during interphase of the cell cycle. This heterochromatinized chromosome is referred to as the *sex chromatin body*, (*Barr body*) seen only in the female (Figs. 1-18 and 1-19). Only one X chromosome is needed to help direct the activities of the cell (euchromatin), the other is simply turned off (heterochromatin). Different X chromosomes are active in different cells in the normal female. For example, in one cell the paternal X may be operative and the maternal X heterochromatinized; whereas in another cell the opposite may be true. Thus, the normal female is a mosaic with respect to which of her X chromosomes is active in any particular cell in her body. Since the

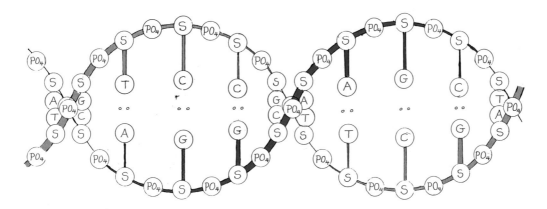

Fig. 1-24. Diagramatic and simplified version of the DNA double helix. The helix is composed of two polynucleotide chains linked together by hydrogen between the base pairs.

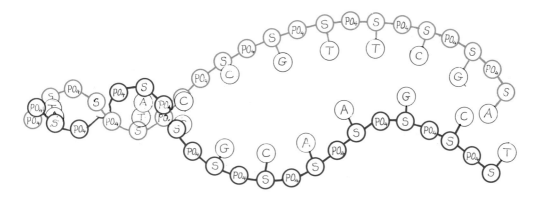

Fig. 1-25. The two polynucleotide chains separate between the base pairs.

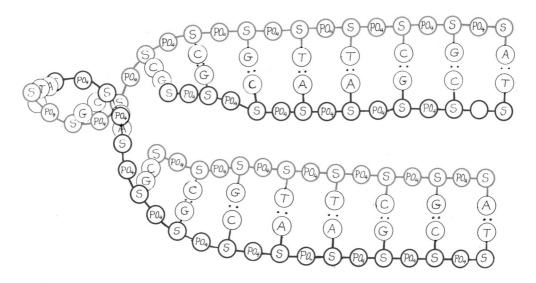

Fig. 1-26. Each of the bases in the two polynucleotide chains are exposed when the strands separate. They now act as templates for the formation of complementary polynucleotide chains.

male has only one X chromosome, it must be kept continually active (euchromatin). Maleness is determined by the Y chromosome, hence the male needs active Y and X chromosomes. He does not have an extra, unneeded X chromosome which can be heterochromatinized, and therefore, the normal male does not show a sex chromatin body.

The sex chromatin body can be recognized in the female as a dark staining dot (in nerve cells) near the nucleolus (Fig. 1-18); against the nuclear membrane (Fig. 1-19); in cells of the stratum spinosum of the epidermis as well as in the middle layer of any stratified squamous epithelium; and as a "drumstick" appendage of heterochromatin in neutrophil white blood cells. Skin biopsies or smears of exfoliated cells taken from oral mucosa serve admirably to identify the true sex in cases of certain types of sexual anomalies (pseudohermaphroditism). While present in every cell of a female, the sex chromatin body can be identified with certainty only in those few nuclei in which it happens to lie in a plane passing through the center of the nucleus and perpendicular to the optical axis of the microscope. In whole cells (neutrophils) the sex chromatin body

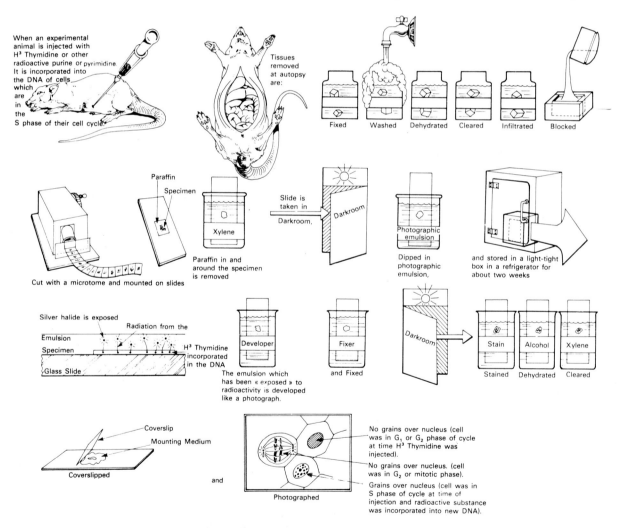

Fig. 1-27. Diagram of autoradiography

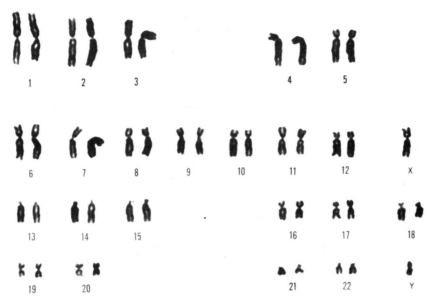

Fig. 1-28, A. Karyotype of a human cell in metaphase stained with routine Giemsa stain. The chromosomes are arranged in order of descending size with morphologically similar chromosomes placed together. Note that some chromosomes, e.g. pairs 1, 2, 3 and 16, have distinctive shapes that permit their individual identification. Most of the homologous chromosome pairs, however, have such a similar appearance that they cannot be differentiated on purely morphological grounds.

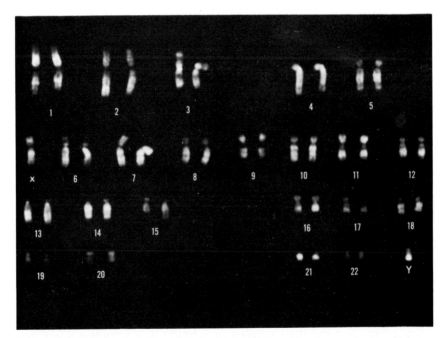

Fig. 1-28, B. Karyotype of same cell stained with quinacrine mustard and photographed with ultraviolet illumination. Each chromosome pair has a unique pattern of fluorescent bands and therefore, can be distinguished from other morphologically similar chromosomes. For example, pair 13 shows more intense fluorescence at the end of the long arm, whereas, pair 14 has more fluorescence near the centromere and only a single band near the end of the long arm. Pair 15 shows moderate fluorescence in the proximal long arm and the distal long arm is dully fluorescent. The intensity of fluorescence for pair 21 is greater than that of pair 22. The end of the Y chromosome is the most brightly fluorescing region of all. (Karyotypes and caption prepared by Dr. Janet D. Rowley, University of Chicago.)

may be folded under the nucleus, making identification impossible.

Mitosis and DNA Replication

The Cell Cycle (Fig. 1-20, A)

The cell cycle may be defined as the orderly sequence of events that occur in the reproductive life of a cell. It may be divided into two major stages, *mitosis* and *intermitotic period* (interphase). The latter may be divided further into three subphases, G1, S and G2. The first subphase, G1 (Gap-1) is the *posmitotic* interval. It lasts from the end of telophase to the beginning of the DNA *synthetic phase* or *S phase*. During the G1 phase, the cell prepares to synthesize new DNA. Before understanding how chromosomes duplicate, it is important to learn something of its chemistry. Each chromosome is a complicated chemical complex of deoxyribonucleic acid (DNA) and proteins, the most important of which are histones. The DNA of the chromosomes consists of a pair of polynucleotides arranged in the form of double helices and hooked together through their base side chains. A *nucleotide* is a compound consisting of a nitrogenous base (purine or pyrimidine), a sugar (ribose or deoxyribose) and a phosphate group (Fig. 1-20).

A *polynucleotide* is a series of nucleotides linked together through their phosphate groups (Fig. 1-21).

DNA is made by joining (via hydrogen bonding) two polynucleotides at their nitrogenous bases (Fig. 1-22).

The sugar in DNA is deoxyribose, and the nitogenous bases are the purines *adenine (A) and guanine (G)*; and the pyrimidines are *thymine (T)* and *cytosine (C)*. The bases are arranged so that adenine always joins with thymine (A-T) and guanine always joins with cytosine (G-C). When they are put together, they may be arranged in a pattern such as depicted in Fig. 1-23. The only additional requirement is to twist the molecule into a double helix (Fig 1-24).

The succession and position of the four bases are subject to permutations. Each sequence of these molecules is, in a way, comparable to a symbol conveying a small portion of "genetic information". "Genes" may be compared to dots and dashes of the Morse code which, when properly arranged in a linear order, are capable of transmitting almost any message among human beings. Much more capable of transmitting information are permutations of position and sequences of pairs of the four nitrogenous bases. They symbolize a practically infinite variety of hereditary characters.

Before mitosis begins, each chromosome must duplicate itself during the S phase of the cell cycle, when the chromosomes cannot be resolved via light and electron microscopy. Duplication probably occurs by unwinding the helical DNA molecule into its two component polynucleotide chains (Fig. 1-25). Separation of the individual strands, exposes the purine and pyrimidine bases so that each of the two polynucleotide chains can act as a template for the formation of a complementary strand of polynucleotides (Fig. 1-26). It is known that each of the bases will react with only one other base. Thus adenine will react only with thymine, and guanine only with cytosine. If a split DNA molecule exposes a sequence of bases: cytosine, cytosine, guanine, thymine, thymine, cytosine, guanine and adenine, it would attract molecules of deoxyribonucleotides with complementary bases, i.e., guanine, guanine, cytosine, adenine, adenine, guanine, cytosine and thymine (Fig. 1-26). The other exposed strand of the originally separated DNA molecule duplicates in a similar manner. In this fashion each of the chromosomes in the nucleus duplicates itself. If a radioactive form of one of the bases is introduced during the S phase, it can be incorporated into the new DNA molecules. Thus when thymidine (thymine attached

MITOSIS AND DNA REPLICATION

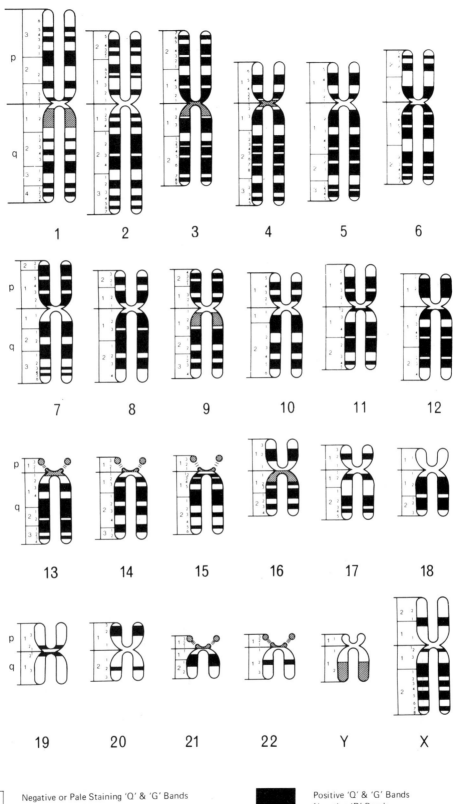

 Negative or Pale Staining 'Q' & 'G' Bands
Positive 'R' Bands

 Positive 'Q' & 'G' Bands
Negative 'R' Bands

Variable Bands

to the sugar) containing, tritium (3H) (tritiated thymidine) the unstable and therefore radioactive isotope of hydrogen, is given to a population of growing cells, those cells in the S phase of the cell cycle incorporate the tritiated thymidine into the newly synthesized DNA. The resulting radioactivity may be detected via autoradiography (Fig. 1-27) or by biochemical techniques such as liquid scintillation counting.

In G_2 (premitotic interval), the cell prepares for the imminent mitosis. The chemical amount of DNA in G_2 is exactly double the amount in G_1. In G_1 the chromosomes are composed of one *chromatid* with one *centromere*. In G_2 the chromosomes are composed of two chromatids held together by one centromere as shown in Fig. 1-16. This morphological arrangement cannot be seen until the chromosomes become visible during mitotic prophase.

Mitosis (Mitotic Cycle) (Fig. 1-16)

Early histologists, studying fixed, stained preparations described four sequential phases of *mitosis: prophase, metaphase, anaphase* and *telophase*. The intervening named phases: *prometaphase, metakinesis, anaphase break* and *reconstruction* attest to continuity of the mitotic cycle.

Prophase commencing when interphase chromosomes contract by coiling (Fig. 1-17 c), accounts for the increased heterochromatin. Concomitantly, centrioles duplicate and migrate to opposite sides of the nucleus to form the *mitotic spindle*.

Fragmentation of the nuclear membrane, permits the shortened chromosomes to engage the spindle machinery. The *nucleolonema* of nucleoli unravels to be dispersed to all chromosomes. Hence it can not be resolved unless special methods are employed.

Metaphase. During spindle development, randomly arranged chromosomes (*prometaphase*) gyrate along the spindle fibers (*metakinesis*) to align at the equatorial plane, midway between opposing centrosomes. The *centromere* of each chromosome is *coupled* to two spindle fibers, one from each centrosome. The composite of chromosomes, spindle and centrosomes is called the *spindle apparatus*. Later in anaphase, contraction of the spindle fibers pulls the chromosome apart at the centromere, to form two *chromatids*. Time lapse microcinematography of single cells in metaphase shows the arms of V-shaped chromosomes waving freely in the gelatinous cytolasm, with their apexes firmly attached to the spindle by centromeres. This "dance of the chromosomes" persists until the centromeres of each chromosome snap apart (*anaphase break*) to signal the end of metaphase and the start of anaphase.

Anaphase. As each centromere snaps apart on cue, the chromosomes "peel" apart from the centromeres to their ends. Identical chromatids become chromosomes of the daughter cells (Fig. 1-16). Contracting spindle fibers pull the daughter chromosomes, centromeres first, toward the centrosomes. At a speed of one µm per minute, anaphase is completed in minutes.

◄
Fig. 1-29 (facing page). A diagrammatic presentation of chromosome bands as observed with quinacrine and special Giemsa staining techniques. This nomenclature was adopted by the Paris Conference (from National Foundation). Note that the short arm of the chromosome is identified by "p", and the long arm by "q". The chromosome arm is divided into regions (up to 4 in the long arm of No. 1) with major landmarks used to separate different regions; each region is further subdivided into bands. The regions and bands are numbered from the centromere to the end of the arm. A particular band is identified by the chromosome, the arm, the region and the band number. Thus 1Q31 refers to the prominent band in the middle of the long arm of chromosome No. 1 (see the first member of pair 1). (Courtesy of Dr. Janet D. Rowley; Human Chromosome Methodology published by Academic Press, Inc.; and the National Foundation.)

Telophase begins when (1) anaphase chromosomes reach the spindle ends, (2) spindle fibers attached to centromeres disappear, (3) the nuclear membrane reassembles around each daughter chromosome group (Fig. 1-16, H), (4) the equatorial cytoplasm undergoes *cytokinesis* (constriction) (Fig. 1-16, G) by microfilaments (plus actin and myosin) under the protoplasmic face of the cell membrane and (5) the nucleolonema reorganizes at specific regions of *nucleolar organizer chromosomes*. Thus, the two *daughter* cells are identical in every aspect. During the *S (synthesis) phase* of the mitotic cycle, the molecular structure of DNA is duplicated for each chromosome. The machinations prevelant from prophase through telophase serve to *split* the chromosomes into chromatids and distribute identical genetic information to each cell.

Reconstruction. Reconstruction is the initial step in the G_1 (*Gap 1*) *Phase* for the daughter cells. The morphologic features of the *mother cell* are reestablished and the cells resume their functions. Cells repeatedly lost from the epidermis, intestinal epithelium and bone marrow are thus continually replaced to maintain the integrity and viability of the tissues. Some cells may exit the G_1 Phase and function for long periods of time without dividing in a state called G_0. Changing stimuli and environment may drive the cell back into the mitotic cycle to provide new cells. On the other hand, some tissues (i.e., cardiac muscle and nerve) repeat the mitotic cycle during development until an optimum number of cells in achieved. At that time, genes controlling cell division are *turned off,* and the cells exit the mitotic cycle to enter the G_0 phase. Subsequent loss of these cells (by disease, injury or oxygen deprivation), will not stimulate a reparative series of mitoses to replace the dead cells. In their place scar tissue (connective tissue) grows in.

Chromosome Karyotypes. The grouping of human chromosomes according to morphologic characteristics is called *karyotyping*. The techniques involve: (1) stimulating cells in tissue culture to enter mitosis, (2) arresting cells in metaphase with antimitotic drugs to disrupt the mitotic spindle, (3) immersing cells in a hypotonic solution causing the chromosomes to spread out and (4) analyzing stained chromosomes according to length of the chromosome arms and position of the centromeres. A *normal* count shows 22 pairs of autosomes and 1 pair of sex chromosomes (XX, female; XY, male). This procedure distinguishes 7 *Groups*: *A, B, C, D, E, F,* and *G*. Pairs 1-3 of *Group A* are large *metacentric* chromosomes in which the centromere is located midway so that each chromosome has essentially equal arm lengths. Pairs 4-5 (*Group B*) are large *submetacentric* chromosomes possessing eccentric centromeres so that one arm is longer than the other. Pairs 6-12 and X (*Group C*) are medium sized submetacentric chromosomes. Pairs 13-15 of *Group D* are large *acrocentric* chromosomes, i.e. one long arm and one minuscule arm bearing a satellite (tiny knob). *Group E* has a pair of no. 16 small metacentric chromosomes, and two pairs (nos. 17, 18) of small submetacentric chromosomes. Pairs 19-20 of *Group F* are small metacentric chromosomes, and Pairs 21-22 (*Group G*) are short acrocentric chromosomes with satellites. The *Y* sex chromosome (Male) of *Group G*, lacks a satellite.

New methods involving banding patterns provide better characteristics to identify human chromosomes. Prominently stained bands subdivide the chromosome arms into well defined regions (*Paris Conference on Standardization in Human Cytogenetics*, The Williams and Wilkins Co., Baltimore. 1973). Banding patterns reveal many structural details and features of chromosomes that were unknown. The chromosomes of man have been classified into seven groups. Such charts are used to detect chromosomal abnormalities in certain congenital diseases. In most individuals with Down's syndrome, chromosome number 21 is present in triplicate rather than in duplicate (Fig. 1-30). This

Fig. 1-30. Group 21-22 chromosomes from a trisomy 21. (Preparation and photography by Drs. L. Scheving and S. Tsai, University of Arkansas for Medical Sciences.)

syndrome, is properly named *trisomy 21*. The term mongoloid is inappropriate and must never be used.

Nuclei possessing more than two complete balanced sets of chromosomes are said to the polyploid (euploid). Specifically, a cell with 3n chromosomes is triploid, one with 4n is tetraploid, etc. Among adult liver cells, diploid, tetraploid, octaploid and dodecaploid nuclei occur. Liver cells seem to show a proportional relationship between cell volume and the total number of chromosomes. In a large liver cell the polyploid chromosomes may be combined in one giant nucleus or distributed among two or more smaller nuclei. Figure 1-15 shows a few liver cells with nuclei of various sizes. A cell is said to exhibit aneuploidy if its chromosome number deviates from the basic number (n) or from exact multiples of n. For example, a cell with one extra chromosome of a homologous pair of chromosomes is *trisomic* (2n + 1) (Fig. 1-30). When a cell has a pair of chromosomes present twice, it is *tetrasomic* (2n + 2). When one chromosome of a homologous pair of chromosomes is lacking (2n − 1) the cell is *monosomic*. If both members of a pair of homologous chromosomes are absent (2n-2) the cell is *nullisomic*.

Protein Synthesis

Proteins comprise about half the cellular dry mass, provide cytoskeletal support and participate in development, growth and maintenance of the cell. The molecular events of *protein synthesis*, presented elsewhere, are mentioned to emphasize correlative structure and function. Three significant points are considered: (1) the nucleus contains chromosomes which make *genes* (regions) of the DNA molecules accessible for *transcription* (copying information), (2) the cytoplasm houses organelles that utilize *transcribed information* for protein synthesis and (3) special molecules deliver information from the nucleus to those cytoplasmic organelles (Fig. 1-31).

Synthesis (Fig. 1-31) commences when a molecule of *ribonucleic acid* (RNA) produced by DNA transcription, relays instruction to the cytoplasm. This is *messenger RNA* or *mRNA* which leaves the nucleus via nuclear pores to function in the cytoplasm. The long thread-like molecule of *mRNA* has a *start* and an *end* designated by *codons* or triplets of nucleotides. Thereafter, 63 possible sequences of three nucleotides each make up a total of 64 codons. Two terminal codons specify *end*

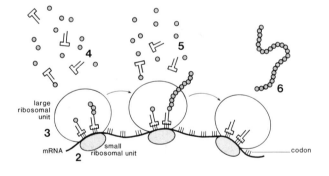

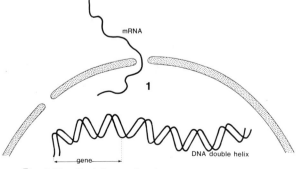

Fig. 1-31. Protein synthesis:

information. The *mRNA* required for synthesis of a specific protein designates the precise sequence of amino acids. Hence, the alignment of single or multiple codons for specific amino acids governs the length and composition of the message encoded on the *mRNA*.

In the next event (Fig. 1-31 [2]), a *small* ribosomal subunit is positioned and bound at a *start* codon on the mRNA, by a molecule of *transfer RNA* (*tRNA*). Then (Fig. 1-31 [3]), a *larger* ribosomal subunit is added, by *tRNA*, to complete the ribosome at the *start* codon. The moving ribosome now begins to assemble a polypeptide at each codon by adding one of 20 amino acids until at the appropriate "stop" codon, the specified polypeptide is formed (Fig. 1-31 [4]). The polypeptide is released when the ribosome reaches a *release* codon. Each amino acid is "recognized" by designated enzymes united to another *tRNA* with the duty of delivering it to the ribosome (Fig. 1-31 [5]).

Numerous ribosomes attached to mRNA form a morphological assemblage resembling a string of pearls (Fig. 1-31 [6]). Such compositions, the *polyribosomes*, are readily detected in transmission electron micrographs either free or fixed to cytoplasmic surfaces of endoplasmic reticulum membranes. Free ribosomal complexes are involved in producing proteins for intracellular use, while those bound to *Er* membranes participate in production of proteins for external use. In the latter case, thread like configurations of protein precursors are injected through transient apertures of the *ER* membrane into the *ER cisternae*. Subsequent processing within the endoplasmic reticulum and the golgi complex prepares the protein for *exocytosis*.

Sizes and Shapes of Cells

According to their functions and locations, the shapes and sizes of cells vary tremendously. Some examples are shown

The average internal cell has, therefore, the shape of a tetrakaidekahedron (Fig. 1-32, B: C); three facets always meet in one edge. Should there be, for a moment, edges in which four facets meet (Fig. 1-32, B: D and G, above), outside pressure (arrows) and slippery surfaces would dissolve any one four-faceted corner into two or four three-faceted corners, sometimes with a new facet being inserted, as seen in Figure 1-32 B: E. The process in which a rhombic dodecahedron is thus transformed into an orthic tetrakaidekahedron is shown in Figure 1-32, B: D-G. Movements and unequal sizes of cells cause variations from the basic type (Fig. 1-32, B: H: human fat cells). Surface cells in stratified epithelia average 11 facets (Fig. 1-32, B: I), while cells of simple epithelia average 8 facets (insert at bottom of Fig. 1-32, B). Since these laws of shape are mechanical in nature, analogous forms are encountered in inanimate masses such as among grains in pure metals and bubbles in a foam (Fig. 1-33). Also the droplets of sebum as well as the cells in sebaceous glands obey these laws of shape (Fig. 1-34).

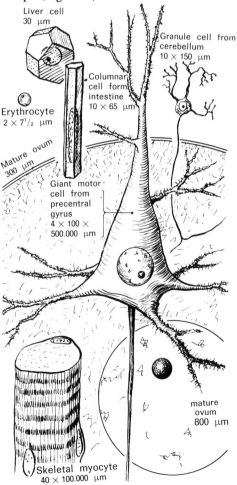

Fig. 1-32, A. Various types of cells, all drawn to the same scale.

42 THE CELL

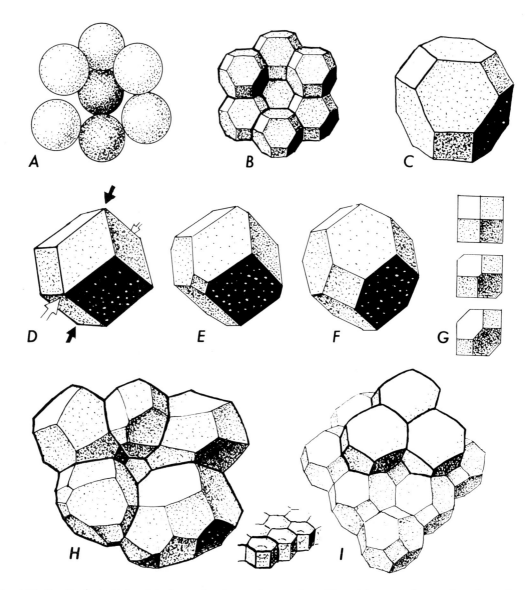

Fig. 1-32, B. Frederic T. Lewis's laws of cytomorphogenesis. (H and I are re-drawn after Proc. Am. Acad. Arts and Sc. 77: 147-187, 1949.)

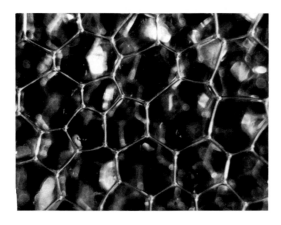

Fig. 1-33. Soap bubbles obey the same laws of shape as do cells. (From Weibel: Morphometry of the Human Lung, Berlin and Heidelberg, Springer, 1963.)

in Figure 1-32, A. The ubiquitous erythrocyte (red blood corpuscle, RBC), a "cell" without a nucleus, with its practically constant diameter of 7.5 μm (in fixed material) serves as an excellent measuring device to estimate cell dimensions in slides and photomicrographs.

Many cells, particularly those with long processes, seem to acquire their shapes by active growth, while many others are forced into shapes by internal and external pressure. Epithelial cells and fat cells, for example, are spherical when their internal pressure (turgor) is higher than that of the environment (Fig. 1-32, B: A). When packed in masses (Fig. 1-32, B: B), they become polyhedral with the average number of facets being 14.

The average internal cell has, therefore, the shape of tetrakaidekahedron (Fig. 1-32, B: C); three facets always meet in one edge. Should there be, for a moment, edges in which four facets meet (Fig. 1-32, B: D and G), outside pressure (arrows) and slippery surfaces would dissolve any one four-faceted corner into two or four three-faceted corner, sometimes with a new facet being inserted, as seen in Figure 1-32 B: E. The process in which a rhombic dodekahedron is thus transformed into an orthic tetrakaidekahedron is shown in Figure 1-32, B: D-G. Movements and unequal sizes of cells cause variations from

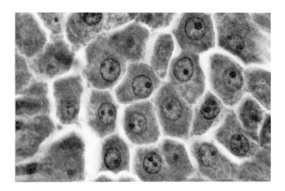

Fig. 1-35. Tangential section of apocrine sweat gland to show that epithelial cells when sectioned appear as polygons with an average number of sides of 6, where 3 cells seem to meet in one point.

the basic type (Fig. 1-32, B: I), while cells of simple epithelia average 8 facets (insert at bottom of Fig. 1-32, B). Since these laws of shape are mechanical in nature, analogous forms are encountered in inanimate masses such as among grains in pure metals and bubbles in a foam (Fig. 1-33). Also the droplets of sebum as well as the

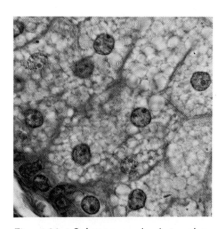

Fig. 1-34. Sebaceous gland to show that the entire cells and the oil droplets in them obey the laws of cytomorphogenesis by mutual pressure.

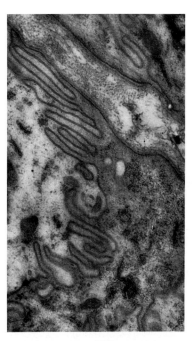

Fig. 1-36. Intercellular interdigitations of gastric surface epithelial cells. Courtesy of Dr. Y. Ito, Harvard Medical School.

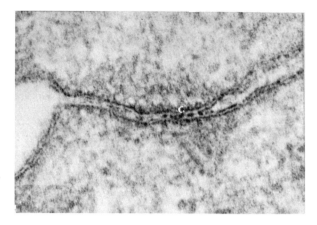

Fig. 1-36, A. Tight junction between two columnar cells, very high resolution EM. Courtesy of Dr. D. Friend, Univ. of Cal. S.F.

Fig. 1-37. Interlocking metal piling in a breakwater at the shore of Lake Michigan.

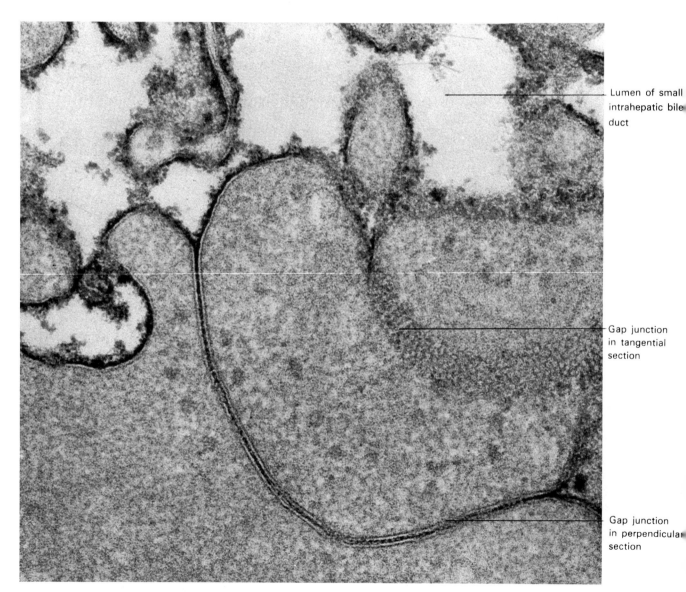

Fig. 1-37, A. Tight junction and gap junction in the wall of a small, intrahepatic bile duct. (Courtesy of Dr. D. Friend, Univ. California at San Francisco.)

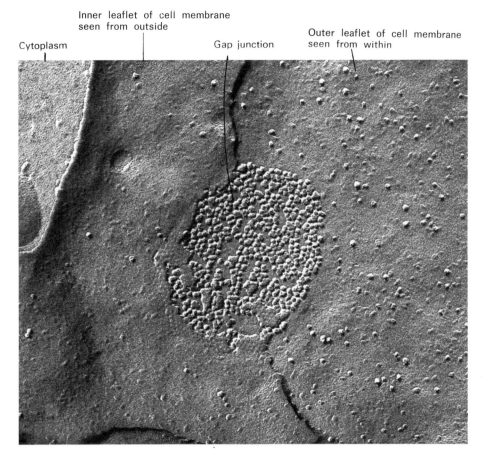

Fig. 1-37, B. Freeze etching of boundary between two cells of the zona fasciculata of rat. (Courtesy of Dr. D. Friend, Univ. California at San Francisco.)

Fig. 1-37, C. Freeze etching of tight junction; from epididymis of rat. (Courtesy of Dr. D. Friend, Univ. California at San Francisco.)

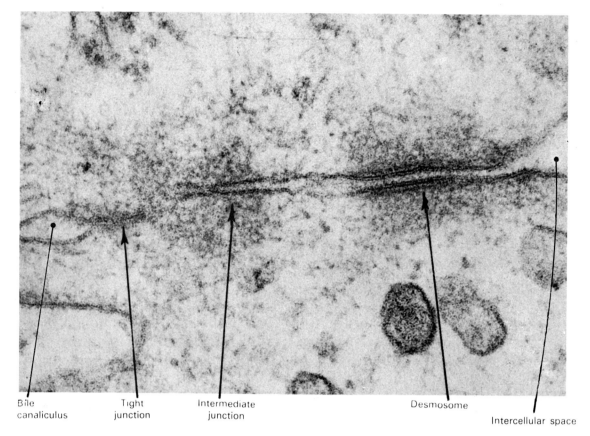

Bile canaliculus   Tight junction   Intermediate junction   Desmosome   Intercellular space

Fig. 1-38. Junctional complex between two liver cells (needle biopsy) from C. Biava, Lab. Invest. *13*: 840-864, 1964.

cells in sebaceous glands obey these laws of shape (Fig. 1-34).

When sectioned, cells in masses (Fig. 1-35) appear as polygons, six being the average number of sides. Three sides converge in every corner of such a section. Four-sided corners are unstable. A corner in a section corresponds to an edge in space where three cells meet.

Connections Between Cells

Cells may just stick to each other by virtue of the adhesiveness of the cell membranes, as among mesenchymal cells; or they may be fastened to each other by dove-tailing superficial projections and depressions (Fig. 1-36). The arrangement is very similar to dove-tailing and tongue-and-groove joints in cabinet work. Interlocking metal pilings in foundations of large buildings or breakwaters at the seashore (Figs. 1-37 and 2-19), employ the same mechanical principle.

Cells in masses often are fastened to each other by adhesive surface specializations: the desmosome (macula adherens), the intermediate junction (zonula adherens) and the tight junction (zonula occludens).

The *desmosome* or *macula adherens* is a spot attachment of limited extension between two cells. One-half of the desmosome (hemidesmosome) is a structural configuration belonging to one cell, while the other hemidesmosome of the macula adherens belongs to the adjacent cell. The intercellular space, between the two hemidesmosomes (about 25 nm wide) contains a layer of "intercellular cement". The protoplasmic face of the plasmalemma of

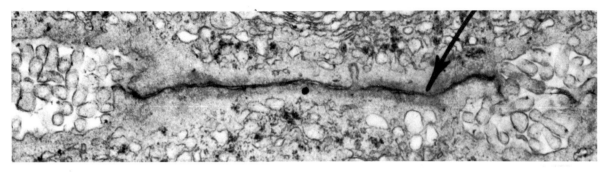

Fig. 1-39. Longitudinal section of tight junction connecting two liver cells along a bile canaliculus. (From C. Biava, Lab. Invest. *13*: 840-864, 1964.)

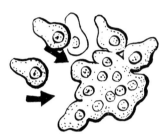

Fig. 1-40. A syncytium.

Fig. 1-41. A plasmodium.

each hemidesmosome is composed of some dense material into which many fine cytoplasmic filaments (tonofilaments) approach, and loop back like hairpins. The macula adherentes are found distributed at irregular intervals over the lateral boundaries of a variety of cells.

An intermediate junction, *zonula adherens*, may surround entire cells in a band-like fashion (Fig. 1-38) or may be of limited extension. In the area of the intermediate junction the external surfaces of adjacent plasmalemmas are about 20 nm apart. Tight junctions (zonula occludentes) are formed by linear arrangement and fusion of protein particles (Fig. 1-37, C). In very high resolution EM, a tight junction is shown in Fig. 1-36A, 1-37A).

When all three types of adhesive mechanisms occur together, they collectively are called a *junctional complex* (Fig. 1-38). Only rarely does a section pass parallel to the zonula occludens so that its continuity can be seen. Such a section is shown in Fig. 1-39.

Another type of cell junction has been identified. Polygonal projections from adjacent cells may approach each other in an alternating arrangement which results in partial occlusion of the intercellular space between the adjacent cells. This type of junction is a *nexus* (gap junction). Because it is an area of decreased electrical resistance, communication between cells is possible and hence the proposed name *macula communicans*. (Fig. 1-37, A and B).

The surface of a cell is covered with a material called the *glycocalyx*, a complex of carbohydrates associated with lipids (glycolipids) or proteins (glycoproteins). The basement membrane or basal lamina is a specialized form of glycocalyx. The glycocalyx is particularly well developed on the surface of the epithelial cells of the gastrointestinal tract. There it serves not only to protect the cell from many proteolytic and mucolytic enzymes, but also

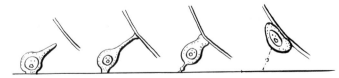

Fig. 1-42. One type of ameboid locomotion.

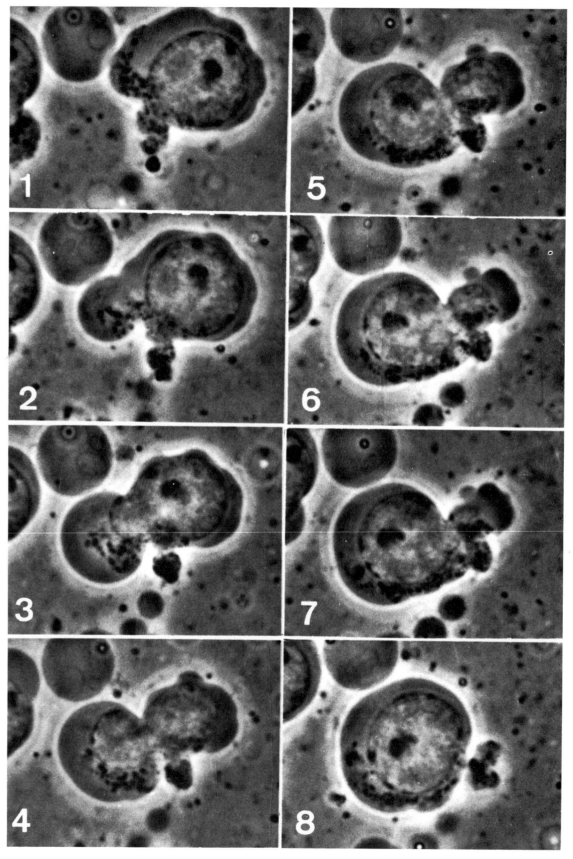

Fig. 1-43. Another type of ameboid motion such as executed by primordial germ cells. Selected frames from a motion picture on primordial germ cells by Drs. R. Hayashi and R.J. Blandau, Department of Biological Structure, University of Washington.

to act as a barrier to large particles or molecules, while permitting smaller substances to pass through it.

Symplasms

Most cells possess only one nucleus, but cells with two or several nuclei may be found (i.e., some liver cells, surface cells of transitional epithelium, skeletal muscle cells). A large body of cytoplasm containing many nuclei is called a symplasm.

A syncytium is a symplasm which has arisen from the confluence of many individual cells (Fig. 1-40).

A plasmodium is a symplasm which has developed from a single cell, the nucleus of which has divided repeatedly, while the cytoplasm failed to divide (Fig. 1-41).

It is assumed that "foreign-body-giant-cells", osteoclasts and skeletal muscle cells are syncytia. The external layer of the trophoblast arises early as a syncytium, but continues to grow as a plasmodium.

Movements

All cells perform movements. In some, the movements are slight, while in others they are rather intense. Muscle cells, by contracting, bring about movements of whole parts of the body. Cilia on the free surface of cells propel liquids. Several kinds of cells, however, are endowed with the ability of individual locomotion. A spermatozoon, for example, swims by means of its flagellum (tail).

Amebism-Ameboidism

Many cells of connective tissue, as well as the white blood cells, maneuver by ameboid motions. They extend temporary processes (pseudopodia); (slender and long in the case of mesenchymal cells, broad and short in the case of leucocytes and macrophages) which adhere to fibers or to other cells. A pseudopodium can draw the cell body toward the point of its attachment (Fig. 1-42). Another form of ameboid motion is illustrated in Figure 1-43. A cell simply thrusts its body through the tissue sending forward a very broad pseudopodium into which the nucleus moves. The cytoplasm from the trailing side of the cell then is drawn in. These cells do not attach themselves to existing structures but "elbow their way" through the tissue.

Phagocytosis

Certain cells, such as macrophages (histiocytes) and leukocytes, can engulf particles by invaginating the adjacent cell membrane and surrounding them with a smooth-walled vesicle, the phagosome (Fig. 1-44). If these particles consist of organic material, the phagocytic cells (phagocytes) can digest them (see lysosomes). Phagocytosis by some cells occurs so rapidly that the particle is seen completely within the cell practically as soon as it has touched the cell surface.

Fig. 1-44. Phagocytosis.

Pinocytosis

Some cells and plasmodia, such as the components of the embryonic trophoblast, mesenchymal cells, macrophages, endothelial cells and, perhaps, the podocytes in the kidney, "drink" by surrounding drops of liquid with flat, spoon-shaped pseudopodia (Fig. 1-45) or by producing cuplike depressions in their surface. This process

Fig. 1-45. Macro-pinocytosis.

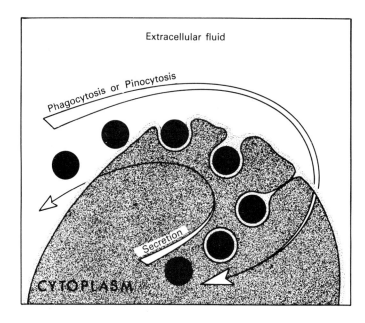

Fig. 1-46. An interpretation of phagocytosis and of merocrine secretion as developed by Dr. Stanley Bennett.

is known as *pinocytosis* and is seen in cells in tissue culture as viewed by time lapse micro-cinematography.

On a submicroscopic scale, depressions appear in the cell membrane transporting small amounts of extracellular fluid into the interior of the cell. The process is called *micropinocytosis*, and the little depressions are called *micropinocytic vesicles*. The importance of this process lies in the transport of materials across cells, especially across endothelial and mesothelial linings. In general, the uptake of extracellular materials is known as endocytosis; whereas the discharge of intracellular material is exocytosis. The material transported across cells is called a "transudate".

Figure 1-46 illustrates the mechanism of phagocytosis, pinocytosis and secretion.

The cell membrane is considered selectively adhesive to specific molecules which, by their chemical affinity to the surface membrane, are adsorbed to it. Adhesion may be due to mechanical irritation. The cell membrane forms a depression around the particle. The depression becomes a pocket which is pinched off in the interior of the cell. Lysosomes release enzymes which dissolve the little intracellular balloon of membrane and digest the particle or molecules whose substances are now within the cell without having traversed its membrane.

The process of secretion occurs by exocytosis (merocrine secretion). Molecules of the substance to be secreted, accumulate in a drop around which a membrane forms. This subsequently joins the cell membrane, and the droplet is discharged (left arrow in Fig. 1-46).

Transformation of Cells

The immediate offspring of the fertilized ovum (zygote) which result from the initial six divisions are roughly alike in structure and size, but thereafter diverge in appearance and behavior. The cells pass through many developmental stages before assuming their final location, form and function. Attainment of this stage is called differentiation. When differentiation is completed, a cell can be classified as a liver cell, a renal tubular ell, a lymphocyte, a muscle cell, etc. Nerve cells, once differentiated, retain their specific character and position throughout life.

However, there are many cells which undergo drastic changes of character. For example, the mucous cells in the intestinal glands proliferate so rapidly by mitotic division that they are pushed upward onto the free surface. After attaining a position on the villi, they lose their secretory function, acquire microvilli and become absorptive in function, but eventually lose their foothold and die. Epithelial cells of the embryonic thymus are transformed into lymphocytes. Connective tissue cells in the capsule of the suprarenal gland divide continuously throughout life. Their offspring grow and become zona glomerulosa cells which in turn are converted to zona fasciculata cells. Finally they become zona reticularis cells and die. In every pseudostratified epithelium we find a layer of small, basal cells (replacement cells). When a superficial cell dies, a basal cell moves up to take its place.

In very young embryos (4 to 8 mm in length), mesenchymal cells can differentiate into liver cells; and later (17 mm until middle childhood), the intrahepatic duct system develops by transformation of liver cells into ductal epithelial cells. The above are examples of irreversible (one-way) transformation.

2...
Epithelium

A continuous layer of cells in mutual contact which covers a surface or lines a lumen or cavity is called an *epithelium* (Fig. 2-1). In certain organs, such as the liver and many endocrine glands, the relation to surfaces is obscure or absent; but

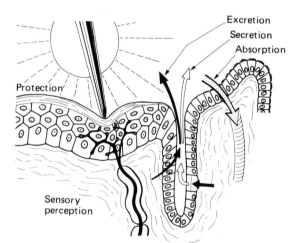

Fig. 2-2. Functions of epithelium.

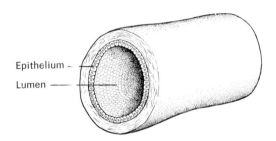

Fig. 2-1. An epithelium.

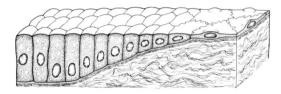

Fig. 2-3. Simple epithelium.

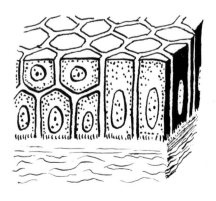

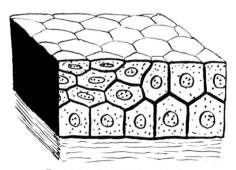

Fig. 2-4. Stratified epithelium.

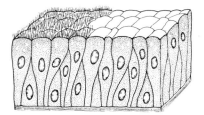

Fig. 2-5. Pseudostratified epithelium.

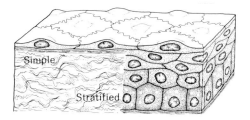

Fig. 2-6. Squamous epithelium.

the intimate association of the cells offers adequate evidence of their derivation from an epithelium and indeed justifies their description as epithelial cells. The narrow spaces between epithelial cells usually contain an intercellular "cement" which plays a role in holding the cells together. An epithelium in the strict sense of the word is attached to a *basement membrane* rich in glycoproteins.

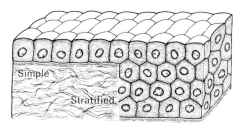

Fig. 2-7. Cuboidal epithelium.

The basement membrane, a product of the epithelial cells, may be characterized by its structure and composition. Structurally it presents as a uniform continuous layer of extracellular matrix produced by the epithelial cells. Via electron microscopy, two distinct laminae (layers) may be resolved: (a) the *lamina lucida (rara)*, an electron lucent, 10 nm thick layer in contact with the base of the epithelial cells, and (b) the *lamina basalis* (*densa*), an electron dense layer (20 to 50 nm thick), separated from the epithelial cells by the lamina lucida (Figs. 2-20, 2-21). The basement membrane consists of (a) collagenous glycoproteins (type IV collagen) and non-collagenous glycoproteins (laminin and entactin), and (b) a heparan sulfate proteoglycan. Basement membranes frequently are anchored to underlying connective tissues by a *lamina fibroreticularis*. Unlike the basement membrane, this lamina is produced by connective tissue fibroblasts. Delicate reticular fibers, embedded in a protein polysaccharide ground substance, are characteristic of this layer. In the older light microscopic literature, the basement membrane was considered to be, what electron microscopy has differentiated as

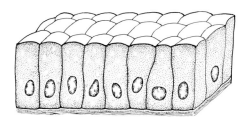

Fig. 2-8. Columnar epithelium.

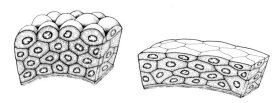

Fig. 2-9. Transitional epithelium, relaxed on the left, stretched on the right.

the lamina, lucida lamina basalis and lamina fibroreticularis.

The base of the epithelial cells attached to the basement membrane is termed the basal or proximal surface. The opposite free or apical surface, usually lines a space, such as the cavity of the stomach, or it faces the air in which we live. The space enclosed by epithelium such as the wall of a tube is called a *lumen*. All epithelia (except for endocrine epithelial cells) are avascular and therefore depend on the underlying connective tissue for metabolic support. Both nutrients and metabolites must cross the basement membrane. Epithelium is comparable to a stone wall or tile floor; the stones or tiles represent cells, and the mortar or mastic represents *intercellular substance* (Fig. 2-1), and the basement membrane.

The cellular linings (*mesothelium*) of body cavities, of blood and lymph vessels (*endothelium*) and the secretory cells of endocrine glands are classified as epithelia (Fig. 2-10).

Epithelium protects, secretes, excretes, absorbs and aids in reception of stimuli (Fig. 2-2).

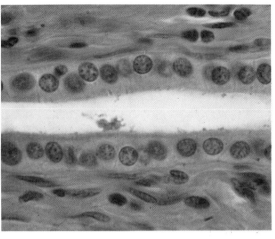

Fig. 2-11. Simple cuboidal epithelium from small bile duct, H&E. 1800 ×.

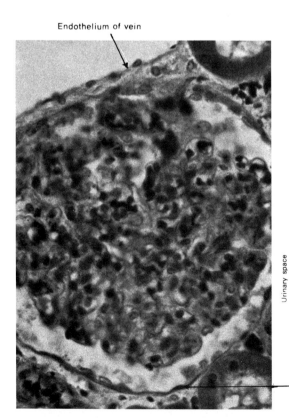

Fig. 2-10. Simple squamous epithelium (glomerular capsule in kidney), Azan. 750 ×.

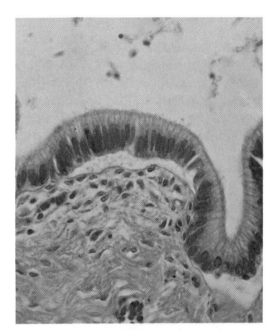

Fig. 2-12. Simple columnar epithelium from gall bladder, Safranin and Picric acid. 1400 ×.

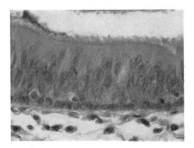

Fig. 2-13. Pseudostratified ciliated epithelium from bronchus, H&E. 650 ×.

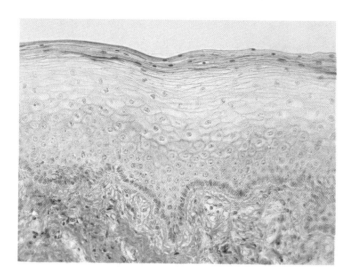

Fig. 2-14. Stratified squamous epithelium from vagina, Azan. 350 ×.

Classification of Epithelia

1. BY NUMBER OF LAYERS:

Simple epithelia consist of a single layer of cells which vary in height (Fig. 2-3), but this height is practically constant in each specific type.

Stratified epithelia consist of 2 or more layers of cells (Fig. 2-4).

Pseudostratified epithelium consists of one layer of cells, most of which are elongated. While every cell is anchored to the basement membrane, not all reach the luminal surface. In pseudostratified epithelium, the nuclei characteristically are located at various levels (Fig. 2-5).

2. BY SHAPE OF SURFACE CELLS:

In *squamous* epithelium, cells with lens shaped nuclei forming the free surface are flat. If the cells are extremely thin, they bulge where the nuclei are located much like a layer of sunnyside fried eggs (Fig. 2-6).

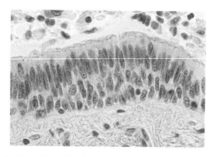

Fig. 2-16. Stratified columnar epithelium from duct of sublingual gland, H&E. 750 ×.

The surface cells of *cuboidal* epithelium, although seldom in the shape of true cubes, have a form of isodiametrical prisms, i.e., they are about as tall as they are wide. Their nuclei are approximately spherical (Fig. 2-7).

Columnar epithelium consists of tall, prismatic cells. The nucleus is oval in most cases and may be polarized toward the basement membrane (Fig. 2-8).

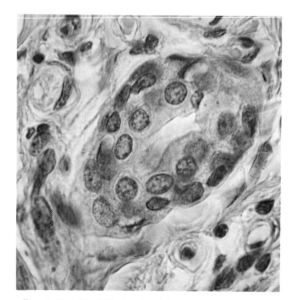

Fig. 2-15. Stratified cuboidal epithelium. Duct of sweat gland, Goldner. 1900 ×.

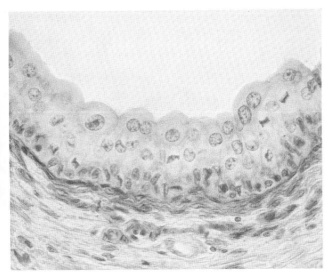

Fig. 2-17, A. Transitional epithelium, contracted from almost empty urinary bladder, Azan. 750 ×.

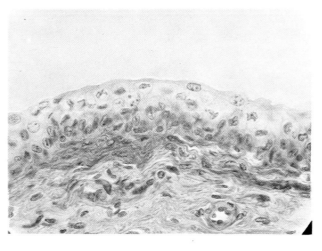

Fig. 2-17, B. Transitional epithelium, extended from full urinary bladder, Azan. 750 ×.

Surface cells of *transitional* epithelium change shape in response to conditions of stretching and relaxation. The cells appear convex on the free surface, when viewed via light microscopy. Via electron microscopy, the luminal surface of the surface plasmalemma is somewhat scalloped. Cells are flexible and slippery and yet adhere to each other (Fig. 2-9). The surface cells are larger than those in the deeper layer, and are frequently *binucleated* or *polyploid*.

Specific Epithelia

To define a specific type of epithelium, two adjectives are usually used: the first refers to the number of layers (simple stratified), the second describes the shape of the surface cells (squamous, cuboidal or columnar). However, one adjective suffices for pseudostratified epithelium which is always columnar and for transitional epithelium which is always stratified.

The known epithelia are the following:
SIMPLE SQUAMOUS EPITHELIUM (Fig. 2-10) - examples: external (parietal) layer of the renal glomerular capsule, lining of the thin segment of the nephric loop, lining of the pulmonary alveoli, endothelium and mesothelium.

SIMPLE CUBOIDAL EPITHELIUM (Fig. 2-11) - examples: small bile ducts, most kidney tubules, loop of the nephron, salivary ducts, thyroid follicles.

SIMPLE COLUMNAR EPITHELIUM (Fig. 2-12) - examples: gallbladder, cervical canal, intestine.

PSEUDOSTRATIFIED EPITHELIUM (Fig. 2-13) - examples: respiratory passages (except smaller bronchioles), duct of the epididymis.

STRATIFIED SQUAMOUS EPITHELIUM (Fig. 2-14) - examples: epidermis, esophagus, vagina.

STRATIFIED CUBOIDAL EPITHELIUM (Fig. 2-15) - only known example in adult man: ducts of sweat glands.

STRATIFIED COLUMNAR EPITHELIUM (Fig. 2-16) - example: male urethra (only in patches), short portions of excretory ducts of lacrimal and salivary glands.

TRANSITIONAL EPITHELIUM (Fig. 2-17) - only known example in animal kingdom: urinary passages.

Since there are three adjectives to identify an epithelium as to the number of layers and four to designate it as to the shape of its surface cells, 12 potential classes of epithelia could exist; although examples have not been found for a few of them (Fig. 2-18).

58 EPITHELIUM

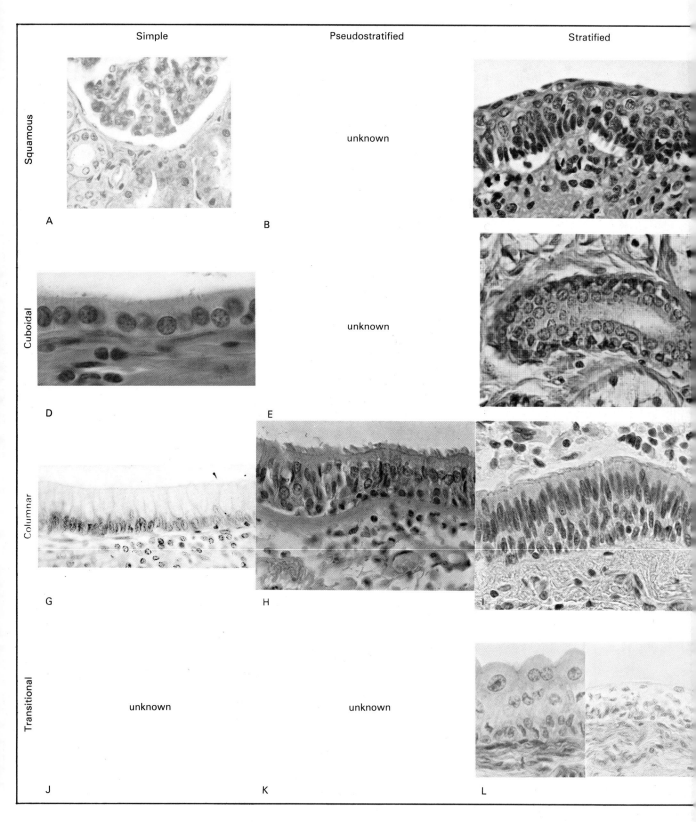

Fig. 2-18. A chart of known and unknown epithelia. A: glomerular capsule; C: vaginal epithelium of infant; D: small bile duct; F: duct of sweat gland; G: colon; H: bronchus; I: duct of sublingual gland; L: urinary bladder.

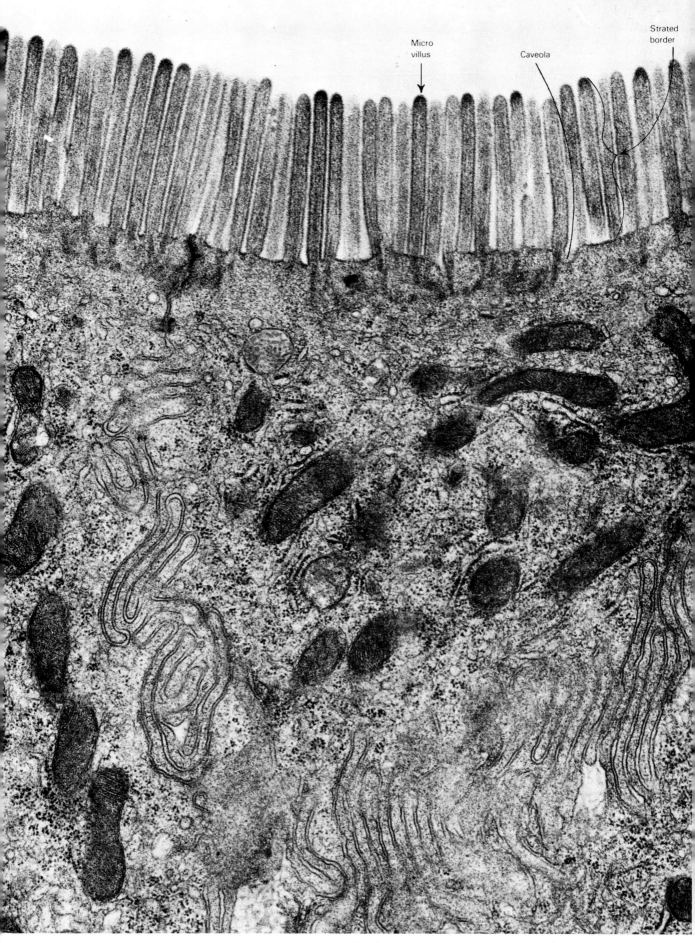

Fig. 2-19. Intestinal cells with striated border. Courtesy of Dr. W. C. Forssmann, Univ. of Heidelberg.

Surface Specializations of Epithelia

The free surface of an epithelium is adapted and modified to its function. Many passages lined by epithelium simply conduct a material, and the surface of their cells appear smooth. Examples: oral cavity, esophagus, excretory ducts of salivary glands, ureter.

The *striated border* of light microscopy (Figs. 2-19 and 2-20) consists of cylindrical, cytoplasmic projections termed *microvilli*, all about 2μm long. The cores of mi-

Fig. 2-20. Intestinal cells with striated border.

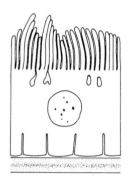

Fig. 2-21. Cell from the proximal convoluted tubule of the kidney showing brush border, caveolae (pinocytic vesicles) and basal infoldings.

crovilli are filled with fine parallel actin filaments anchoring them to the basal cytoskeleton (terminal web). Interaction between these actin filaments and myosin of the terminal web produces a complicated pumping movement. Microvilli serve to increase the surface area of cells involved in absorption. A glycoprotein "fuzzy" coat, the glycocalyx, projects from the plasmalemma as an extension of integral protein. The glycocalyx protects the luminal surface of cells and provides a barrier to prevent certain substances from entering and at the same time selecting colloidal particles and substances in solution for absorption. *Pinocytic vesicles* may be found in the valleys between the microvilli. Examples: bile ducts, gallbladder, intestine.

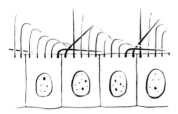

Fig. 2-22. Ciliary movement.

A *brush border* of light microscopy (Fig. 2-21), not unlike a striated border, consists of microvilli possessing a dense, axial core. The microvilli of "brush borders" are much taller than those of striated borders, more closely packed and of unequal length. Those nearer the center of each cell are tallest, and those at the periphery are shorter, presenting a resemblance to the tuft of a paint brush. Example: proximal convoluted tubules of the kidney. Although individual microvilli are best visualized via electron microscopy, the over-all effect of a striated or brush border can be recognized with the light microscope.

Cilia (Fig. 2-22), rhythmically whipping surface projections, move a blanket of mucus along moist epithelial surfaces. They are cytoplasmic extensions anchored in the cells by *basal corpuscles* derived from centrioles. The ciliary motion, in wavelike fashion, progresses over the entire ciliated epithelial surface, regardless of cell boundaries. During the rapid downstroke, a cilium, except for its basal segment, remains rigid; but it is flexed during the slower recovery upstroke (much like the stroke of a bull-whip). Ciliary motion is autonomous (not controlled by the nervous system), and may be studied in fragments of respiratory epithelium in tissue culture, whipping at about 1000 strokes per minute. A singular cilium such as the tail of a sperm or the moving projection of a cuboidal cell in the rete testis is called a *flagellum*. Throughout the animal kingdom and the lower plant kingdom up to the Cycadeae (palm-like gymnosperms), every cilium (Fig. 2-23) and every flagellum contains two axial and nine peripheral double filaments. Figure 2-24 distinguishes between the taller and wider cilia (5 to 10 μm in length and 0.2 μm in diameter) and shorter, narrower microvilli (2 μm in length) in longitudinal section. Examples: uterine tubes, respiratory passages.

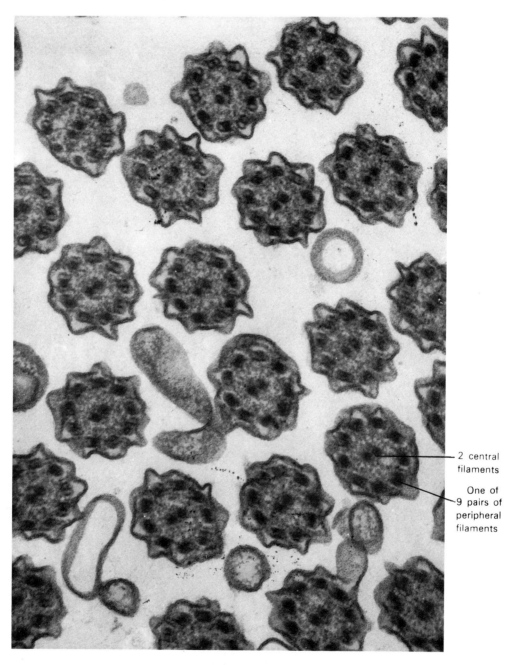

Fig. 2-23. Cross sections of cilia and of microvilli from nasal respiratory epithelium. Courtesy of Dr. D. H. Matulionis, University of Kentucky.

Stereocilia are long, thin, sometimes branched non-motile microvilli arranged around the free edges of cells like fences (Fig. 2-25). In the spiral organ they have a special orientation to assist in transmitting auditory sensations (Fig. 17-14). They do not have an internal structure characteristic of true cilia, but they resemble the microvilli of the nose (Fig. 2-24). Example: epididymis, inner and outer hair cells of the spiral organ (of Corti).

Terminal bars is an old light microscopic term for the entire junctional complex of electron microscopy (Fig. 1-4, A, Fig. 11-38). They seal the luminal ends of cells together, separating the lumen from the intercellular spaces (Fig. 2-26, also Figs. 1-37, A and C; 1-38 and 1-39). Examples:

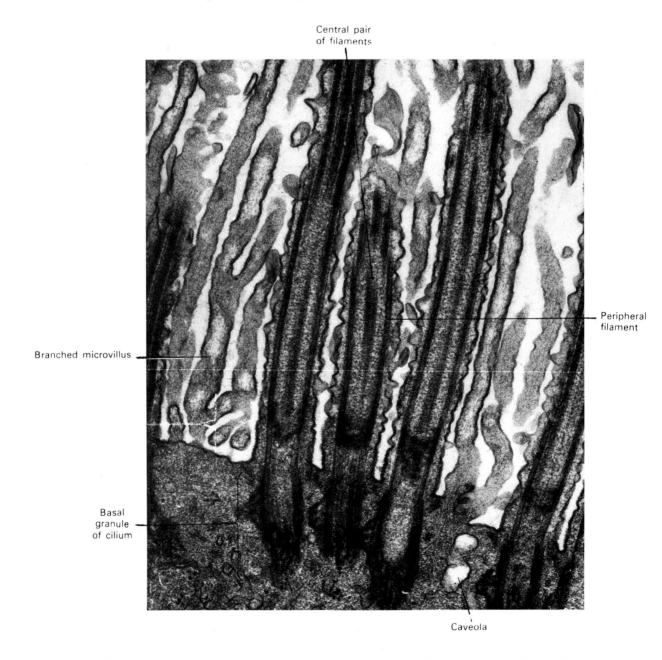

Fig. 2-24. Cilia and branched microvilli from nasal respiratory epithelium. Courtesy of Dr. D. H. Matulionis, University of Kentucky.

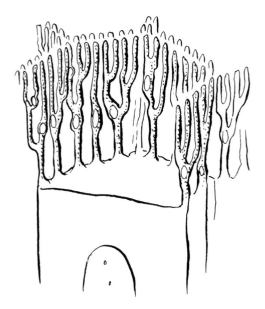

Fig. 2-25. Stereocilia of the epididymis.

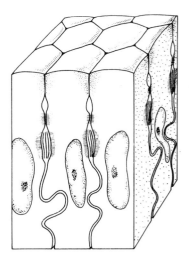

Fig. 2-26. Junctional complex.

intestinal columnar epithelium, liver parenchyma along the bile canaliculi.

There are three types of STRATIFIED SQUAMOUS EPITHELIUM:

Non-cornified, stratified sguamous epithelium (Fig. 2-27, A) is found on moist and soft surfaces, such as the lips, vagina, esophagus and the fossa navicularis of the male urethra. Toward the surface, the cells accumulate a watery iiquid within them and present a mucous appearance. Desquamated, they yield a slimy mass.

Where moderate friction occurs, such as in oral and esophageal mucosa and often in facial epidermis, the *ordinary type of stratified sguamous epithelium* is encountered (Fig. 2-27, B). As cells flatten toward the surface they are shed while alive, as evidenced by their still stainable nuclei (Fig. 2-27, B).

Cornified, stratified squamous epithelium (Fig. 2-27, C) occurs where mechanical protection and protection against desiccation is needed, such as over most of the

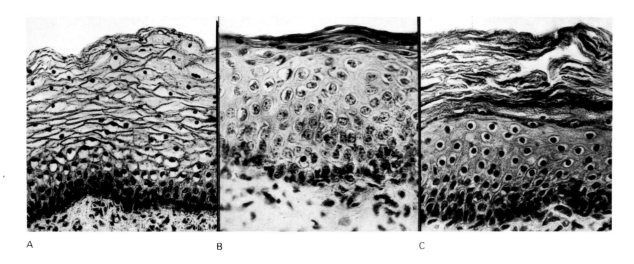

A B C

Fig. 2-27. The three types of stratified, squamous epithelium: A, mucous type (vagina); B, ordinary type (esophagus); C, keratinized type (epidermis).

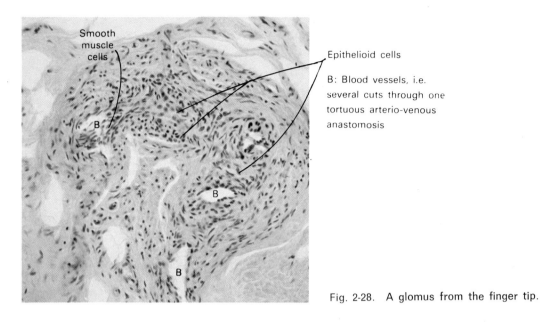

Fig. 2-28. A glomus from the finger tip.

epidermis, on the gums and hard palate. The surface layer consists of dead, cornified cells made of *keratin* (= horn) and is called *stratum corneum*. *Keratinization* is described more fully in Chapter 11. Keratinized epithelium is absent on most continuously moist surfaces in man such as in the oral cavity and the esophagus. Nevertheless, the epithelium of the hard palate in man is slightly keratinized and is usually sloughed when subjected to very hot foods.

Glandular Epithelium

Some epithelial cells are specialized for secretion. Some of these are located in the lining epithelium (intraepithelial glands); others are located in the wall of a tubular structure (intramural glands). Epithelial glands such as the liver and pancreas maintain a tubular connection with their epithelial progenitor but are located away from the duodenal wall (extramural glands). These epithelial structures form a variety of glands (see Chapter 8).

Epithelial-Like Tissues

Some organs, composed largely of polyhedral cells not in direct contact with any space, resemble epithelium and are often described as epithelial. Most endocrine glands exhibit such structure.

Some smooth muscle cells surrounding arteriovenous anastomoses, the vas afferens of the renal glomerulus and the vessels in a glomus (Fig. 2-28), are modifed to resemble epithelial cells. They are called *epithelioid* cells.

3...
Connective tissue

Connective tissue is the supporting, binding and packing tissue of the body. Its supportive, binding and wrapping function is readily seen in Figure 3-1 where it is stained blue. With few exceptions, it is characterized by having relatively few living cells in a greater proportion of intercellular material (matrix or ground substance and fibers). Fig. 3-2 shows the major components of connective tissue serving as a packing material for blood capillaries which provide nourishment and oxygen. Connective tissue is classified according to:
A. The intercellular material
   1. nature of the ground substance
   2. type of fibers (when present)
   3. abundance and arrangement of these fibers
B. The types of cells.

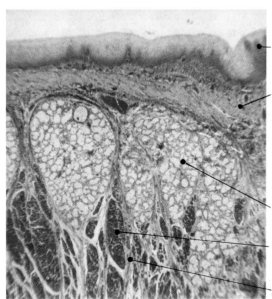

Fig. 3-1. Connective tissue (blue) of soft palate, Azan. 55 ×.

— Epithelium

Areolar or loose connective tissue permits restricted gliding of the surface which it fastens gently to the underlying organs such as

Salivary glands and

Muscles.
To tie these deeper structures firmly together, dense irregular connective tissue is organized into flat septa which form capsules all continuous with each other

THE CONSTITUENTS OF CONNECTIVE TISSUE

The Intercellular Material

hyalin

GROUND SUBSTANCE. Ground substance is an amorphous, optically homogeneous material that varies from a liquid to a thick, gel-like state. Because it has the same refractive index as water, it is relatively invisible in fresh preparations and does not stain well with the routine dyes. It has been identified histochemically as a colloidal solution of mucopolysaccharides. Mucopolysaccharides may be either sulfated or nonsulfated. Sulfated mucopolysaccharides include chondroitin sulfates a-, -b, -c, keratosulfate and heparin. Nonsulfated mucopolysaccharides include chondroitin and hyaluronic acid. Hyaluronic acid is a substance capable of increasing its viscosity by polymerization. The enzyme, hyaluronidase, changes it back to a more fluid state by hydrolysis. Changes in the viscosity of the ground substance can bind water in the tissue spaces, prevent the spread of infections and influence the metabolic activity of the cells. Sulfated mucopolysaccharides form particularly firm gels that act as cement substances.

The "fixed" connective tissue cells (that is those which do not exhibit intense locomotion) help to maintain the composition of the ground substance. Connective tissue fibers are found within the ground substance. Electrolytes, nutrients, fluids, metabolites, etc., must diffuse through ground substance on their way between

Fig. 3-2. Stereogram of connective tissue showing its components. 1: elastic fiber; 2: blood vessel; 3: collagen fiber; 4: fat cells; 5: lymphocyte; 6: histiocyte = macrophage; 7: mast cell; 8: reticular fibers; 9: membrane; 10: reticular cell; 11: fibroblast; 12: monocyte; 13: plasma cell; 14: mesenchyme cells.

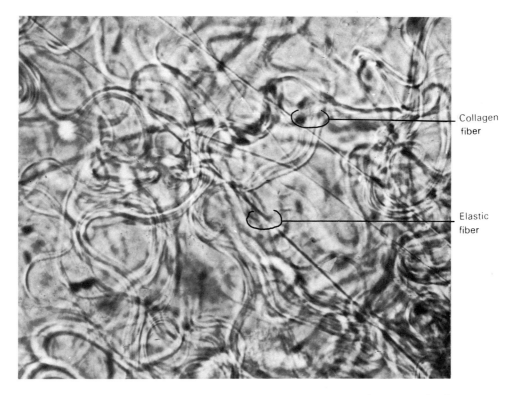

Fig. 3-3. Fresh, subcutaneous connective tissue of rat. Not stained, but visualized by narrowing the condenser diaphragm.

the cells and the walls of capillaries.

CONNECTIVE TISSUE FIBERS. Three kinds of fibers are found in connective tissue: collagenous fibers consist of an albuminoid called *collagen* which yields gelatin *Collagen Fibers* (Figs. 3-2, 3-3 and 3-4) are long, white, slightly wavy, non-elastic fibers which vary in thickness from 1 to 100 μm and are composed of bundles of parallel non-branching fibrils about 0.1 μm in diameter. The fibrils do not branch, but bundles of them may leave one fiber and join another giving the fiber a branched appearance. A *cement substance* holds the fibrils together by forming a coating around the entire fiber. Chemically, collagenous fibers consist of an albuminoid called *collagen* which yields gelatin when boiled. They will dissolve in strong acids and alkali or in gastric and pancreatic juices. When the fibers are immersed in a weak acid, the cement substance dissolves, the fibers swell and the fibrils separate. Under the electron microscope, these fibrils in most cases exhibit a cross banding with a periodicity of approximately 64 nm (Fig. 3-4). This periodicity can be altered experimentally (Fig. 3-5). Collagen fibers stain with Van Gieson's picrofuchsin, eosin and other acid dyes and show a positive birefringence (see Appendix 1).

Parallel fibers of tropocollagen molecules align and overlap one another by one quarter lengths to produce the characteristic 64 nm banding pattern. Two nm thick tropocollagen molecules develop from three polypeptide alpha chains which are coiled into a right handed helix. Each alpha chain of three amino acids repeats along its 280 nm length. Of the three amino acids, the first can be of any kind except lysine or proline. The second amino acid could be either proline or lysine, but the third must always be glysine. This makes it possible for the existence of a variety of collagen types, of which at present 5 are well known: TYPE I, found

Fig. 3-4. Collagen fibers from human skin showing bands spaced 64 nm. (Electron micrograph by Dr. Jerome Gross, M.I.T. and Harvard Medical School, from Scientific American, May 1961.)

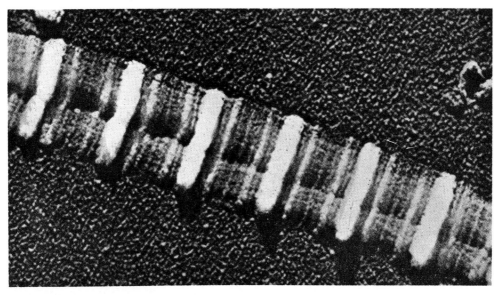

Fig. 3-5. "Fibrous long spacing" from a collagen produced in vitro by adding glycoprotein to an acid solution of native collagen. Period spacing: 280 nm. (Same source as Fig. 3-4.)

in loose and dense connective tissue; TYPE II characteristic of cartilage, the notochord, and vitreous humor; TYPE III may be identified in the dermis, loose areolar connective tissue, walls of vessels and the placenta; TYPE IV found in the basement membrane (basal lamina); TYPE V, also found in the basement membrane but may not be confined here.

Elastic Fibers. These are long, yellow, thin, and electron-optically homogeneous threads (Figs. 3-2, 3-3, 3-6 and 3-7). In sections prepared for light microscopy they are usually straight branching fibers which appear highly refractive. The phenomenon of refractility can be demonstrated in tissue sections, by moving the fine adjustment of the microscope back and forth. Unlike the strong and non-extensible collagen fibers, they can be stretched to about one and one-half times their original length. When cut, they tend to snap into twisted or coiled figures (Figs. 3-2 and 3-6). Chemically, the fibers consist of *elastin*, an albuminoid, which is very resistant to boiling, acids and alkali. Trypsin, however, will digest elastin rapidly. Although routine stains will not differentiate

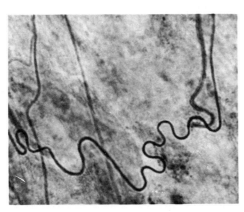

Fig. 3-6. Elastic fiber, curled as if recoiled upon cutting, Resorcin fuchsin stain.

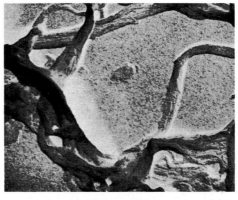

Fig. 3-7. Elastic fibers visualized by electron microscopy after shadowing. Courtesy of Dr. G. F. Bahr, University of Lund.

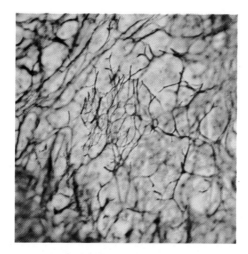

Fig. 3-8. Reticular fibers in a lymph node. Silver impregnation. 750 ×.

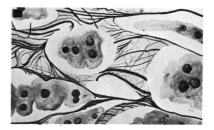

Fig. 3-10. Reticular fibers spanned by collagen membranes from adrenal cortex.

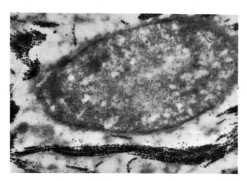

Fig. 3-9. Reticular fibers impregnated with silver, just formed by a fibroblast from a chick embryo. (Electron micrograph by Prof. Friedrich Wassermann, Argonne National Laboratory.)

Fig. 3-11. Fiber-reinforced plastic sheet.

these fibers very well, orcein and resorcin-fuchsin stain them intensely.

Reticular (Argyrophil) Fibers. These fibers branch and join each other to form an extensive, delicate, fibrous network called a reticulum (Figs. 3-2, 3-8 and 3-9). Often, instead of branching and joining they cross each other, like wires in fences. Because of their small diameters (0.2 to 1.0 μm) and resistance to dyes, the fibers are difficult to demonstrate in routine sections. When treated by special silver techniques, however, they are blackened by the deposition of metallic silver, hence the name argyrophilic or "love of silver". This is demonstrated in Figure 3-9.

Reticular nets frequently are found in association with basement membranes and often continuous with collagen fibers. Electron microscopic studies reveal that they have the same 64 nm periodic cross banding as collagen fibers, and the two are chemically similar. Unlike collagen, reticular fibers are not birefrigent, perhaps because of the difference in thickness.

FIBER FORMATION (FIBRILLOGENESIS). The *fibroblast* synthesizes polypeptides at the polyribosomes of the endoplasmic reticulum. The polypeptides are transported to the surface of the cell via the golgi complex and golgi derived vescicles. Macromolecules of tropocollagen, 280 nm in width, are released at the surface of fiber forming cells (fibroblasts). Tropocollagen is composed of 3 polypeptide alpha chains coiled in a right handed helix. The tro-

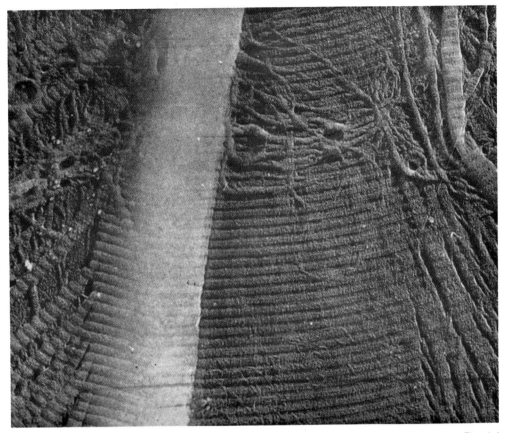

Fig. 3-12. A sheet of collagen precipitated in vitro from a solution. (Same source as Fig. 3-4 and Fig. 3-5.)

pocollagen fibrils polymerize into collagen fibrils exhibiting the 64 nm periodicity. The collagen fibrils increase in thickness by apposition to a point characteristic of the anatomical regions.

MEMBRANOUS CONNECTIVE TISSUE. Very thin membranes, which can be stained with the same dyes as collagen, occur normally in the zona glomerulosa of the adrenal cortex and form most of the stroma of the male breast. In sections they appear as thin, often wavy lines. These membranes frequently span the spaces between the meshes of argyrophil fiber networks (Figs. 3-10).

Membranous connective tissue has a structure similar to reinforced plastic sheets (Fig. 3-11). Electron microscopy suggests that connective tissue membranes may be sheets of extremely thin collagen fibrils, similar to a film of collagen formed in vitro (Fig. 3-12).

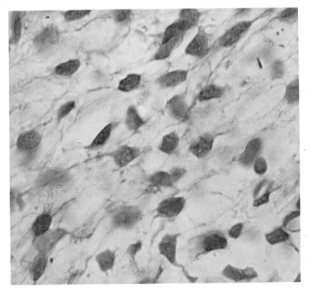

Fig. 3-13. Mesenchyme from 24 mm human embryo, H&E. 1900 ×.

72 CONNECTIVE TISSUE

Fig. 3-14, A and B. Fibroblasts from axillary skin, from the same specimen, Azan. 1900 ×.

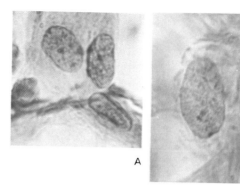

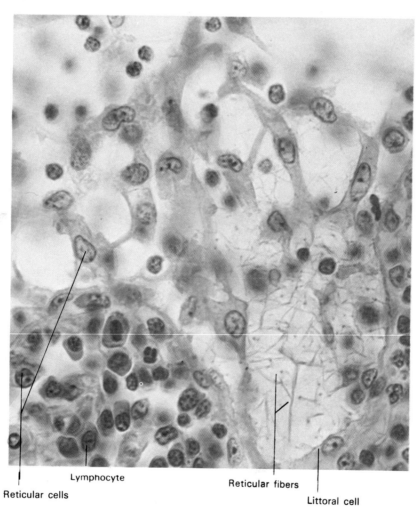

Lymphocyte

Reticular cells

Reticular fibers

Littoral cell

Fig. 3-15. Reticular cells in a medullary sinus of a lymph node. Some reticular fibers can also be seen. 1900 ×.

Cells of Connective Tissue

MESENCHYMAL CELLS. Mesenchmyal cells are primitive cells found chiefly in young embryos. They have moderately large spheroid to angular nuclei with fine chromatin granules, and one or more nucleoli (Fig. 3-13). Their cytoplasm is drawn out into processes that frequently give the cell a stellate appearance. Cell outlines are rather distinct. The processes of one cell adhere to those of neighboring cells, but contact points are temporary since mesenchymal cells maintain their individuality at all times and move about. The death of one cell does not affect those around it. Some investigators believe that primitive mesenchymal cells persist in adult tissue where they "stand ready" to differentiate into fibroblasts, macrophages, parenchymal cells of the adrenal cortex, and perhaps many others. Some maintain that de-differentiation of more mature cells also contributes to their number in the adult.

FIBROBLASTS (Figs. 3-14, A and B). Fibroblasts, in all adult connective tissues, are instrumental in the production of the intercellular ground substance and of the precursors of the fibers. Fibroblasts are recognized by large, oblong, faintly staining nuclei with very fine, dustlike chromatin granules. The preponderance of euchromatin gives morphological evidence of fibroblastic protein synthetic activity. Fibroblasts, not synthesizing actively, have nuclei with more heterochromatin, and hence may be termed fibrocytes. Like mesenchymal cells they are stellate shaped, and their cytoplasm rarely can be made out in the light microscope. They vary greatly in size (compare Figs. 3-14, A and B). When located in dense connective tissue, such as in tendons or in the cornea, the fibroblasts are compressed and elongated; and their nuclei are small and dense (Fig. 3-31).

MACROPHAGES (Fig. 3-16). Macrophages are voracious phagocytes with small isodiametric, darkly staining nuclei showing a few large, heterochromatin granules. The cytoplasm is slightly granular, and may exhibit pseudopodia extending from fairly distinct cell outlines. Some of these cells, fixed macrophages (*macrophagocytus stabilis*), are attached, while others, free macrophages (*macrophagocytus nomadicus*), maneuver in the intercellular matrix engulfing bacteria, cell debris and other foreign

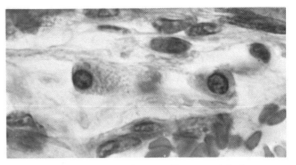

Fig. 3-16. Two macrophages (histiocytes) from human skin, same specimen as 3-14, Azan. 1900 ×.

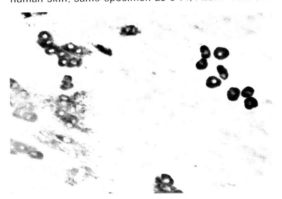

Fig. 3-16, A. Mast cells stained with toluidine blue, 50 ×.

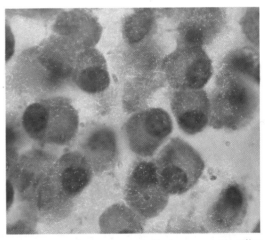

Fig. 3-17. Normal plasma cells in a synovial villus in rheumatoid arthritis, photographed from a slide by Dr. V. Cameron, Ottawa, H&E. 1900 ×.

material. The fixed macrophages were called "histiocytes". Monocytes from the blood stream may transform into macrophages after entering the tissue spaces. Phagocytic and lysosomal activity in macrophages is seen via electron microscopy in Fig. 1-7.

74 CONNECTIVE TISSUE

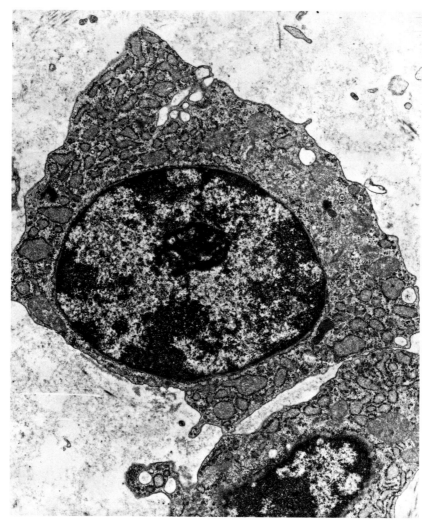

Fig. 3-18. Plasma cell from lamina propria mucosae of stomach. EM by Prof. Wolf G. Forssmann, University of Heidelberg.

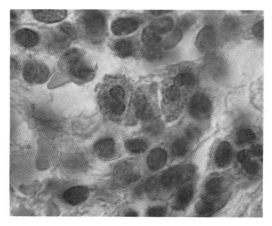

Fig. 3-19. Three eosinophil leukocytes in the lamina propria mucosae of the stomach, H&E. 1900 ×.

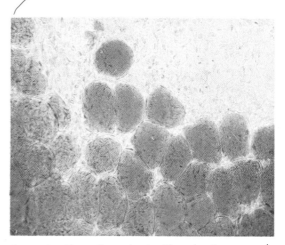

Fig. 3-20. Fat cells stained with sudan in a spread of rat subcutaneous tissue. 600 ×.

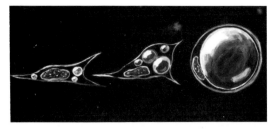

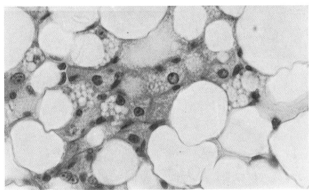

Fig. 3-21, A and B. Developing fat cells.

TISSUE BASOPHILS (mast cells) (Fig. 3-16, A) are large cells with small, pale-staining nuclei. Their cytoplasm is filled with many course basophilic granules which are *metachromatic*; that is, they have the property of concentrating certain dye substances such as toluidine blue, which is light blue in its dilute form and deep purple when concentrated. Thus, when tissue basophils are stained with dilute toluidine blue, their cytoplasmic granules acquire a deep purple hue. Because these granules are water soluble, they are not preserved in routine sections. Tissue basophils can be found along the course of small blood vessels. They elaborate *heparin*, an anticoagulant; and *histamine*, a vasodilator which also increases permeability of capillaries. SEROTONIN, an epinephrine-like substance, is found in tissue basophil granules of some nonhuman species, but not in man. It is generally agreed that blood basophilic granulocytes do enter the tissue to become tissue basophils. There is some evidence to suggest that tissue basophils (mast cells) may originate from pericytes, relatively undifferentiated cells associated with blood vessels (c.f., Chapter 9).

PLASMA CELLS (Fig. 3-17). Plasma cells

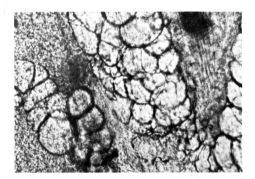

Fig. 3-22. Fresh connective tissue of rat with fat cells.

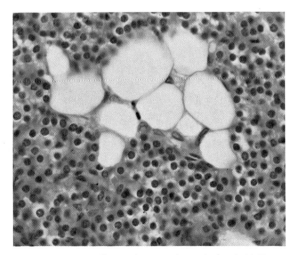

Fig. 3-23. Fat cells in the parathyroid gland, H&E. 750 ×.

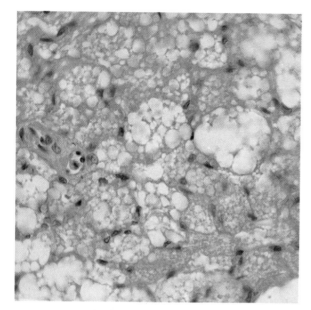

Fig. 3-24. "Brown" periadrenal fat, H&E. 750 ×.

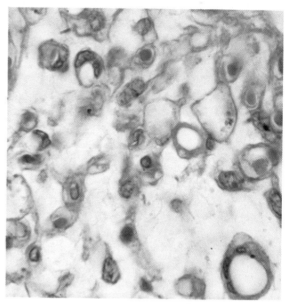

Fig. 3-25. Adipose tissue of terminal cancer patient, H&E. 1600 ×.

imp. possess small, eccentrically placed nuclei, and heterochromatin clumps frequently attached to the nuclear membrane (cartwheel nucleus). The cytoplasm is intensely basophilic due to granular endoplasmic reticulum except for a clear area next to the nucleus that contains the golgi complex and centrosome. These cells are infrequent in normal connective tissue, except in the lamina propria of the stomach where they are most numerous. They are seen in hemopoetic tissues and are prominent in areas of chronic inflammation. This is one of the cell types responsible for the

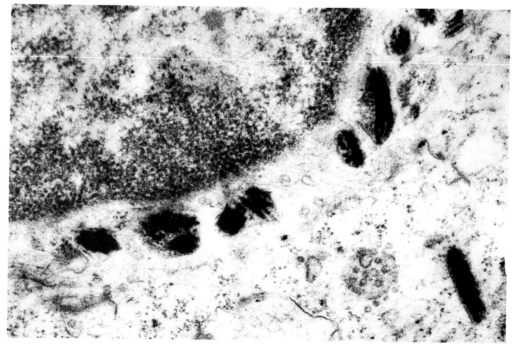

Fig. 3-26. Melanosomes in a melanocyte of a hamster. Courtesy of Dr. Tabashi Nakai, Chicago Medical School.

production and storage of antibodies. The typical structure of the plasma cell is well seen in the electron micrograph Fig. 3-18. Plasma cells are derived from B-lymphocytes. They are non-motile and live only a few days.

LEUCOCYTES (white blood cells). Leucocytes may wander into connective tissue from the blood stream. Lymphocytes, monocytes and eosinophils are occasionally seen (Fig. 3-19). Polymorphonuclear neutrophils are common in areas of acute inflammation. Wandering leucocytes resemble those seen in blood smears (see Chapter 5), but they appear smaller, exhibiting their true shape; while in blood smears the originally spherical leucocytes are artificially flattened, thus appearing larger.

ADIOPOCYTE (fat cell). Adipocytes occur singly or in groups (Fig. 3-20) between the connective tissue fibers. They originate from fibroblast-like cells that accumulate fat in vacuoles (Fig. 3-21, A and B), and coalesce to form progressively larger ones until the entire center of the cell is occupied by a single vacuole. In the fresh state, fat cells resemble soap bubbles (Fig. 3-22). Some lying free are spherical, while others packed together closely have their abutting surfaces flattened (Figs. 3-20 and 3-23). In routine histological preparations, the fat is dissolved out of the cells by the clearing agents, so that a fat cell looks very much like a signet ring if the plane of section passes through its nucleus.

In certain areas, as in the neighborhood of the kidney and of the adrenal gland, fat cells undergo continuous cycles of depletion and repletion (Fig. 3-24). This special kind of adipose tissue is known aa *brown fat*. It is characterized by multivacuolated cells whose individual fat droplets are not coalesced. Brown fat is more frequently found in infants than in adults.

During severe diseases and in starvation, fat cells will become chronically depleted of fat (Fig. 3-25).

PIGMENT CELLS. Pigment cells (see Chapter 11) contain pigments of specific true or structural color. The best known among them is *melanin*, a dark brown or black pigment. A melanin granule (melanosome) is a tiny pile of submicroscopically thin lamellae (Fig. 3-26). The cells are known as melanophores, or *melanocytes*. Melanocytes contain an enzyme, tyrosinase, which is involved in the synthesis of melanin-containing units, the melanosomes. Tyrosinase is the cellular component which will oxidize dihydroxyphenylalanine (dopa reagent) forming a black reactive product. Mature melanin granules do not contain tyrosinase and therefore do not give a positive dopa reaction. In Caucasians, the pigment granules fail to darken unless irradiated with ultraviolet rays (suntan). In all races the melanin granules darken in certain cells found in the lamina fusca and tunica vasculosa of the eye. Examples of pigment cells are shown in Figs. 1-3, 11-13 11-16 11-17 and 11-18. *Guanophores* contain guanine crystals. These cells are found in the irides (plural of iris) of blue-eyed persons. *Lipophores*, frequent in lower vertebrates, but rare in man[1], contain the yellow, orange or red pigment called lipochrome.

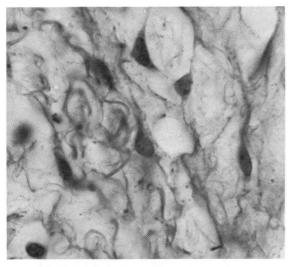

Fig. 3-27. Mucous connective tissue from umbilical cord of a 48 mm embryo. 1900 ×.

[1] Except for those rare individuals with green irides, where they may lie in front of the guanophores.

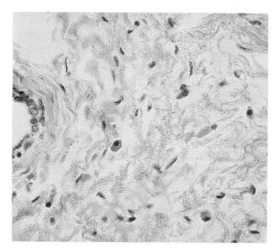

Fig. 3-28. Areolar (loose) connective tissue from mamma of nullipara, H&E. 1000 ×.

TYPES OF CONNECTIVE TISSUE

Mesenchyme

Mesenchyme, composed of a network of branching mesenchymal cells, is embryonic connective tissue, distinguished by the lack of fibers in watery ground substance (Fig. 3-13).

Mucous Connective Tissue

Mucous connective tissue (Wharton's jelly) is found in the umbilical cord and in the cores of placental villi. It is probably nothing more than a specialized type of mesenchyme. It has fewer cells but a more gelatinous, viscous ground substance than mesenchymal tissue, as well as a few very fine fibers in the matrix (Fig. 3-27). Sometimes it is called *myxomatous* tissue.

Reticular Connective Tissue

Reticular connective tissue is the most primitive connective tissue in the adult. It consists of a network of reticular cells and a latticework of very thin, reticular fibers (Fig. 3-15). Some of the reticular cells are attached to the fibers (*fixed cells*), while others lie free between the fibers (*free cells*). This tissue is characteristic of hemopoetic (Chapter 5) and lymphoid organs and tissues, and forms an integral part of the reticuloendothelial system of the body (Chapter 10).

Areolar (or Loose) Connective Tissue
(Figs. 3-1, 3-3, and 3-28)

This tissue is composed of a loose, irregular arrangement of collagen, elastic and reticular fibers in a gelatinous ground substance of low viscosity. The collagen fibers predominate.

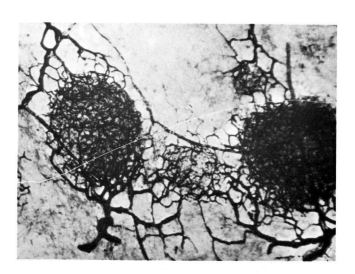

Fig. 3-29. Injected plexus of primary fat lobules of a 22 cm fetus. Courtesy of Prof. A. Dabelow, University of Mainz.

All cells of adult connective tissue, except reticular cells, can be found in areolar tissue. Figure 3-3 is a photomicrograph of artificially edematous, unstained, superficial subcutaneous connective tissue of a rat. In the fresh state, areolar connective tissue is a colorless, transparent, glistening soft mass. It acts as a packing and binding material throughout the body, forming parts of the superficial and deep fascias, the *stroma* of organs and packing for vessels and nerves. The dissection of gross specimens is largely a matter of separating this tissue to better visualize the named structures. All exchanges between the blood vessels and *parenchyma* of organs must occur through areolar tissue. (The cells which perform the function of an organ constitute the *parenchyma*; whereas the connective tissue which provides support for parenchymal cells is called the *stroma*.)

Adipose (Fat) Tissue

This is a type of connective tissue in which adipocytes predominate. Certain areas of the body are particularly prone to contain fat (e.g., renal sinus, ischiorectal fossa). Spherical capillary plexuses, each supplied by its own artery and drained by its own veins, arise in these areas during fetal life, even before fat deposition begins (Fig. 3-19). A fat lobule develops in the territory of each spherical plexus and grows until adjacent lobules almost touch each other; remaining separated, however, by fibrous septa. In subcutaneous tissue, these septa are known as *retinacula cutis*. The orientation of many lobules in a single plane results in fatty layers (e.g., panniculus adiposus). Fat lobules act as pressure absorbers (Chapter 11) and as storage organs. Although it is customary to consider fat merely as an inactive substance, in reality it is a physiologically dynamic tissue and the lipid within cells has a relatively short turnover time.

Dense Irregular Connective Tissue

This is similar to areolar connective tissue except that collagen fibers are coarser and more densely packed; hence, the

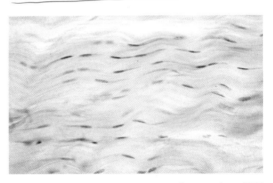

Fig. 3-31. Longitudinal section of a tendon, H&E. 525 ×.

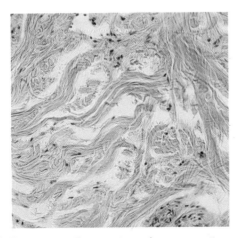

Fig. 3-30. Dense, irregular connective tissue from dermis, H&E. 280 ×.

Fig. 3-32. Ligamentum nuchae. Longitudinal section. Van Gieson stain. 375 ×.

Fig. 3-33. Ligamentum nuchae, cross section Van Gieson stain: the elastic fibers are stained yellow, collagen fibers red. 1900 ×.

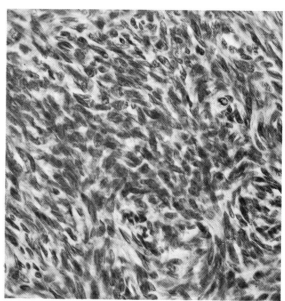

Fig. 3-34. Dense, cellular connective tissue from cortex of ovary, H&E. 833 ×.

spaces between fibers are smaller, and the tissue has a firmer, more resistant consistency (Fig. 3-30). It is found in the dermis of the skin, the capsules of certain organs, the dura mater and in deep fascias.

Dense Regular Connective Tissue

Tendons, aponeuroses and ligaments are sites of dense regular connective tissue. *Tendons* and *aponeuroses* connect muscles to bones. Tendons are made of parallel bundles of thick collagen fibers (Fig. 3-31), while aponeuroses are composed of parallel fibers arranged in crisscrossing layers. Fibroblasts in interrupted rows between the fibers are the predominant cells of the intercellular material. In cross section these cells exhibit stellate forms between the large polygonal sections of collagen fibers.

Ligaments connect bones to bones and resemble tendons histologically. Elastic (yellow) ligaments differ, however, by virtue of the fact that the parallel bundles are composed of predominantly of thick elastic fibers separated and tied together by small amounts of areolar connective tissue (Figs. 3-32 and 3-33). Both tendons and ligaments are poorly vascularized.

Dense Cellular Connective Tissue
(Fig. 3-34)

Dense cellular connective tissue is a mass of closely packed, fusiform cells, arranged as if in whirling streams. This type occurs in the cortex of the ovary and in the basal layers of the endometrium.

4...
Skeletal Tissue

The vertebrate body is supported by an internal skeleton of cartilage and of bone. Both the word "cartilage" and the word "bone" have dual meanings. A cartilage is a specific named piece of the skeleton, such as the thyroid cartilage or the right, third, costal cartilage. Cartilage (without an article) is a building material or a kind of tissue. Similarly, a bone, such as the humerus or the patella, is a specific, skeletal piece; while bone (without an article) is the material of which named bones are made.

The skeleton of a six-week-old human embryo consists entirely of models of bones made of cartilage. For example, even at this early state the hip bones (acetabulum and femur, Fig. 4-1) have characteristic adult shapes; but are made entirely of cartilage.

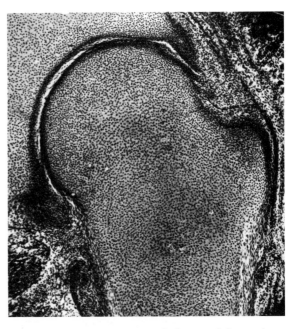

Fig. 4-1. Cartilaginous acetabulum, and femoral head, neck and greater trochanter of a 6-week-old embryo.

Astonishingly enough, although cartilage appears earlier in the human embryo than bone, bone is the more ancient of the two tissues. The earliest known fossil vertebrates, the extinct ostracoderms (animals more primitive than fishes) possessed a dermal armor made of true bone almost identical in structure with the membranous bones of our own skull. It is unknown at which geological time cartilage appeared, but it is probably a later acquisition. In sharks (chondrichthyes), bone has almost disappeared, except for the dermal scales. The use of bone in the construction of an internal skeleton seems to have been an innovation of bony fishes (osteichthyes).

82 SKELETAL TISSUE

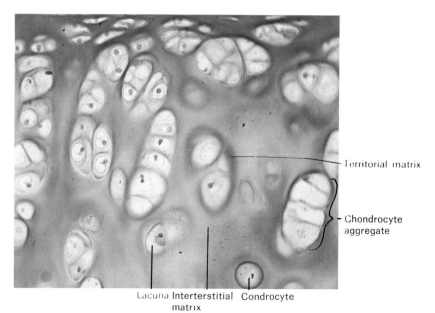

Fig. 4-2. Hyaline cartilage from a canine humeral head, H&E. (375 ×). Courtesy, Dr. Jo Ann C. Eurell, Hahnemann University.

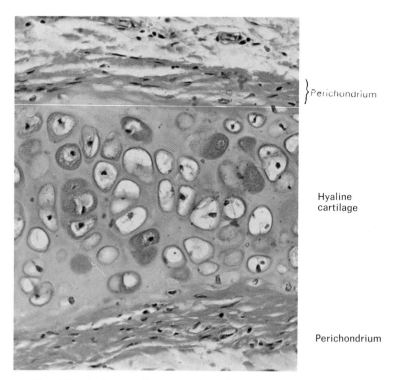

Fig. 4-3. Hyaline cartilage from bronchus, H&E.

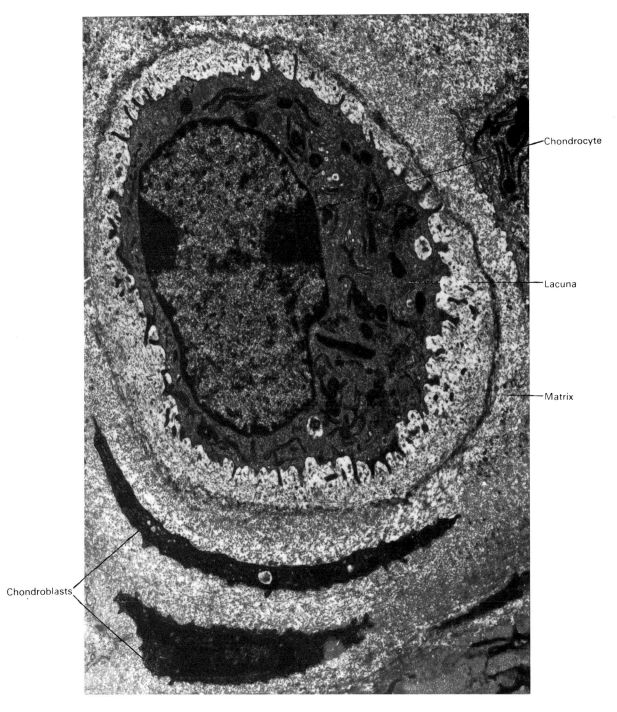

Fig. 4-4. Peripheral area of hyaline cartilage (mouse). Courtesy of Dr. D.H. Matulionis, University of Kentucky. (11,000 ×).

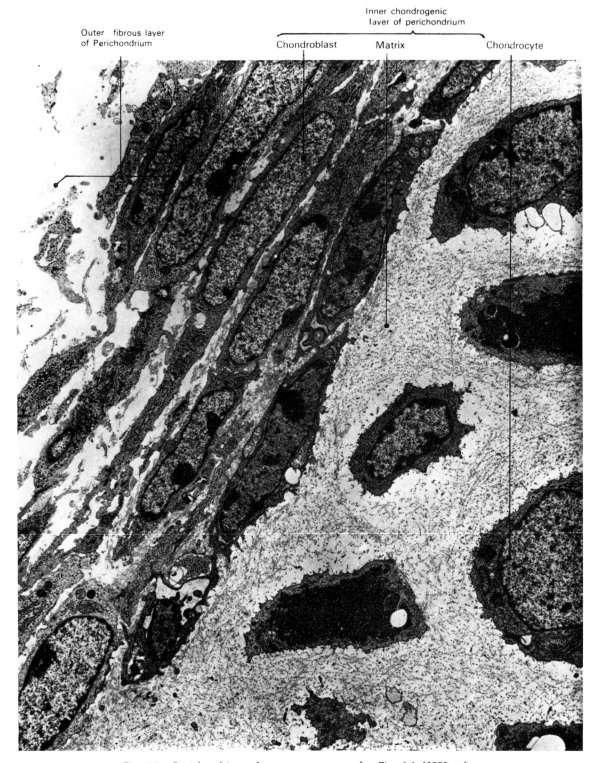

Fig. 4-5. Perichondrium of mouse, source as for Fig. 4-4. (3550 ×)

CARTILAGE

Cartilage is flexible, slightly elastic and can be cut easily with a knife. Its unique biomechanical properties make cartilage a key element of the skeletal system.

The cells of cartilage are the *chondroblasts* and *chondrocytes* (Figs. 4-2. 4-3 and 4-4). Chondroblasts actively secrete matrix to form mature cartilage and then are transformed into mature chondrocytes. Chondrocytes are characterized by a large, central, spherical nucleus with one or more nucleoli. The space occupied by a

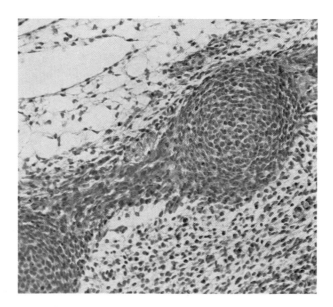

Fig. 4-6. Cross section of rib in the early precartilage stage. Note also the myoblasts which form the intercostal muscles adjacent to the cartilage. Mouse embryo, H&E. (200 ×). Courtesy, Dr. Jo Ann C. Eurell, Hahnemann University.

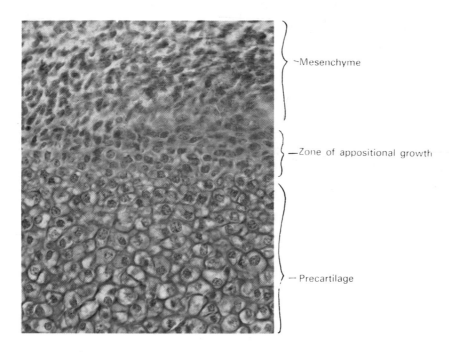

Fig. 4-7. Precartilage of a vertebral body. 31 mm embryo, H&E. (600 ×).

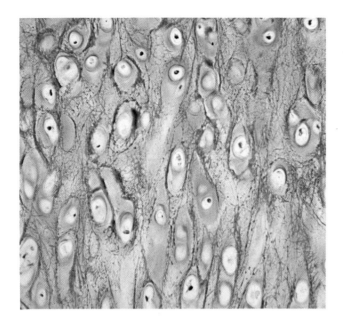

Fig. 4-8. Elastic cartilage of external auditory meatus, hematoxylin and orcein. (390 ×).

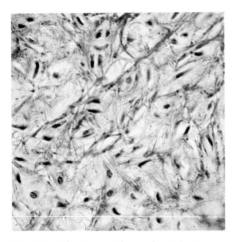

Fig. 4-9. Elastic cartilage of epiglottis, hematoxylin and orcein. (550 ×).

chondrocyte is a *lacuna*. Some lacunae contain more than one cell, and these cell groups are referred to as *chondrocyte aggregates* (cell nests, isogenous groups). After death, the chondrocytes often shrink, but the lacunae retain their shape and are visible in tissue preparations.

The chondrocytes are surrounded by a *matrix* consisting of a homogeneous *ground substance* in which fibers are embedded. The ground substance contains proteoglycans composed of a protein core and covalently bound side chains of glycosaminoglycans such as chondroitin-4 and 6-sulfate, keratan sulfate, and hyaluronic acid. The proportion of the side chains in cartilage varies with anatomic location and age. The fibers of the cartilaginous matrix include Type II collagen and elastic fibers. Fiber distribution and quantity varies with cartilage type.

Most cartilages are surrounded by *perichondrium* (Figs. 4-3 and 4-5), a connective tissue envelope composed of an outer *fibrous layer* and an inner cellular, *chondrogenic layer*. Perichondrium is not present on articular surfaces.

Nerves and blood vessels are not present in adult cartilage. Nutritive materials must diffuse from capillaries in the perichondrium through the matrix to reach the cells. The end products of chondrocyte metabolism return to the the blood by the reverse route. Calcium deposits may form in the matrix of older cartilages and impair diffusion of nutrients. If this calcification process occurs, chondrocytes will die and the cartilage will deteriorate.

Formation of Cartilage

In the embryo, mesenchymal cells group together to form masses shaped like primordial bones and cartilages (Fig. 4-6). The cells lose their characteristic processes, round up, and deposit small quantities of matrix between one another (Fig. 4-7). This process occurs during the *precartilage* stage and the cells are called *chondroblasts*.

Chondroblasts manufacture matrix in the following manner: amino acids are synthesized into protein chains in the rough endoplasmic reticulum, the proteins are transported to the golgi complex where carbohydrates synthesized into glycosaminoglycans are combined with the proteins, and the resulting proteoglycans then are secreted at the cell surface. This newly-formed ground substance combines with fibers produced by the chondroblasts and matrix is formed.

Growth of Cartilage

There are two types of cartilage growth: *interstitial growth* and *appositional growth*.

INTERSTITIAL GROWTH. Interstitial growth is found in young cartilage and involves (a) repeated mitotic division of chondrocytes with subsequent deposition of matrix between daughter cells (Fig. 4-2) and (b) increase in size of the cartilage by continued deposition of matrix between existing chondrocytes. During interstitial growth, recently divided cells may group together as chondrocyte aggregates.

APPOSITIONAL GROWTH. In appositional growth, chondroblasts in the chondrogenic layer of the perichondrium retract their processes and secrete a layer of matrix around themselves. These cells and their lacunae are characteristically flattened during early stages (Fig. 4-5). An abundance of collagen fibers, relative to the amount of proteoglycans in an area where the matrix is newly formed, gives the area a less basophilic quality. Appositional growth occurs simultaneously with interstitial growth early in cartilage development, and later is the primary type of cartilage growth after interstitial growth ceases.

Once secretion of matrix is complete, the chondroblasts transform into chondrocytes.

Types of Cartilage

Cartilage is classified according to the distribution and quantity of fibers in the matrix.

HYALINE CARTILAGE (Figs. 4-2 and 4-3). Hyaline cartilage is composed of a feltwork of extremely delicate collagen fibers embedded in an amorphous ground substance. The collagen fibers are not visible in ordinary preparations, because their re-

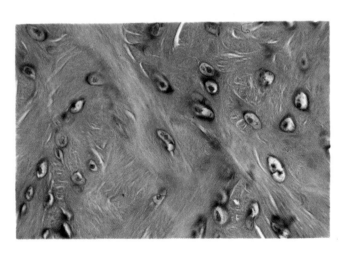

Fig. 4-10. Fibrocartilage, H&E. (370 ×). Courtesy, Dr. Jo Ann C. Eurell, Hahnemann University.

88 SKELETAL TISSUE

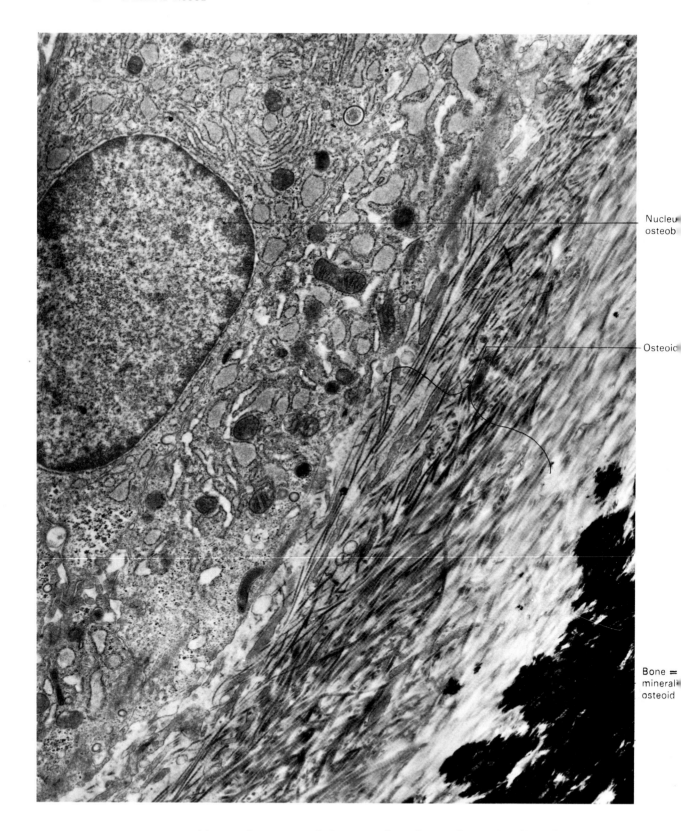

Fig. 4-11. An osteoblast producing osteoid. Courtesy of Dr. S. W. Whitson, Southern Illinois University. 5800 ×.

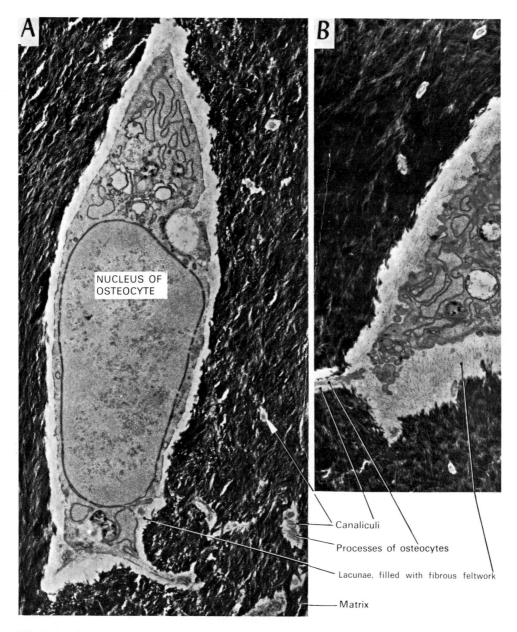

Fig. 4-12. Osteocytes of mouse. (Electon micrographs by Prof. Friedrich Wassermann, Argonne National Laboratory).

fractive index approaches that of the ground substance. The matrix immediately surrounding the chondrocytes and exhibiting a high degree of basophilia is the *territorial matrix* or capsule (Fig. 4-2).

The less basophilic areas surrounding the territorial matrix is designated *interterritorial matrix*. Hyaline cartilage, the most common type of cartilage, can be found in the nose, most parts of the larynx, trachea, bronchi and on articular surfaces.

ELASTIC CARTILAGE (Figs. 4-8 and 4-9). Elastic cartilage consists of an amorphous matrix which contains a dense network of branching elastic fibers which is more dense in the interior than near the perichondrium. The elastic fibers can be demonstrated by special stains which impart a yellow color to gross samples of the tissue. The cartilage of the external ear, auditory tube, epiglottis and parts of the larynx (tips of arytenoids, corniculate and cunei-

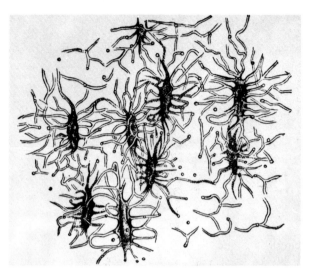

Fig. 4-13. Bone lacunae and canaliculi, occupied, during life, by the branched osteocytes (Rollett, 1872).

form cartilages) is of the elastic type (see Chapter 13).

FIBROCARTILAGE (Fig. 4-10). In fibrocartilage, groups of chondrocytes are separated from each other by thick bundles of collagen fibers. The cells are sometimes arranged in characteristic rows. Fibrocartilage is acidophilic in contrast to the basophilia of elastic and hyaline cartilages. This type of cartilage lacks a perichondrium. The annulus fibrosus of the intervertebral disc, the symphysis pubis, the menisci of the knee, and the articular disc of the sternoclavicular joint are formed of fibrocartilages.

Bone

Bone is a dense hard connective tissue, unique for its lightweight structure and high tensile strength with some elasticity. *Calcium ions* and other minerals are stored and released as required by this dynamic living tissue. Throughout life, bone is renewed and remodeled in response to homeostasis, mechanical stress, aging and disease.

Like many connective tissues, bone cells and fibers are embedded in an organic ground substance. Bone differs via a composition of 50% inorganic salts and about 50% collagen fibers Type I (Fig. 4-11). Minerals deposited as amorphous calcium phosphate form *hydroxyapatite crystals* (hydrated calcium phosphate). Slender crystalline rods, up to 60nm long and 3 to 6 nm wide, align along oriented collagen fibers.

Of four bone cell types: (1) *osteoprogenitor cells*, (2) *osteoblasts*, (3) *osteocytes* and (4) *osteoclasts*; the first three are considered identical but in different functional states. *Osteoprogenitor cells* function as *stem cells* with mesenchymal potencies to become osteoblasts, osteocytes, fibroblasts or hemopoetic cells. On bone surfaces they "wait" to function in reorganization or repair of bone.

OSTEOBLASTS (Fig. 4-11) produce *osteoid*, a protein composed of Type I collagen, proteoglycans and glycoproteins. During bone formation, osteoblasts interdigitate to form a quasi-cuboidal epithelium. A rich RER and a prominent golgi complex diminishes when the *osteoblast* reverts to osteoprogenitor status or becomes entrapped in bone matrix as an *osteocyte* (the definitive bone cell).

OSTEOCYTES (Fig. 4-12) reside in and conform to the *lacunae* (spaces) in bone matrix. Multiple *canaliculi* (microscopic tunnels), between adjacent lacunae accomodate cell processes (Fig. 4-13) specialized to communicate with processes of other osteocytes via *nexuses*. Nearby capillaries provide nutrients and remove wastes via diffusion through the canaliculi. Evidence implicate osteocytes as assisting osteoclasts in removing calcium from perilacunar bone to circulating blood.

OSTEOCLASTS (Figs. 4-14, A and B) are conspicuous throughout life on bony surfaces being reshaped by resorption. These "bone breakers" are huge branched cells with up to 50 spherical nuclei, occupying *erosion lacunae* (Howship's). The branches possess ruffled borders with actin containing motile processes. Branches with ruf-

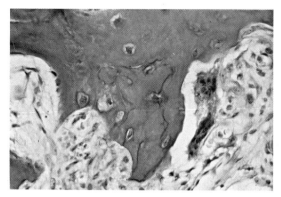

Fig. 4-14A. Two osteoclasts (righ side of picture) from the femur of a human fetus, H&E.

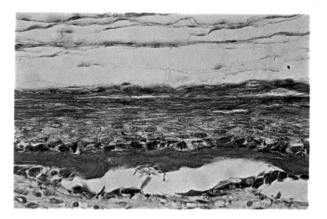

Fig. 4-15. Periosteum on the outer surface of a developing bone. Masson trichrome, (370 ×). Courtesy, Dr. Jo Ann C. Eurell, Hahnemann University.

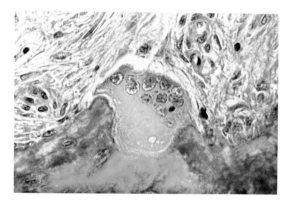

Fig. 4-14B. An osteoclast in an erosion lacunae. Human biopsy from the compact bone of the tibia at an osteolytic focus due to hyperparathyroidism secondary to renal disease. Undecalcified, plastic embedded semithin section. Goldner stain. (400 ×). Courtesy of Dr. Hanns Plenk, University of Vienna.

fled borders increase the surface area so extensively that when one part of the cell rests, the others are active. Calcium crystals trapped between folds and ruffled borders and within intracytoplasmic vacuoles, implicate these cells in bone resorption. The contention that *osteoclasts* develop from fusion of several osteoblasts is questioned. Current studies consider fusion of *blood derived monocytes* as osteoclast precursors. Numerous mitochondria account for cytoplasmic acidophilia, while cell processes filled with lysosomes test positively for acid phosphatase.

BONE LINING CELLS. Flat, elongated cells with spindle-shaped nuclei cover most of the surfaces of bone in the adult skeleton. These cells are thought to be derived from osteoblasts which have ceased their activity. The bone lining cells may divide and differentiate into additional osteoblasts, or serve as barrier cells between the canalicular system and the interstitial fluids.

PERIOSTEUM. Bones are surrounded by a sheath of dense, irregular connective tissue, the *periosteum* (Fig. 4-15). This tissue is firmly attached to the underlying bone by *perforating fibers* (Sharpey's fibers). The periosteum, like the perichondrium, has an outer *fibrous layer* and an inner

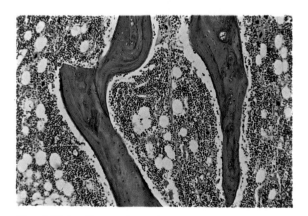

Fig. 4-17, A. Spongy bone from a canine humeral head, H&E. (90 ×). Courtesy, Dr. Jo Ann C. Eurell, Hahnemann University.

92 SKELETAL TISSUE

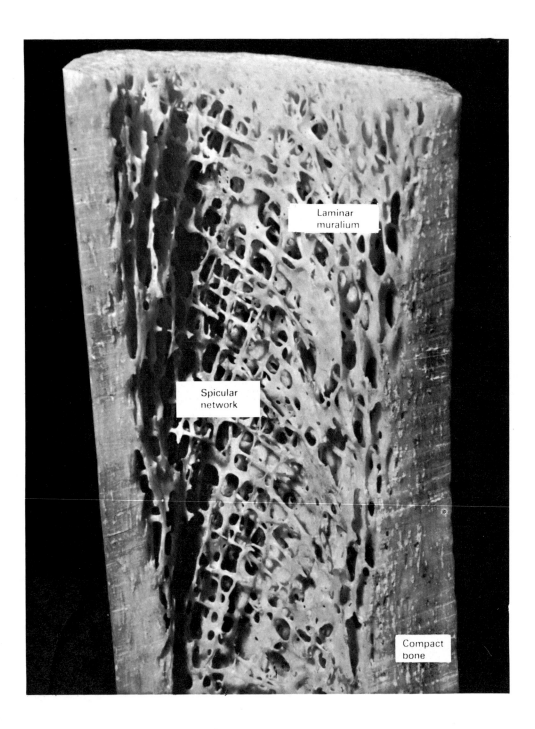

Fig. 4-16. Shaft of a tibia, cut open.

cellular *osteogenic layer*. Periosteum is not found on the surfaces of bones which are covered with articular cartilage, or where tendons and ligaments insert.

ENDOSTEUM. The *endosteum*, another layer of connective tissue similar to periosteum, lines the marrow, soft tissue cavities and osteons of bone.

Classification of Bone

STRUCTURAL CLASSIFICATION OF BONE. Structurally, there are two kinds of bone: spongy and compact (Fig. 4-16). In most adult bones, both types are present. *Spongy bone* is sometimes referred to as *trabecular bone*. Compact bone forms the dense shell which is continuous with the internal, spongy framework. In flat bones such as the bones of the skull, the internal, spongy framework is called *diploe*.

Spongy Bone (Figs. 4-16 and 4-17, A, B). Spongy bone is a muralium of thin laminae and rods called *trabeculae*. Adult mammalian trabeculae are composed of lamellae of mineralized osteoid demarcted by regularly arranged lacunae containing osteocytes. Canaliculi from the lacunae penetrate adjacent lamellae. The surfaces of the trabeculae are covered by endosteum.

Compact Bone (Figs. 4-18 and 4-19). Compact bone is composed of lamellar bone arranged in concentric cylinders around a *central canal* (Haversian canal). These bony subunits are called *osteons* (Haversian systems). Within individual lamellae of an osteon, the collagen fibers of the matrix are helically arranged around the axis of the central canal. The direction of the helix alternates approximately ninety degrees between adjacent lamellae. The central canal contains nutrient vessels, nerves, and connective tissue. Transverse or oblique *perforating canals* (Volkmann's canals) connect the central canal of one osteon with the central canal of an adjacent osteon, the periosteum, or the *medullary (marrow) cavity*. The outer boundary of the osteon is delineated by a "cement line" of mineralized matrix devoid of collagen.

The outer and inner surfaces of compact bone are formed by lamellae which run parallel to the surface and cover the deeper osteons. The *outer circumferential lamellae* are covered with periosteum, while the *inner circumferential lamellae* are covered with endosteum. Within the osteonal bone are *interstitial lamellae* which fill in cavities between adjacent osteons.

Bone Marrow. In the later stages of fetal development, the cavities of long bones contain active hemocytopoetic tissue called red bone marrow. In the adult, the *red bone marrow* of long bones is replaced gradually by marginally active hemocyto-

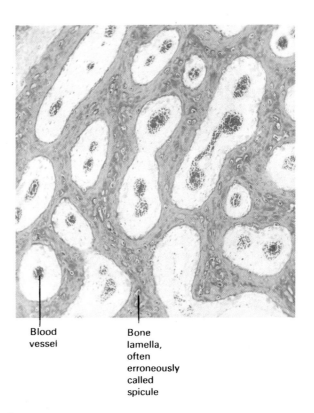

Blood vessel

Bone lamella, often erroneously called spicule

Fig. 4-17, B. "Spongy" bone, also called cancellous bone, is a muralium osseum. From mandible of 6-months old fetus, Azan. 185 ×.

94 SKELETAL TISSUE

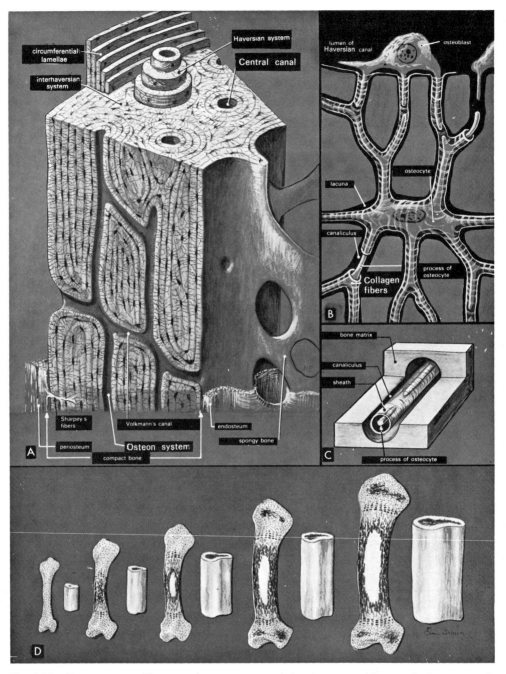

Fig. 4-18. Stereogram to illustrate the structure and development of bones. A: Structure of compact bone; B: an osteocyte in its lacuna and canaliculi; C: a canaliculus with processes of two adjacent osteocytes; D: growth of a long bone.

poetic tissue which contains large amounts of fat, called *yellow bone marrow*.

Red bone marrow occurs less frequently in the adult skeleton than in the fetal skeleton and is restricted to: (a) the interspicular and interlaminar spaces of the spongy bone within the epiphyses of long bones, (b) vertebral bodies, (c) ribs, (d) the sternum, and (e) certain bones of the skull. See Chapter 5 for a more complete description of the bone marrow.

CLASSIFICATION BY MODE OF DEVELOPMENT. *Osteogenesis* is the formation of bone, and *ossification* is the process by

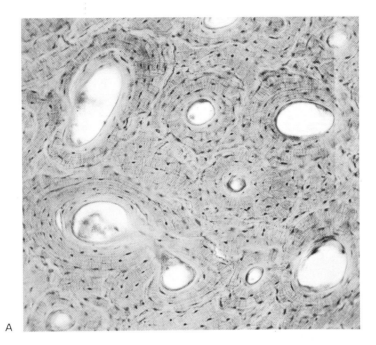

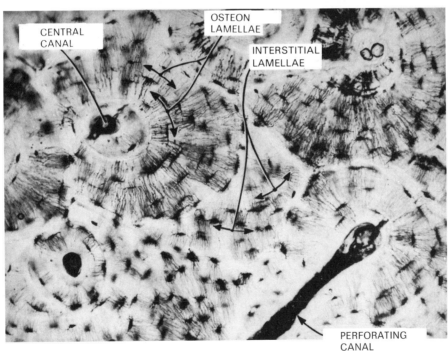

Fig. 4-19, A + B.

which any non-osseus tissue is transformed into bone. Bone formation occurs by two major modes: intramembranous and endochondral ossification.

Membranous ossification (Fig. 4-20) is a process whereby bone develops directly in well vascularized mesenchymal tissues. Mesenchymal cells differentiate into osteoprogenitor cells, and then to osteoblasts which form an *osteoid complex*. The fate of the osteoblast was considered previously. The resulting network of trabecular

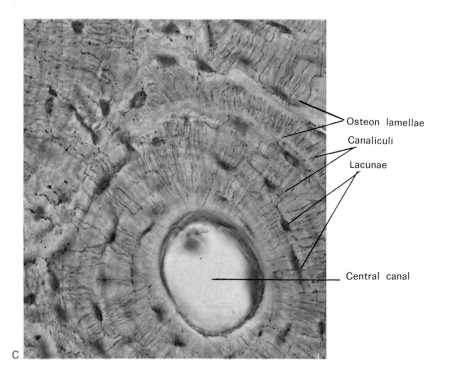

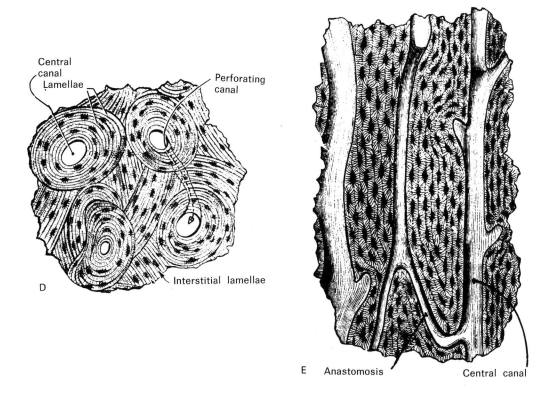

Fig. 4-19. Compact bone, ground, dry preparations. Air in the lacunae and canaliculi makes them visible. Lamellae are located between what appear to be rows of lacunae. The canaliculi (dark, very thin lines) traverse the lamellae. A-D are transverse sections, E is a longitudinal section. A: 220 ×; B: 500 ×; C: 860 ×. D and E: from Rollet, 1872.

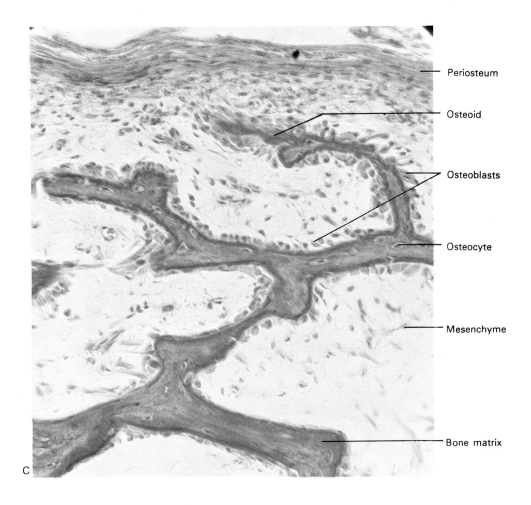

Fig. 4-20. Intramembranous ossification of bone, Masson's trichrome.

bone is known as the *primary spongiosa* (*os membranaceum reticulofibrosum*).

In areas where compact bone will ultimately form, bone deposition continues until all the space between trabeculae of the primary spongiosa is filled in with irregular osteons, the *os membranaceum lamellosum*. If the primary spongiosa is the precursor of spongy bone, the thickening of trabeculae stops and the spaces fill with hemocytopoetic tissue.

Secondary cartilage forms from mesenchyme as the articular cartilage of intramembranous bones involved in movable joints. This cartilage develops after the bone is established.

The frontal and parietal bones, mandible, maxilla and parts of the occipital and temporal bones form by intramembranous ossification.

Endochondral ossification (Fig. 4-21). Bones which develop from pre-existing hyaline cartilage models of the future adult bone, form by endochondral ossification. In the center of the cartilaginous model, the chondrocytes hypertrophy while the surrounding territorial matrix calcifies. The chondrocytes die and the calcified matrix fragments. The remaining calcified cartilage cores provide a framework for the formation of primary spongiosa. Mesenchymal tissue, osteoblasts and periosteal vessels invade from the periphery of the cartilage model. Invading osteoblasts align along the calcified cartilage cores and

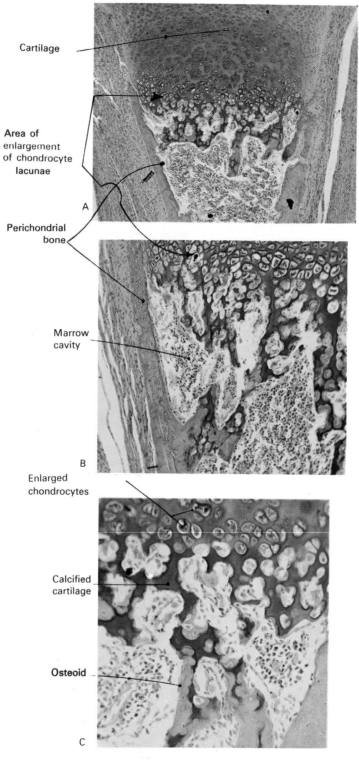

Fig. 4-21. Endochondral ossification in human finger phalanges. Hematoxylin without acid alcohol and eosin. A: 100 ×; B: 166 ×. C: 300 ×.

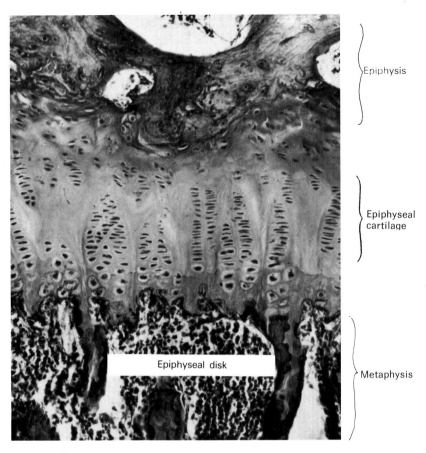

Fig. 4-22. The epiphyseal cartilage of a long bone.

form osteoid which is subsequently mineralized.

The formation of long bones such as the tibia involves both intramembranous and endochondral ossification. Initially, the perichondrium forms a bony ring around the cartilaginous model by intramembranous ossification. The periochondrium is then referred to as periosteum. Concurrently, the hyaline cartilage within the bony model becomes primary osseous trabeculae by endochondral ossification. The initial primary osseous trabeculae located in the developing diaphysis of the long bone is referred to as the *primary ossification center*. This center expands toward the epiphyses at each end of the long bone.

Continued growth in length of a long bone is carried out by the *epiphyseal cartilage*[1], a disc of specialized hyaline cartilage which separates the epiphysis from the metaphysis. The epiphyseal cartilage consists of five zones: (1) the reserve zone, (2) the proliferating zone, (3) the zone of hypertrophy (4) the resorbing zone, and (5) the zone of ossification. These zones can be seen in Fig. 4-22. Interstitial growth of the epiphyseal cartilage occurs with addition of cells by division of the chondrocytes in the proliferating zone, and secretion of matrix by these cells. The cells in the hypertrophying zone continue to secrete matrix, and they gradually enlarge and accumulate glycogen. The vertical matrix of the chondrocytes in the resorbing zone becomes calcified, and the surrounded cells die. The uncalcified trans-

---

[1] The "epiphyseal cartilage" is correct terminology from the Nomina Anatomica, 5th edition. "Physeal cartilage" has been proposed as an alternative because this cartilage is involved in formation of metaphyseal bone, and is not considered part of the epiphysis.

verse septae of matrix are resorbed, and vertical spicules of calcified cartilage are left. In the zone of ossification, invading capillary loops and perivascular connective tissue with osteoprogenitor cells establish bone formation on the calcified catilage spicules. The calcified cartilage cores with newly deposited bone (woven) on their surfaces are referred to as *primary osseous trabeculae* (primary spongiosa).

In the epiphysis, a *secondary center of ossification* forms similar to the process for the primary center of ossification. The secondary center expands to fill the epiphysis except for the area of *articular cartilage* (Fig. 4-23). At skeletal maturity, when closure of the epiphyseal cartilage occurs, the bone of the primary and secondary centers becomes continuous.

Eventually, the primary osseous trabeculae of both intramembranous and endochondral bone will be replaced by *secondary osseous trabeculae* (secondary spongiosa). Osteoclastic activity will remove the woven bone and cartilaginous cores of the primary osseous trabeculae while osteoblastic activity forms new lamellar bone on other trabecular surfaces.

Osseous tissue, in long bones such as the humerus, increases in width by appositional activity of osteoprogenitor cells residing under the periosteum. These differentiate into osteoblasts which lay down concentric bony lamellae. The wall of the metaphysis does not increase in relative thickness, because endosteal bony layers are resorbed and remodeled at a rate equalling subperiosteal deposition.

The formation and resorption of bone plays an active role in bone growth, modeling and remodeling. Bone *growth* is the mechanism by which bones increase in size. Bone *modeling* is the process by which bones are reshaped from the initial cartilage model into the mature bone. *Remodeling* is the renewal of mature bone through a continual process of resorption and formation. The activities of bone formation and resorption are influenced by hormones such as parathyroid hormone, calcitonin, estrogens and prostaglandins, and numerous other factors, such as vitamins.

JOINTS

Three main categories of joints are recognized: (1) fibrous joints, (2) cartilaginous joints and (3) synovial joints.

Fibrous joints. The fibrous joints are classified by the tissue involved in the articulation. *Syndesmoses* are formed by the connection of two bones by a ligament, such as exists between the tibia and the fibula (Fig. 4-24). *Sutures* are connected by a thin band of connective tissue such as the sutures of the skull (Fiq. 4-25). The *gomphosis* is a specialized joint which anchors teeth to bone.

Cartilaginous joints. Bones joined together by hyaline cartilage are termed *synchondroses*. An example would be the junction of the ribs and sternum. The junc-

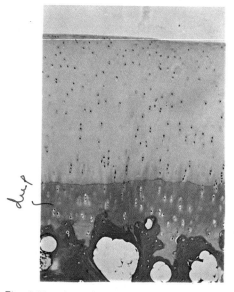

Fig. 4-23. Articular cartilage of a canine humeral head, H&E. (90 ×). Courtesy, Dr. Jo Ann Eurell, Hahnemann University.

tion of bones by fibrocartilage is termed a *symphysis*. The intervertebral discs and the pubic symphysis are such joints (Fig. 4-26).

Synovial joints (Fig. 4-27). The articular surface of bones in synovial joints is covered with hyaline cartilage. The *articular cartilage* has three zones (Fig. 4-23): (1) superficial, (2) intermediate and (3) deep. Cells in the superficial zone are discoidal while those in the intermediate zone are spheroidal; the cells of both zones are randomly arranged. Spheroidal cells of the deep zone are arranged in columnar groups. The portion of the deep zone adjacent to subchondral bone is calcified cartilage. The calcified cartilage is separated from the uncalcified cartilage of the deep zone by basophilic staining *tidemark*.

The boundaries of the joint are formed by a *joint capsule* which is composed of an outer *fibrous layer* and an inner *synovial membrane* (Fig. 4-28). The synovial membrane is folded or villous in structure, and it has connective tissue which varies from areolar to adipose to dense in arrangement. The synovial cells adjacent to the joint lumen are macrophage or fibroblast-like with intermediate forms also present. The cells secrete components of

Fig. 4-25. A suture, an example of a fibrous joint.

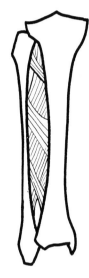

Fig. 4-24. A syndesmoses, an example of a fibrous joint.

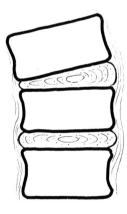

Fig. 4-26. The intervertebral discs, examples of cartilaginous joints.

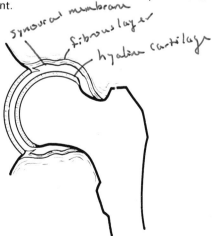

Fig. 4-27. A synovial joint.

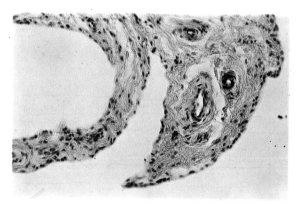

Fig. 4-28. Synovial membrane from an equine carpal joint, H&E. (64 ×). Courtesy, Dr. Jo Ann C. Eurell, Hahnemann University.

synovial fluid and are phagocytic. The joint capsule is well vascularized (Fig. 4-29).

Synovial joints may have extra-capsular or intra-articular ligaments. These ligaments are dense, regular connective tissue. In addition, intra-articular menisci of fibrocartilage may be present in the space between opposing articular surfaces.

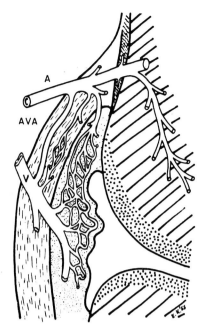

Fig. 4-29. Vascularization of a joint capsule. A = arteriole; V = venule; AVA = arteriolar-venous anastomosis. (From Ekkehard Kleiss, Microcosmos, 47: 1957.)

5...
Blood

Blood, classified as a connective tissue, is composed of cellular elements suspended in an intercellular fluid matrix, called *plasma*.

If an anticoagulant such as heparin is added to a test tube of freshly drawn blood (to prevent clotting), and this mixture is centrifuged, the cells are forced to the bottom to form a semi-solid pack. The volume-percentage of erythrocytes which comprise most of these packed cells is known as the *hematocrit*, the determination of which is an important clinical test. The supernatant fluid, the plasma (Fig. 5-1), is a straw-colored liquid that contains proteins, fats, carbohydrates, inorganic salts, hormones, etc. One of the plasma proteins is *fibrinogen*, which is essential for blood clotting. If blood is allowed to stand before centrifuging, the

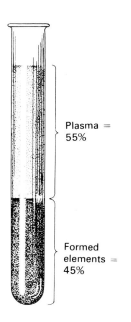

Fig. 5-1. Centrifuged blood.

Fig. 5-2. Erythrocytes and erythrocytic aggregates (rouleaux formations) (Alexander Rollet, 1872).

fibrinogen in the plasma is converted to *fibrin* which, in turn, entangles the blood cells (*clotting*). When this clotted mass is centrifuged, the cells entrapped in the fibrin mesh are forced to the bottom. The supernatant fluid, consisting of all the elements of the plasma that remain after the blood has clotted, is the *serum*.

The formed elements of blood include: erythrocytes (red blood cells, RBC), leucocytes (white blood cells, WBC) and thrombocytes (platelets).

Erythrocytes

Erythrocytes are highly differentiated cells, which in humans and most adult mammals lack nuclei (Fig. 5-2). A unique configuration, like biconcave disks, provides the erythrocyte far more surface area than if it retained its developmental spherical shape. In human embryos, erythrocytes are nucleated up to the gestational age of seven weeks. By eleven weeks, no nucleated erythrocytes remain in the peripheral blood. Healthy adult human males average 5.4 million erythrocytes per cubic millimeter of whole blood, while a healthy adult female has about 4.5 million. The cytoskeleton (stroma) of the erythrocyte is a network of protein strands holding, in their meshes, a dark red pigment called hemoglobin, which loosely combines with oxygen from air inspired in the lungs. The oxygenated hemoglobin (*oxyhemoglobin*) confers a bright red color to the blood. It is estimated that there are some 280 million molecules of hemoglobin confined in each erythrocyte. If the hemoglobin mole-

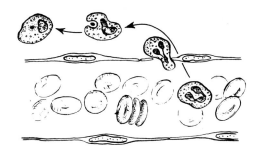

Fig. 5-4. Diapedesis.

cules were free in the plasma, they would leak through the endothelium of the blood vessels and be lost in the urine. When the blood reaches the capillaries, oxygen is released, changing oxyhemoglobin to hemoglobin. Carbon dioxide, released from cells, is transported back to the lungs where it is released and exhaled (Fig. 5-3). Whereas all the oxygen was carried by the hemoglobin molecule, only 23 percent of the carbon dioxide is returned by hemoglobin as *carbaminohemoglobin*. The greater portion is carried in the blood plasma as biocarbonate ion. Although erythrocytes average 7.5 μm in diameter, they are extremely flexible and thus can be deformed so that they may be carried through capillaries as narrow as 4 μm in diameter. Erythrocyte often stack themselves in piles as *erythrocyte aggregates* (Rouleaux) (Fig. 5-2).

Leucocytes

Leucocytes are true cells (each with its own nucleus) capable of ameboid movement. There are from 5,000 to 9,000

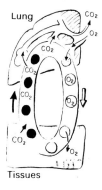

Fig. 5-3. Gas exchange.

WBC's in a cubic millimeter of whole blood. Leucocytes can leave the blood stream by passing between the endothelial cells of capillaries (see Chapter 9) or by piercing temporary holes through them (*diapedesis*). Thus, they can reach a tissue space where they wander about by ameboid movement and engulf bacteria and particulate matter (Fig. 5-4). Certain leucocytes with characteristic specific granules in their cytoplasm, are the *granulocytes*. Others which do not possess conspicuous granules, are the *agranulocytes*.

AGRANULOCYTES. The agranulocytes include lymphocytes and monocytes.

Lymphocyte. A small lymphocyte (Fig. 5-5) has a dark-staining, spheroid nucleus that occupies almost the entire cell. The chromatin pattern is dense and predominantly heterochromatic. Only a small shell of faintly basophilic cytoplasm is visible. A few small, irregularly shaped, non-specific *azurophilic* granules are seen occasionally in the cytoplasm. Two larger sizes of lymphocytes may be classified: the medium and large lymphocytes. These cells possess more cytoplasm, while the nuclei have additional euchromatin. Though small, medium and large lymphocytes are categorized, it should be understood that in reality there exists a gradual transition of the three forms.

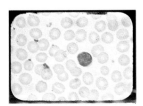

Fig. 5-5. Lymphocyte, Wright's stain.

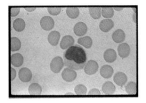

Fig. 5-6. Monocyte, Wright's stain.

Monocyte. A monocyte (Fig. 5-6), the largest agranulocyte, has a large, often slightly indented nucleus that stains lighter than that of the lymphocyte. The pattern of chromatin is a combination of euchromatin and heterochromatin. In a Wright's stained blood smear, the cytoplasm usually appearing grayish-blue in color, is greater in quantity than that of a large lymphocyte. As in the lymphocyte, a few non-specific azurophilic granules may be seen. When a monocyte leaves the blood stream and enters a tissue space, it usually transforms into a macrophage. Under pathological or demanding conditions, several monocytes may fuse to form a syncytium known as a foreign body giant cell.

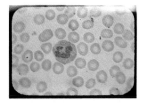

Fig. 5-7. Polymorphonuclear neutrophils, Wright's stain.

GRANULOCYTES. The granulocytes are recognized by lobular nuclei which assume many shapes and by specific granules in the cytoplasm, which permit classification by staining properties into neutrophilic, acidophilic and basophilic granulocytes.

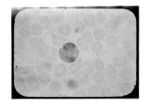

Fig. 5-8. Acidophil. Wright's stain.

Neutrophilic granulocyte. A polymorphonuclear (nuclei with a variety of shapes) neutrophilic granulocyte (commonly referred to as a PMN or "poly", Fig. 5-7), has a nucleus constricted into 2 to 5 lobes connected by thread-like nuclear bridges. The cytoplasm contains small, evenly dispersed, light-staining granules which contain proteolytic enzymes (lysosomes). This is the most common of all the leucocytes in peripheral blood.

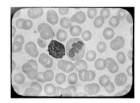

Fig. 5-9. Basophil (left), neutrophil (right). Wright's stain.

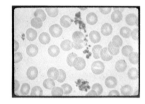

Fig. 5-10. Thrombocytes (Blood platelets) Wright's stain.

Younger cells with one elongated or two lobes are called juvenile neutrophilic granulocytes ("bands" or "stabs") (Fig. 5-12). The "stab" is a nucleus that folds across itself like a saber. One percent of the neutrophils seen in normal blood are bands, but during an infection their percentage in peripheral blood increases to reflect the greater demand. In 3% of the neutrophils from females, a chromatin sex corpuscle (Barr body, sex chromatin) can be seen. Functionally, neutrophils are attracted by chemotaxis (directional affinity) to areas of inflammation and bacterial proliferation, where they actively phagocytose bacteria and other foreign particles. The neutrophil itself dies in this process.

Acidophilic granulocyte (eosinophil). An acidophilic granulocyte (Fig. 5-8) has a nucleus composed of two to three lobes connected by broad bridges. The cytoplasm is filled with large, coarse, round, acidophilic, refractive granules. These cells increase in number in peripheral blood and at an antigenic inflammatory site in response to hyperimmune or allergic reactions such as: parasitic infestations, infectious diseases and certain degenerative lesions such as polyarteritis nodosa.

Basophilic granulocyte. A basophilic granulocyte (Fig. 5-9) has an irregular, light staining, usually horseshoe-shaped nucleus. The cytoplasmic granules are variable in size and shape and are intensely basophilic. If they are numerous, they may obscure the nucleus. The basophilic granules contain *heparin,* important in liproprotein metabolism and in hemostasis. However, basophils per se, have no recognized relationship to anticoagulation. The *histamine* content of basophils is significant in total basophil counts. A count of 50 per cubic millimeter is a sign of allergic sensitization, whereas a count below 20 per cubic millimeter regularly accompanies allergic reactions.

Thrombocytes

Thrombocytes (blood platelets) (Fig. 5-10) are very minute (2 to 4 μm in diameter), irregularly shaped fragments of megacaryocytes that tend to clump together in blood smears. They have fine basophilic thrombocytic granules, the *granulomeres,* and light-staining peripheries, the *hyalomeres.* Thrombocytes tend to aggregate at sites of injury to blood vessels and thus promote *hemostasis.* They also assist in the formation of *thromboplastin,* a factor essential in the conversion of prothrombin to thrombin. A reduction in thrombocyte number or the presence of abnormal platelets leads to poor or no clot forma-

Table 5-1. Sizes and Frequencies of Blood Cells

	Diameter of cells in blood smears in μM	Number of Cells per mm³	Percentage among Leukocytes	Typical Pathologic Condition in Which Number Is Increased
Erythrocytes	Average: 7.5	4,500,000 to 5,500,000	0	Polycythemia rubra
Leukocytes	6-15	5,000 to 9,000	100	Leukemia
Lymphocytes	6-12	1,000 to 2,700	20-30	Whooping cough
Monocytes	12-15	150 to 720	3-8	Tuberculosis
Neutrophils	9-12	3,000 to 6,750	60-75	Acute pus-forming infections
Acidophils	10-14	100 to 360	2-4	Allergies and parasitic infections
Basophils	8-10	25 to 90	0.5-1.0	Chicken pox

tion. In a cubic millimeter of blood, there are about 250,000 platelets.

The sizes and frequencies of the various kinds of blood cells are shown in Table 5-1. The sizes refer to artificially flattened cells in smears. If white blood cells are counted consecutively and identified in a normal blood smear (differential count), the percentages of each type expected would be those shown in Table 5-1.

Hemocytopoesis

Hemocytopoesis (blood cell formation) occurs in *hemopoetic tissues* via differentiation of mesenchymal cells into colonies of spherical basophilic cells, the *blood islands*. In yolk sac and body stalk mesenchyme of two week embryos, hemocytopoesis produces nucleated erythrocytes. At 6 weeks, definitive *erythroblasts* (erythrocyte precursors) appear in liver to produce enucleated erythrocytes. In the second month, the first *granular leukocytes* and *megacaryocytes* appear in liver and then spleen. *Bone marrow* appears in the four month fetus and as its function increases, that of the liver and spleen decreases. Successive hemopoetic sites (yolk sac, liver, spleen and bone marrow) are seeded by self-duplicating, pluripotential *hemocytoblasts*. The hemocytoblast is also referred to as the *Pluripotential Hemocytopoetic Stem Cell* (*PHSC*) or *Colony-Forming Unit* (*CFU*). [While acceptable, this text uses terms recommended by the *Nomina Histologica*]. Hemocytopoesis in red bone marrow, is termed *medullary hemocytopoesis*. Should demands for blood cells exceed the capacity of bone marrow, the liver and spleen may resume latent potentialities via *extramedullary hemocytopoesis*.

Bone marrow structural components

Bone marrow has a stroma of reticular fibers associated with reticular cells (Fig.

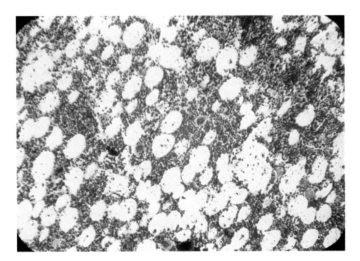

Fig. 5-11. Bone marrow section. H. & E.

5-14) and interstices occupied by hemocytopoetic cells and adipocytes. The quantity of adipocytes classifies marrow as *red* or *yellow*. Red (*active*) marrow contains up to 75% adipocytes, while yellow (*inactive*) marrow exceeds that amount. The color red is attributed to the hemoglobin accumulating in developing erythrocytes. Generally *red marrow* is localized in flat bones. In adolescents it also occupies the proximal epiphyses of the humerus and femur. Red marrow predominates in the fetus, but with time, *yellow marrow* accumulates until both are equal. Aging, disease and starvation stimulate marrow to become gelatinous. The name *myeloid tissue* alludes to an abundance of adipocytes derived from reticular cells.

Marrow is pervaded by *venous sinuses* of nonfenestrated endothelial cells joined by junctional complexes. The lumina widths vary from 50 μm to 75 μm. A poorly defined basal lamina of scattered flocculent material surrounds the venous sinuses. An outer supporting *adventitial* coat contains long branching reticular cells which can give rise to neighboring adipocytes. The *adventitia* shields or exposes external endothelial surface areas in response to toxins, hormones or cells attempting to gain access to the venous sinuses. Generally 40% to 60% of the surface is covered by adventitia.

Bone marrow vasculature is described elegantly by DeBruyn (Fig. 5-14). A *nutrient artery* pierces periosteum and bone and arborizes into *arterioles* which distribute blood to *periosteal capillaries* and *venous sinuses*. The latter drain into a *central marrow vein* which conducts blood to the exiting *vena comitans*.

Mature cells enter the closed circulation via a *transcellular* route: (1) a new blood cell compresses the exposed *external surface* of an endothelial cell, (2) which fuses to its *internal surface*, to (3) produce a *transient migration pore* through which the blood cell maneuvers and (4) the pore seals behind the cell. Permanent endothelial openings, or spaces between adjacent endothelial cells usually are not present.

Hemocytoblasts proliferate three cell lines committed to produce *erythrocytes*, *granulocytes* and *megacaryocytes*. The respective processes are *erythrocytopoesis*, (Fig. 5-13), *granulocytopoesis* (Fig. 5-12) and *megacaryocytopoesis* (Fig. 5-15). *Thrombocyte* (platelet) formation from megacaryocytes is termed *thrombocytopoesis* (Figs. 5-18, 19A, 19B).

Erythrocytopoesis produces erythrocytes by differentiation of erythroblasts in three

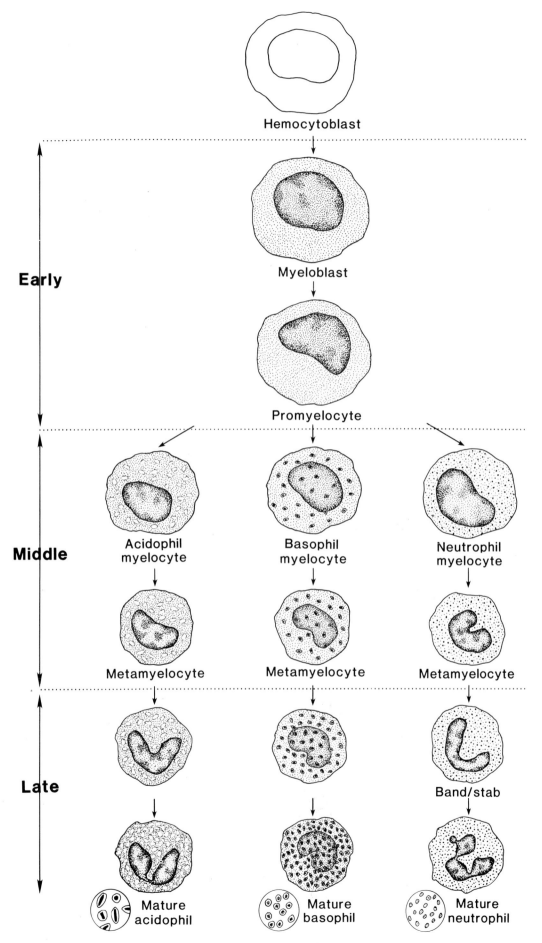

Fig. 5-12. Granulocytopoesis.

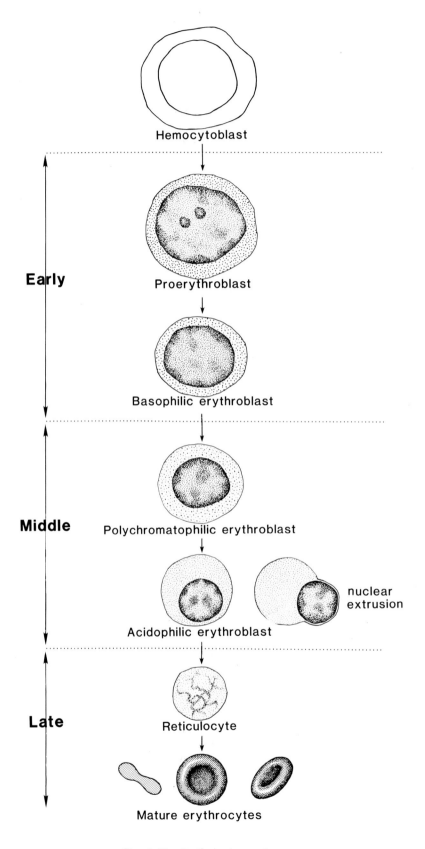

Fig. 5-13. Erythrocytopoesis.

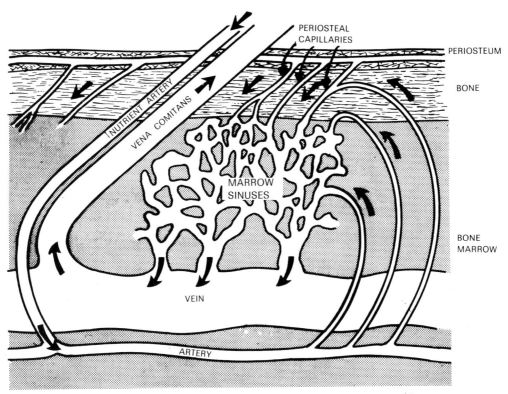

Fig. 5-14. Vasculature of bone marrow, based on the work of Dr. P.P.H. DeBruyn, The University of Chicago. (Ref. DeBruyn, 1981, Seminars in Hematology, vol. 18: 179-193).

stages: *early, middle* and *late*. In the *early stage* the *proerythroblast* appears with a large euchromatic nucleus surrounded by a rim of basophilic cytoplasm. This active cell, with 24 nucleoli, elaborates free polyribosomes. A series of mitoses produces several smaller *basophilic erythroblasts*, with deep cytoplasmic basophilia due to the accumulation of polyribosomes which synthesize the protein hemoglobin. Nuclei of dense heterochromatin minus nucleoli, indicates cessation of ribosomal production. Dividing basophilic erythroblasts yield the middle stage *polychromatophilic erythroblasts*.

The *middle stage* smaller polychromatophilic erythroblasts exhibit less basophilia than their progenitors because acidophilic hemoglobin accumulates in the cytoplasm. When the hemoglobin content exceeds that of polyribosomes, increasing acidophilia (via eosin-azure dyes) produces a reddish tint, hence the name *acidophilic erythroblast*. After cessation of hemoglobin production, the nucleus is extruded.

Late stage enucleated *erythrocytes* enter the circulation when mature. The diffusely basophilic appearance of immature erythrocytes indicates that they entered the circulation prior to completing synthesis of hemoglobin. Frequently called *reticulocytes*, they are useful indicators of erythrocyte formation in patients suffering from anemia of in individuals at high altitudes.

The fate of erythrocytes depends on cell membrane structure. Loss of sialic acid residues are detected in the spleen, which removes aging (120 day old) erythrocytes to reclaim iron. Degraded hemoglobin transported to the liver by *plasma transfer-*

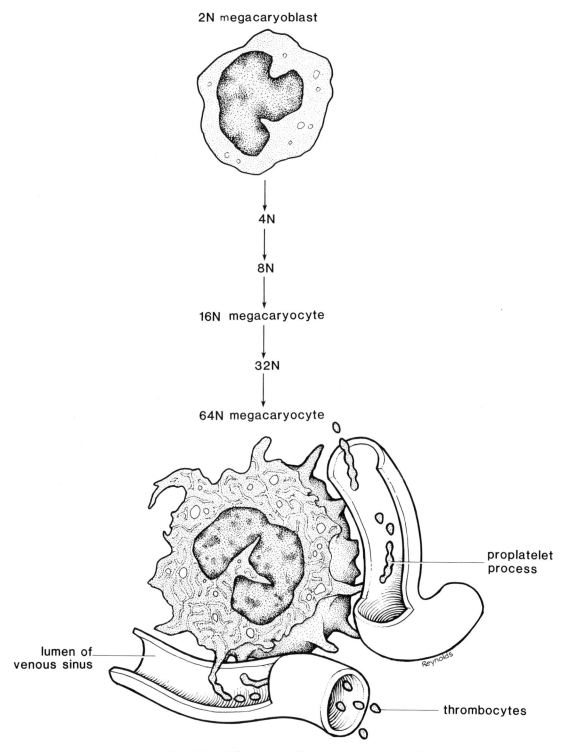

Fig. 5-15. Different cell kinetic compartments of red bone marrow.

rin is converted to bilirubin. The hormone *erythropoetin* maintains erythrocytes in circulation and stimulates marrow to increase production.

Granulocytopoesis is patterned after erytrocytopoesis (Fig. 5-12). In the *early stages*, the *myeloblast* derived from the hemocytoblast, gives rise to *promyelocytes* with *non-specific* granules. *Middle stage myelocytes* and *metamyelocytes* accumulating *specific granules* characteristic of *acidophils* (eosinophils), *basophils* and *neutrophils* retain the reproductive capacities of their progenitors. *Late stages* are forms of immature and mature *granulocytic neutrophils, basophils* and *acidophils*. Nuclear segmentation and increased heterochromatin indicate decreasing nuclear activity and diminished protein syntesis. Granulocytes enter the venous sinuses as do the erythrocytes. *Monocytes* and *granulocytes* share the myeloblast as the common stem cell, as evidenced by marrow transplantation experiments in mice. Special techniques must be used to distinguish monocytes from myeloblasts.

Megacaryocytopoesis (Fig. 5-15) occurs as the result of several nuclear divisions (*karyokinesis*) and nulcear fusion without cytoplasmic division (*cytokinesis*) in the

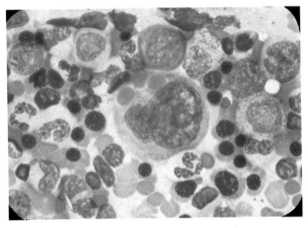

Fig. 5-16. Young megacaryocyte in a bone marrow imprint. Note phagocytosis of erythrocytes.

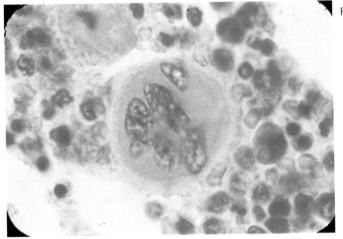

Fig. 5-17. Mature megacaryocyte. Wright's stain.

megacaryocytoblast. These activities result in the formation of a giant cell up to 70 μm in diameter, possessing 8 to 16 times the amount of DNA in a normal cell. *Demarcation membranes* (Fig. 5-18) form cytoplasmic channels isolating *prothrombocytic processes* which penetrate transient migration pores of the marrow venous sinus endothelium (Fig. 5-19A, Fig. 5-19B). Constrictions of intravascular processes fragment off as *thrombocytes* (platelets). A hormone, *thrombopoetin*, stimulates megacaryocyte formation and release of thrombocytes, which function in the clotting process.

Myeloid cells (erythrocytes, granulocytes and monocytes) mature within bone marrow prior to release into the circulation. *Lymphocytes*, however, develop in bone marrow but complete their maturation elsewhere, viz., *B-Lymphocytes* in the spleen and *T-Lymphocytes* in the thymus.

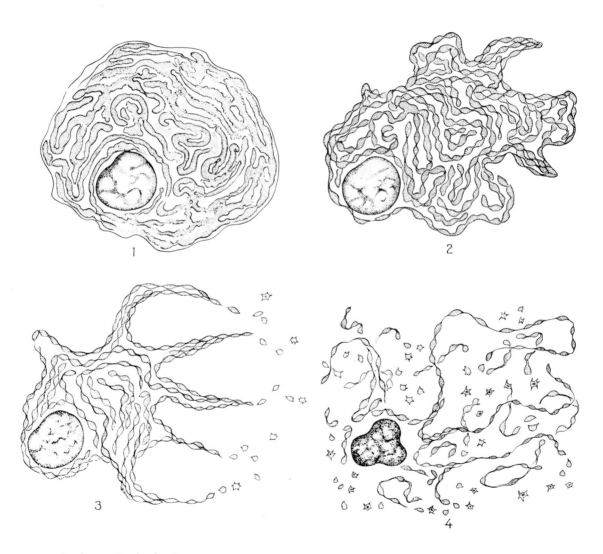

Fig. 5-18. Stages in platelet formation and liberation of platelets. Platelet demarcation membranes appear in the cytoplasm (1-2). These membranes fuse with one another creating a series of cytoplasmic cylinders around the nucleus. Granules are found within these cylinders. Projections appear which progressively elongate giving the cell an octopus-like appearance (3). The enlongated filaments break up into platelets. (From *Living Blood Cells* and *Their Ultrastructure* by Dr. Marcel Bessis, Springer-Verlag, 1973.)

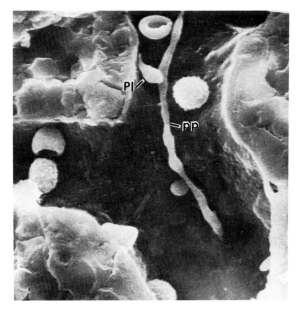

Fig. 5-19A. Scanning electron micrograph of the marrow sinus wall. Note the intraluminal segment of a prothrombocytic process (PP) with periodic constrictions along its length. PL = teardrop shaped platelet. Courtesy of P.P.H. DeBruyn, Seminars in Hematology, 18: 179-193, 1981.

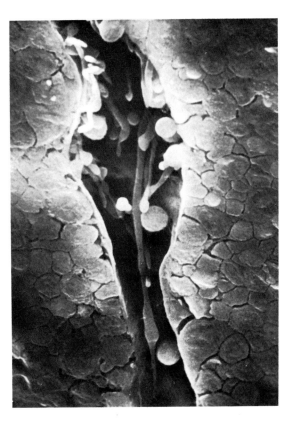

Fig. 5-19B. Scanning electron micrograph of an exposed marrow sinus with a dense cluster of prothrombocytic processes. Courtesy P.P.H. DeBruyn, Seminars in Hematology, 18: 179-193, 1981.

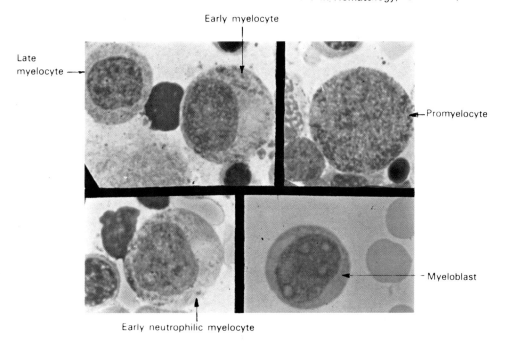

Fig. 5-20. Bone marrow cells, Wright's stain. Courtesy of Dr. D. Bainton. University of California at San Francisco.

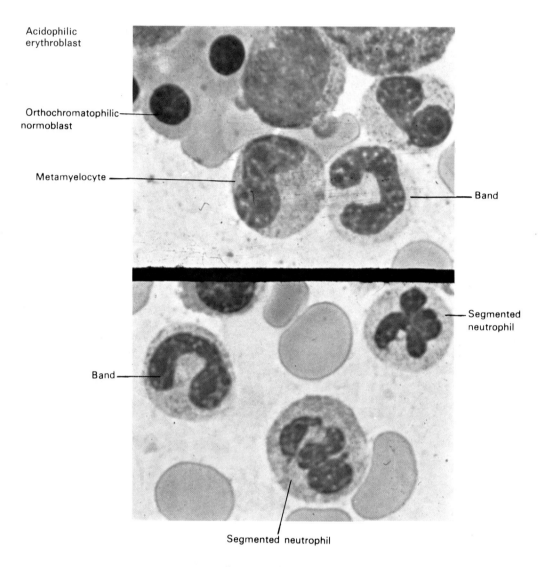

Fig. 5-21. Bone marrow cells, Wright's stain. Courtesy of Dr. D. Bainton.

6...
Muscle

Muscle tissue is composed of cells specialized for contraction and relaxation and held together by connective tissue.

Contractility is a general property of protoplasm. Without it, ameboid movement and even cell division would be impossible. In multicellular organisms, contractility, although present in every cell, is most highly developed in muscle cells. The most ancient cells so specialized are the myeoepithelial cells of coelenterates. These are epithelial cells, both in the ectoderm and the endoderm (Fig. 6-1) which possess a basal, subepithelial extension containing myofibrils. In higher animals, cells specialized for contraction are located deep to epithelium.

There are three different kinds of muscle, classified as shown in Table 6-1. The 1983 *Nomina Histologica* deletes terms such as "sarcolemma", "sarcoplasm" and "sarcosome" in favor of the tautologous *plasmalemma*, *cytoplasm* and *mitochondrion*, respectively.

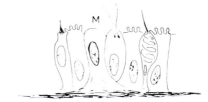

Fig. 6-1. A myoepithelial cell (M) in the ectoderm of a coelenterate.

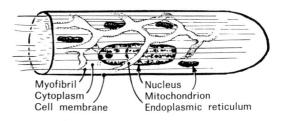

Fig. 6-2. A generalized muscle cell.

Table 6-1

Kinds of muscles

Non-striated muscle	1. Smooth muscle tissue (e.g. in the wall of gut and blood vessels	Involuntary (cannot be controlled by the will)
Striated muscle (striped)	2. Cardiac muscle tissue (e.g. in the heart)	
	3. Skeletal muscle tissue (e.g. in the biceps brachii)	Voluntary (can be controlled by the will)

Non-Striated Muscle Tissue

Non-striated *myocytes* (smooth muscle cells) may occur singly (as around sweat glands), in isolated bundles of fibers (e.g., arrector pili muscles of skin) or in well-organized layers (e.g., tunica media of blood vessels). The smooth myocyte, an individual cell (Fig. 6-3), is often spindle-shaped but may be branched. It possesses a single, centrally located, oval nucleus with fine chromatin granules and one or more nucleoli. Myofibrils cannot be seen easily with the light microscope in non-striated myocytes.

The fringed ends of many non-striated myocytes (Fig. 6-3) facilitate their attachment to connective tissue. The myoepithelial cells of sweat glands are frequently classified with the non-striated myocyte (Figs. 6-3 and 6-4).

When a bundle of non-striated myo-

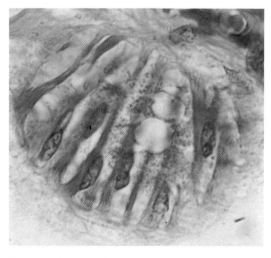

Fig. 6-3. Tangential section of a sweat gland showing isolated smooth muscle cells. Each has one nucleus and ends in fringes. At the right one sees a branched smooth muscle fiber. Goldner stain, 1900 ×.

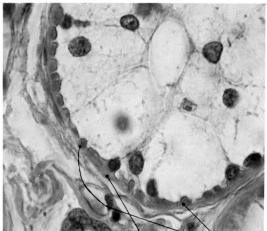

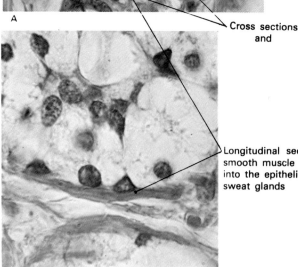

Fig. 6-4. Eccrine sweat gland surrounded by smooth muscle cells, erroneously called "myoepithelial". Goldner stain, 950 x. A, Muscle cells seen in cross section. B, One such cell with its nucleus, longitudinally cut.

Cross sections and

Longitudinal sections of the smooth muscle cells pressed into the epithelium of sweat glands

NON-STRIATED MUSCLE TISSUE

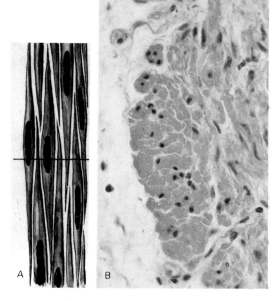

Fig. 6-5. The nuclei of smooth muscle cells are located near their centers. Since in closed bundles the cells are staggered, a nucleus is encountered only in a few sections through individual muscle cells. A, Explanatory diagram. B, Musculature of vesicular gland, H&E. 700 ×.

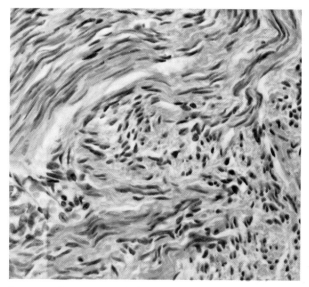

Fig. 6-6. Smooth muscle bundles in the wall of the uterus. 750 ×.

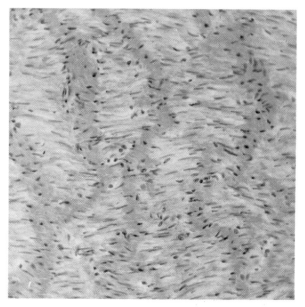

Fig. 6-7. Waves of contraction in muscularis of cecum, H&E. 360 ×.

cytes is cut transversely, those sections which pass through a nucleus are widest (Fig. 6-5). Where non-striated myocytes are irregularly arranged as in the uterus, one finds longitudinal, oblique and transverse sections through them (Fig. 6-6).

Contraction of non-striated myocytes occurs under the influence of autonomic nerves (see Fig. 12-54) and passes in a relatively slow, sustained, wavelike fashion over entire muscle bundles (Fig. 6-7).

Where two non-striated myocytes come into apposition, the plasmalemma may form specialized zones of contact, the nexuses (Fig. 6-8) (gap junctions) as well as interdigitations (Fig. 6-9). These junctions provide intimate union of the cells involved and probably facilitate the transmission of impulses between the cells.

Non-striated myocytes in longitudinal section from the esophagus are shown in Figure 6-10. When examined by transmission electron microscopy, a unique characteristic of smooth muscle, the *insertion corpuscle* (area densa) is visible. These dense areas occur throughout the cytoplasm and along the inner surface of the plasmalemma. Also, numerous flask-like pits or invaginations of the plasmalemma are present in the peripheral cytoplasm. These invaginations resemble micropinocytotic vesicles and are termed *superficial vesicles* (caveolae) (Fig. 6-10). Superficial vesicles may be

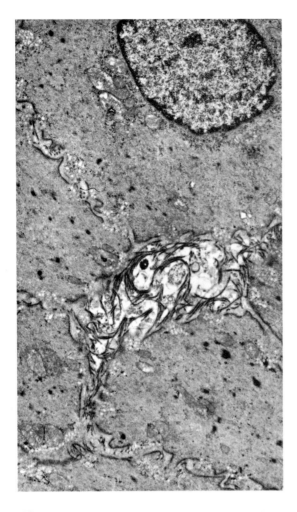

Fig. 6-8. Smooth muscle cells of vesicular gland with collagen fibers (left of center of picture) and nexuses. 4800 ×. Courtesy of Dr. G. Aumüller, University of Heidelberg.

Fig. 6-9. Smooth muscle cells from vesicular gland. Note interdigitations and nexuses. ▼ 700 ×. Same source.

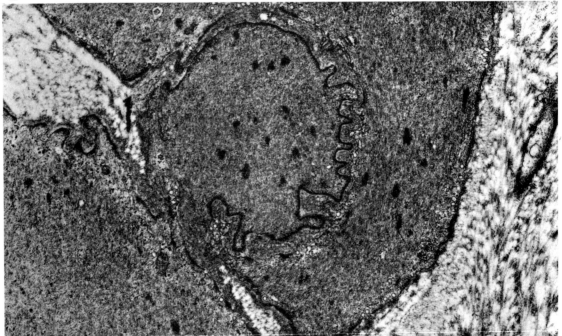

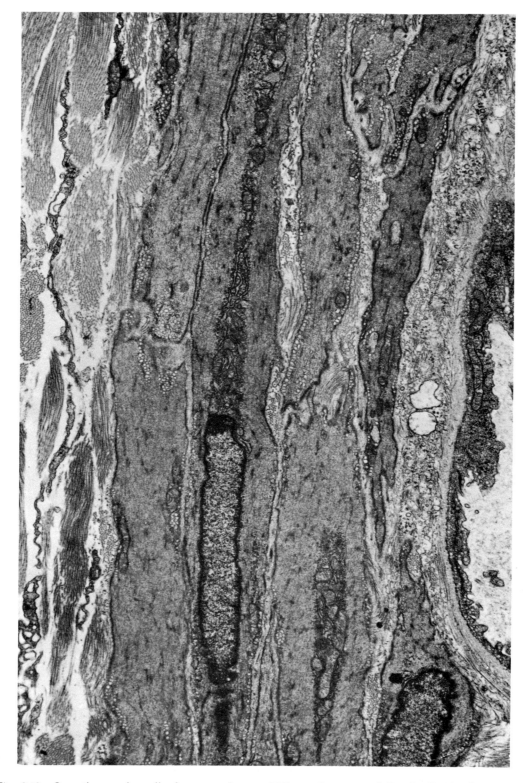

Fig. 6-10. Smooth muscle cells from esophagus, 9400 ×. Courtesy of Dr. D. H. Matulionis, University of Kentucky.

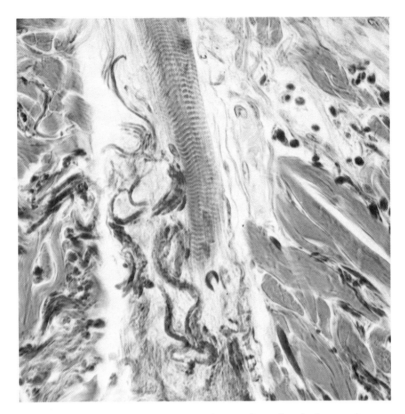

Fig. 6-11. Attachment of a skeletal muscle cell of the m. levator palpebrae superioris to connective tissue of the upper lid. Note fringed, microtendinous continuations of the myofibrils. Azan, 950 x.

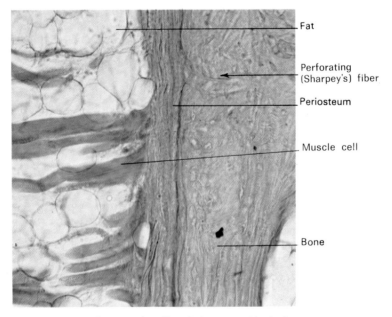

Fig. 6-12. Attachment of cells of the m. orbicularis oculi (left to the periosteum) to the lamina papyracea of the ethmoid bone (right), 350 x.

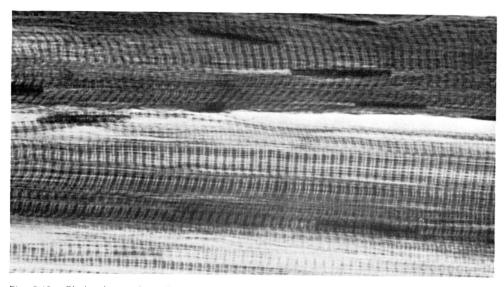

Fig. 6-13. Skeletal muscle cells stained with iron hematoxylin. Courtesy of Leo Massopust.

involved in moving substances into the cell or transmitting impulses into the cell.

Skeletal Muscle Tissue

Striated skeletal myocytes occur in bundles which may be attached to facial skin (Fig. 6-11), bone (Fig. 6-12), or cartilage by tendons, aponeuroses or connective tissue. The individual myocytes are multinucleated, with pale, vesicular nuclei dispersed along their length at the periphery just under the plasmalemma (Figs. 6-2, 6-13). The striated skeletal myocyte, therefore, has the character of a syncytium. Although the diameter of each individual myocyte is approximately the same throughout its length (Figs. 6-12 and 6-13), various myocytes vary considerably in diameter (Figs. 6-14 and 6-15). Skeletal myocytes are usually cylindrical with circular cross-sections when lying alone (Fig. 6-14). When present in organized muscle bundles, the myocytes are so densely packed that, by mutual compression, they become prismatic, having polygonal cross-sections (Fig. 6-15).

In the scanty cytoplasm of the skeletal myocyte, there are embedded numerous longitudinally oriented myofibrils, each composed of regularly alternating bands or segments with distinctive optical properties (Fig. 6-16):

The light band is the *isotropic disk* (I disk) (singly refractive).

The dark band is the *anisotropic disk* (A disk) (birefringent).

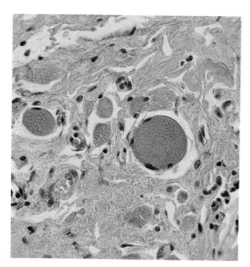

Fig. 6-14. Skeletal muscle cells in cross section in the external sphincter ani, where they are not tightly packed and reveal their primary, cylindrical shape. Note also different sizes and superficial location of nuclei.

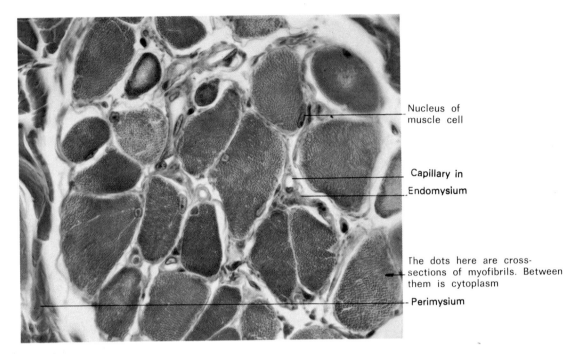

Fig. 6-15. Sagittal section of m. orbicularis oris. Goldner stain, 1000 ×.

The extremely thin *telophragma* or *Z line* appears to bisect the I disk.

A *zona lucida (H band)* is the lighter center of the dark A disk which is bisected by the darker *mesophragma (M band)*, barely resolved with the light microscope.

Myofibrils are divided by the Z lines into segments called *myomeres* (old name = sarcomere). Identical myomeres of myofibrils occur side by side, in register, thus

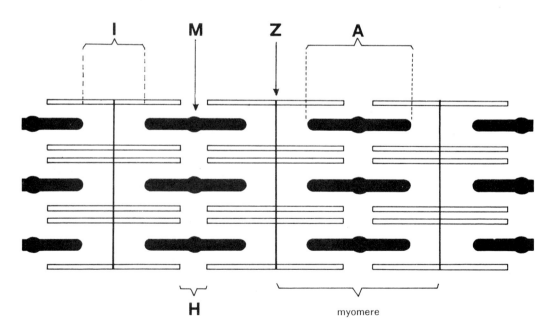

Fig. 6-16. Diagram of myofibril.

causing the characteristic cross striations seen in longitudinal sections.

Electron microscopy reveals that each myofibril, in turn, possesses smaller longitudinal *myofilaments* which, in turn, consist of molecular chains. The relative size and structure of the components of a myocyte are summarized in Figure 6-17. Figure 6-18 shows the ultrastructure of skeletal myocytes.

Two major types of myofilaments have been described: (1) thick myofilaments of *myosin* and (2) thin myofilaments of *actin*. Myosin myofilaments extend from the end of an A band to the other. The actin myofilaments extend from the Z line through the I band and into the adjacent part of the A band (up to the H band) (Fig. 6-18 A).

Cross sections through the H and M bands show only thick myofilaments. A transverse cut through the I band shows only thin myofilaments. Sections through the terminal region of the A band show both thick and thin myofilaments arranged in a hexagonal pattern with a thick myofilament in the center of each hexagon (Fig. 6-18, C). Figure 6-20 attempts to illustrate the events during relaxation, moderate contraction and extreme contraction.

Each thin actin myofilament is a helix of two strands of *F actin (fibrous actin)*. F actin strands are actually polymers of *G actin monomers* which are smaller globular units. Two additional proteins, *troponin* and *tropomyosin*, are associated with the thin myofilaments. The thick myosin myofilaments are composed of light and heavy *meromyosin* subunits which have globular head regions to serve as crossbridges between the thick and thin myofilaments.

Agranular endoplasmic reticulum tubules of skeletal myocytes (no longer called the sarcoplasmic reticulum) surround the myofibrils, and end as dilated *terminal cisternae* at the A-I junction or the Z line (Fig. 6-21). In this region, a transverse tubule is interposed between two terminal cisternae. This arrangement of a centrally located transverse (or T) tubule with agranular endoplasmic reticulum terminal cisternae (on either side of the T-tubule) is known as a *triad*. The

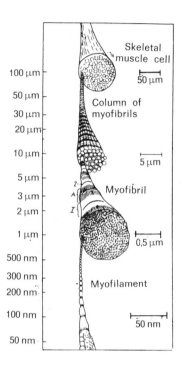

Fig. 6-17. Stereogram of a skeletal muscle cell according to Prof. F. Buchthal, University of Copenhagen.

transverse tubules are invaginations of the surface plasmalemma. Thus, the space in a transverse tubule is actually an extracellular apace. The transverse tubule-terminal cisternae system provides a morphological arrangement whereby an impulse (an action current) may travel over the outer plasmalemma, down the T-tubes to cause the release of calcium ions from the endoplasmic reticulum, thus providing coordinated activity of all myofibrils.

Figure 6-22 shows the T-tubules (black) and the endoplasmic reticulum (dark grey) simultaneously. This combined demonstration occurs only in very rare, fortunate preparations.

During partial contraction, the thin ac-

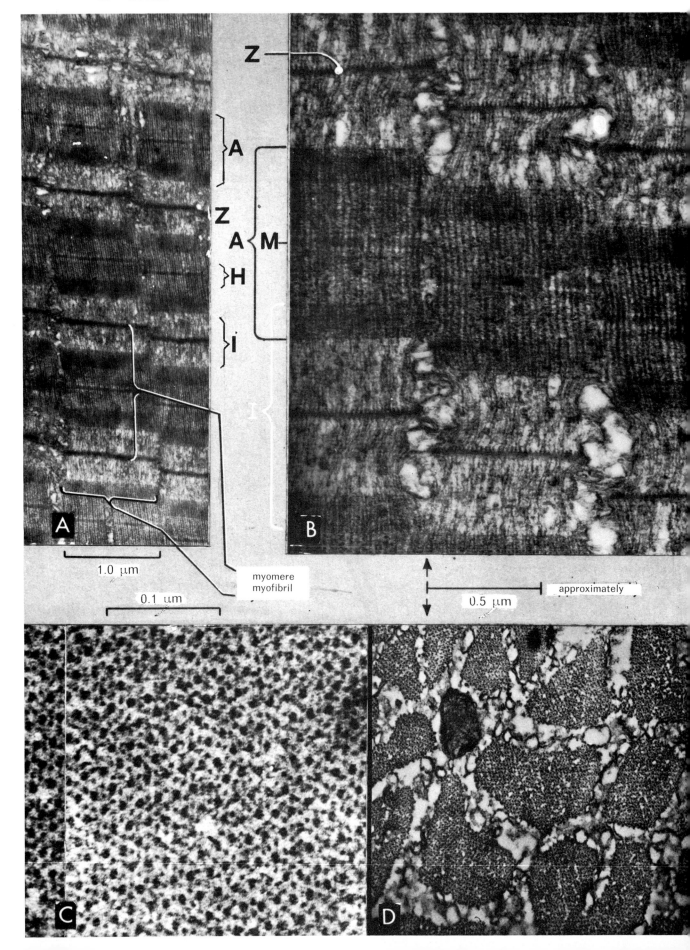

tin myofilaments slide between the thick myosin myofilaments (Figure 6-20), thus shortening the myomeres. The I disk is eliminated when contraction has approximated 50% of the resting length (Fig. 6-19). The H band may also disappear at this time. The A disk remains unchanged during the cycles of normal contraction and relaxation. During extreme contraction, the A disk may shorten; and a contraction band of dark material appears along the Z line (clumping of actin myofilaments against the Z line).

Calcium released from the agranular endoplasmic reticulum during the spread of the *wave of depolarization* from the transverse tubules is necessary for initiating interdigitation of the actin and myosin myofilaments.

Contraction of skeletal myocytes occurs under the influence of motor nerves. Each somatic motor nerve axon ends on a multinucleated mass of cytoplasm, pushing the plasmalemma inward. This connection, shown in Figure 6-23, is known as the *motor end-plate* or as the *neuromuscular synapse* (see *also* Figs. 7-13 and 7-14). When the nerve is destroyed, paralysis and muscular atrophy result.

A "muscle" (Figs. 6-15, 6-24 and 6-25) is a bundle of fascicles of skeletal myocytes tied together by connective tissue and supplied with blood vessels and nerves. *Endomysium* in contact with the myocyte plasmalemma is a thin delicate connective tissue layer composed of fine reticular fibers and some collagenous and elastic fibers. It is this layer that carries the capillaries which nourish the myocytes (Figs. 6-15 and 6-24). Bundles or fascicles of myocytes are surrounded by a thicker layer of connective tissue, the *perimysium*; and these fascicles are held together by a common envelope of connective tissue, the *epimysium*, also known as the fibrous sheath of the muscle or the *muscle fascia*.

The energy of muscle contraction is transmitted to a tendon, periosteum or some other fibrous structure. The ending of a skeletal myocyte is best seen in facial muscles which end in the dermis of the skin. The myofibrils seem to spread like fringes of an old cut-off rope and adhere to collagen fibers of the dermis (Fig. 6-11), and the plasmalemma blends with the connective tissue. In the muscle-tendon junction, the myocytes taper to conical or

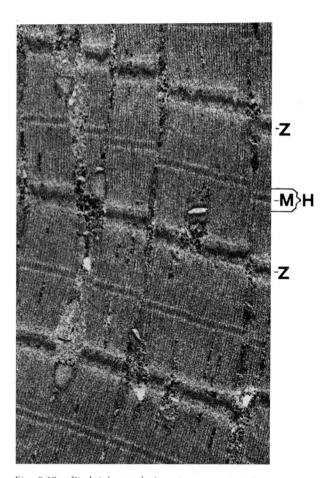

Fig. 6-19. Skeletal muscle in extreme contraction. Courtesy of Dr. C. N. Sun, University of Arkansas for Medical Sciences.

◄

Fig. 6-18. Sections of skeletal muscle cells, very slightly contracted. A and B longitudinal. C and D transverse. In C we note the zone where the myosin and the actin filments overlap. Courtesy of Drs. Carlson and Knappeis, University of Copenhagen. See also Fig. 6-20, B.

rounded ends. The endomysium surrounding each myocyte becomes continuous with the comparable connective tissue investments (*endotendinium*) of the collagen fibers in the tendon. Perimysium and epimysium blend with *peritendinium* and *epitendinium* respectively.

Myosatellite cells, with scanty cytoplasm and elongated nuclei, are flattened against the surface of the myocyte and are enclosed in the surface coat (glycocalyx or basal lamina) of the myocyte. The myosatellite cells may play a major role in the regeneration of skeletal muscle.

Cardiac Striated Muscle Tissue

The structure of the heart muscle is shown stereographically in Figure 6-26.

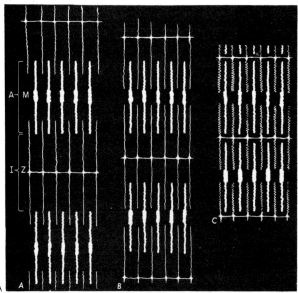

Fig. 6-20. A: Diagram of muscular contraction. B: Stereogram of the arrangement of actin (yellow) and myosin (blue) filaments in skeletal muscle in the A band region of a myomere. Note the hexagonal position of the actin filaments around the myosin filaments.

Cardiac muscle is found in the *myocardium* of the heart walls and in the proximal portions of large conducting arteries. The myocardium, formerly conceived to be a syncytium, is a network composed of individual, slightly branched cardiac myocytes, each possessing a single, centrally located nucleus. Figure 6-27 shows an un-

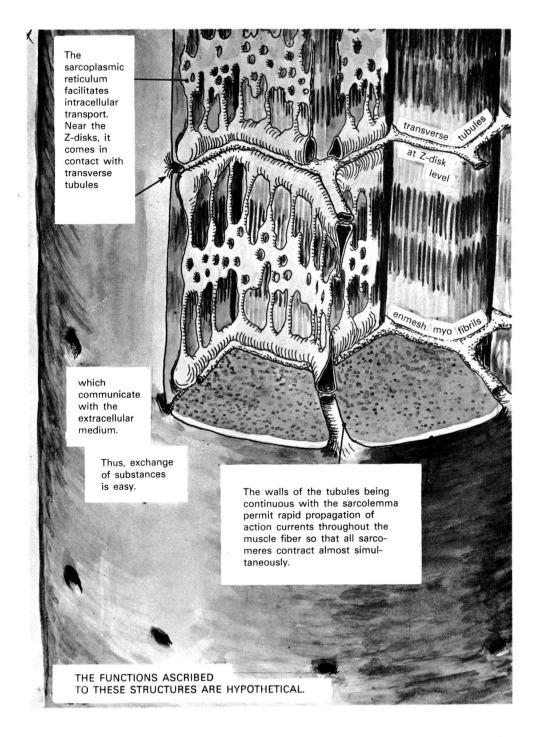

Fig. 6-21. Stereogram of the ultrastructural relationships of the plasmolemma, endoplasmic reticulum and T. tubule system in skeletal muscle of the frog.

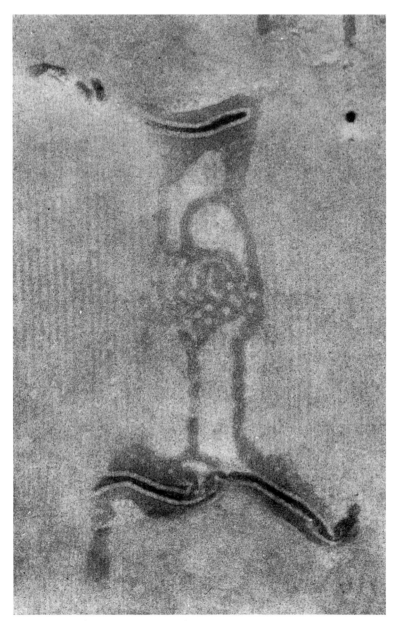

Fig. 6-22. The T tubules (black) and the endoplasmic reticulum (dark grey), filled with lanthanum. Rat diaphragm. Courtesy of Drs. R. A. Waugh and J. R. Sommer, Duke University.

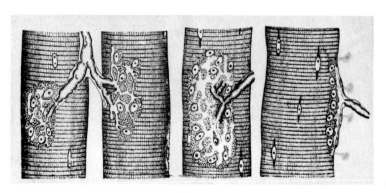

Fig. 6-23. Neuromuscular synapses (motor end plates) as seen in fresh, unstained muscle cells (J. Arnold, 1872).

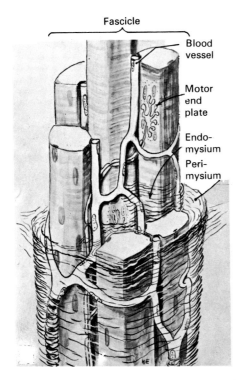

Fig. 6-24. Construction of a muscle fascicle.

stained, isolated cardiac myocyte teased out of a papillary muscle from a human cadaver.

The ends of the cardiac myocytes are studded with very small, slender projections and indented by depressions (Figs. 6-26, B and 6-28). By means of these jagged surfaces, adjacent cardiac muscle cells are joined together. These junctional boundaries, known as *intercalated disks*, are unique to cardiac muscle. Easily observable in fresh pieces of heart and in iron hematoxylin preparations, the intercalated disks may be difficult to see in routine slides, since the mounting medium masks their refractive properties.

In certain regions of the intercalated disk, a desmosome-like arrangement, the *macula adherens*, is apparent in EM preparations. The extensiveness of this arrangement is responsible for the term "fascia adherens". In other regions, usually where the myocytes meet laterally, the connections form nexuses or gap junctions.

Cardiac myocytes, rich in cytoplasm, contain many mitochondria with numerous cristae and few delicately striated myofibrils. In comparison to skeletal muscle, the A bands are not as conspicuous in cardiac muscle; thus the Z bands are particularly prominent (Fig. 6-28).

In the ventricles of the heart the myocytes are short and thick, in the atria they are more slender.

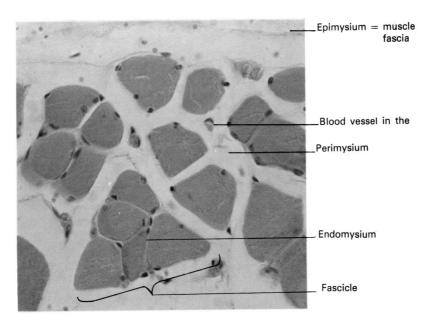

Fig. 6-25. Construction of a skeletal muscle, H&E. 750 ×.

132 MUSCLE

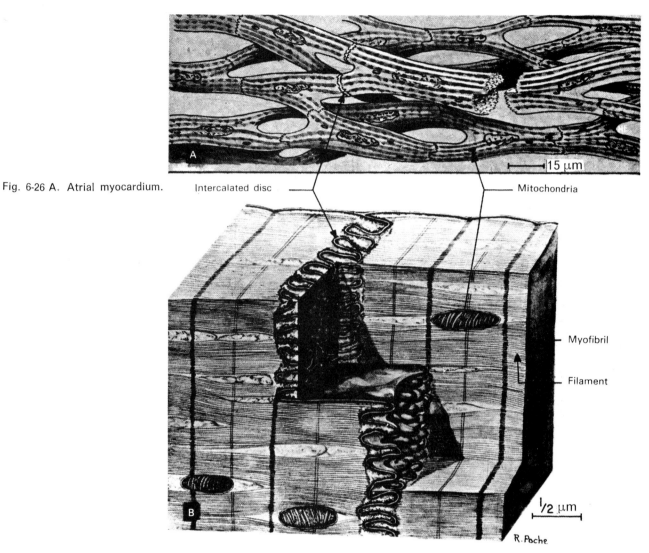

Fig. 6-26 A. Atrial myocardium.

Fig. 6-26, B. An intercalated disc. Stereogram by Dr. R. Poche (from Poche und Lindner, Z. Zellforsch. *43*: 104-120, 1955).

Fig. 6-27. Isolated cardiac muscle cell from papillary muscle of a human, embalmed cadaver, obtained by teasing with needles, unstained.

For a comparison of skeletal and cardiac muscle see Figure 6-29).

The transverse tubule-endoplasmic reticulum system is similar to that of skeletal muscle except that the T-tubules are associated with only one terminal cisterna at any one point. The resulting association is a *dyad* as opposed to the triad of skeletal myocytes. The T-tubules in skeletal myocytes are located at the A-I junction of the myomere, while in cardiac muscle, they occur at the Z line.

Cardiac myocytes are autonomous, i.e., they have an innate rhythmic contraction, as seen in tissue culture. Autonomic nerves to the heart can speed up or slow down contractions, but can neither initiate nor stop them.

The cardiac myocytes are surrounded by a delicate net of fine reticular and collagenous fibers corresponding to the endomysium of skeletal muscle.

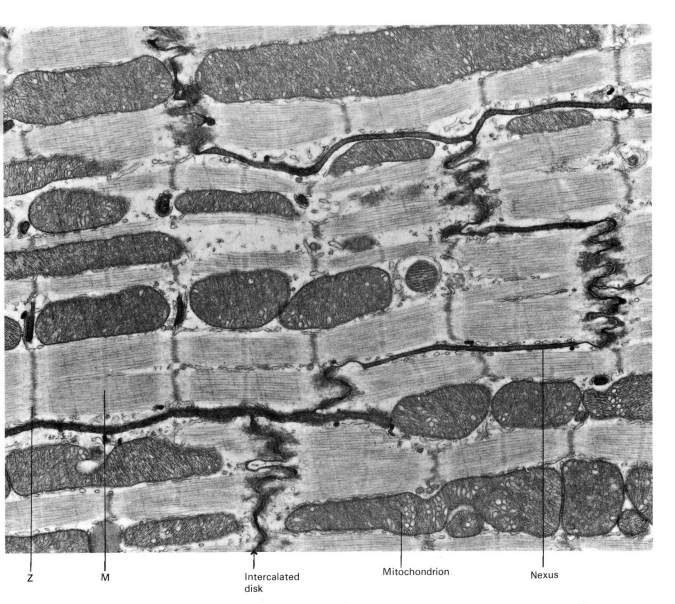

Fig. 6-28. Cardiac muscle of bat, 20,000 ×. (From Hagopian, Anversa and Nunez, Anat. Rect. *178*: 599, 1974.)

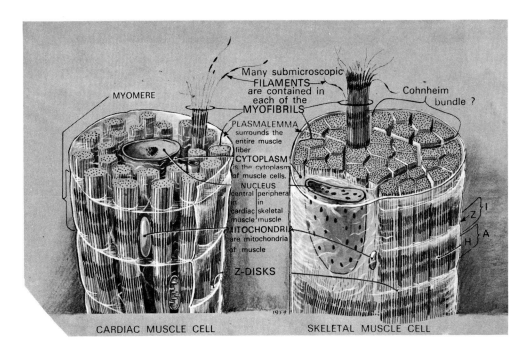

Fig. 6-29. Comparison of cardiac and skeletal muscle fibers.

7...
Nervous tissue

NERVE CELLS AND THEIR MUTUAL RELATIONSHIPS

The nerve cell or neuron (Figs. 7-1, 7-2 and 7-3) is the structural and functional unit of nervous tissue, specialized for re-

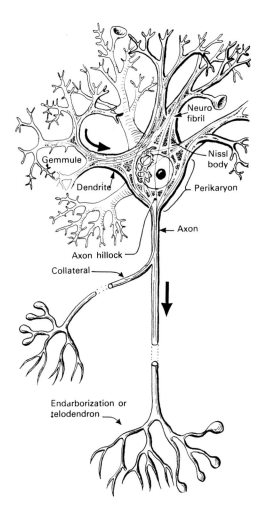

Fig. 7-1. Diagram of a nerve cell (neuron).

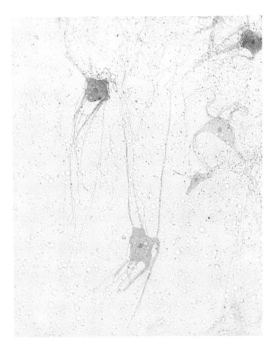

Fig. 7-2. Spread of multipolar nerve cells from spinal cord. Borax carmine, 200 ×.

136 NERVOUS TISSUE

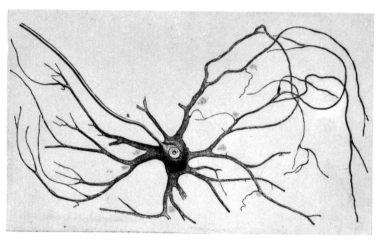

Fig. 7-3. Motor cell from ventral horn of spinal cord. The axon runs toward the upper left. All other processes are dendrites. Schweiger-Seidel, 1872.

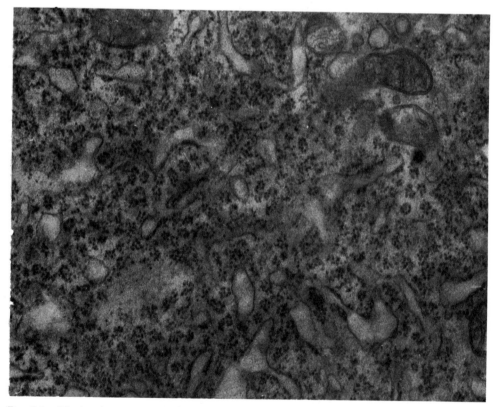

Fig. 7-4. Nissl substance = rough-surfaced endoplasmic reticulum plus many free ribosomes. Courtesy of Dr. Arthur Hess, University of Utah.

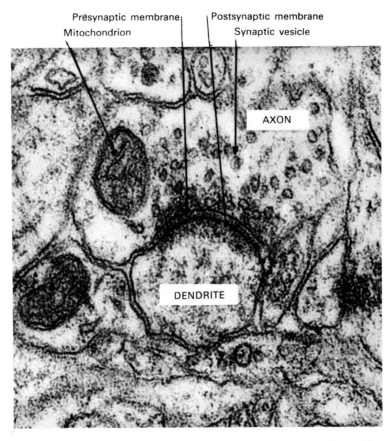

Fig. 7-5. An axo-dendritic synapse in the striate cortex of a rabbit. (From H. van der Loos, Z. Zellforsch. *60:* 815-825, 1963.) For better understanding see Fig. 7-17.

ception of stimuli and for conduction and transmission of impulses.

Nerve Cell or Neuron

The nerve cell or neuron (Figs. 7-1, 7-2 and 7-3) consists of a cell body (*corpus neuroni*) from which one or more cytoplasmic processes arise. The processes are known as *nerve fibers*. They may be short or very long, and a sensory neuron in man may extend for up to 1.5 m. Processes which conduct impulses toward the cell body are *dendrites*. The single process which carries an impulse away from the cell body is the *axon*. Irregularly shaped masses of basophilic *chromatophilic substance* (*Nissl*) are scattered through the cytoplasm of the cell body and dendrites.

Electron microscopy (Fig. 7-4) reveals that the chromatophilic substance consists of dense, rough endoplasmic reticulum and dense clusters of free ribosomes. This substance is absent from all axons and from the peripheral processes of unipolar neurons. The chromatophilic substance is absent from that small region of the cell body which gives rise to the axon. This region is the *axon hillock (colliculus axonis)*. Examination of nerve cells via electron microscopy reveals neurofilaments and neurotubules coursing through the cell body and its processes. Groups of neurofilaments seen at the light microscopic level are called *neurofibrils* (seen best in silver impregnations). Mitochondria are present in the nerve cell bodies, axons and in dendrites. The golgi complex is present in all cell bodies.

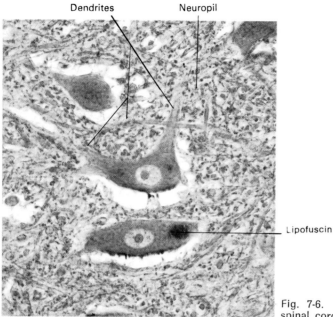

Fig. 7-6. Multipolar neurons from ventral horn of spinal cord. Bodian silver stain, 750 ×.

The cell body has a vesicular nucleus with a predominance of euchromatin, signifying a marked degree of RNA synthesis (Fig. 1-1). This accounts for the numerous ribosomes of the chromatophilic substance. In neurons of females, the sex chromatin may be seen as a nucleolar satellite (Fig. 7-6) (see chapter 1).

Dendrites branch richly and often are studded with very small side branches, called *gemmules*.

The axons are relatively smooth throughout most of their course, but may give rise to side branches called *collaterals*. Axons and their collaterals usually terminate in profuse branches, the *end arbori-*

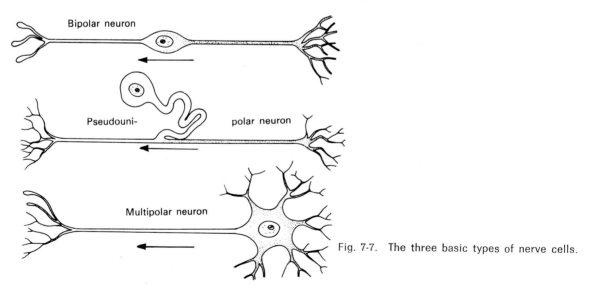

Fig. 7-7. The three basic types of nerve cells.

Fig. 7-8. Various types of multipolar nerve cells, photographed at the same magnification by Prof. Clement Fox, Marquette University. A: Pyramidal cell from motor cortex; B: Purkinje cell from cerebellum; C: Golgi type II cell from cerebellum; D: granule cell from cerebellum; E: cells from reticular formation; F: double pyramidal cells from hippocampus. In A and F some blood vessels are blackened. Please do not confuse them with parts of nerve cells!

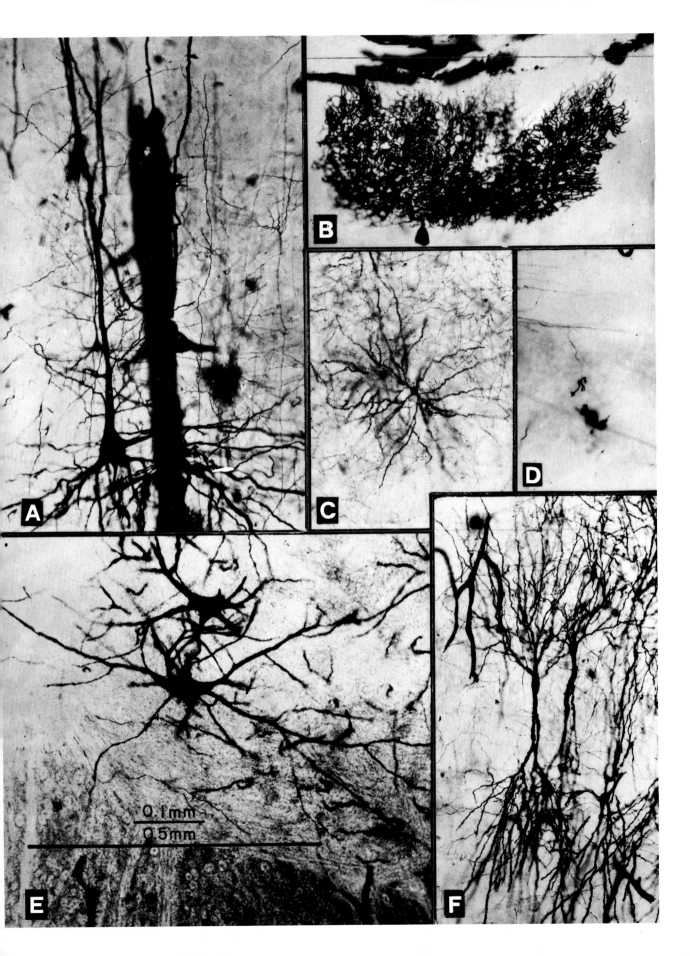

zations or *telodendria*. The specific morphological ending of an axon telodendron is the *terminal*. The axon terminals contain tiny *synaptic vesicles* ranging from 40-50 nm in size (Fig. 7-5). Terms such as *axolemma* (for plasmalemma) and *axoplasm* (for cytoplasm) are deleted from the current edition of the *Nomina Histologica*. The acceptable terms *plasmalemma* and *cytoplasm* are tautologous.

Lipofuscin pigment, accumulating in aging neurons (Fig. 7-6), probably consists of undigested residues from the lysosomal system.

Types of Nerve Cells

The three basic types of nerve cells are shown in Figure 7-7. There are multipolar, bipolar and unipolar (pseudo-unipolar) neurons.

Multipolar nerve cells (Figs. 7-1, 7-2, 7-3, 7-6, 7-7) are of many different sizes and shapes, each type characteristic for a specific region of the nervous system. They possess many dendrites and one axon.

Figure 7-8 shows several multipolar nerve cells which were impregnated with silver by the golgi method and photographed at the same magnification (scale at lower left). Multipolar cells predominate in the central and autonomic nervous systems.

Bipolar nerve cells possess one dendrite and one axon. This type is confined to olfactory epithelium, retina and ganglia of

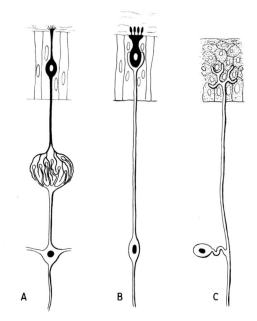

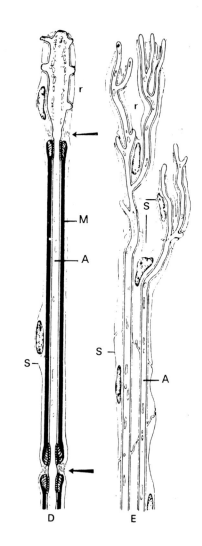

Fig. 7-9. Three basic types of receptors. A: Primary sense cell (example: olfactory cell). B: Secondary sense cell (example: hair cell of ear). C: Sensory nerve cell (all pseudounipolar cells of dorsal root ganglia). D and E: Ultrastructure of some sensory nerve endings (S = neurolemmocyte (Schwann)). Arrows = neurofiber (Ranvier) nodes, M = Myelin; A = Axis cylinder, D: with myelinated fiber, E: with non myelinated fibers. (From K. H. Andres and M. v. Düring, Morphology of Cutaneous Receptors, in Handbook of Sensory Physiology, Vol. II: 3-28, Springer-Verlag. Heidelberg, 1973).

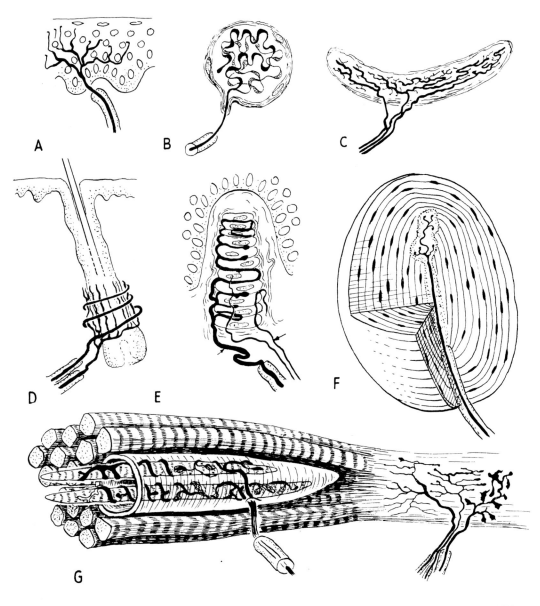
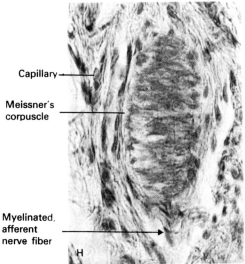

Fig. 7-10. General somatic sensory nerve endings and their functions. A: Free nerve ending (pain?); B: end bulb (of Krause) (cold?); C: Organ of Ruffini (heat? more likely pressure); D: hair root (light touch and wind); E: tactile corpuscles (of Meissner) (touch and perhaps pain); F: lamellated corpuscle (Vater-Pacini) (pressure and perhaps translatory motion); G: left, muscle spindle (muscular contraction); right, musculo-tendinous organ (of Golgi) tension. A, B, C, D and E are exteroceptors; F and G are prorpioceptors H: (Meissner's) tactile-corpuscle in the urinary bladder, Azan.

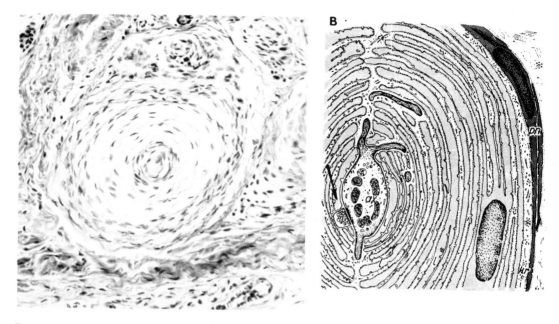

Fig. 7-11. A: Lamellated (Vater-Pacini) corpuscle from the palm of an infant. Goldner stain, 725 ×. B: Diagram of ultrastructure of a lamellated pressure receptors. (From Andres and v. Düring, Handbook of Sensory Physiology, Vol. II: 3-28, Spinger-Verlag, Heidelberg, 1973).

the stato-acoustic nerve (eighth cranial nerve).

Unipolar (pseudo-unipolar) *nerve cells* have one T-shaped process. The leg of the T emerges from the cell body, and the cross bar of the T consists of a peripheral and a central process. The peripheral process ends in a specialized receptor; the central process ends in the central nervous system. The unipolar nerve cell is the only known type of neuron present in sensory ganglia of the spinal and cranial nerves (except the eighth) and in the mesencephalic nucleus of the trigeminal nerve.

Receptors

Receptors are activated by stimuli originating (a) outside the body (*exteroceptors*), (b) in the viscera (*interoceptors*), (c) in the musculoskeletal system and (d) in organs

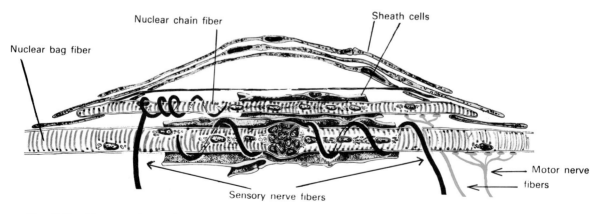

Fig. 7-12. Diagram of muscle spindle. Courtesy of Dr. H. Schmalbruch, University of Copenhagen.

specialized to register position and motion (*proprioceptors*). Stimuli can be received by three basically different arrangements as illustrated in Figure 7-9. A *primary sensory nerve cell*, located at the periphery, receives a stimulus and transmits it directly to the central nervous system (Fig. 7-9, A) (examples: olfactory cells, rods and cones of retina). A *secondary sensory nerve cell* is a specialized epithelial cell

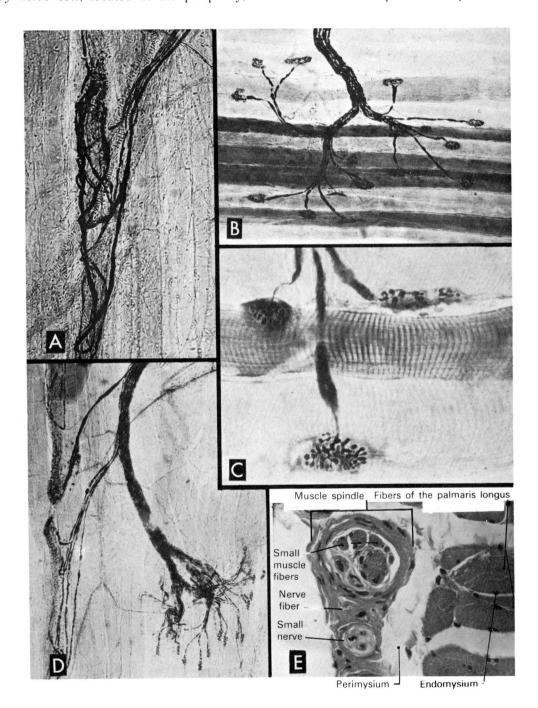

Fig. 7-13. Some nerve endings. A: muscle spindle; B and C: motor end plates; D: a mixed nerve with fibers leading to working (extrafusal) muscle fibers (right) and others to and from two muscle spindles (left); E: muscle spindle in palmaris longus. (Courtesy of Leo Massopust St. Louis Univ.)

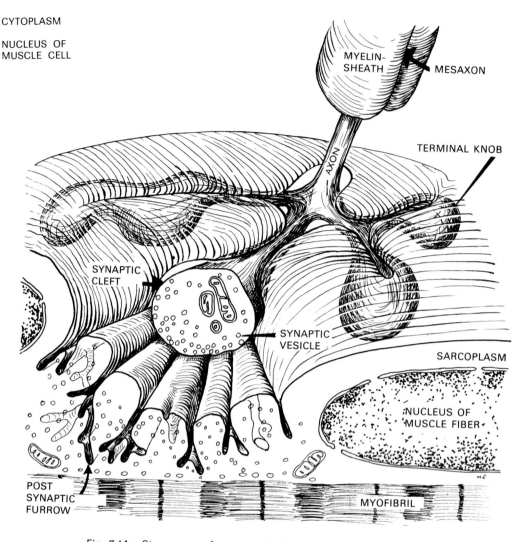

Fig. 7-14. Stereogram of motor end plate (neuromuscular synapse).

which receives a stimulus but depends on a nerve cell to conduct the impulse to the central nervous system (Fig. 7-9, B) (Examples: hair cells in the ear, taste cells in the tongue). A *sensory (afferent) nerve cell* is a neuron whose cell body is located at great depth in the body and whose long peripheral process (dendrite) receives the stimulus (Fig. 7-9, C) and whose axon transmits the stimulus to the central nervous system (example: unipolar cells in dorsal root ganglia). "Special" receptors are located in specialized sense organs (nose, ear, eye). The various "general" receptors are so called because of their general distribution in the body. Their structure and functions must be studied in Figure 7-10. A section through a lamellated corpuscle is seen in Fig. 7-11.

Of special interest among receptors is the muscle spindle (Fig. 7-10 left and Fig. 7-12) which has a motor and a sensory component. It is a bundle of small specialized muscle fibers (nuclear bag fibers and nuclear chain fibers) in a sheath of flattened cells derived from connective tissue. Collaterals from motor axons cause contraction of the spindle fibers simultaneously with the contraction of the working muscle fibers, while peripheral processes of unipolar cells in dorsal root ganglia register that contraction. See also Figures 7-13, A, D and E.

Effectors

Effectors are the organs which respond to stimuli. They include muscles and glands, which are set into action by the endings of axons of motor cells. Motor (*efferent*) endings to glands and to smooth myocytes are delicate end arborizations, those to skeletal myocytes are the highly organized *motor end plates*, shown in Figures 6-23, 7-13, B, C and D (right) and in Figures 7-14 and 7-15. In a motor end-plate, the knob-like terminations of the axon fit into depressions (*primary synaptic clefts*) in the skeletal myocytes. The motor end-plate is sometimes called a *neuromuscular junction* or a *neuro muscular synapse*. In the enlarged knob-like *boutons terminaux* are mitochondria and many *synaptic vesicles*. The myocyte plasmalemma, in addition to being indented by the *boutons terminaux* of the end arborization shows numerous oblong infoldings which sometimes branch (*secondary synaptic clefts* or *junctional folds*) (Figs. 7-14 and 7-15).

When the axonal action potential reaches the area of terminal bulb (knob) of the axon, the neurotransmitter *acetylcholine* is released. The acetylcholine crosses the synaptic cleft and stimulates receptor sites in the post-synaptic plasmalemma (sarcolemma) of the myocyte which becomes increasingly permeable to small ions. The ion flux causes a *depolarization* of the myocyte plasmalemma. When the depolarization reaches a threshold value, an *action potential* is generated in the myocyte plasmalemma which propagates along this membrane. It travels into the interior of the myocyte via the transverse tubule smooth endoplasmic reticulum (sarcoplasmic reticulum), and muscle contraction occurs. The released acetylcholine is rapidly destroyed by the enzyme *acetylcholinesterase* located on or in the myocyte plasmalemma. This enzymatic mechanism limits the duration of the neuromuscular interaction to a brief response. After a short *refractory period*, the process may be repeated.

Synapses

Neurons communicate constantly via electrical impulses and specialized chemical messengers called *neurotransmitters*. The more than 50 billion neurons create a complex, but orderly and repetitive circuitry at sites called *synapses*. *Presynaptic* neurons release the *neurotransmitter* into a space called the *synaptic cleft*. Upon reaching the *postsynaptic* neuron the neurotransmitter initiates another electrical impulse. This direct mechanism of transmission involves conversion of an electrical signal to a chemical one, and back to an electrical signal. Classification of *morphological synapses* is based on size and shape of neuron synaptic vesicles. Although *cholinergic* and *adrenergic* synapses are the first and best known *chemical synapses*, others are considered in neuroscience courses. Improvements in EM technology suggest the existence of an *electrical synapse*, per existence of gap junctions in pre- and postsynaptic neuron membranes.

Cholinergic synapses, or synapses that release the neurotransmitter acetylcholine have electron lucent synaptic vesicles (via electron microscopy). *Adrenergic synapses*, or synapses which release noradrenalin, contain synaptic vesicles which have an electron dense core (via electron microscopy).

A nerve fiber (axon or dendrite), when stimulated, will carry the impulse in both directions from the point of stimulation. It is only the polarity at the synapse which controls the direction of impulse transmission. From among the various kinds of synapses, five varieties are shown in Figure 7-16.

Throughout the gray matter of the spinal cord and of the brain, very delicate den-

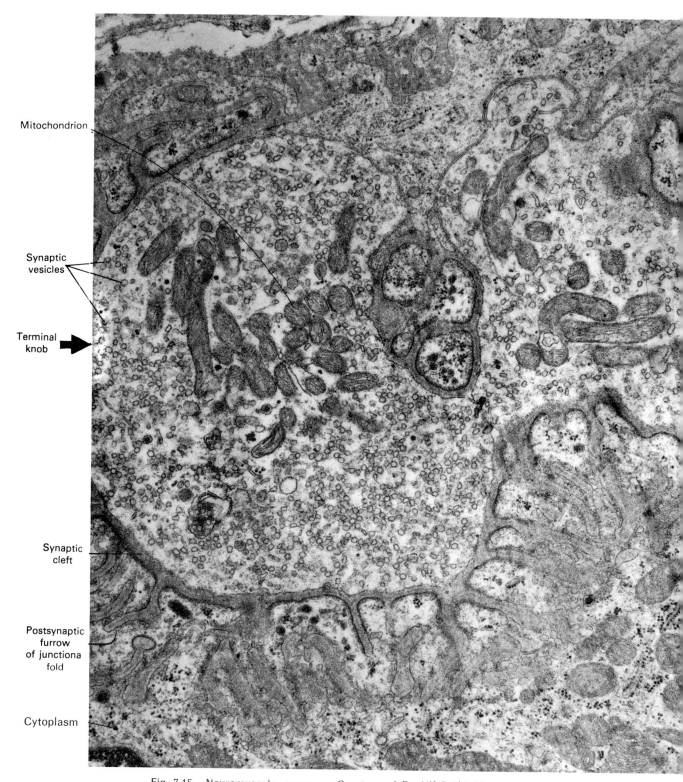

Fig. 7-15. Neuromuscular synapse. Courtesy of Dr. Ulf Brühl, University of Heidelberg.

drites interlace with equally delicate endings of axons. Such a region of interlacing nerve cell processes (including gliocyte processes) is called the *neuropil* (Fig. 7-6 and Fig. 7-17). A synapse at which an axon passes its impulse to a cell body is an axo-somatic synapse; one at which it transmits its impulse to a dendrite is an axo-dendritic synapse, and one at which it passes its impulse to another axon is an axo-axonic synapse. Figure 7-17 shows the ultrastructure of some types of synapses in the central nervous system. At every synapse the cell membranes of both neurons (which are contiguous, but not continuous) are relatively thick, about 20-30 nm in width, and are separated by the *synaptic cleft*. Both cells may be connected across the cleft by little threads of glycoprotein (Fig. 7-5). At other places, short rods, which arise from the dendrite and which bear a little knob, protrude halfway into the synaptic cleft. There is an accumulation of *synaptic* vesicles in the axon near the presynaptic membrane. Synapses may be located at the ends of axons and at the ends of dendrite branches, i.e., where the two ends almost touch; but they can also be located all along the winding course of a fiber. Synapses may consist of a passing contact *(synapse en passage)*. These possibilities are illustrated in Figures 7-16 and 7-17. A climbing fiber in the cerebellum for example, is an axon that winds around a dendrite of a purkinje cell like a vine around a trellis. In such a situation, it synapses with that dendrite all along its winding course. Terminal branches of "basket cells" form baskets around the cell bodies of the Purkinje cells. At such places the synapse is as long as all these basket fibers together.

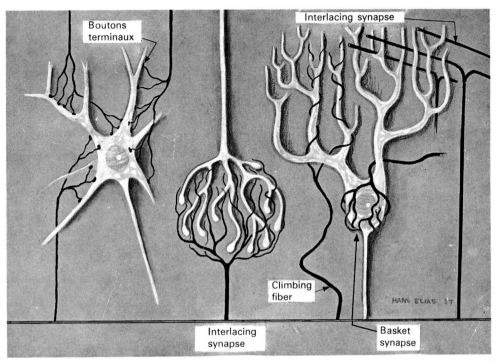

Fig. 7-16. Various kinds of synapses. *Left*, As in the ventral gray column of the spinal cord and in the somatic motor nuclei in the brain stem, boutons terminaux make axo-dentritic and axo-somatic contacts. *Center*, Glomerulus in olfactory bulb. *Right*, Climbing fiber, and interlacing synapse (axo-dendritic) and basket synapse (axo-somatic) around a piriform neuron (Purkinje cell).

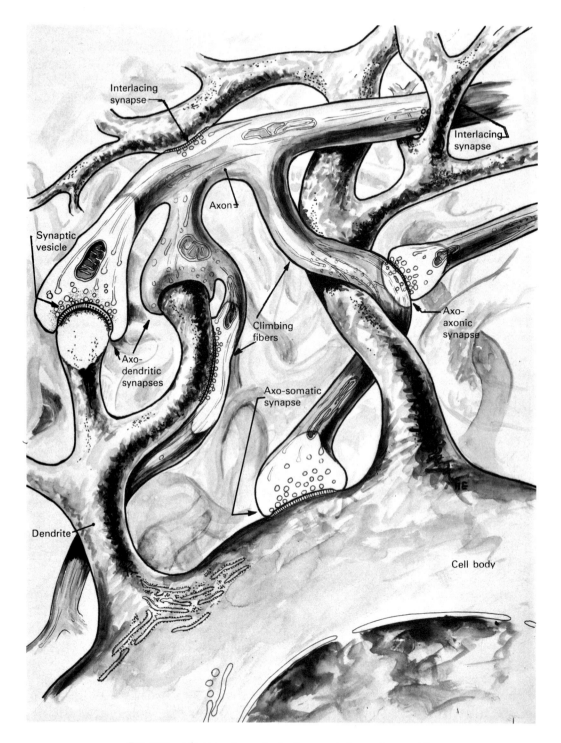

Fig. 7-17. Ultrastructure of plexiform substance (neuropil).

MICROANATOMY OF THE NERVOUS SYSTEM

Central Nervous System

The central nervous system, i.e., the brain and spinal cord, receives incoming stimuli to be classified, interpreted and, then, either stored for future reference or acted upon immediately.

NEUROGLIA. The central nervous system is a large mass of nerve cells and nerve fibers supported by a special kind of connective tissue called *neuroglia*.

The nervous elements are supported by *neuroglial cells* (glial *cells*). The glial cells and their processes fill most of the space between the nervous elements. The glial cells are classified into *astrocytes* (star cells), *oligodendroglia* and the *microglia* (Fig. 7-18).

Astrocytes are stellate glial cells with many cytoplasmic processes. *Protoplasmic astrocytes* with a multitude of processes are found principally in the gray matter around nerve cell bodies (in the neuropil of the central nervous system (CNS), i.e., the brain and spinal cord).

Fibrous astrocytes have fewer, straighter and longer processes when compared to protoplasmic astrocytes. Fibrous astrocytes are found principally in the white matter (myelinated nerve fibers).

Both the protoplasmic and fibrous astrocytes have processes which adhere to blood vessels by perivascular feet.

Oligodendroglia. These glial cells have only a few, small processes. Their nuclei, smaller, denser and rounder than those of astrocytes, are surrounded by a relatively thin rim of cytoplasm, having the appearance of fried eggs. This resemblance should be helpful in distinguishing oligodendroglial cells from the astrocytes. Oligodendroglia may be further classified according to their position, e.g., perivascular, perineuronal and interfascicular (in white matter).

Microglia. The microglial cells are small cells which may have long angular cytoplasmic processes. They have phagocytic properties when the nervous tissue is injured or damaged. Although the specific origin of microglial cells is unknown, they may arise from perivascular pericytes in the CNS or from mesenchymal cells which have migrated into the CNS. Some investigators feel they may represent blood monocytes in the CNS; others that they may be true glial cells.

EPENDYMA. The central canal and the ventricles are lined by a single layer of central gliocytes, ciliated epithelial cells, the *ependymal cells* (Fig. 7-19). In the spinal cord and lower medulla oblongata, ependymal cells are columnar with basal ends that taper into long processes which traverse the entire thickness of the cord.

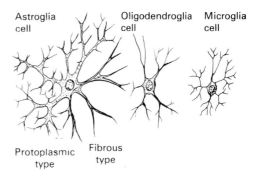

Fig. 7-18. Various types of neuroglia cells.

At the surface, these processes spread and contribute to the formation of the *external limiting membrane*, a delicate feltwork of fine glial fibers and connective tissue fibers.

In the ventricles of the brain, the ependyma varies in structure from columnar to squamous, as shown in Figure 7-19.

CHOROID PLEXUS. In the early embryo, when the neural groove deepens and the neural folds meet to close the neural tube; a portion of the amniotic cavity (i.e., a part of the outer world) is captured to form the central canal of the spinal cord, ventricles and aqueduct of the brain. These are filled at first with amniotic fluid and later dis-

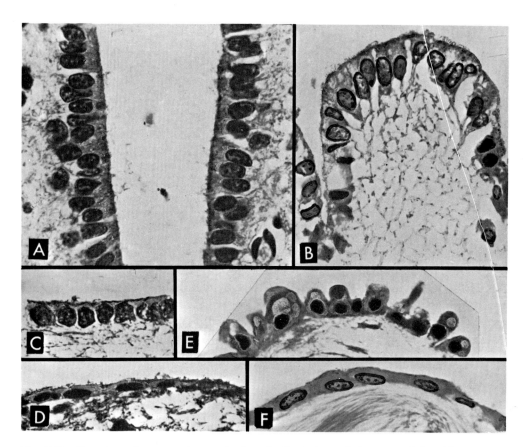

Fig. 7-19. Ependyma from one human brain. A, Central canal, lower medulla oblongata. B, Stria terminalis. C, Roof of lateral ventricle. D, Wall of lateral ventricle. E, Choroid plexus. F, Tela chorioidea of the fourth ventricle.

placed by cerebrospinal fluid which is very similar to plasma in its composition. The ependymal cells which line the ventricular system form a true epithelium.

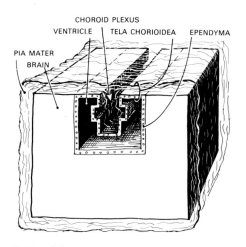

Fig. 7-20. Schematic stereogram illustrating the relationship of the tela chorioidea and of the choroid plexus to a ventricle of the brain.

The brain may be compared to a thick-walled box (Figs. 7-20 and 7-21) lined by a single layer of ependymal cells and wrapped in pia mater. Brain tissue, i.e., gray and white matter, form the bulk of the wall. This thick-walled box has a very thin lid in which brain tissue is absent, and where ependyma and pia mater are in direct contact, the *tela choroidea* is formed. The tela choroidea is invaginated so that at certain places a great mass of fringed folds, the *choroid plexus*, hangs into the ventricles. The core of these folds and fringes consists of pia mater which is richly vascularized. The covering is ependyma (Figs. 7-21 and 7-22). The ependymal cells at this location contain many vacuoles related to their secretory activity (Fig. 7-19, E). The choroid plexus maintains and produces the cerebrospinal fluid which can flow into and fill the subarach-

CENTRAL NERVOUS SYSTEM 151

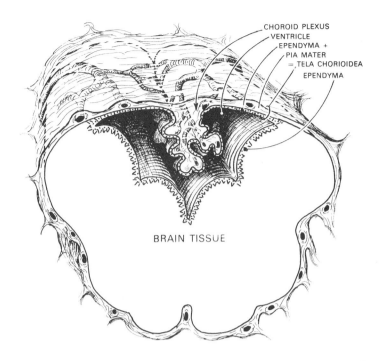

Fig. 7-21. Tela chorioidea and choroid plexus of the fourth ventricle.

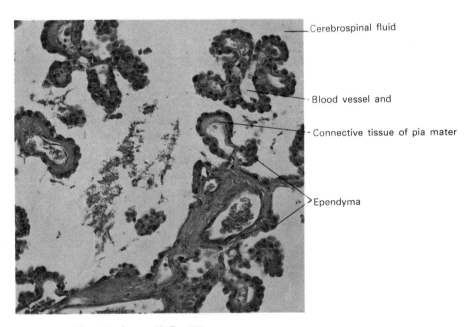

Fig. 7-22. Choroid plexus, H&E. 375 ×.

noid apace through three openings in the tela choroidea of the fourth ventricle. Thus cerebrospinal fluid provides a liquid medium which fills the cavities of the central nervous system and surrounds it so that it is suspended in a hydrostatic cushion.

GRAY AND WHITE MATTER. The cell bodies of neurons of the central nervous system are assembled either in cortical lay-

152 NERVOUS TISSUE

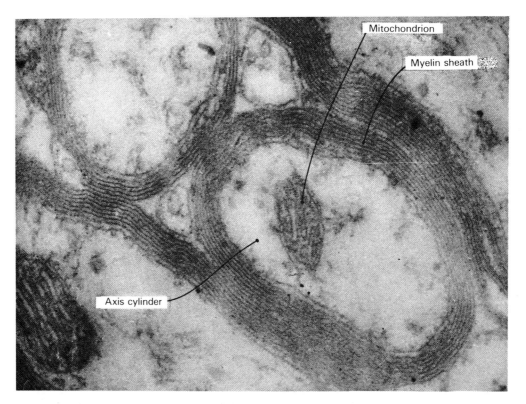

Fig. 7-23. Myelinated nerve fibers from the spinal cord of the hamster. (Electron microgram by Dr. H. Hager, University of Munich.)

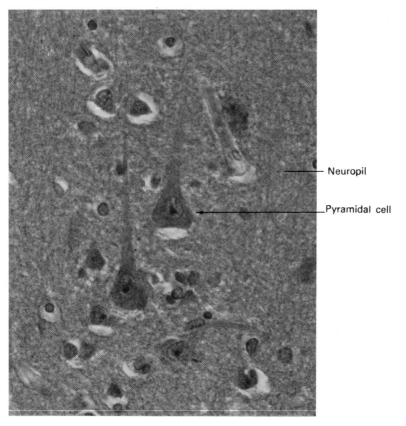

Fig. 7-24. Some pyramidal cells in the cerebral cortex, H&E. 1000 ×.

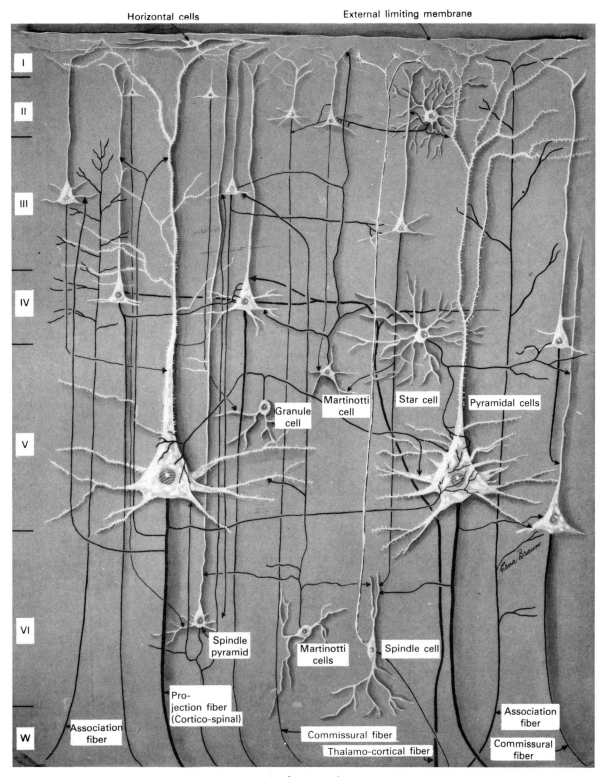

Fig. 7-25. A diagram of the cerebral cortex, its layers and various cell types. Axons are drawn black.

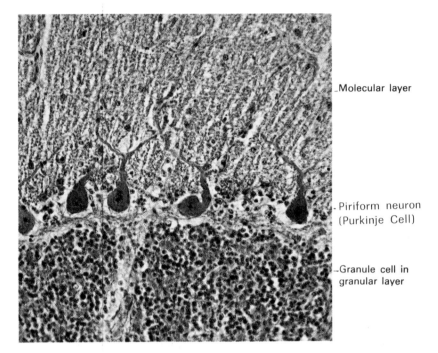

Fig. 7-26. Cerebellar cortex, Hansen's iron hematoxylin, orange G and eosin. 400 ×.

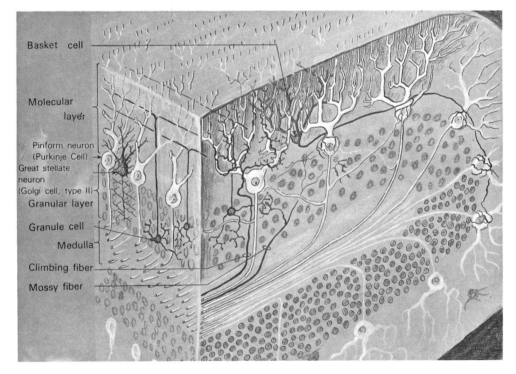

7-27. Layers and cells in the cerebellar cortex.

ers or in relay centers, referred to as "nuclei" or cell columns. In and near these centers, synapses take place. Together, the cortical layers and the "nuclei" form the gray matter (Fig. 7-6, 7-24 and 7-26) composed of nerve cell bodies and neuropil (molecular or plexiform substance). The mixture of neuropil and masses of nerve cells is due to the fact that the dendrites of the latter are usually short and highly ramified. Axons of neurons whose cell bodies are far away or nearby, synapse with other neurons within the gray matter.

Many nerve cells have short axons

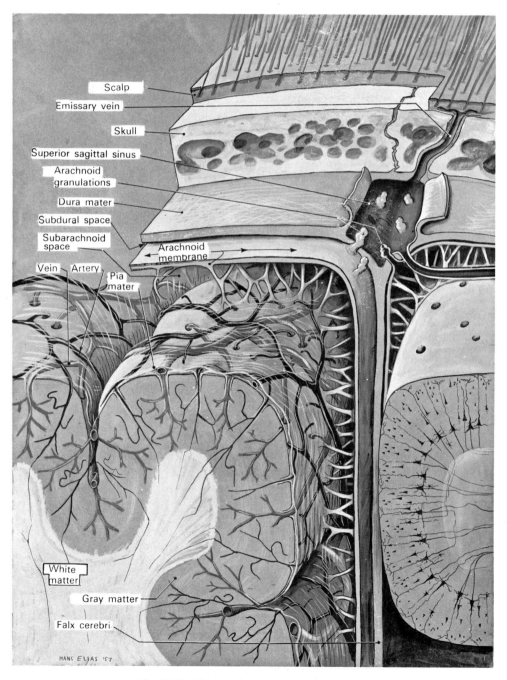

Fig. 7-28. The cerebral cortex and the meninges.

156 NERVOUS TISSUE

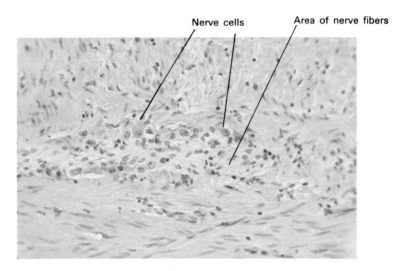

Fig. 7-29. Intramural ganglion of (Auerbach's) myenteric plexus in the muscularis of the colon, H&E. 375 ×.

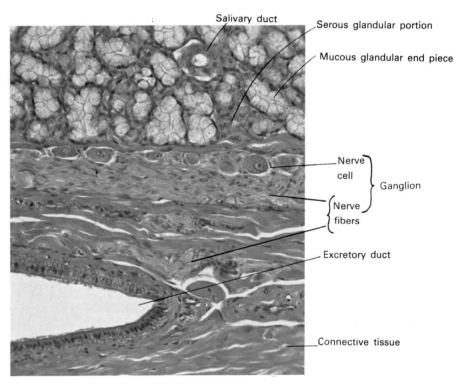

Fig. 7-30. Intramural ganglion in the submandibular gland, H&E. 400 ×.

which arise and terminate within their own "nucleus" of origin (golgi type II cells, horizontal cells, amacrine cells, bipolar cells of the retina, Martinotti cells and others). Others have long axons which conduct impulses to distant locations. The latter (projection, commissural and association fibers) are insulated by sheaths of a glistening, fatty substance called *myelin*. Collectively such myelinated nerve fibers in the central nervous system constitute the white matter. The electron microscope

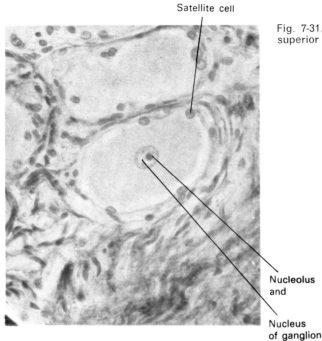

Fig. 7-31. Nerve cell with satellite cells in the superior cervical ganglion, Azan. 380 ×.

shows a myelin sheath to be formed of almost concentric layers (Fig. 7-23). These layers actually spiral around the nerve fibers. They are, in fact, extremely attenuated parts of oligodendroglia cells which have wrapped themselves around the fiber.

Each area of gray matter has a characteristic structure which distinguishes it from other portions of the central nervous system. One such area, the ventral gray column of the spinal cord, is shown in Figure 7-6. Figure 7-8 shows a great variety of nerve cell types, characteristic for specific regions of gray matter.

CEREBRAL CORTEX. Sections of the cerebral cortex reveal a series of ill-defined layers which vary from area to area. These layers are designated by numbers and names. Figure 7-25 and Table 7-1 present generalized synopses of the "cyto-architectronics" of the cerebral cortex. There are great regional variations.

The cerebral cortex is characterized chiefly by the presence of pyramidal cells (Figs. 1-18, 7-8, A left and 7-24). This cell has, indeed, the shape of a very tall pyramid whose long axis is perpendicular to the surface of the brain. It has one large apical dendrite which branches moderately and many basal dendrites fanning out from its base. The axon arises from its base and is directed toward the white matter. Pyramidal cells vary greatly in size, the smallest being located near the pia mater (in "layer II"), the largest are deeper and located in "layer V". Figure 7-25 shows the various types of pyramidal cells and other cells present in the cortex.

CEREBELLAR CORTEX. The *cerebellar cortex* is organized around a branching core of white matter (medulla or arbor vitae). Its architecture is shown in Figs. 7-26, 7-27, and the parts found in various layers are listed in Table 7-2.

MENINGES. The *meninges* (singular, *meninx*) are membranes which surround, nourish and protect the brain and spinal cord. They are the *pia mater, arachnoid mater* and the *dura mater* (Fig. 7-28).

Pia Mater. The pia mater is a delicate vascular membrane of areolar connective tissue immediately applied to the brain and spinal cord.

TABLE 7-1. CEREBRUM

Layer Number	Name of Layer	What Is in It	What Terminates in It
I	Molecular or plexiform	Horizontal cells and their axons.	Dendrites of pyramidal and fusiform cells, axons of small neurons (Martinotti cells).
II	Outer granular	Many small pyramidal cells. Their axons end in deeper cortical layers; some enter the white matter.	Axons of deep small neurons (Martinotti cells).
III	Outer pyramidal	Medium-sized pyramidal cells, whose axons pass to the same layer, deeper layers or into the white matter; stellate cells with axons to the same and deeper layers, and small neurons (Martinotti cells) whose axons are directed upward.	Axons of small neurons (Martinotti cells), association fibers.
IV	Inner granular	Stellate cells with short axons terminating in this layer, some in deeper layers.	Horizontal fibers, which are terminations of thalamocortical fibers, are found at this level in the post central gyrus.
V	Inner pyramidal	Large pyramidal cells, whose axons pass into the white matter with collaterals to the same and more external cortical layers. In the precentral gyrus, these cells are the giant cells (of Betz). Their axons terminate in the ventral gray column of the spinal cord. Stellate and granule cells with short axons.	Axons from other cortical layers.
VI	Fusiform	Fusiform (spindle) cells whose axons enter the white matter and small neurons (Martinotti cells) whose axons go to more superficial layers.	Axons from outer layers and association fibers.
VII	White matter	Axons to and from cortex.	Nothing.

TABLE 7-2. CEREBELLUM

Name of Layer	What Is in It	What Terminates in It
Molecular layer (outer layer)	Dendrites of piriform cells (Purkinje) basket cells and their dendrites. Great stellate neurons (Golgi type II) and their dendrites.	Climbing fibers, T-shaped axons of granule cells.
Piriform layer (middle layer)	Perikarya of piriform (Purkinje) cells whose axons run to the dentate, emboliform and fastigial nuclei.	Axons of basket cells making axosomatic synapses, returning collaterals of (Purkinje) cells.
Granular layer (inner layer)	Granule cells.	Axons of Great stellate neurons (Golgi type II) from molecular layer. Mossy fibers.
White matter (arbor vitae)	Axons to and from cerebellar cortex.	Nothing.

Projections of the pia mater extend into the brain and spinal cord carrying arteries and veins into their substance (Fig. 7-28).

Arachnoid Mater. The arachnoid mater, a sheet of delicate areolar tissue separated from the pia by the *subarachnoid space*, is connected with the pia mater by delicate trabeculae. The subarachnoid space contains cerebrospinal fluid and is lined everywhere by flat "mesothelial" cells.

Dura Mater. The dura mater (the outer mennix), is a thick sheet of dense fibrous connective tissue separated from the arachnoid by the very narrow *subdural space*. In the skull, the *dura mater encephali* is adjacent to the internal periosteum of the cranium. In the vertebral canal, the *dura mater spinalis* is separated from the perios-

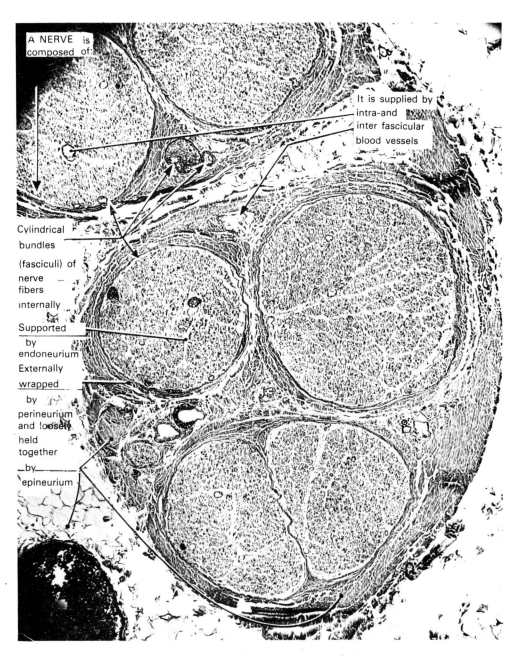

Fig. 7-32. A small portion of a cross section of the sciatic nerve.

teum by an *epidural space* which contains fat, small arteries and a well-developed venous plexus. The *falx cerebri* and the *tentorium cerebelli* are extensions of the cranial dura mater. *Venous sinuses* are contained within the cranium between layers of dura mater encephali. Cerebral veins (Fig. 7-28, upper right) drain into dural sinuses, and cerebrospinal fluid filters into sinuses from the subarachnoid space through *arachnoid granulations*.

Peripheral Nervous System

GANGLIA. A ganglion (Figs. 7-29 and 7-30) is an accumulation of nerve cell bodies outside the central nervous system. Each nerve cell body (Figs. 7-31 and 1-1) is surrounded by a capsule of flat *satellite cells*. Ganglia are aurrounded and supported by connective tissue, and form pathways for traversing nerve fibers.

Ganglia of the *autonomic nervous*

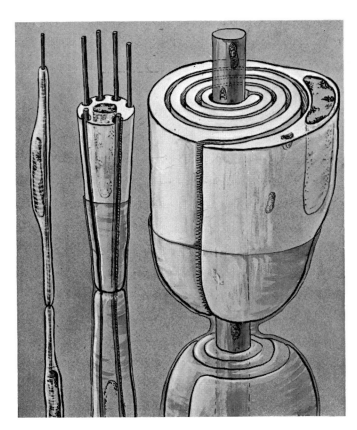

Fig. 7-33. Relation of peripheral nerve fibers to the neurolemmocyte (Schwann cell). *Left and center,* Non-myelinated nerve fibers. *Right,* Myelinated nerve fiber.

system containing multipolar neurons (Figs. 7-29, 7-30 and 7-31) are relay stations in which synapses occur. Axons and dendrites of neurons located in other parts of the nervous system synapse within these ganglia or traverse without synapsing.

Spinal ganglia and the *sensory ganglia* of cranial nerves, except the eighth, harbor the nerve cell bodies of unipolar (pseudounipolar) neurons. Synapses are not known to occur in them, nor do synapses occur in the ganglia of the eighth nerve (the vestibular and the cochlear ganglia), which contain bipolar nerve cells. In routine histological sections, it is difficult to distinguish a spinal ganglion from a sympathetic chain ganglion (Fig. 7-31).

NERVES. A *nerve* is a bundle (*fascicle*) of nerve cell fibers (or a bundle of several bundles of nerve cell fibers) held together by layers of connective tissue as shown in Figure 7-32. A large nerve is composed of cylindrical bundles (fasciculi) of nerve cell fibers. Each nerve cell fiber is wrapped in a very delicate areolar connective tissue, called *endoneurium*. Several of these wrapped nerve cell fibers form a bundle (fascicle) which is wrapped externally by *perineurium*, a sheath of relatively dense connective tissue. The entire nerve (several fascicles) is wrapped by *epineurium*. A

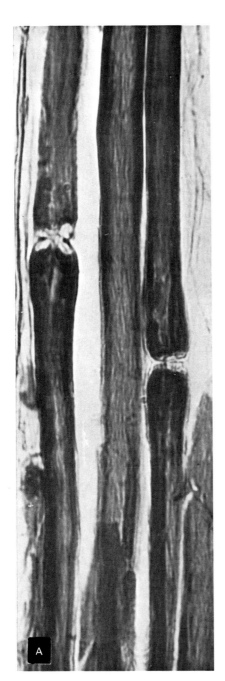

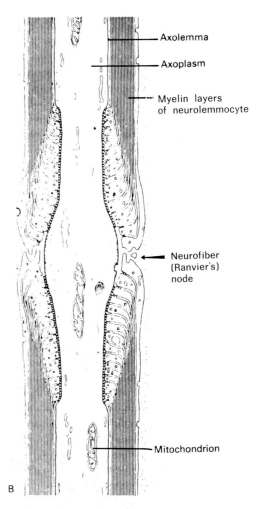

Fig. 7-34. A: Myelinated nerve fibers stretched out on a slide after osmic acid treatment. Two of the fibers in this picture show constrictions neurofiber nodes (of Ranvier). They are the boundaries between neurolemmocytes (Schwann cell). B: Diagram of a longitudinal section through a peripheral myelinated nerve fiber (peripheral process of a pseudounipolar cell in a spinal ganglion at the level of a neurofiber node (arrow). (From K. H. Anders and M. v. Düring, Morphology of Cutaneous Receptors, in Handbook of Sensory Physiology, vol. II: 3-28, Springer-Verlag, Heidelberg, 1973).

nerve is supplied by intra- and interfascicular blood vessels. *Nerves do not usually contain nerve cell bodies.*

Except for certain, highly specialized cranial nerves, nerves contain both sensory and motor fibers, as is demonstrated in Figure 7-13, D, which shows a small nerve carrying sensory fibers from two muscle spindles (left) and motor fibers to at least 15 neuromuscular end-plates (right).

Peripheral Nerve Fibers. These are the axons of motor (efferent) neurons or peripheral processes of sensory (afferent) neurons. They are called *axis cylinders.* Each peripheral nerve fiber is wrapped in a *neurolemmal sheath* (*Schwann's sheath*) composed of several elongated *neurolemmocytes (Schwann cells)* aligned as shown at the left in Figure 7-33. Frequently, when several nerve cell fibers are embedded in a single neurolemmocyte sheath (Fig. 7-33, middle), they are said to be unmyelinated. The thickest nerve fibers are surrounded by neurolemmocytes (Schwann cells) which have wrapped themselves many times around a nerve cell fiber (Figs. 7-33, right, 7-34, 7-35 and 7-36). Such fibers are called *myelinated fibers.* In myelinated fibers the neurolemmocytes are separated from each other by *neurofiber nodes (nodes of Ranvier)* (Fig. 7-34). In the neurofiber nodes, one neurolemmocyte interdigitates with the next. There is one neurolemmocyte between two nodes. Osmium tetroxide, a substance which simultaneously fixes and stains fat, makes it possible to detect myelin, the lipid found in the plasmalemma. In light microscopy one can distinguish this myelin sheath easily after osmium tetroxide treatment (Figs. 7-34, A and 7-35). Since osmium tetroxide is a post-fixative for electron microscopy, myelin stands out very well in EM pictures.

Thinner axis cylinders, particularly those of the autonomic nervous system and many visceral afferent fibers, are surrounded by neurolemmocytes with minimal wrapping. These, the so-called *non-myelinated fibers,* are fibers with scanty amounts of myelin presenting an intermediate type (Fig. 7-35). In some autonomic nerves, several axis cylinders may be embedded in a single row of neurolemmocytes (Figs. 7-33, center, 7-38).

Electron microscopy shows that, in myelinated nerve fibers, the neurilemma cell wraps itself around the axis cylinder. It begins with a simple indentation (Fig. 7-39). Initially, two layers of neurolemmocyte membrane, the mesaxon, suspends the axon within the neurolemmocyte. Late stages of this process in which an axis cylinder is surrounded by many layers of myelin are seen in Figures 7-36 and 7-37. It is schematically illustrated in Figure 7-33, right.

In many nerves, myelinated and non-

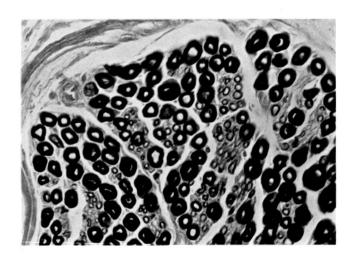

Fig. 7-35. Peripheral nerve showing nerve fibers of different caliber and different degrees of myelinization. Osmic acid.

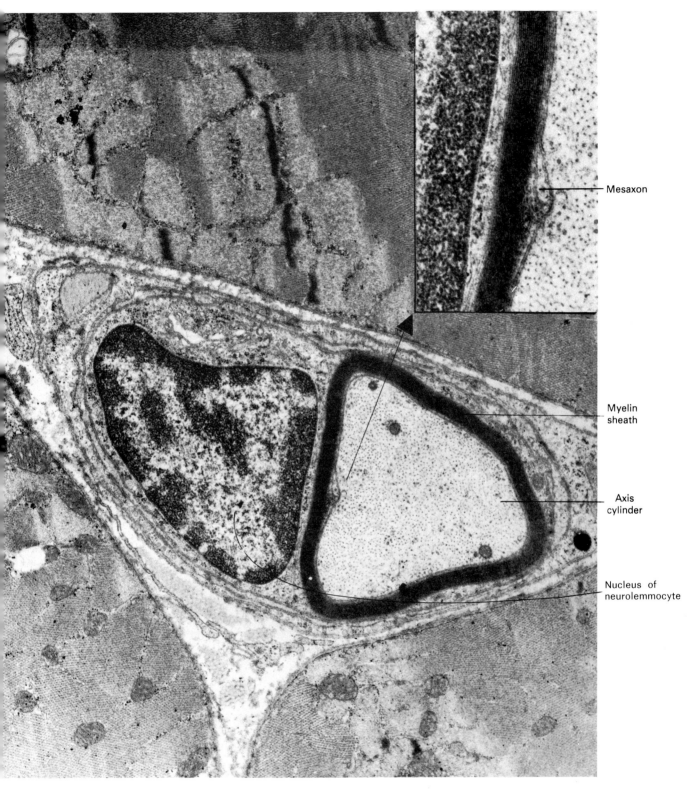

Fig. 7-36. Myelinated nerve fiber. The inset (upper right) shows the mesaxon. Courtesy of Dr. U. Brühl, University of Heidelberg. 19,000 ×.

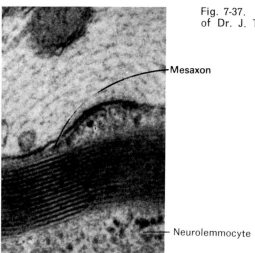

Fig. 7-37. Detail of myelin sheath. 120,000 ×. Courtesy of Dr. J. T. Povlishok, Medical College of Virginia.

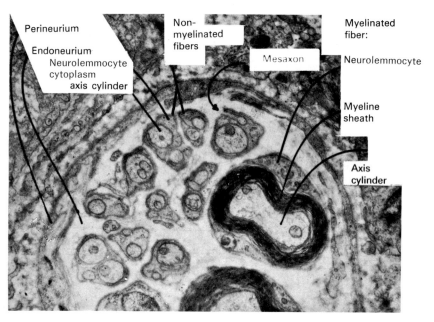

Fig. 7-38. A small mixed nerve from a hamster containig myelinated fibers and non-myelinated fibers. (Preparations and electron micrography by Dr. Takashi Nakai, Chicago Medical School.)

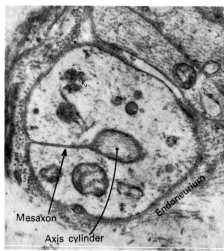

Fig. 7-39 A non-myelinated nerve fiber. (Preparation and electron micrography by Dr. Takashi Nakai, Chicago Medical School.)

myelinated fibers are mixed (Figs. 7-35 and 7-38). The degree of myelinization is determined by the number of times the neurilemmocyte wraps (like a jelly-roll) around the nerve cell fiber.

The myelin sheath is believed to act as an insulator forcing the impulse to "jump" from node to node. This is *saltatory conduction* of an impulse. This type of conduction is much faster (60 meters/second) than that found in relatively non-myelinated fibers (20 meters/second).

As myelin ages, oblique clefts or funnels develop. These are the incisures of the myelin (Schmidt-Lantermann clefts), where the myelin sheath is incompletely compacted, forcing some neurilemmocyte cytoplasm to appear in the cleft.

8...
Organization of the body

The human body is composed of four basic types of tissue: epithelium, connective tissue (which includes osseus tissue and blood), muscle and nervous tissue. Consider the body as a small world surrounded by a larger world from which it must be protected and with which it must communicate. With the exception of one direct opening, the abdominal ostium of the uterine tube, all communication with the outer world occurs through epithelia.

Epithelium, as Figure 8-1, A shows, separates the outer world, from the internal medium of the body, which consists of connective and supporting tissues, blood, nervous tissue, muscles and intercellular substance. The internal medium, in turn, encloses two systems of spaces, as indicated in Figure 8-1, B. These systems are: the cardiovascular system lined by endothelium and the coelom lined by mesothelium. Figure 8-1, C, shows how inward extensions of the outer world lined by epithelium reach deep into the body.

Interaction with the outer world via epithelia, occurs by sensory perception, mechanical responses, diffusion, absorption, secretion and excretion.

Limiting Membranes

The external lining of the body is the integument (Figs. 8-1, C; 11-1 and 11-2), a complex, protective coat, the layers of which are shown on the left side in Figure 8-2. Wherever the body possesses large, externally exposed openings, the outer lining is reflected inward, changing its characteristics from integument to a mucous membrane (Figs. 8-1, C; 8-2, right). When the openings are small (pores), the inward reflected layer serves as the duct of a gland (Fig. 8-1,C). Each layer of the integument has its counterpart in a complete mucous membrane. (Fig. 8-2) However, not all mucous membranes have a full complement of the standard layers.

While integument rests on skeletal parts or on muscles, mucous membranes are supported by connective tissue or suspended in a coelomic cavity (Fig. B-3). In the latter case, the mucous membrane (i.e., intestinal epithelium) is surrounded by connective tissue, smooth muscle and lastly a serous membrane. The serous membrane (*tunica serosa*) consists of two portions: (1) *facing the coelomic cavity* is an outer simple squa-

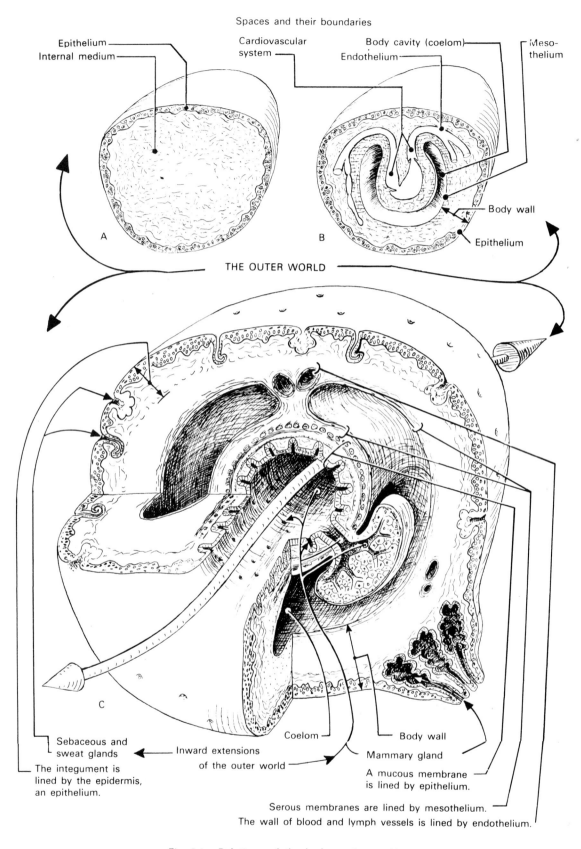

Fig. 8-1. Relations of the body to the world.

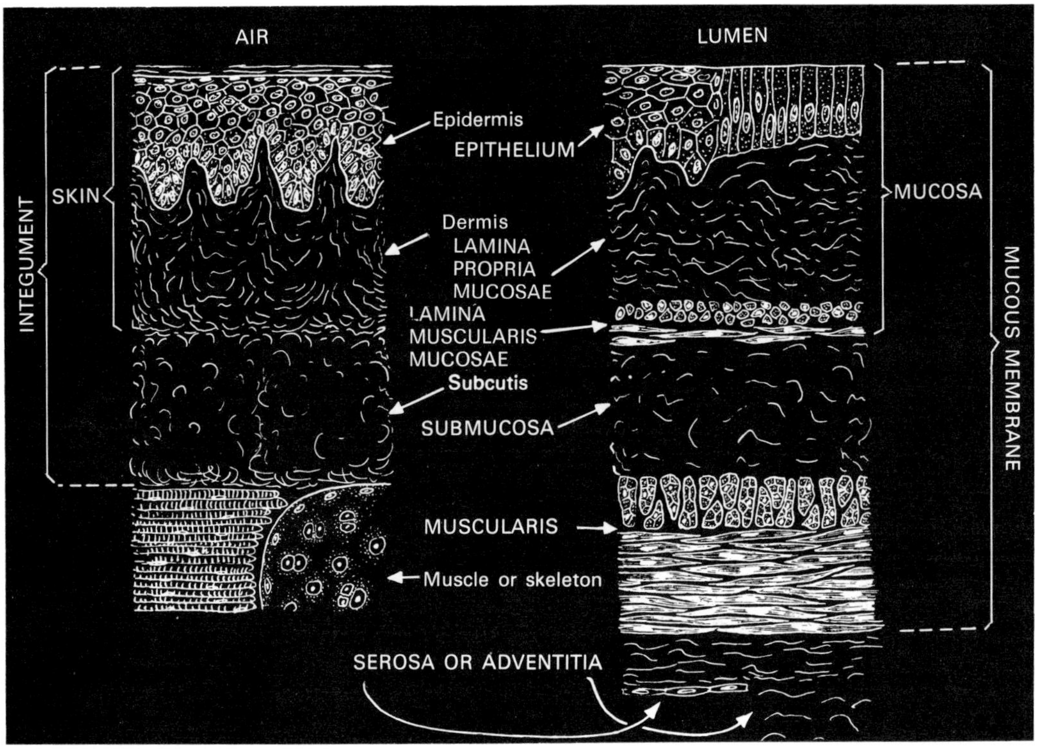

Fig. 8-2. A comparison of integument and mucous membrane.

mous epithelium called *mesothelium*, resting on basement membrane, and (2) an underlying delicate areolar connective tissue layer, the *lamina propria serosa*. Underlying the entire tunica serosa, is another connective tissue layer the *tela subserosa*. If the outer wall of a viscus, does not protrude into the coelom, it can be surrounded by a *tunica adventitia* (Fig. 8-3), a connective tissue layer that blends with adjacent connective tissues. Such organs are described as *retroperitoneal* in location (example: esophagus). Organs which protrude only partially into the coelom, present two surfaces: (1) one facing the coelom, covered by the tunica serosa, and (2) the opposite side protected by *tunica adventitia*. (Fig. 8-3 the ureter). Serous membranes also may cover solid organs such as the spleen (suspended in the coelom by the tunica serosa). The kidney, a solid organ, is located partially retroperitoneally, and thus is covered by both adventitia and serosa. In the spleen and kidney, the covering, whether serous or adventitial is designated a *capsule*.

A complete mucous membrane (tunica mucosa) is composed of (1) an *epithelium* resting on its basement membrane, (2) a layer of loose areolar connective tissue layer, the *lamina propria mucosa*, and (3) a delicate smooth muscle layer, the *laminae muscularis mucosae*. An extreme case of an incomplete mucosa is the lining of the middle ear in which only the epithelium (and its basement membrane) rests directly on periosteum. Most of gastrointestinal tract possesses a complete mucosal lining (Fig. 8-4).

The coelomic cavities, i.e., the pleural, pericardial, peritoneal and scrotal cavities are lined by serous membranes: the pleura (Figs. 8-5, and 8-6), pericardium, peritoneum (Fig. 8-4) and the tunica vaginalis respectively. That portion of a serous membrane in contact with the body wall is its

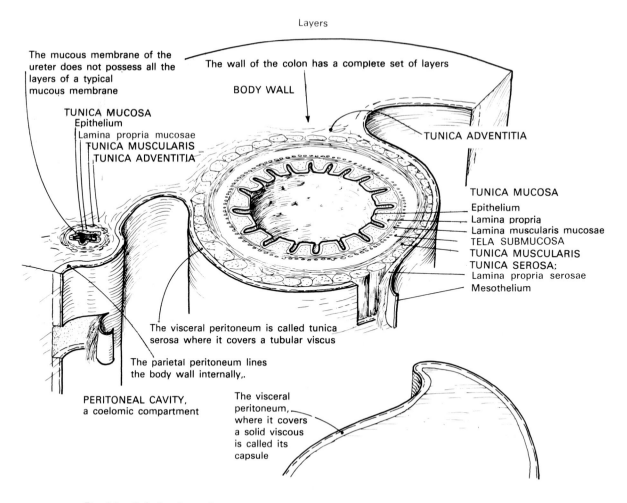

Fig. 8-3. Relationships of various viscera to the body wall and to the body cavity.

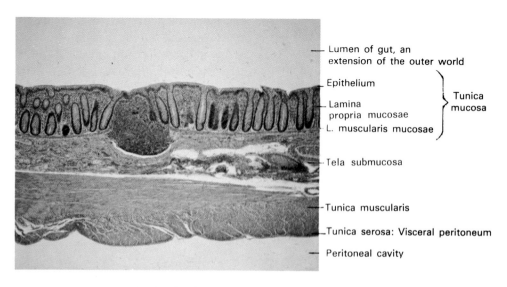

Fig. 8-4. Ventral wall of colon, H&E. 56 ×.

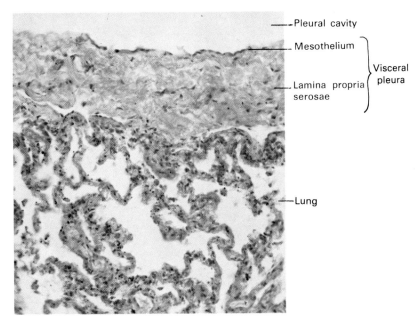

Fig. 8-5. The visceral pleura, H&E. 390 ×.

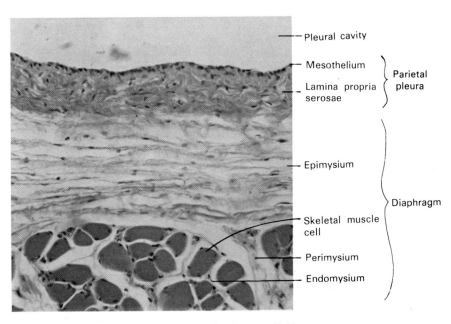

Fig. 8-6. Parietal pleura lining the diaphragm, H&E. 350 ×.

parietal portion, that in contact with an organ is the visceral portion. For example: the serous membrane covering the outer surface of the lung is called the *visceral pleura*, and its reflection lining the internal surface of the thoracic wall is called the *parietal pleura*. The capsule of the free surface of the liver (Fig. 8-7) is a serous membrane (visceral peritoneum) but the side of the liver attached to the diaphragm is an adventitia.

Blood and lymph vessels are lined by a

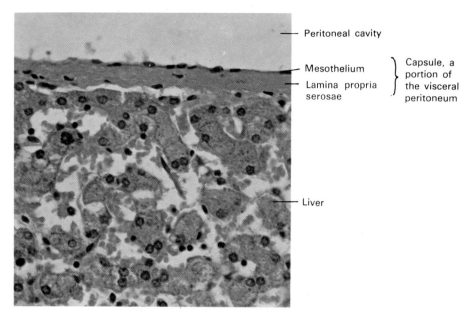

Fig. 8-7. The capsule of the liver, a part of the visceral peritoneum, H&E. 780 ×.

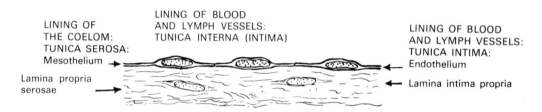

Fig. 8-8. *Left,* Tunica serosa, *Right,* Tunica interna (intima).

tunica intima (Fig. 8-8, right), which consists of endothelium (flat, simple squamous epithelia cells), attached to a basement membrane and supported by a delicate loose areolar connective tissue layer, the tunica interna (intima).

Surface Increase

For absorption, secretion, diffusion and osmosis, increase in surface area confined within a small area, is most essential. Tubular organs (Fig. 8-9 A) may be tortuous (Fig. 8-9, B). Thus the intestine, traversing a distance of 25 centimeters (approximately 9.84 inches), is extended by coiling into a tube 6.25 meters (20.5 feet) long. The ductus epididymis, traversing a distance of 5 cm (2 in.), is coiled to approximately 6.1 m (20 ft) long. Each efferent duct of the testis bridging a distance of 3.2 mm (1/8 in) is 15.2 cm (6 in.) long. The combined length of the seminiferous tubules in one testicle, itself 5 cm (2 in.) long, is about 400 m (1/4 mile).

Branching of tubes further increases the total length and therewith the combined surface area. The combined length of all blood vessels of one person is estimated at 100,000 km (6000 miles) or two and a half times the circumference of the earth.

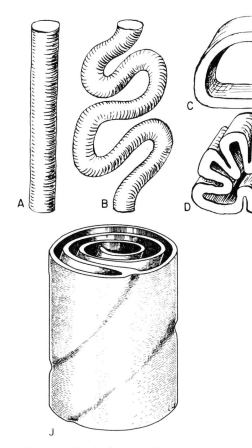

Fig. 8-9. Methods of surface increase.

This enormous length is necessary to give all the capillaries a combined surface area of approximately 2,000 m^2 (1/2 acre).

A sheet or shell (Fig. 8-9,C), to increase its surface, may be thrown into folds (Fig. 8-9,D), as, for example, the cerebral cortex (to accommodate more neurons) or the mucous membrane of the small intestine (to approximately triple its absorptive area). A surface (Fig. 8-9, E), such as the wall of the small intestine, may bear villi (Fig. 8-9.F), increasing the absorbing area again more than sixfold; and the individual epithelial cells may be covered with microvilli (Fig. 8-9, H), increasing the absorptive

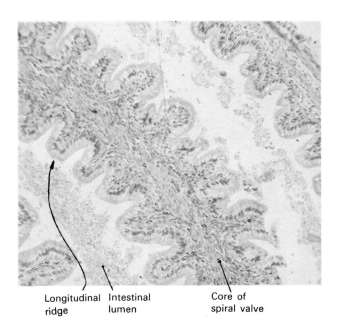

Fig. 8-10. Spiral valve in the intestine of a baby shark. In the adult, the ridges are taller and branched, H&E. 190 ×.

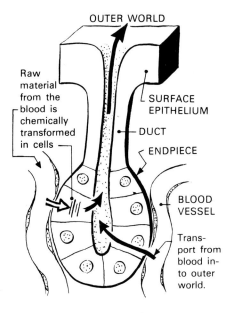

Fig. 8-11. A gland.

surface another sixfold. Thus, the small intestine presents an absorptive surface more than 2,000 times greater than that of a straight tube with a smooth wall traversing the same distance. The "spiral valve" found in the intestine of sharks (Fig. 8-9, J), substantially increases the total surface area by longitudinal ridges (Fig. 8-9, and 8-10).

Increase of secretory surface of glands is accomplished by epithelial invaginations (Fig. 8-9, G). Microvilli and deep invaginations of the cell basal surface increase the reabsorptive and excretory surfaces of kidney proximal convoluted tubule cells tremendously (Figs. 8-9, I, and 14-43, A and B).

Glands

Glands are organs which synthesize substances from raw material delivered to

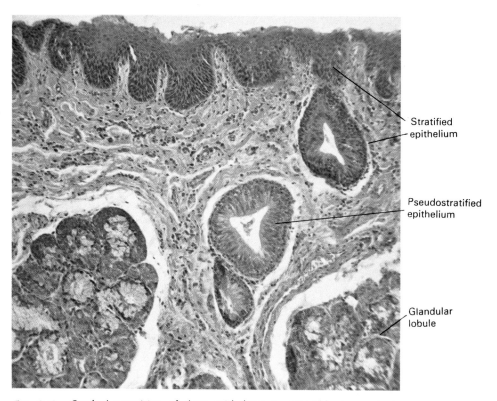

Fig. 8-12. Gradual transition of duct epithelium to resemble increasingly the surface epithelium. Plica ventricularis ("false vocal cord"). H&E, 250×.

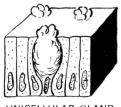
UNICELLULAR GLAND
(Goblet cells)

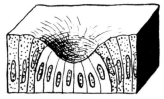
INTRAEPITHELIAL GLAND
(e.g. Tubuli efferentes of testis)

SECRETORY EPITHELIUM
(e.g. surface epithelium of stomach)

Fig. 8-13. Three types of primitive glands.

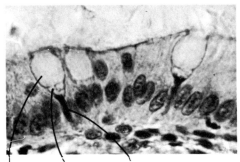

Drop of mucus | Golgi complex | Basal part of goblet cell with nucleus

Fig. 8-14. Goblet cells (unicellular glands) from the colon.

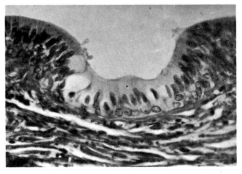

Fig. 8-15. Intraepithelial gland in the excretory duct of the sublingual gland.

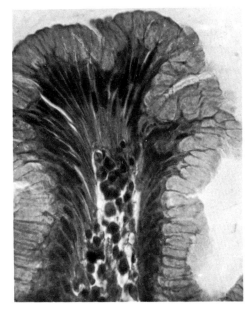

Fig. 8-16. A secretory epithelium. Surface epithelium of stomach.

TUBULUS

ACINUS

ALVEOLUS

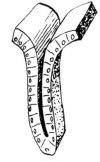
SULCUS

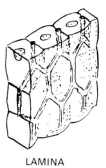
LAMINA

Fig. 8-17. Various forms of glandular end-pieces.

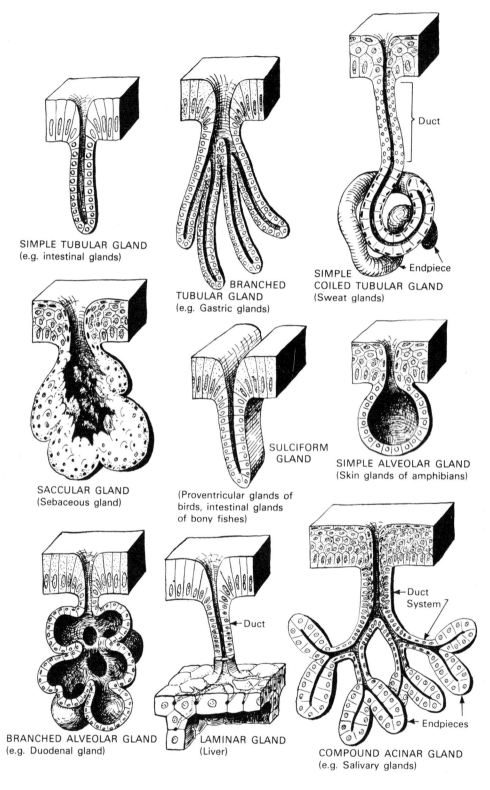

Fig. 8-18. The basic, morphologic types of glands.

GLANDS 177

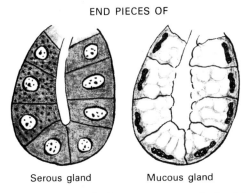

Fig. 8-19. Serous and mucous types of glandular end-pieces.

them by the blood stream and which eliminate the processed substances. The *parenchyma* of glands is derived from epithelium.

EXOCRINE GLANDS. Exocrine glands consist of ducts with terminal, secretory epithelial cells, which produce and discharge secretions into the "inward extensions" (ducts) leading to the outer world (Fig. 8-1,C). In larger glands, a duct carries the secretion to the surface (Fig. 8-11). Toward its point of exit, the epithelium lining the duct of a large gland tends to assume, the structure of the surface epithe-

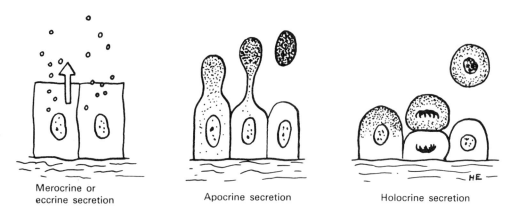

Fig. 8-20. Modes of secretion.

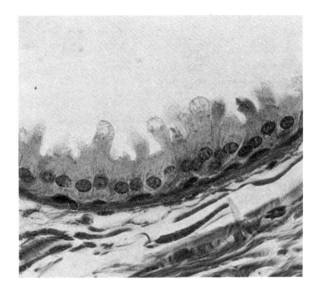

Fig. 8-21. Apocrine secretion in a gland of the external auditory meatus. Goldner stain, 1000 ×.

lium on which it empties. Thus, near their mouth, ducts of large salivary glands are lined by moist stratified epithelium of the oral cavity. (Fig. 8-12). The pancreatic and common bile ducts are lined by simple columnar epithelium, similar to that of the duodenal epithelium.

Intraepithelial Glands. Certain simple glands consist of secretory cells embedded in an epithelial surface (intraepithelial). Among these are unicellular glands (goblet cells), multicellular epithelial glands and secretory epithelia (Figs. 8-13, 8-14, 8-15 and 8-16). Most glands, however, are located deep to the surface epithelium (intramural), but retain a connection via ducts, to the epithelium.

Classification of Glands. Glands may be classified according to shape, type of product and mode of secretion.

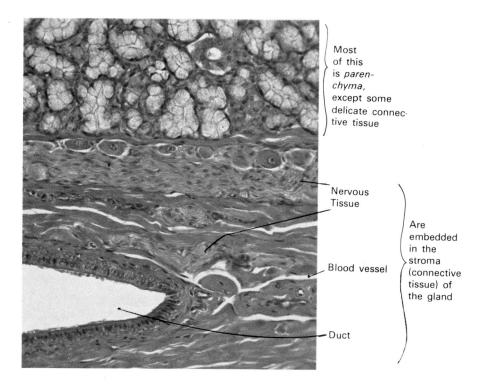

Fig. 8-22. Parenchyma and stroma of a salivary gland. H&E, approx. 350 ×.

Classification of Glands According to Shape. Exocrine glands have secretory end-pieces of various shapes, (Fig. 8-17). The terms *acinus* and *alveolus* frequently are misused. Acinus, in Latin, means a fresh grape torn off its stem with a very narrow cavity in it. In histology, a secretory acinus is an end-piece with a very narrow lumen. An alveolus (Latin: diminutive of alvus = bell) as the name implies has a larger lumen than an acinus.

In Figure 8-18, the various types of glands are shown and named. Combinations of two or more types of end-pieces occur. Sulciform glands (stomach of birds), simple alveolar glands (skin of amphibians) and simple acinar glands (imaginable but not yet found anywhere) have not been found in adult man. Most simple glands and branched glands empty directly onto an epithelial surface others empty through a duct. A gland which empties through a branched duct system is known as a compound gland.

Classification by Type of Product. Glands can be classified according to the

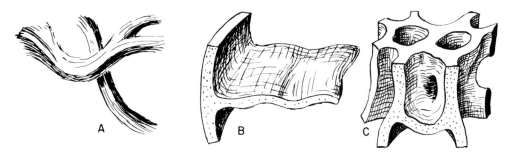

Fig. 8-23. A, Trabeculae. B, Septum. C, Muralium.

physical properties of their secretion (Fig. 8-19). Serous glands produce clear liquids. Usually, their cells have a deeply staining, optically rather homogeneous cytoplasm in which secretory granules may be embedded. Their nuclei are round or oval. Mucous glands secrete a viscous slimy product. The cytoplasm of their cells appears foamy. Frequently, their nuclei are compressed by the accumulation of secretion against the basal cell membrane. Some glands are classified as mixed, that is, they contain both serous and mucous elements.

Classification by Mode of Secretion (Fig. B-20) Electron microscopic studies have revised our concepts of three modes of gland secretion: *holocrine, merocrine* (eccrine) and *apocrine*. These names remain from early light microscopic studies, but are adapted to current ideas. *Holocrine* secretion occurs as a result of the sacrifice of entire cells resulting from lipid accumulation in the cytoplasm. Undifferentiated cells at the periphery of secretory epithelium undergoes mitoses to replace lost cells (examples: sebaceous glands next to hair follicles). *Merocrine* (eccrine) secretion involves release of secretory products, usually a protein, without the loss of any part of the plasmalemma. Secretory granules, produced by granular (rough) endoplasmic reticulum, are packaged by the golgi complex in membranous vesicles. The membrane bound granules migrate to the luminal surface of the cell where it fuses to the plasmalemma. The product is discharged and the membrane which surrounded the secretion granule is incorporated into the plasmalemma, thus conserving the cell membranes (examples: sudoriferous (sweat) glands. *Apocrine* secretion (Fig. 8-28) is that process which releases accumulated globular lipid droplets. Within the cytoplasm, the lipid product is not membrane bound, but as it migrates toward the luminal surface (like protein in merocrine secretion) it forcefully bulges the plasmalemma into the lumen. Soon there is sufficient evagination of the plasmalemma that it appears to surround the lipid droplet. The process is complete when the lipid droplet is pinched off as a membrane bound product. The site of separation seals and heals rapidly. In contrast

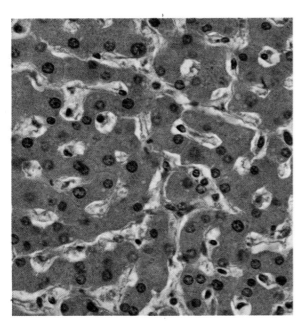

Fig. 8-24. Cellular muralium, walls one cell thick (simplex): liver.

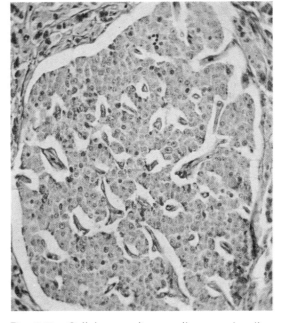

Fig. 8-25. Cellular muralium, walls several cells thick (multiplex): pancreatic island.

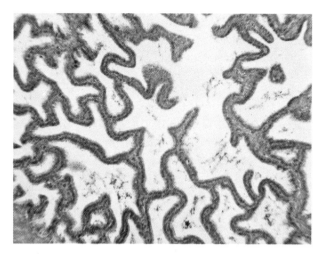

Fig. 8-26. Mucous membrane muralium (vesicular gland).

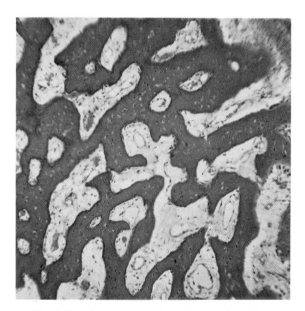

Fig. 8-27. Osseous muralium ("cancellous" bone).

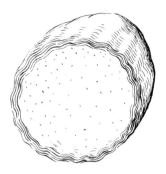

Fig. 8-28. A tunica albuginea surrounding an organ.

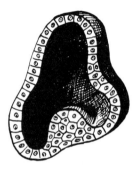

Fig. 8-29. A follicle.

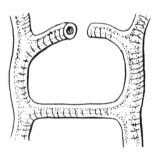

Fig. 8-30. Two anastomoses.

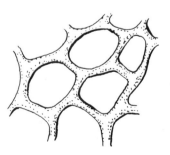

Fig. 8-31. A plexus.

to merocrine secretion, some plasmalemma is lost (example: apocrine sweat glands associated with the hair of the axilla, pubis and circumanal region). The mammary glands secrete milk (protein plus lipids) by merocrine and apocrine secretory modes.

In *cytogenic* glands, i.e., testes and ovaries, entire living, intact cells are discharged.

ENDOCRINE GLANDS. These glands are ductless and discharge their products (hormones) directly into the circulatory system (Chapter 15) usually via merocrine secretion.

Composition of Organs

Specialized cells which perform the specific functions of an organ constitute its *parenchyma*; for example, cardiac myocytes of the heart and hepatocytes of the liver. The parenchyma is supported by connective tissue called the *stroma* of the organ. The stroma not only supports the parenchyma of the organ, but it carries blood vessels, ducts, nerves and ganglia (Fig. 8-22). The organ receives nourishment raw materials and oxygen by arteries, and is drained of waste (and hormones) by veins and lymph vessels. These vessels, together with the capillaries, constitute its vasculature. The activity of the organ is controlled by nerves with ganglia frequently inserted in their course. The nerves and ganglia provide the innervation for the organ.

Structural Forms

Certain structural entities are employed in the construction of various organs, (1) trabeculae (Fig. 8-23, A) act as long beams, (2) a septa (Fig. 8-23, B) form simple walls which divide an organ into compartments, and (3) muralium (Fig. 8-23, C) is a system of complex interconnected walls. Trabeculae, septa and muralia can be composed of cells, fibers, membranes, muscle or bone according to their functions. For example, the trabeculae of the spleen consist of connective tissue fibers and perhaps occasional smooth muscle cells. The most delicate osseous trabeculae consist of bone; the trabeculae carneae and interventricular septum of the heart consist mainly of cardiac muscle; the septula of the testicle consist of dense fibrous connective tissue, and the septum pellucidum of the brain consists of nerve cells, neuropil and neuroglial cells.

A cellular muralium consists of walls built of cells, interconnected like the walls in a building. They may be one cell thick (muralium simplex, Fig. 8-24) as in the adult liver; two cells thick (muralium duplex) as in the liver of infants; or several cells thick (muralium multiplex) as in pancreatic islets (Fig. 8-25) and occasionally in the suprarenal medulla.

A mucous membrane muralium is a system of interconnected walls which, possesses a lamina propria core of connective tissue covered on both sides by epithelium. The seminal vesicle, is an extensive

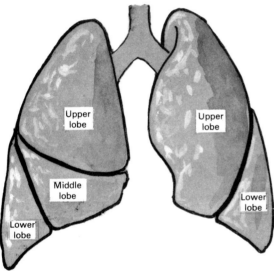

Fig. 8-32. The lungs and their lobes.

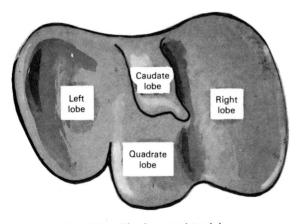

Fig. 8-33. The liver and its lobes.

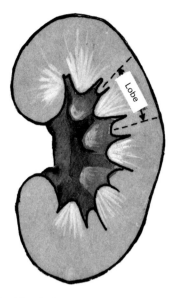

Fig. 8-34. The kidney (cut in half) having its lobes exposed.

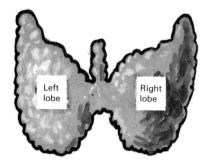

Fig. 8-35. The thyroid gland and its lobes

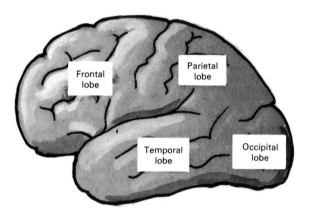

Fig. 8-36. The left cerebral hemisphere and its lobes.

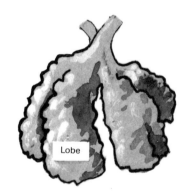

Fig. 8-37. A renal glomerulus and its lobes.

muralium of this kind (Fig. 8-26). With epithelium lining all the individual spaces including the large central cavity. The walls of the gall bladder and cystic duct likewise are constructed on the same principle. The parenchyma of the lung, also a mucous membrane muralium, forms the interalveolar septa.

Other Structural Forms

A *bony (osseous) muralium* consists of intersecting plates and bars of bone. Most trabecular bone (Fig. 8-27) is of this type.

The *tunica albuginea* of the testis (Fig. 8-28) is a nonelastic, retaining envelope made of fibrous connective tissue to maintain a certain internal pressure.

A *follicle* (Fig. 8-29) is a bag, usually without outlet, lined by epithelium.

An *anastomosis* (Fig. 8-30) is an open communication between two *hollow* passages. It is incorrect to speak of "anastomosing" nerves; for nerves are not hollow.

A *plexus* (Fig. 8-31) is a network of tubes or fibers. Blood, lymph vessels and nerves form plexuses in specific locations.

Structural Mechanisms

Students frequently are confused by the terms fibers, fibrils and filaments. Confusion is amplified when equating the term *cell* with fiber in the case of muscle. The latest Nomina Histologica presents a uniform terminology to avoid nomenclature difficulties. Most cells possess *microtu-*

bules and *microfilaments*. Electron microscopy shows microtubules to be actin like proteins averaging 24 nm in diameter and several micrometers long. These proteins, called *tubulins*, form the structural and functional elements of the mitotic spindle, cilia and flagella and also converge on centrosomes to form what is called the *aster*. In non-dividing cells, microtubules are dispersed throughout the cytoplasm. *Microfilaments*, also composed of actin-like proteins, provide the structural framework of cells, but are functionally important in promoting ameboid activity of cells. Aggregates of microfilaments form the *microfibrils* seen with the light microscope. Complicated networks, referred to as *cytoskeleton* have been demonstrated in most cells. Microfilaments are related to, but are of different composition than *myofilaments*, *neurofilaments* and *keratofilaments*. Neurofilaments 7 to 10 micrometers in diameter form the framework of neurons much like microfilaments, but are chemically different. With the light microscope, bundles of neurofilaments stained by various silver methods are identified as *neurofibrils*. In muscle cells, the structural proteins require attention and distinction. The term "fiber" should no longer be equated with a single muscle cell. Designate muscle cells as follows: nonstriated myocyte (smooth muscle cell), skeletal myocyte (skeletal muscle cell) and cardiac myocyte (heart muscle cell). The *myofibrils* are small parallel units about 1 to 3 micrometers in diameter, which in turn are composed of the *myofilaments* responsible for the banding patterns of skeletal and cardiac muscles. The only time that the term *myofiber* should be used is when referring to a series of cardiac myocytes.

In connective tissue it is correct to speak of collagen, elastic and reticular *fibers*.

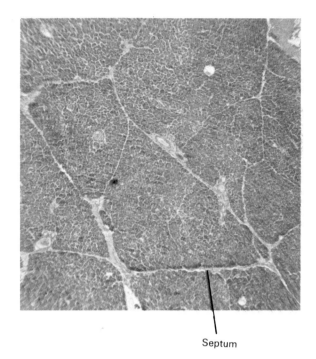

Fig. 8-38. Very low power view of pancreas to show its composition of lobules which are separated by connective tissue septa. H&E, 50 ×.

Lobes and Lobules

A lobe is a relatively large subdivision of an organ or of a corpuscle. For example, one deep fissure divides the left lung into two distinct lobes, two fissures divide the right lung into three lobes (Fig. 8-32). The *liver*, a unified organ in man, has four externally visible protrusions called lobes (Fig. 8-33). The liver of many animals has distinct lobes. The *kidneys* of the newborn are indented by a variable number of externally visible fissures outlining lobes. In childhood these lobes fuse externally; yet, when cutting an adult kidney one can distinguish 10 to 15 lobes (Fig. 8-34). The *thyroid* has three distinct lobes (Fig. 8-35). Each cerebral hemisphere is indented by only one large, constant fissure. Nevertheless, one divides it conveniently into four lobes (Fig. 8-36). Even the very small renal glomerulus is divided by two to six deep fissures into three to seven lobes (Fig. 8-37). The nucleus of a polymorphonuclear neutrophilic leukocyte (PMN) is divided into two to five lobes connected by strands of nuclear material (Fig. 5-7).

Each lobe of an organ is subdivided into lobules. Thousands or even millions of lobules are contained in certain organs. Lobules may be just recognizable with the naked eye. In the salivary glands, the pancreas, the thymus, and in masses of fat, lobules are separated from each other by connective tissue septa (Fig. B-38); but in the human liver and kidney, lobules are arbitrarily defined territories whose external surface boundaries cannot be readily discerned.

9...
Cardiovascular system

The cardiovascular system consists of the tubes devoted to the transportation of blood within the internal medium. The blood itself serves for the transport of substances, as well as leucocytes and erythrocytes, between the various portions of the body.

Actual exchange of matter occurs through the walls of the capillaries. Arteries and veins are conduits to and from the capillary bed, and the heart pumps the blood through these vessels. Lymph vessels, described with the defense system, drain the tissue spaces of excess fluid.

BLOOD VESSELS

Arteries carry blood from the heart; *veins* return it to the heart. *Capillaries* connect arteries with veins. The walls of blood and lymph vessels consist of the following layers, from inside outward:

Tunica interna (intima)
   endothelium (simple squamous cells)
   Lamina intima propria (areolar connective tissue). With advancing age, the thickness of the lamina intima propria increases.

Internal elastic membrane (elastic membrane or network of elastic fibers and bands)

Tunica media (mixture of mostly circular smooth myocytes and some elastic fibers or cardiac myocytes mixed with elastic, fenestrated membranes).

External elastic membrane (feltwork of elastic fibers)

Tunica externa (adventitia) (areolar connective tissue).

(These layers are shown in Fig. 9-7, F, G and H.)

Capillaries

Figure 9-1 deals with capillaries in sequential order: (A) Blood is carried through arteries and arterioles into the capillary bed, and returns through venules and veins to the heart. The artery is narrower and has more smooth myocytes in its wall than the vein. (B) The *endothelial cell* is the building stone for capillaries. (C) In section, this squamous cell shows a bulge around the nucleus. (D) As part of a tube, the cell must be curved. (E) Two endothelial cells fit together to form an average capillary. (F) This is a photomicrograph of a spread rat omentum, showing a

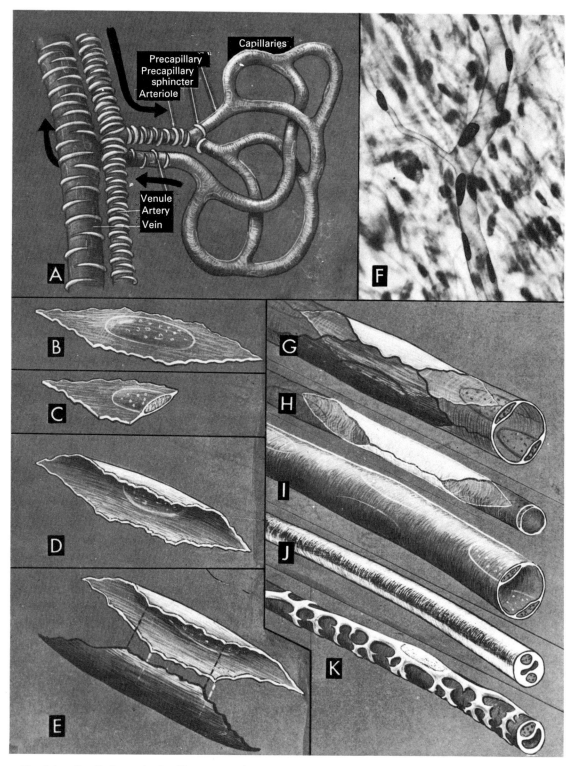

Fig. 9-1. Capillaries. A, Capillary net and its connections. B, An endothelial cell, as if flattened. C, An endothelial cell, cut in half. D, An endothelial cell as a component of a cylindrical capillary wall. E, Two opposing endothelial cells. F, An average-sized capillary runs vertically through the field; a very narrow capillary arises from it at the left. (From spread of greater omentum of rat.) G, An average-sized capillary. H, A narrow capillary surrounded by a single endothelial cell. I, A capillary as it would appear without silver-impregnated cell boundaries. J, Endothelial contraction narrowing the capillary. K, A pericyte (Rouget cell) surrounding a capillary and constricting it.

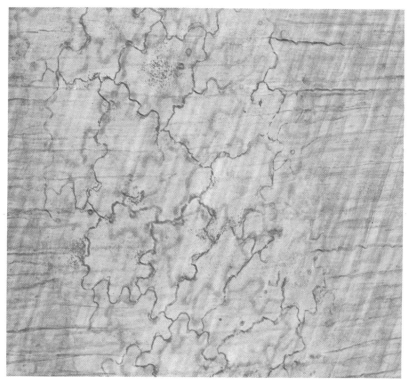

Fig. 9-2. Spread of rat mesentery impregnated with silver to show the scalloped outlines of endothelial cells of a lymph vessel, 1000 ×.

capillary of average caliber ascending almost vertically through the field. This capillary is surrounded at each level by two endothelial cells with deeply stained nuclei that stand out clearly. The cytoplasm of the endothelial cells, forming an extremely thin layer, is so transparent that the entire capillary resembles a very delicate glass tube. Just above the center of Figure 9-1, (F) a much narrower capillary arising from that just examined, courses toward the upper left. This side branch is so thin that one endothelial cell suffices to form its wall throughout its visible length. The nucleus of this single endothelial cell is seen at the upper left as an oval spot. (G) Endothelial cell boundaries can be revealed by silver impregnation. This is shown very well in a lymph vessel, because this lymph vessel is spread flat (Fig. 9-2). In blood vessels, the cell boundaries stain equally well (Fig. 9-3) but are seen less clearly, because these vessels remain more or less cylindrical. Continuing with the examination of Figure 9-1 (H) very thin capillaries (such as the smaller one at the left in F) are surrounded by a single endothelial cell. (I) Usually the capillary wall (endothelium) is extremely thin and permeable to electrolytes and other substances of small molecular size. (J) When the capillary is narrowed, its wall is thick. Such narrowing is probably due to the blockage of blood flow whereupon the capillary empties and the endothelial cells assume shape as if liberated from intracapillary pressure. Blockage of blood flow can be the result of a blood cell stuck in the entrance to a tiny capillary, or to an embolus or to a hypothetical precapillary sphincter. (K) At intervals along the length of the capillary, irregularly shaped, isolated cells reside within the basement membrane surrounding the capillary endothelial cell. These cells, the *pericytes* are non-contractile, relatively undifferentiated

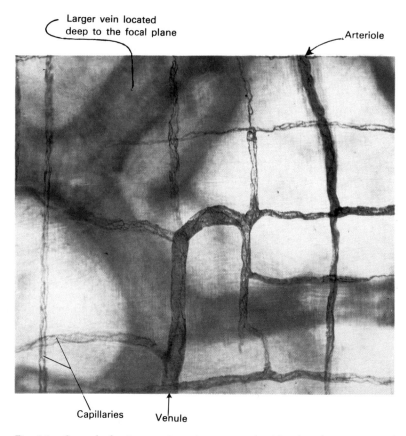

Fig. 9-3. Spread of rat mesentery, impregnated with silver by injection of silver nitrate into a mesenteric artery to show boundaries of endothelial cells, as well as net-like arrangement of capillaries. 500 ×.

cells with the potential to differentiate into a variety of different types, even smooth muscle. Pericytes have erroneously been equated with Rouget cells. In 1875, Rouget reported the existence of contractile cells on vessels now recognized as precapillary arterioles (metaterioles) in the nictitating membrane of the frog eye. The cells described are recognized now as smooth myocytes.

A capillary, in cross section, often appears like a signet ring. This is especially evident in an electron micrograph (Fig. 9-4). Many capillaries possess a continuous, endothelial wall (Figs. 9-4 and 9-5, A). Where two endothelial cells meet in continuous endothelia, junctional complexes occur at intervals. Transendothelial transport is a process by which substances enter or leave the vessel lumen by passing (1) between adjacent endothelial cells or (2) through the entire cell, i.e., plasmalemma, cytoplasm and plasmalemma again, and in addition the basement membrane (Fig. 9-4). Diffusion, pinocytosis followed by exocytosis (ptyocytosis) are instrumental in transendothelial transport (Fig. 9-5, A). Pinocytic vesicles, intracellular vacuoles and exocytotic vesicles are found at the boundaries of and within the cytoplasm of endothelial cells. The passage of such droplets may take but a tiny fraction of a second. Therefore, this process may account for the transport of much liquid.

Three specialized capillary vessels have been recognized via E.M. studies:. (1) non-fenestrated, (2) fenestrated and (3) sinusoids and sinuses (Figs. 9-4, 9-5, 9-6). The *non-fenestrated continuous capillaries* are composed of endothelial cells held tightly together by gap-junctions and desmosomes. *Superficial vesicles (caveolae)*

Fig. 9-4. A capillary from the vocalis muscle, 10,000×. Courtesy of Dr. Ulf Brühl, University of Heidelberg.

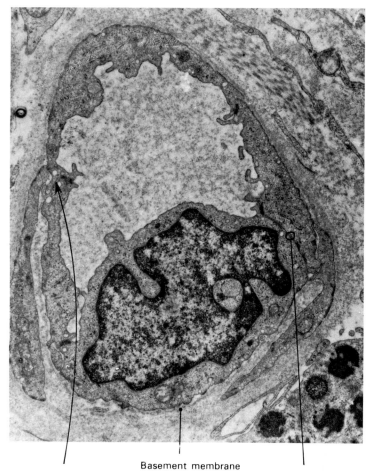

Basement membrane
Cell boundaries with junctional complexes and interdigitations

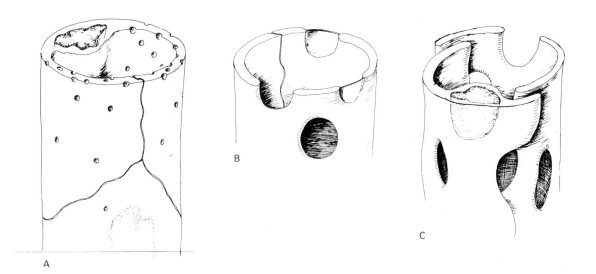

Fig. 9-5. The three basic types of capillaries. A: With continuous endothelium and micropinocytosis. B: With fenestrated endothelium. Note that some fenestrae are open, others are closed by diaphragms. C: With discontinuous endothelium. This type is called a sinusoid.

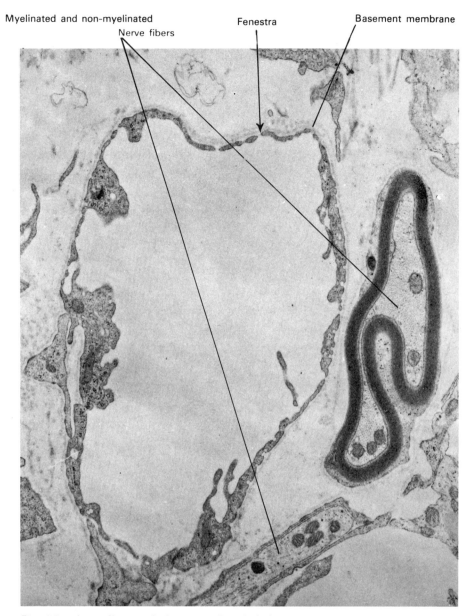

Fig. 9-6. A capillary with fenestrated endothelium, also a myelinated and a non-myelinated nerve fiber. Courtesy of Dr. J. T. Povlishock, Medical College of Virginia.

pock the outer and luminal plasmalemma surfaces of the endothelial cells. One variety, with a 200 nm thick wall, is associated with muscle tissues (smooth, akeletal and cardiac) and parts of the ovary and testis. Another, 100 nm type is characteristic of blood-tissue barriers. In these capillaries, true zonula occludens (tight junctions) also occur, but superficial vesicles are rare. *Fenestrated capillary endothelial cells* have 60 to 100 nm diameter fenestrations (windows or pores). One variety, with thin *diaphragms* closing the fenestrations, occurs in tissues requiring rapid fluid exchange (GI mucosa, kidney peritubular plexuses, ciliary body of the eye and choroid plexus). Fenestrated endothelial cells *lacking diaphragms* occur in kidney glomerular capillaries. Open, but larger fenestrations

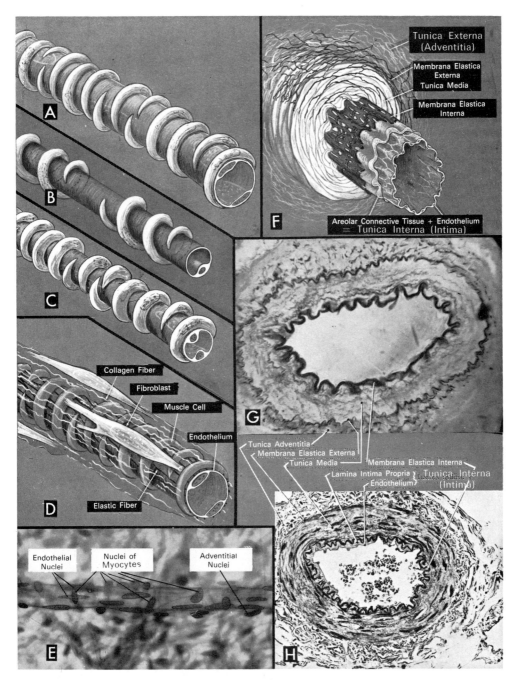

Fig. 9-7. Arterioles and small arteries. A, A typical arteriole. B, An arteriole surrounded by muscle fibers which spiral around the endothelial tube. C, Contraction of the muscle fibers constricts the arteriole. D, The same arteriole as shown in C, now complete with tunica adventitia. E, An arteriole from a spread of rat omentum. F, The layers of a typical artery. G, Section of a small artery stained with resorcin fuchsin. H, A section similar to that shown in G, stained with hematoxylin and eosin.

are encounterd in *sinusoidal endothelial cells* of endocrine glands, carotid and aortic bodies, but basement membranes associated with these capillaries are thin and attenuated. *Sinusoids* are adapted for retaining large quantities of blood under low pressure. Cell junctions are rare, causing gaps to appear between opposing edges of adjoining endothelial cells. The basement membrane is scanty or absent in the sinusoidal endothelium occuring in liver and bone marrow. *Splenic sinuses* are characterized by non-fenestrated, elongated endothelial cells arranged parallel to the lumen like the staves of a barrel and retained by threads of reticular fibers

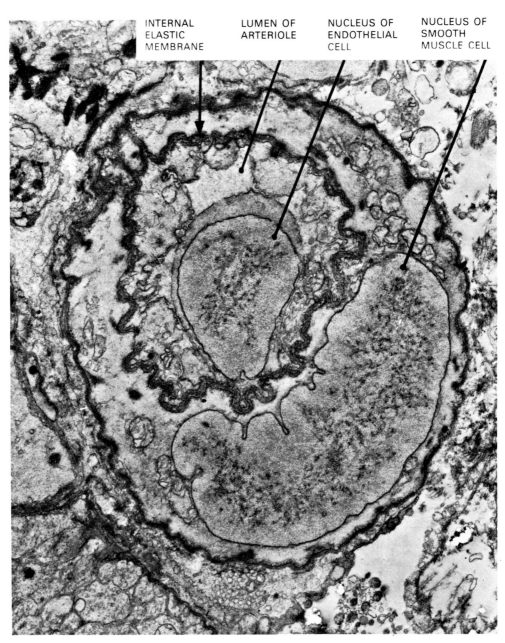

Fig. 9-8. An arteriore from the adenohypophysis of a gerbil. (Electron microgram by Dr. E. Millhouse, Chicago Medical School.)

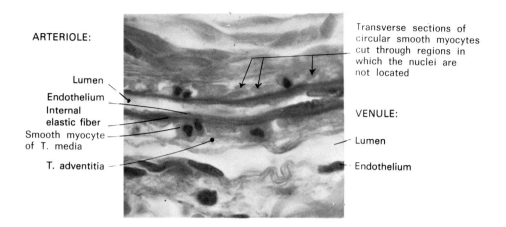

Fig. 9-9. Axial, longitudinal section of arteriole and venule. Note that the myocytes of the Tunica media (which run circularly) are cut transversely and that in most of their cross-sections the nucleus is not seen. For an explanation, see chapter 6. The scanty elastic tissue is colored black by silver impregnation. It consists, in a vessel as small as this, of longitudinal fibers. 1875 ×.

analogous to the barrel hoops. Arterioles deliver blood and venules drain these sinuses.

At the capillary bed, fluid may exit the lumen of the capillary through the endothelial cells into the tissue spaces. This tissue fluid bathes tissue cells to supply them with oxygen and nutrients. More tissue fluid leaves the arterial side of the capillary bed than is returned to the blood at the venous side. The excess tissue fluid is drained via the lymphatic system. In general, the relationship between the hydrostatic and colloidal osmotic pressures determines the movement of fluid in and out of a capillary bed. High hydrostatic pressure forces fluid out of the capillaries on the arterial side, Decreased blood pressure and a high colloidal osmotic pressure draws tissue fluid into the blood at the venous side of the capillary bed. The high colloidal osmotic pressure results from a loss of fluid and small molecules at the arterial side which produces a hemo-concentration of large molecules in the blood. Excess fluid (*edema*) in the tissue spaces may result from a number of causes: (1) lymphatic obstruction prevents removal of excess fluid, accumulating in the interstitial areas; (2) venous stenosis decreases normal return of fluid at the venous side of the capillary bed; (3) high arterial pressure promotes the formation of an excess amount of tissue fluid at the arterial side of the capillary bed; (4) malnutrition causes a decrease in the concentration of colloidal molecules in the blood which results in a colloidal osmotic pressure that is too low at the venous side of the capillary bed for normal movement of tissue fluid into the blood; and (5) lack of skeletal muscle movement results in failure to compress veins so that they empty.

Arteries

The arteries, vessels which transport blood away from the heart toward capillary beds, arise by repeated branching, from the aorta and the pulmonary trunk. *Conducting arteries* (pulmonary trunk, aorta, common carotids, subclavian, renal, internal iliac and femoral arteries) are distinguished from *distributing arteries* which arise from them. The structure of arteries in general can be studied best in medium size, distributing arteries.

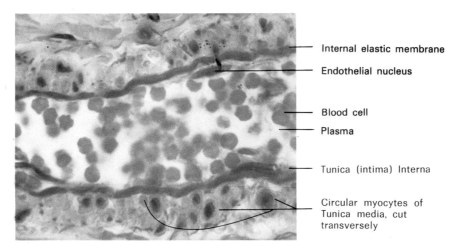

Fig. 9-10. Axial, longitudinal section through a small artery. 1875 ×.

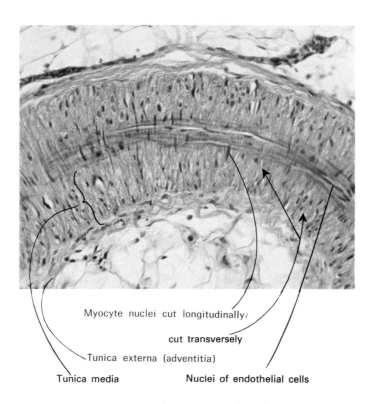

Fig. 9-11. Longitudinal section of medium sized artery, tangential to the endothelium and the innermost muscle layer. Galacyanin and Picric acid. 375 ×.

Small- and Medium-sized Arteries. Figure 9-7 demonstrates arteries of microscopic size. The *arteriole* consists of endothelial cells surrounded by a basement membrane plus a monolayer of spiraling smooth myocytes forming the tunica media (Fig. 9-7, A). An electron micrograph of an arteriole is seen in Figure 9-8. One smooth myocyte may spiral several times around an arteriole (Fig. 9-7, B). Contraction of the myocytes narrows the arteriole lumen (Fig. 9-7, C). Although they are modestly developed, all layers of a complete blood vessel are present in arterioles. Fibroblasts and collagen fibers comprise the tunica externa and sparse smooth myocytes represent the tunica media. A few delicate, longitudinal elastic fibers make up the internal elastic membrane (Fig. 9-7, D), while the endothelium and its basement membrane represents the tunica intima. Elongate nuclei of endothelial cells and fibroblast nuclei of the tunica externa parallel the longitudinal axis but smooth myocyte nuclei spiral circularly (Fig. 9-7, E). Therefore, in axial longitudinal sections through arterioles and arteries (Figs. 9-9 and 9-10), smooth myocytes are cut transversely or obliquely while endothelial cell nuclei are cut lengthwise. In a longitudinal section through a medium sized artery, slightly off center, the smooth myocyte nuclei run prependicular to endothelial cell nuclei (Fig. 9-11, A). Figure 9-11, B, shows a small artery and its vena comitans (companion vein). The disposition of smooth myocytes in the tunica media of the artery is seen. In the wall of the vein, there are three isolated smooth myocytes. The perirenal brown fat can also be seen.

In *medium-sized arteries* layers common to all blood and lymph vessels are clearly recognizable and may be studied when comparing Figure 9-7, F, G and H, with the layers listed above. In Figure 9-12, all layers can be idenified.

Figure 9-13 represents the structure of large arteries to be compared in G with that of smaller ones.

ELASTIC ARTERIES. The tunica media of the large, *conducting arteries* (aorta, the pulmonary, common carotid and subclavian arteries) contain many fenestrated elastic membranes, connected by elastic bands (Fig. 9-13, A). Smooth myocytes, attached by elastic cell tendons connect the elastic membranes in the aorta (Fig. 9-13, D). When the myocytes contract, the elastic membranes become wavy, the vessel lumina are narrowed resulting in an increase in blood pressure (Fig. 9-13, E). The internal elastic membrane, in contrast lacks fenestrations, and myocytes attach to membranes of conducting arteries by two methods (Fig. 9-13, F). The larger the artery, the more prominent is the elastic tissue of its wall, but in smaller arteries the muscular component is more prominent (Fig. 9-13, G). An intimate interplay exists between the passive elastic tissue and the contractile myocytes in the arterial wall.

The relation of myocytes to elastic membranes is best exemplified in the aorta of hoofed animals (Fig. 9-14, A). The muscle fibers, when contracted, establish tension

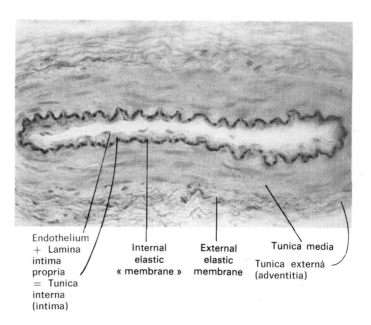

Fig. 9-12. Transverse section of a relatively small, collapsed artery. The internal elastic membrane in this size artery consists of longitudinal, elastic fibers. Since the artery is cut transversely, these fibers are cut crosswise too and appear as black dots. Goldner stain and silver. 675 ×.

Fig. 9-13. A, B and C, Aorta of infant. D and E, Attachment of muscle fibers to the elastic membranes of the tunica media of the aorta. F, Attachment of myocytes to the internal elastic membrane. G, Musculature (white) and elastic tissue (black) in arteries of various sizes.

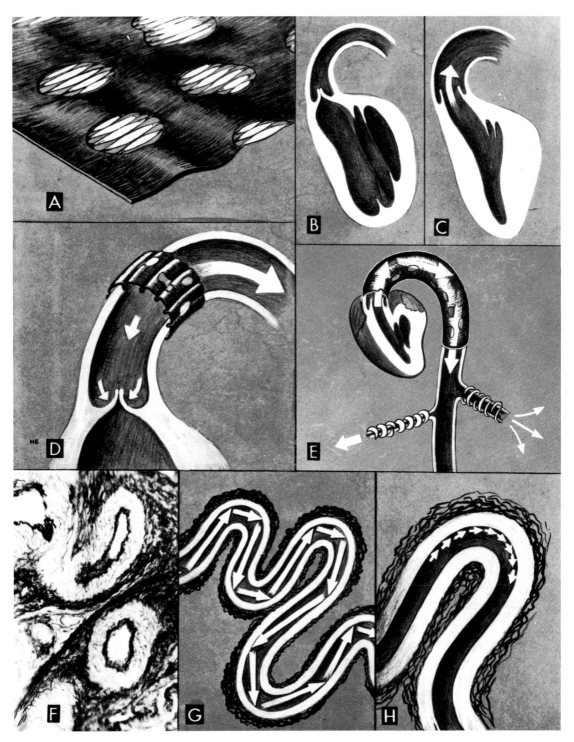

Fig. 9-14. Mechanics of blood propulsion. (See text.)

198 CARDIOVASCULAR SYSTEM

in the elastic membranes. Figure 9-14, B to E, illustrate the mechanism of blood propulsion. Blood is set in motion by the pumping action of the heart in which expansion of the heart chambers (diastole; B) alternates with contraction (systole, C). During diastole, the lumina of the aorta and pulmonary arteries are narrow and the semilunar valves are closed (Fig. 9-14, B). Systole propels blood through the semilunar valves causing dilatation of the conducting arteries (fig. 9-14, C). The elastic membranes rebound and, by their contraction, maintain systolic pressure during car-

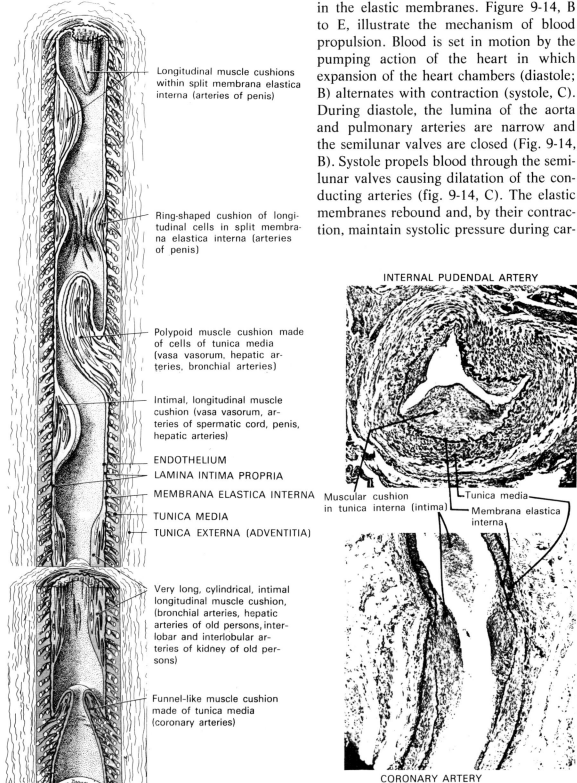

Fig. 9-15. Intra-arterial muscle cushions.

Fig. 9-16. Intimal muscle cushions.

diac diastole, thereby forcing most of the blood forward (large arrow in Fig. 9-14, D). Backward flow (small arrows) forces closure of the semilunar valves. The total energy required for blood propulsion is furnished by ventricular systole (Fig. 9-14, E). Blood flow toward the periphery is aided by the elasticity of the passive components of the arterial wall. The musculature in distributing arteries controls only their caliber but contributes little to propulsion. Activity of smooth myocytes determines the volume of blood admitted to each region and controls blood pressure proximally and distally (Fig. 9-14, E).

The tunica externa (adventitia) of tortuous arteries, like coiled arteries in the uterus, contains much elastic tissue (Fig. 9-14, F). This elastic tissue reflects onrushing blood around bends so that it reaches peripheral points with undiminished velocity (Fig. 9-14, G). The reflection of the blood flow is not as jagged as demonstrated in the diagram in Figure 9-14, G, but is relatively smooth (Fig. 9-14, H).

Intimal Muscle Cushions. Longitudinal, smooth muscle bundles are found inside many arteries. Figure 9-15 gives a summary of the various organizational patterns

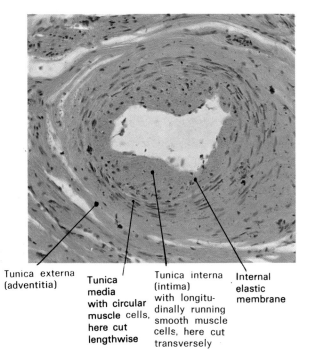

Fig. 9-17. Large helicine artery, H&E. 350 ×.

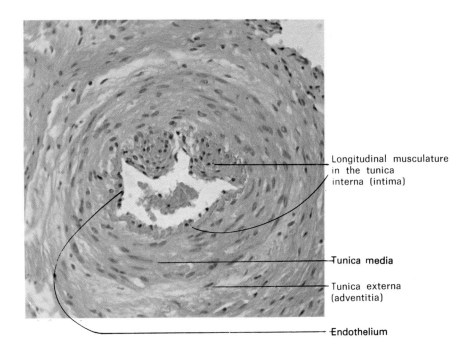

Fig. 9-18. Small helicine artery, H&E. 500 ×.

and occurrence of these muscle masses. They may form continuous layers, encircling the entire lumen for a considerable length as in the renal interlobar arteries of some elderly people or more frequently, they form small, local thickenings, called intimal muscle cushions (Figs. 9-15, 9-16, 9-17 and 9-18), or rings. Funnel-like and polypoid projections of muscle masses of the tunica media may project into the lumen of arteries. These internal muscle cushions can narrow or temporarily occlude a vessel. In organs such as the kidney and the liver, intimal muscle cushions often develop with advancing age. In bronchial arteries they occur regularly and in the internal pudendal arteries and their branches, they play an important role in the erection mechanism of the penis (Chapter 18). Intimal muscle cushions and medial muscle polyps are very common in coronary arteries. Occasionally, intimal musculature in arteries of the gastric tunica submucosa is associated with peptic ulcers. Contraction of intimal muscle cushions in the chorionic branches of the umbilical arteries at birth prevents blood loss in the new-born child.

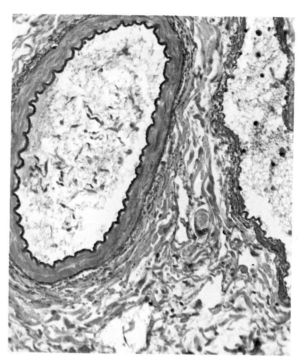

Fig. 9-19. An artery (left) and a vein (right). 375 ×.

Veins

Veins are organized like arteries, but commonly their walls are thinner, their lumina are wider; and, as a rule, intravenous pressure is lower than arterial pressure (Fig. 9-19). This must be so, because blood flow in arteries is the result of cardiac, systolic pressure; blood flow in veins is due, to a great extent, to negative pressure in the thorax (suction).

Often, a vein (*vena comitans*) which drains a small territory accompanies the artery which supplies that territory (Fig. 9-19). Usually it is easier to distinguish large, thin-walled veins from smaller, thicker-walled arteries (Figs. 9-1, A; 9-19

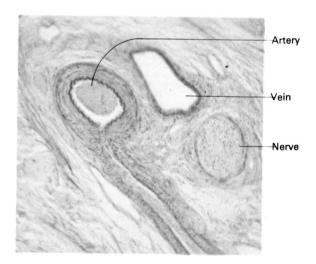

Fig. 9-20. A "neurovascular bundle" consisting of a nerve, an artery (which is seen branching) and a vein. Note sphincter around the base of the arterial branch. (Use a magnifer to see the sphincter.) Goldner and resorcin-fuchsin.

and 9-20). Even on a very small scale the distinction between arterial, venous and lymphatic vessels is based primarily on comparative wall thickness (Fig. 9-21, A & B).

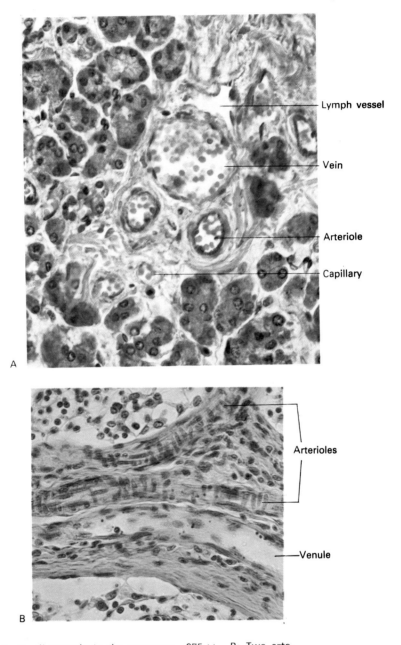

Fig. 9-21. A: Small vessels in the pancreas. 875 ×. B: Two arterioles and one venule in a lymph node, Azan, 630 × (longitudinal view).

Venules which collect blood from capillaries are similar in structure to capillaries, except for their larger diameter, i.e., their walls consist of endothelium without a muscular coat (Fig. 9-9 and Fig. 9-21, B). Proceeding toward the heart, circularly arranged smooth myocytes, develop a thin tunica media around the veins which gradually increases in thickness. Many veins, particularly in the extremities, are provided with valves which prevent backflow (Fig. 9-22, B, C and D). Venous blood flow in the head and trunk is controlled by expansion of the thorax during inspiration. Expansion produces a partial vacuum which draws blood into the atria of the heart similar to the mechanism by which air is inspired into the lungs. In the extremities, however, venous blood flow is aided by lateral pressure of skeletal muscle (Fig. 9-22, E) and perhaps by smooth muscle contractions of the tunica media (Fig. 9-22, F).

Variations from the basic structure of veins occur at specific locations. The veins in extremities have a tunica media as thick as that of arteries, capable of resisting hydrostatic pressure (Fig. 9-22, B).

The large veins in the abdominal cavity (inferior vena cava, renal veins and suprarenal veins) possess in their tunica externa powerful longitudinal musculature (Fig. 9-23, A) as well as longitudinal elastic fibers. The mechanical advantage of this structure is illustrated in Figure 9-24. The longitudinal muscle cells (white arrows) and the longitudinal elastic fibers (black arrows) in the tunica externa of the vena cava inferior and the suprarenal veins provide strong longitudinal traction capable of counteracting the transverse pressure of the viscera, which would otherwise cause these vessels to collapse. This traction is effective in the inferior vena cava because this straight vessel is anchored at both ends. The longitudinal musculature of

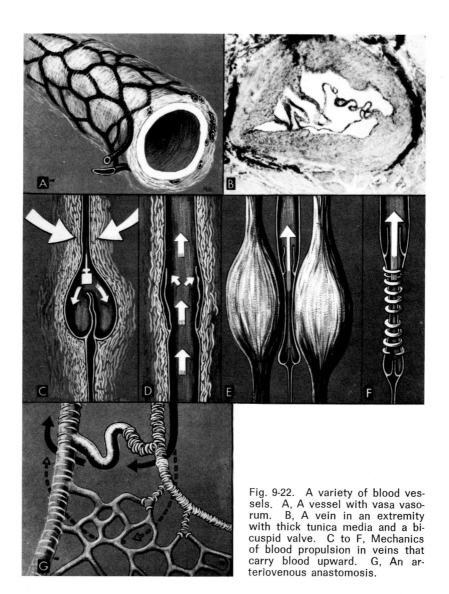

Fig. 9-22. A variety of blood vessels. A, A vessel with vasa vasorum. B, A vein in an extremity with thick tunica media and a bicuspid valve. C to F, Mechanics of blood propulsion in veins that carry blood upward. G, An arteriovenous anastomosis.

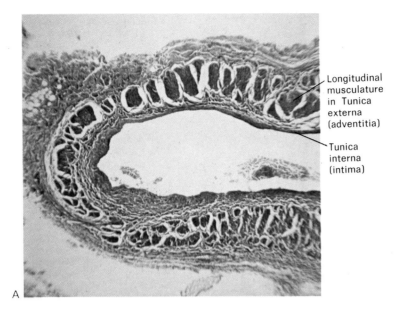

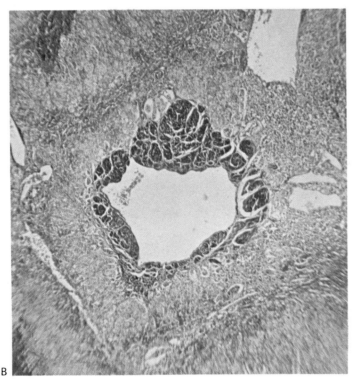

Fig. 9-23. A: Transverse section of suprarenal vein. Goldner stain, 60 ×. B: Central vein of adrenal gland. Goldner stain, 70 ×.

large abdominal veins extends into the central vein of the suprarenal gland (Fig. 9-23, B), and to the small veins in the suprarenal medulla.

Vasa Vasorum

The walls of larger blood vessels, drained by small veins, are nourished by

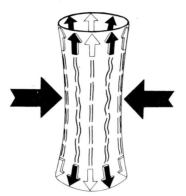

Fig. 9-24. Mechanics of the wall of large, abdominal. veins. Muscle: white. Elastic tissue: black.

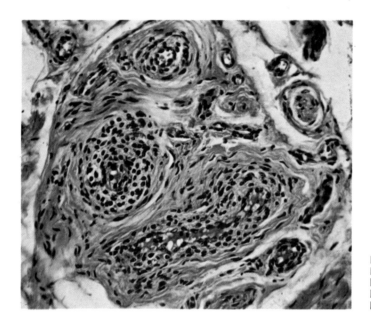

Fig. 9-25. Glomus from finger tip. Epithelioid cells stained with Astra blue. Preparation and photography by Prof. J. Staubesand, University of Freiburg, i.B.

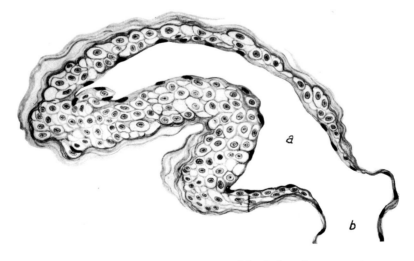

Fig. 9-26. An arterio-venous anastomosis (blood flow from a to b) with epithelioid cells. (From Giuseppe Conti, Annales d'Anat. Path. 3: 5-32, 1958.)

small nutrient arteries and the *vasa vasorum* (Fig. 9-22, A).

Arteriovenous Anastomoses

An arteriovenous anastomosis (*AVA shunt*, Fig. 9-22, G) is a direct connection between an artery and a vein, through which blood can be diverted from the capillary bed when the pre-capillary sphincters are contracted.

When an arteriovenous anastomosis is closed by musculature on the arterial side, blood flows freely through the capillary

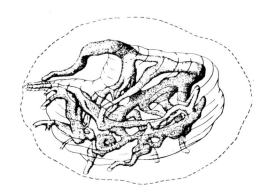

Fig. 9-29. A coccygeal glomus. (Graphic reconstruction by Staubesand: Acta Anat. 19: 105-344, 1953.)

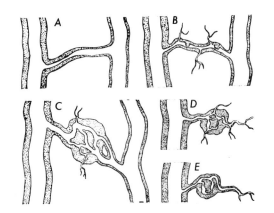

Fig. 9-27. Arteriovenous anastomoses and glomera. (From J. Staubesand: Verh. Anat. Ges. 49, 1951.)

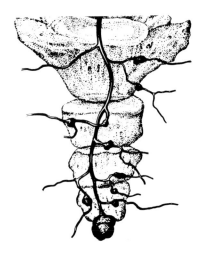

Fig. 9-30. The middle sacral artery and glomera. (From Staubesand:: Acta Anat. 19: 105-344, 1953.)

bed, but when the AVA sphincter opens and the pre-capillary sphincters are closed, blood is diverted from the capillary bed. This happens often when the hands and feet are exposed to extreme cold. The blood is thereby protected from excessive cooling, but the tips of the fingers and toes become numb. Some arteriovenous anastomoses are very tortuous, and many of the smooth myocytes in their tunica media assume an epithelioid appearance (Fig. 9-26) which can be differentiated by special stains (Fig. 9-25).

Glomera

If an AVA branches extensively and forms vascular networks as seen in Fig-

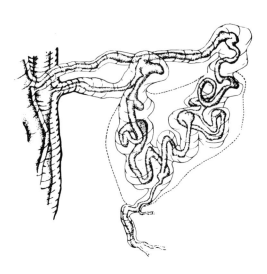

Fig. 9-28. An arteriovenous anastomosis with epithelioid wall. (From Staubesand: Acta Anat. 19: 105-344, 1953.)

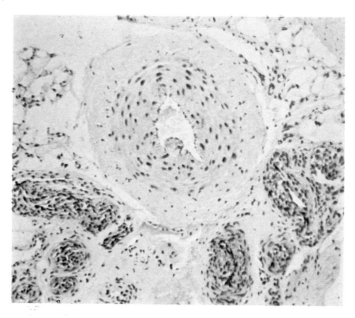

Fig. 9-31. A glomus consisting of several windings of one or several vessels whose epithelioid cells are stained with Astra blue, while the muscle cells in the nearby artery remain unstained. Preparation and photography by Prof. J. Staubesand, University of Freiburg, i.B.

ures 9-27, 9-28 and 9-29, and the epithelioid cells predominate, it is called a *glomus* (plural: glomera).

A glomus is a convolution of arterioles in the walls of which smooth myocytes are modified into *epithelioid cells* (Figs. 9-25 and 9-26).

Glomera, innervated by delicate free nerve endings, are assumed to be unconscious receptor organs which register blood pressure and the chemical composition of the blood.

Glomera may be inserted in arteriovenous anastomoses (Fig. 9-27, C) or they may drain into the capillary bed of the surroundings (Fig. 9-27, D and E; 9-29).

Glomera are frequent in the finger tips, (Figs. 9-25 and 11-3) the glans penis and alongside branches of the middle sacral artery (Fig. 9-30). The largest glomus in the body is located at the bifurcation of the common carotid artery. Pressure on this organ may induce drowsiness and even general anesthesia. This knowledge was used in ancient times as a means of anesthesia during surgery. Hence the name carotid, from *Kapaelv* = to induce sleep.

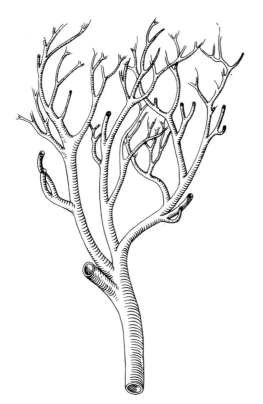

Fig. 9-32. An end-artery. (Redrawn from Staubesand: Verh. Deutsch. Ges. Kreislaufforsch. 22: 263-367, 1956.)

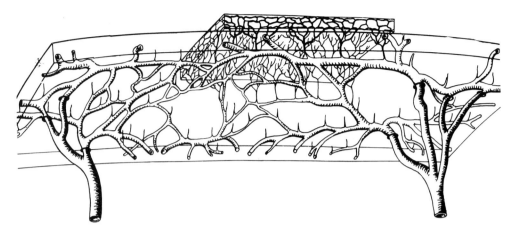

Fig. 9-33. Superficial and deep capillary plexus in the skin. (Redrawn from Staubesand: Verh. Deutsch. Ges. Kreislaufforsch. 22: 263-267, 1956.)

Simple arteriovenous anastomoses exist (Fig. 9-27, A). Other AV anastomoses have epitheliod walls (Fig. 9-26) capillary outlets, and intimal cushions along their course (Figs. 9-27, B, and 9-28).

The glomus in Figure 9-29, arising from the small artery at the left in Figure 9-30, discharges blood into several capillaries. The epithelioid cells can be differentially stained with Astra blue while the real myocytes remain unstained as seen in Figure 9-31. Glomera function, in part, as unconscious sensory organs to register blood pressure and chemical composition of blood. Evidence suggests that glomera may release hormone-like materials that provide for adjustment of pressure and composition of the blood.

End-Arteries versus Arterial Plexuses

Arteries in large, three-dimensional organs, such as salivary glands, pancreas, kidney, liver, spleen, lung, etc., tend to be *"end-arteries"* (Fig. 9-32), i.e., each branch is independent of its neighbors, just as are the branches of most trees. This means that they do not anastomose with one another within that organ.

In flat sheets such as skin, serous and mucous membranes (but not the retina), arteries form anastomotic networks or *plexuses* arranged in a deep, wide-meshed layer and a superficial layer of finer meshes and thinner vessels (Fig. 9-33). In the capsule of the suprarenal gland, the arterial plexus forms but one layer.

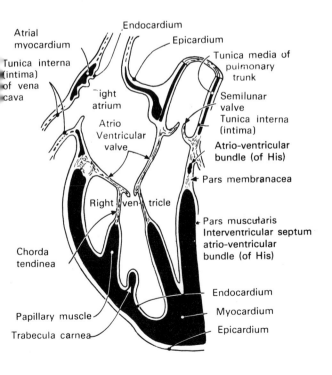

Fig. 9-34. Diagram of the heart.

HEART

The kinetic energy for arterial blood propulsion is created by the rhythmic contractions of the *heart* (Fig. 9-34), a living pump of intricate construction. It is suspended in the *pericardial cavity* which is part of the coelom. The luminal lining of the heart is the *endocardium* (homologue of the tunica interna), its muscular component is the *myocardium* (homologue of the tunica media), and the outer lining which borders on the pericardial cavity is the *epicardium* (homologue of the tunica externa). The heart may be considered a very large, specialized blood vessel with the coats of the vascular wall continued into the heart (Fig. 9-34). The parenchymal elements of the heart, i.e., the myocardium, receive support from dense connective tissue structures collectively known as the *cardiac skeleton*. The components of

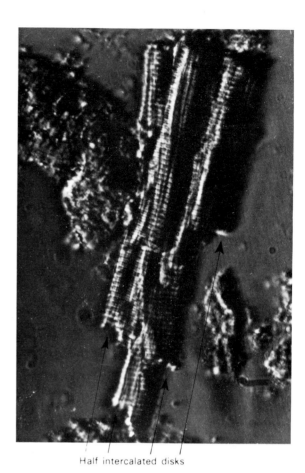

Fig. 9-36. Cardiac muscle cell isolated by trypsin. Courtesy of Dr. Joachim R. Sommer, V. A. Hospital, Durham, N. C.

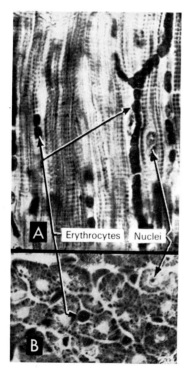

Fig. 9-35. A: Longitudinal section of myocardium. Erythrocytes appear black. B: Transverse section of myocardium. Iron hematoxylin.

the cardiac skeleton are the *annuli fibrosi, trigona fibrosa* and the *septum membranaceum*. The annuli fibrosi are fibrous rings of dense connective tissue encircling the two atrioventricular openings, and the pulmonary trunk. The trunk of the aorta is situated between the three annuli fibrosi. The fibrous ring of the aorta is composed of three portions: (1) a left fibrous trigone (between the bicuspid orifice and the aorta), (2) the right fibrous trigone (between the two atrioventricular rings and the aorta, and (3) the thickened conus ligament, between the pulmonary trunk and the aorta. The remainder of the cardiac skeleton is formed by the septum membranaceum, an extension of the right fibrous trigone descending into the interventricular septum. The cardiac skeleton forms the major attachment sites for cardiac myofibers.

Epicardium

The epicardium (*visceral pericardium*) invests the heart. Its connective tissue layer is continuous with both the tunica externa of the great vessels and with the tela subserosa of the *parietal pericardium*. The epicardium is a serous membrane whose mesothelium is continuous with that of the parietal pericardium. The tela subepicardiaca consists of areolar connective tissue containing abundant adipocytes (Fig. 9-44), especially around the blood vessels, and harbors the intramural parasympathetic ganglia of the heart.

The slight amount of albuminous fluid (pericardial fluid) within the pericardial cavity, between the visceral and parietal layers of the smooth serous membrane, facilitates frictionless motion. In cases of inflammation (pericarditis, which can be due to trauma, cancer, tuberculosis, uremia, etc.), the motions of the heart may become painful. The pericardial cavity can be filled with fibrin, and adhesions between pericardium and epicardium may ensue.

Myocardium

The myocardium, (c.f., Chapter 6) continuous with the tunica media of the blood vessels, is a three-dimensional network of short, slightly branched, cardiac myocytes.

Capillaries parallel the muscle cells (Fig. 9-35) in a delicate connective tissue stroma supporting each cardiac myofiber.

Fig. 9-37. Fatty degeneration of individual cardiac muscle cell after ligation of a small branch of the coronary artery of a dog. Courtesy of Drs. Yokoyama, Jennings and Wartman, Northwestern University. Hematoxylin and Sudan III.

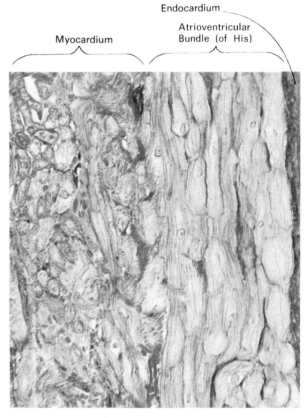

Fig. 9-38. Longitudinal section of bundle of His in the left ventricle. Azan, 420 ×.

Fig. 9-39. Conduction myofiber (Purkinje) of the atrioventricular bundle on the interventricular septum, Azan, 820 ×.

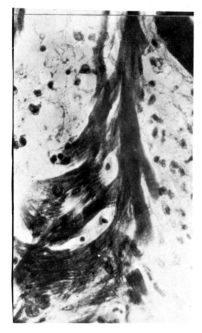

Fig. 9-40. One of the fringed endings of the bundle of His as seen in a tangential section of subendocardium.

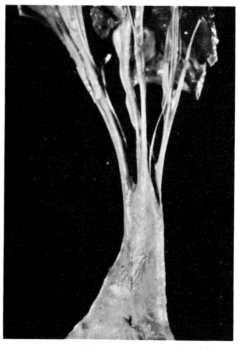

Fig. 9-41. Papillary muscle and tendinous cords.

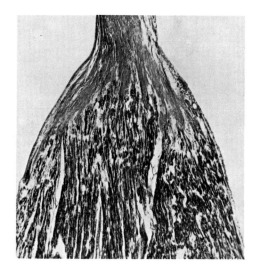

Fig. 9-42. Papillary muscle and tendinous cord.

dog. Shortly after ligation, individual myocytes begin to degenerate (Fig. 9-37). In tissue cultures of fetal mouse heart, individual cardiac myocytes pulsate, each with its own rhythm, provided they do not touch. Upon contact, a myofiber forms and synchronous or wave-like contractions begin. This is possible because cardiac myocytes must possess some moderate impulse conducting ability.

Since the myocardium consists of individual myocytes, each endowed with the ability of autonomous, rhythmic contraction, it is indeed necessary that their activity be coordinated. This is accomplished effectively by the *cardiac conduction system* an impulse-conducting system of the heart consisting of the sinuatrial (S-A)

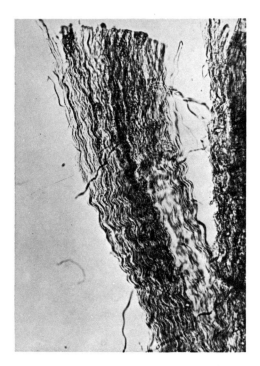

Fig. 9-43. Chorda tendinea, teased.

Cardiac myocytes can be isolated by teasing or by trypsin digestion (Fig. 9-36). Phase microscopy reveals myofibrils, A-bands, Z-lines, and the scalloped cell boundary which forms one-half of an intercalated disk. The individuality of cardiac myocytes can be confirmed by ligation of a branch of a coronary artery in a

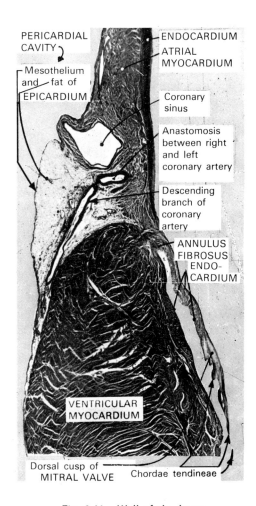

Fig. 9-44. Wall of the heart.

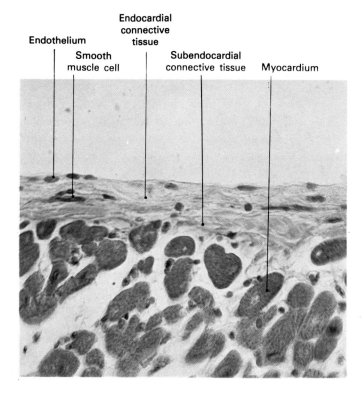

Fig. 9-45. Endocardium, Goldner stain. 750 ×.

node, atrioventricular (A-V) node and the atrioventricular bundle. The S-A node is located in the subepicardial connective tissue where the superior vena cava joins the right atrium. The initiation of the stimulus for contraction occurs in the S-A node, which passes an excitation wave over both atria to the A-V node. The A-V node resides in the tela subendocardialis (connective tissue) of the right atrium near the termination of the coronary sinus. Special nodal myofibers of the A-V node leave as cardiac conduction myofibers (Fig. 9-38). The large conducting myocytes are characterized by a few myofibrils located around the periphery, more cytoplasm, and more glycogen than regular cardiac myocytes (Fig. 9-39). The A-V bundle terminates (Fig. 9-40) by contacting progressively smaller cardiac conduction myocytes until there is little different between conduction myocytes and cardiac myocytes of the ventricles. The conduction fibers' rate of impulse conductions is greater than that of the regular cardiac myofibers. The smaller conduction myofibers have been observed to contract and yet appear like specialized conducting myocytes.

The heart walls and the greater part of the *interventricular septum* consist mainly of myocardium. The luminal surfaces of the ventricles are recognized by irregular muscular ridges and bridges termed *trabeculae carneae*. In addition to their role in myocardial contraction, they serve to shunt cardiac impulses quickly across the ventricle.

A papillary muscle (Figs. 9-41, bottom and 9-42) is a conical projection of myocardium connected to one or more *chordae tendineae* (Figs. 9-41 and 9-42). The *tendinous cords* attached to the cusps of the atrioventricular valves consist mainly of strong, wavy, collagen fibers (Fig. 9-43).

The *atrioventricular valves* are re-duplications of endocardium covering a dense core of collagen fibers continuous with the cardiac skeleton (Fig. 9-44) and isolated smooth myocytes.

Endocardium

The endocardium (Fig. 9-45) lining of the heart, is a continuation of the tunica interna of blood vessels. It consists of an endothelium supported by loose connective tissue of the *stratum subendotheliale*. This connective tissue layer contains some smooth myocytes which adapt the endocardium to the size of the chambers and keep the internal surfaces of the heart smooth during systole. Endocardium lines the entire heart inside, so that it covers the myocardium, trabeculae carneae, chordae tendineae, valves and the membranous septum. The endocardium is bound to the myocardium by the layer of subendocardial connective tissue. Cardiac conduction myofibers are located in the subendocardial layer.

10...
Defense system

The name "defense system" designates collectively all organs, cells and tissues which aid in the defense against foreign substances termed antigens. The following belong to the defense system: cells capable of phagocytosis, cells which produce humoral and cellular immunity, lymph vessels, sinusoids, lymphoid tissues and lymphoid organs.

Usually the word *lymphatic* refers to conduits for lymph, while *lymphoid* is preferred for tissues and parts of organs which harbor and produce lymphocytes. Often both words are used synonymously.

Reticuloendothelial System

The reticuloendothelial system (Fig. 10-1) is an integral part of the defense system, and, like the defense system as a whole, is scattered throughout the body. Components of the reticuloendothelial system (RE system) are encountered in connective tissues, all lymphoid structures and the liver. The RE system is composed of the *phagocytic macrophaqes, reticular cells* and *stellate macrophages* (Kupffer cell). Two forms are encountered, the *fixed macrophage* (histiocyte) and the *nomadic* or *wandering macrophage*. The nomenclature

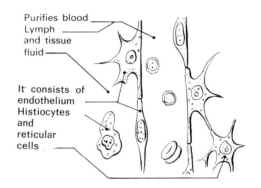

Fig. 10-1. Components of the reticuloendothelial system.

committee elected to limit the use of reticular cell to the stem cell of reticular connective tissue. Thus, reticular cells attached to reticular fibers together constitute the stroma of lymphoid tissue and organs. Reticular cells are structurally intermediate between mesenchyme and epithelial cells. Large vesicular nuclei (very little heterochromatin and large amounts of euchromatin) possess at least one prominent nucleolus. Stromal reticular cells in lymph nodes and spleen originate from mesoderm, while those of the thymus (the *epithelioreticular* cells) are derived from endoderm of the third and fourth pharyngeal pouches.

Endothelial cells line the venous sinuses

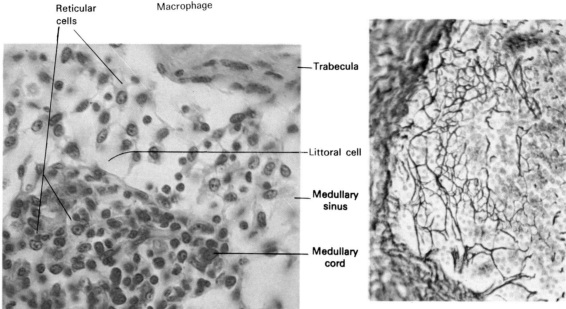

Fig. 10-2. Reticular cells in a medullary sinus of a lymph node, Azan. 940 ×.

Fig. 10-3. Reticular fibers in the subcapsular sinus of a lymph node, Foot silver impregnation 640 ×.

of the spleen, sinusoids of lymphoid tissue, liver and bone marrow. The structural modifications seen between neighboring endothelial cells in the RE system (e.g., continuous, fenestrated and discontinuous endothelium) reflect their functional role.

For example, the endothelium of the hepatic sinusoids has: (1) gaps between processes of adjacent endothelial cells, (2) fenestrae without diaphragms and (3) a discontinous basal lamina. These modifications ease passage of blood-borne agents into the liver perisinusoidal spaces where they are engulfed by stellate macrophages (Kupffer).

Plasma Cells

Plasma cells are distributed in all connective tissues (see Chapter 3). Quite numerous in the lamina propria mucosae of the stomach, these cells produce circulating or humoral antibodies. Since they possess no phagocytic abilities, they are not considered members of the RE system.

Plasma cells are characterized by an eccentric nucleus which has a clock-face or cartwheel chromatin pattern. Ultrastructurally they show extensive development of the granular endoplasmic reticulum and the golgi complex.

TISSUES AND ORGANS OF THE DEFENSE SYSTEM

Reticular tissue is the chief supporting tissue of the defense system.

It is composed of a network of reticular cells (Fig. 10-2) and reticular fibers (Fig. 10-3). Unfortunately, it is practically impossible to stain reticular cells and reticular fibers on the same slide. The functional components of the defense system are: diffuse lymphoid tissue, solitary lymph nodules, aggregated lymph nodules (Peyer's patches), lymph nodes, spleen and the lymphoepithelial organs (tonsils, thymus, and appendix).

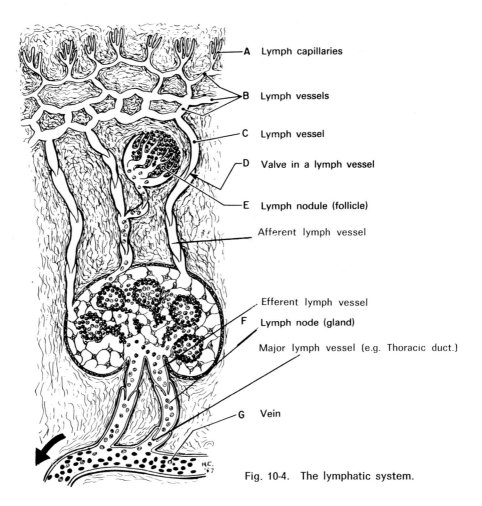

Fig. 10-4. The lymphatic system.

Lymphatic and Lymphoid System

The lymphatic system (summarized in Fig. 10-4) is a system of vessels and nodes in which lymph streams from the periphery toward larger lymphatic vessels and ultimately to the greater veins. It begins with numerous, blind-ending *lymph capillaries* (A) which take up excessive tissue fluid, the *lymph*. Since some lymphocytes recirculate between blood, tissue fluid and lymph, these cells may even occur in the terminal lymph capillaries. Lymph flows slowly through vessels of increasing caliber (B,C), many of which are provided with valves (D). Lymphocytes produced in solitary lymph nodules (E), are added to the lymph which is filtered through lymph nodes (F), where more lymphocytes and antibodies are added. In the diagram (Fig. 10-4), efferent vessels are drawn as if they empty into a vein. Please understand this to signify that the efferent lymphatic vessels join progressively larger lymphatic vessels which ultimately empty into veins (G). All lymphatic vessels ultimately terminate in two large lymphatic vesaels, the thoracic duct and the right lymphatic duct. The right lymphatic duct collects lymph from the upper right portion of the body (half the head and neck, and the right upper extremity) and empties into the junction of the right internal jugular and subclavian veins. The thoracic duct, the largest lymphatic vessel in the body, receives blood from the remainder of the body and empties at the junction of the left internal jugular and subclavian veins.

Figure 10-5 shows how lymph is produced and removed. An artery (1) supplies

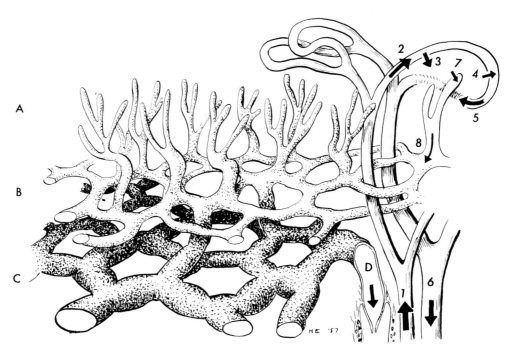

Fig. 10-5. Origin of lymph.

blood to a region of the body, and distributes it through blood capillaries (2). Diffusion takes place through the endothelial walls through which not only oxygen and carbon dioxide pass, but fluid components of the blood as well. In this manner, nutritive substances and hormones (as solutes) reach the tissues. Diffusion of fluid and solutes out of a blood capillary, is shown by arrow number 3. Some of the tissue fluid laden with waste products returns into the blood capillaries by diffusion (4) and streams (5) back into the venous return (6). Since the effect of arteriolar hydrostatic pressure supplied by cardiac systole (3) is greater than the effect of venous oncotic pressure (4), more fluid exits the blood vascular system than can return.

Continuous accumulation of interstitial fluid could lead to *edema* (swelling) if the lymphatic system did not provide a means for the return of tissue fluid into the blood stream (see Chapter 9). Excess fluid enters into the lymph capillaries (7) and henceforth is called *lymph*. From the lymph capillaries (A) lymph flows (8) into a peripheral or superficial plexus (B) of delicate lymph collecting vessels. From there it drains into a deep plexus (C) of coarser lymph vessels, which drains through

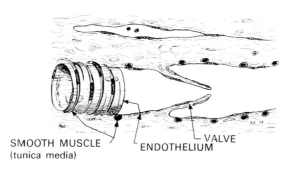

Fig. 10-6. Typical lymph vessels.

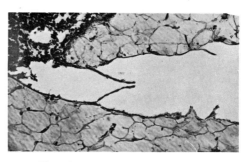

Fig. 10-7. A small lymph vessel.

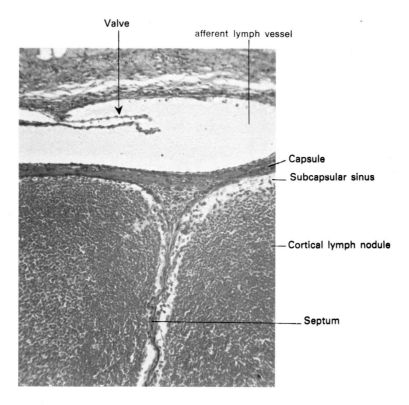

Fig. 10-8. Afferent lymph-vessel to a lymph node, Azan. 185 ×.

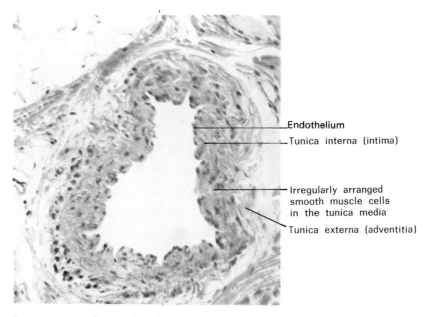

Fig. 10-9. Lymph vessel in the axilla, Goldner stain. 335 ×.

lymph vessels (D) of ever increasing caliber.

LYMPH VESSELS. Lymph vessels (Figs. 9-21, A, 10-6, 10-7 and 10-8) are constructed according to the same histological plan as blood vessels; but, except for the endothelium, the layers of their walls are usually much thinner than in veins. Lymph vessels, with a weak tunica media, are often collapsed (Fig. 10-6, upper right) and practically invisible. The feeble tunica media of large lymph vessels (Fig. 10-6, lower left) initiates propulsion of lymph through the valves by slow rhythmic contractions. Assisting in this process are respiratory movements, and the massaging effect of surrounding muscles. While most lymph vessels possess very thin walls (Figs. 9-21, A and 10-8), there are some mainly in the extremities, the axilla and the spermatic cord, whose walls are so thick that it is possible to confuse them with veins or even arteries (Fig. 10-9). The walls of such lymph vessels consist mainly of dense, fibrous connective tissue. Smooth muscles, if present, are arranged irregularly in obliquely criss-crossing bundles within this connective tissue.

The thoracic duct (Fig. 10-10) possesses longitudinal smooth muscle in the intima. It has a delicate, incomplete internal elastic membrane, and a moderately developed tunica media of circularly arranged smooth myocytes.

Lymph vessels can be demonstrated by injection of plastic materials. Figure 10-11 gives examples of such preparations.

DIFFUSE LYMPHOID TISSUE. Diffuse lymphoid tissue (Fig. 10-12) is, in essence, lymphocytes suspended in a stroma of reticular fibers and reticular cells. It is widely distributed in mucous membranes, particularly in those of the gastrointestinal (gut associated lymphoid tissue [GALT]) and respiratory (bronchiolar associated lymphoid tissue [BALT]) systems. In the diffuse lymphoid tissue of the lamina propria mucosae of the gastrointestinal tract, plasma cells, macrophages and fibroblasts may outnumber the lymphocytes; and acidophil granulocytes are numerous.

SOLITARY LYMPH NODULES. Solitary lymph nodules (lymph follicles) (Figs. 10-13, 10-14 and 12-1) are rather sharply defined spheroids of dense, lymphoid tissue (the *corona*), often possessing a less dense

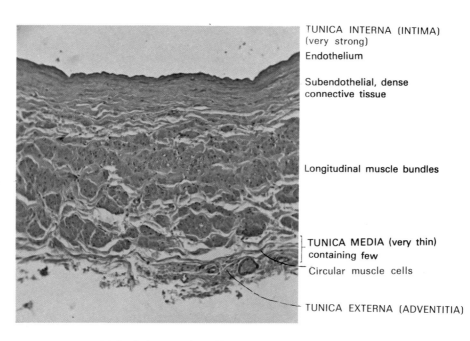

Fig. 10-10. Wall of thoracic duct, H&E. 185 ×.

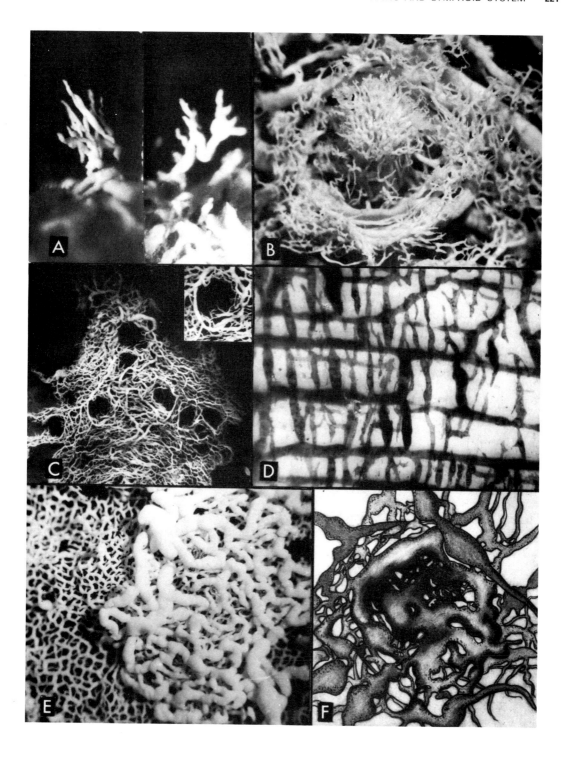

Fig. 10-11. Latex-injected lymph vessels. A, In filiform papillae. B, In and around circumvallate papilla. C, Deep lymphatics of tongue (posterior third). D, In diaphragm. E, Superficial (left) and deep (right) plexus in the colon of the cat. F, Around solitary nodule in the colon of three year old boy. (Preparation and photography by Dr. G. Ottaviani and co-workers: Ateneo Parmense, *23*, Suppl. and *26*, Suppl. 1952 and 1954.)

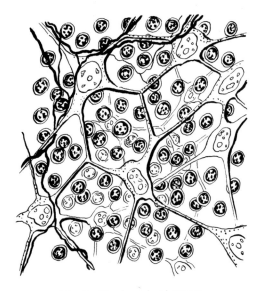

Fig. 10-12. Lymphoid tissue.

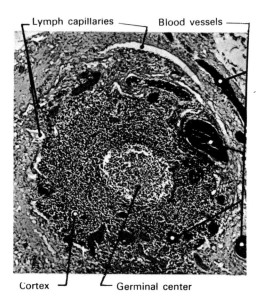

Fig. 10-13. Lymph nodule from uvula.

light-staining central region, the *"germinal center"*.

Many solitary lymph nodules are found in the mucous membrane of the intestine (Figs. 10-14 and 10-15). The germinal center of a lymph nodule cannot be seen in sections that do not pass through the middle of the lymph nodule.

Germinal centers are not present until after birth; but during the childhood years, when many infections are encountered, they are very prominent. Animals born and raised in germ-free environments, do not develop germinal centers.

Germinal centers contain stellate reticular cells fixed to reticular fibers, and free macrophages or reticular cells that have moved away from their reticular fibers. A germinal center is usually surrounded by a dark corona of small lymphocytes, often arranged in concentric layers (Fig. 10-14). Figure 10-15 shows the distribution of lymph nodules and vessels in the wall of the large intestine.

AGGREGATED LYMPH NODULES: Aggregated lymph nodules are lymph nodules collectively embedded in masses of diffuse lymphoid tissue. They occur in tonsils and in the ileum; at the latter location they were called *Peyer's patches* (Fig. 10-24, A).

LYMPH NODES. Lymph nodes (lymph glands) (Figs. 10-4, 10-16, 10-17 and 10-18, also 10-2, 10-3 and 10-18) are encapsulated lymphoid organs through which lymph slowly streams. In its typical form, a lymph node has roughly the shape of a bean. It is surrounded by a dense, fibrous *capsule* which blends with the surrounding connective tissue. *Trabeculae* of

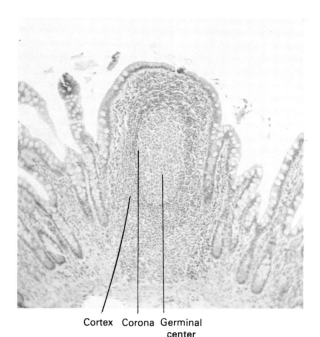

Fig 10-14 Lymph nodule in an intestinal villus, H&E. 180 ×.

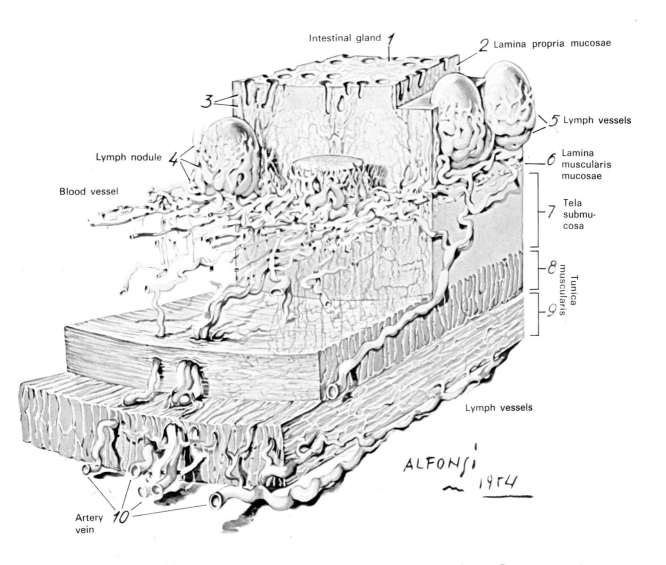

Fig. 10-15. Lymph nodules and lymph vessels in the colon (from Ottaviani, Ateneo Parmense, 1954).

dense, fibrous tissue arising from the capsule support the organ internally. A slight depression in the capsule is the *hilus*. The organ is pervaded by reticular connective tissue. The meshes of this reticulum are filled with densely packed lymphocytes, except for the sinuses. Thus, the lymph node possesses a highly cellular cortex and a moderately cellular medulla.

The *cortex* contains typical *lymph nodules* and between them a diffuse zone of lymphocytes (termed the *paracortical region*) (Fig. 10-17). Thick cords of densely packed lymphocytes, the *medullary cords*, extend from the cortex into the *medulla*. The rest of the organ is occupied by *sinuses* bridged by reticular cells and fibers with meshes wide enough to permit flow of lymph.(Fig. 10-2). Immmediately under the capsule is the *subcapsular sinus* (Figs. 10-8 and 10-17). This is continuous with the paratrabecular sinuses running along the trabeculae and themselves continuous with the very wide *medullary sinuses*. Lymph enters the lymph node through several *afferent lymph vessels* (Figs. 10-8 and 10-17) which pierce the capsule to enter into the subcapsular sinus. Lymph passes from the subcapsular sinus to the paratrabecular sinuses and then to the medullary sinuses. Several *efferent lymph vessels* emerge from the hilus (Fig. 10-16).

224 DEFENSE SYSTEM

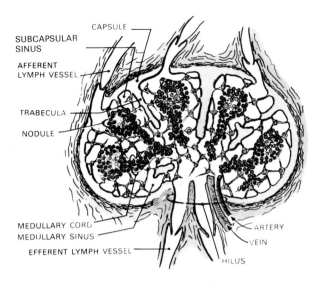

Fig. 10-16. Diagram of a lymph node.

The sinuses of a lymph node are lined by endothelial cells. Several examples are shown in Figures 10-1 and 10-19. Some reticular cells bridge the sinuses. This arrangement provides for a slowing and redirection of lymph flow, thereby increasing the percolation of lymph through the node.

As the lymph flows through the sinuses, more lymphocytes are added to it. One such lymphocyte is seen penetrating, by diapedesis, through the layer of endothelial cells in Figure 10-19.

The lymph node is supplied with blood by arteries and drained by veins (Figs. 10-16 and 9-21, B). These pass through the

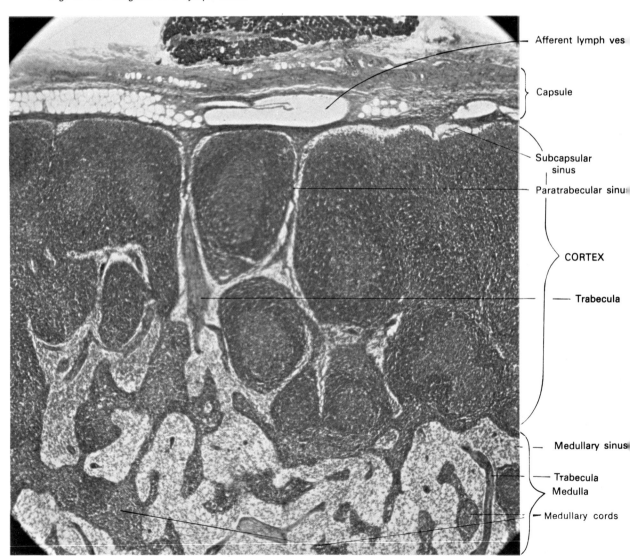

Fig. 10-17. Peripheral portion of a lymph node, Azan. 100 ×.

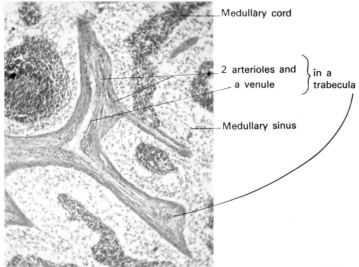

Fig. 10-18. Medulla of lymph node, Azan. 160 ×.

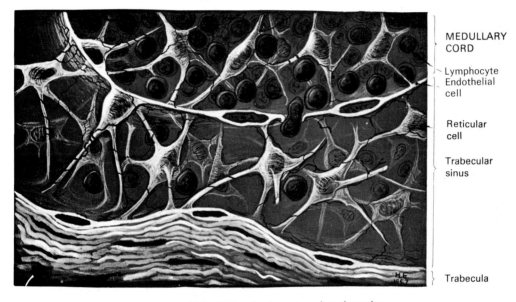

Fig. 10-19. A sinus in a lymph node.

Fig. 10-20. Surface view of lingual tonsil.

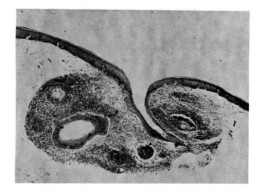

Fig. 10-21. Crypt of lingual tonsil with lymphoid tissue.

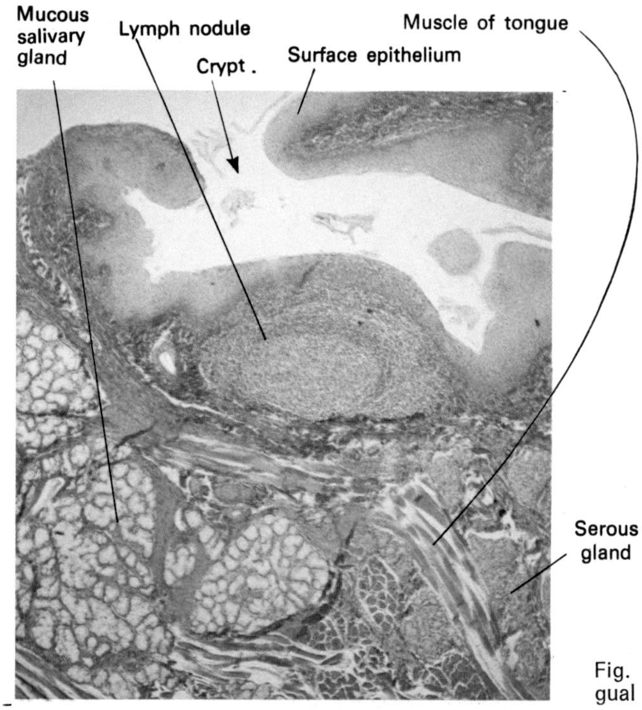

Fig. 10-22. A small portion of a lingual tonsil, H&E. 55 ×.

hilus and break up into capillaries within the lymph node without, however, communicating with its sinuses.

In young individuals, most lymph nodes exhibit the architecture just described. With advancing age, the lymph nodes become more and more amorphous masses of dense, lymphoid tissue.

Lympho-endothelial Organs

TONSILS. A tonsil is a mass of aggregated lymph nodules and lymphoid tissue closely associated with an epithelium. Tonsils are characterized by folds or depressions (*crypts* in the surface epithelium (Figs. 10-20, 10-21 and 10-22, lingual tonsil), around which *aggregated lymph nodules* are grouped. Lymphocytes penetrate the epithelium disrupting much of the epithelium (Figs. 10-23 10-24 and 10-25) which is, however, continuously regenerated. After emerging from the epithelium, these lymphocytes, are called *salivary corpuscles*, as soon as they appear on the surface. The *pharyngeal tonsil (adenoid)* is covered with pseudo-stratified, ciliated, columnar epithelium (Fig. 10-25). Folds rather than crypts characterize the pharyngeal tonsilar mucosa. The *lingual* and *palatine tonsils* (Figs. 10-21, 10-22, 10-23 and

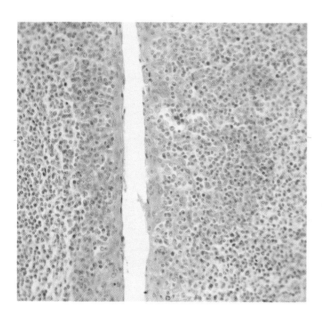

Fig. 10-23. A crypt of a lingual tonsil lined by stratified squamous epithelium invaded and partially digested by lymphocytes, H&E. 375 ×.

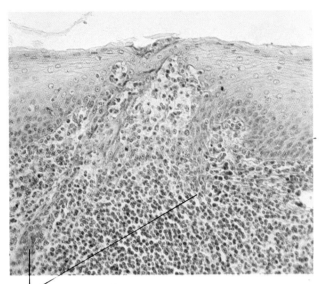

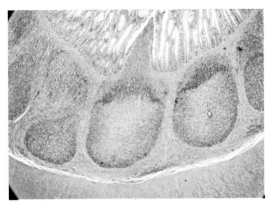

Fig. 10-24. A. Aggregated lymph nodules (Peyer's patch) in ileum, a lymphoepithelial organ. 30 ×.

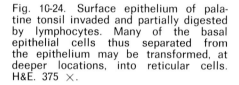

Basal cells separated from the epithelium, perhaps to be transformed into reticular cells

Fig. 10-24. Surface epithelium of palatine tonsil invaded and partially digested by lymphocytes. Many of the basal epithelial cells thus separated from the epithelium may be transformed, at deeper locations, into reticular cells. H&E. 375 ×.

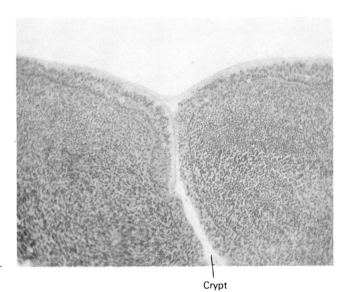

Fig. 10-25. Pharyngeal tonsil, Azan. 190×.

Crypt

10-24) are covered by stratified, squamous epithelium. Minor tonsils may occur in the auditory tube, in the crypts of foliate papillae, and in the laryngeal ventricle (Figs. 12:11, G and 13-14).

ILEUM and VERMIFORM APPENDIX. The ileum and the vermiform appendix may be considered as lympho-epithelial organs, since they harbor aggregated lymph nodules deep to their epithelium. They are sometimes called intestinal tonsils (see Fig. 10-24, A).

THYMUS. The thymus (Figs. 10-26 and 10-27) is a lobulated, lymphoid organ surrounded by a fibrous *capsule*. Its *medulla*, lightly staining and branched, is covered by a very dense lymphoid *cortex*. The medulla contains aggregates of degenerating reticular cells, the *thymic corpuscles (Hassall's corpuscles)*, with a hyaline, keratin-

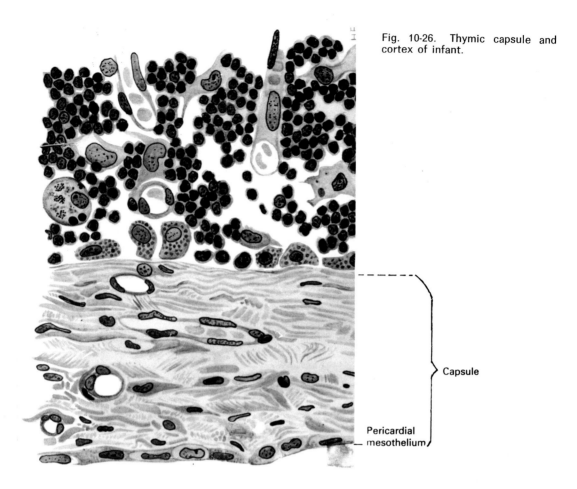

Fig. 10-26. Thymic capsule and cortex of infant.

like core (Fig. 10-28). This structure, reminiscent of "epithelial pearls", is probably due to the derivation of these cells from oral epithelium.

With advancing age the organ involutes,

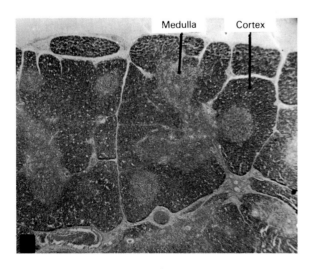

Fig. 10-27. Thymus of infant.

that is, slowly the parenchyma is replaced by loose connective tissue and fat (Fig. 10-29) and the lobulation becomes more pronounced.

The stromal cells of the thymus originate from pharyngeal endoderm in the embryo. In late fetal life, practically all the lymphocytes for the body are produced in the thymus. Many of these lymphocytes migrate to the other lymphoid organs and tissues during early perinatal life where they multiply in specific zones called *thymus dependent areas*. Thus, neonatal thymectomy results in a severe deficiency in the development of other lymphoid organs.

The thymus also produces several hormone-like substances (e.g., thymosin, thymopoetin, thymic humoral factor, etc.) capable of effecting the development of other lymphoid tissues. Lymphyocytes derived from the thymus are referred to

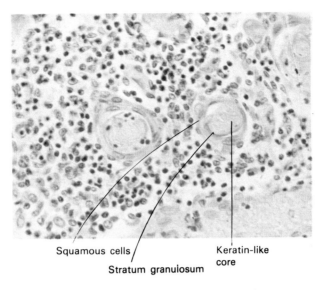

Fig. 10-28. Thymic corpuscles, H&E. 750 ×.

as "T" lymphocytes which play a primary role in cellular immune reactions (which include transplantation immunity or graft rejection mechanisms).

Lymphocytes derived from the bone marrow (in mammals) are termed "B" lymphocytes, because they are analogous to those from the bursa of Fabricius, a cloacal lymphoepithelial organ of birds (Fig. 10-30). In brief, "T" stands for thymus and "B" stands for bone marrow or bursa. "B" lymphocytes, after antigenic stimulus, may differentiate into plasma cells which synthesize and release antibody which enters the bloodstream and circulates (humoral antibody) in search of the complementary foreign antigen.

Neonatal thymectomy (experimental) or thymic aplasia (clinically known as Di George's syndrome) results in a deficient to absent development of the T-lymphocyte system. In this situation there is a loss of the delayed hypersensitivity reactions including the graft rejection mechanism. In these animals or patients, foreign organs or tissue grafts are not rejected. The humoral antibody (B-cell) system is not appreciably destroyed in these situations.

B-lymphocytes and T-lymphocytes can be found in separate zones of lymphoid tissue. For example, in the lymph node, B-lymphocytes occur primarily in the lymphoid nodule of the cortex (Fig. 10-17), in association with macrophages and a particular type of antigen-presenting cell called a follicular-dendritic cell. T-lymphocytes occur primarily in the paracortical zone of the cortex (Fig. 10-17), in association with macrophages and a particular type of antigen presenting cell called an interdigitating cell.

Spleen

In addition to performing the general functions of the defense system, the spleen breaks down erythrocytes, facilitates the differentiation of macrophages and is a reservoir for erythrocytes and thrombocytes. It is a large hemolymph gland inserted into the blood stream. It is surrounded by a fibrous *capsule* containing *smooth muscle cells*, and internally supported by *fibromuscular trabeculae*, which form a network connected with the capsule (Fig. 10-31). It is filled with reticular connective tissue and venous sinuses lined with elongated endothelial cells. Portions of the reticular

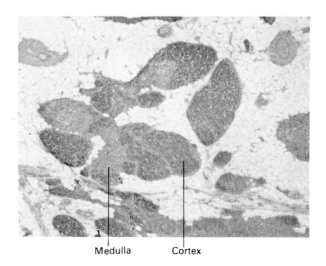

Fig. 10-29. Thymus of 60-year-old man, H&E. 55 ×.

Medulla   Cortex

tissue packed with lymphocytes are called *splenic corpuscles* which together constitute the *white pulp*. Unlike other lymph nodules, the splenic corpuscles are oblong and branching (Fig. 10-33). Those portions of the organ occupied largely by the venous sinuses and filled with erythrocytes constitute the *red pulp* (Figs. 10-32 and 10-33). *Central arteries* (Fig. 10-34) course in the branching splenic corpuscles, but shortly after emerging from the white pulp, divide into straight branches, the *penicillar arteries*. Each penicillar artery has a *penicillar arteriole*, an *ellipsoid arteriole* (sheathed by an indistinct investment of reticular cells and macrophages) (Fig. 10-35) and a *terminal capillary*. The terminal capillaries open directly into the venous sinuses (closed splenic circulation) or into the interstitial spaces (open splenic circulation). In the latter case, the arterial blood maneuvers through the red pulp by percolating through the interstitium and then passing between the fenestrations of the endothelial cells lining the sinuses. As a RBC passes, it is "inspected" by the endothelial cell; and any abnormal granule or com-

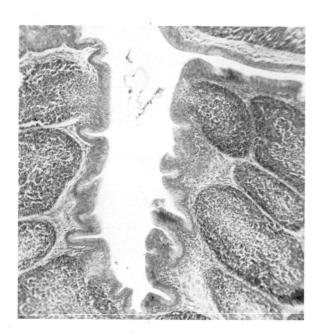

Fig. 10-30. Fabrician bursa of chick, H&E. 185 ×.

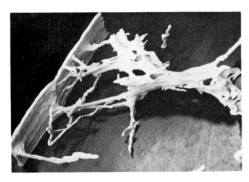

Fig. 10-31. Reconstruction from serial sections of the capsule and of some trabeculae in the spleen. (By Prof. L. Martin, University of Fribourg, Switzerland.)

filtration, the filtrates vary with the tissue, i.e., tissue fluid for lymph nodules, lymph for lymph nodes and blood for the spleen.

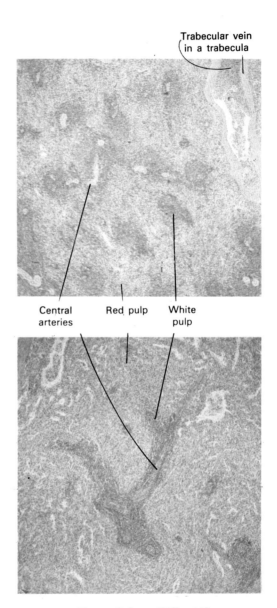

Fig. 10-32. Above: Spleen, H&E. 112 ×.

Fig. 10-33. Below: Spleen. A splenic corpuscle is cut longitudinally. Thus, one can verify the fact that it branches, H&E. 112 ×.

ponent such as a Howell-Jolly body (see Chapter 5) is removed.

Investigators still disagree about the circulation of blood through the spleen. Some subscribe to a "closed circulation theory" which states that the blood never leaves the enclosed endothelium-lined spaces. Others believe in an "open circulation" where terminal capillaries arising from the sheathed arteries open directly into the red pulp rather than into venous sinuses. Finally there are those who postulate that both open and closed circulation occur under different physiological circumstances.

The *fusiform endothelial cells* of the venous sinuses are elongated and oriented longitudinally. Reticular (argyrophilic) fibers are arranged spirally around sinuses. Electron micrographs reveal that the walls of the venous sinuses of the spleen permit passage of blood cells between the fusiform endothelial cells (Figs. 10-36, 10-37, 10-38 and 10-40). Figure 10-39 summarizes the splenic structure.

Although various lymphoid tissues have the same basic structural plan and perform

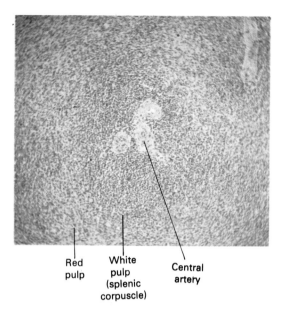

Fig. 10-34. A splenic corpuscle with a few branches of its central artery. The corpuscle is part of the white pulp. It is white in the fresh state due to the presence of numerous lymphocytes. Since their nuclei occupy most of the cells, the white pulp appears blue when stained with hematoxylin. H&E, 385 ×.

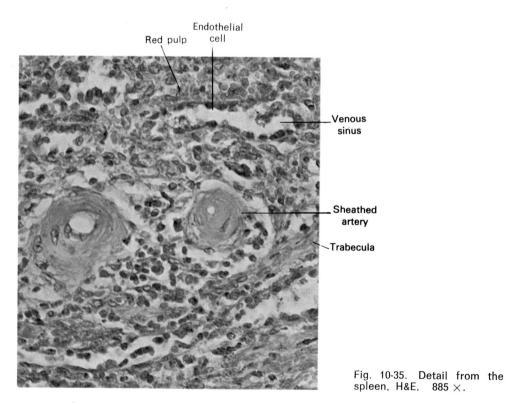

Fig. 10-35. Detail from the spleen, H&E. 885 ×.

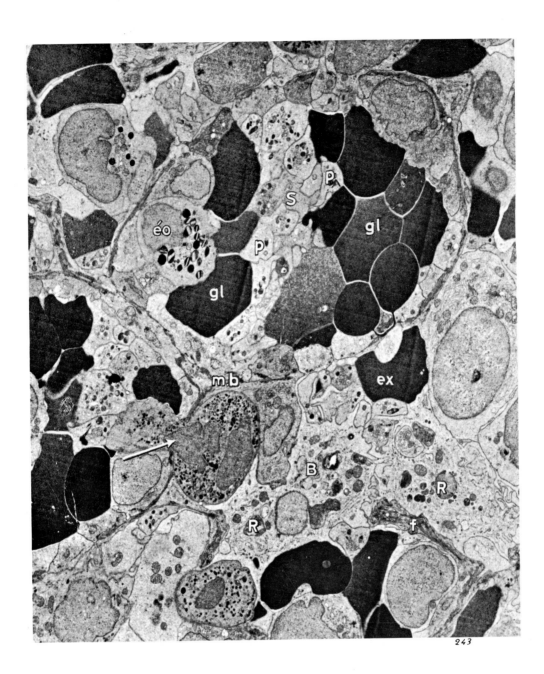

Fig. 10-36. Red pulp of spleen of mouse. B = basophil polymorphonuclear leukocyte; eo = eosinophil polymorphonuclear leukocyte; ex = extravascular erythrocyte; f = fibroblast; gl = erythrocytes; mb = basement membrane of venous sinus; p = platelets; R = reticular cells; S = space of venous sinus; *arrow* = a neutrophil polymorphonuclear leukocyte passing through the wall of a venous sinus by ameboid motion. (Electron microgram from G. Simon and R. Pictet: Acta Anat. 57: 163-171, 1964.)

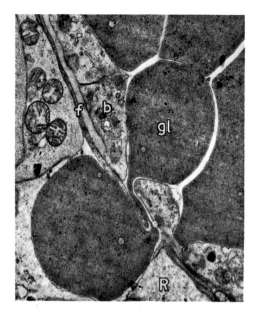

Fig. 10-37. An erythrocyte (gl) passing through the wall of a venous sinus. b = littoral cell; f = process of fibroblast; R = reticular cell. (Electron microgram from G. Simon and R. Pictet: Acta Anat. 57: 163-171, 1964.)

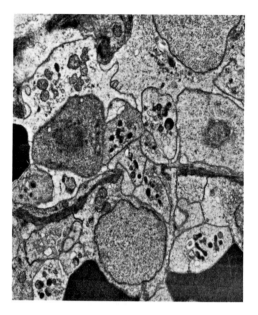

Fig. 10-38. Platelets passing through the wall of a splenic venous sinus. (Electron microgram from G. Simon and R. Pictet: Acta Anat. 57: 163-171, 1964.)

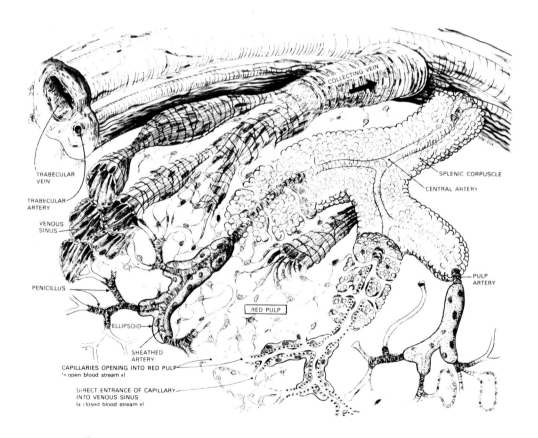

Fig. 10-39. Stereogram of the spleen. Illustration of the concepts of both closed and open circulation.

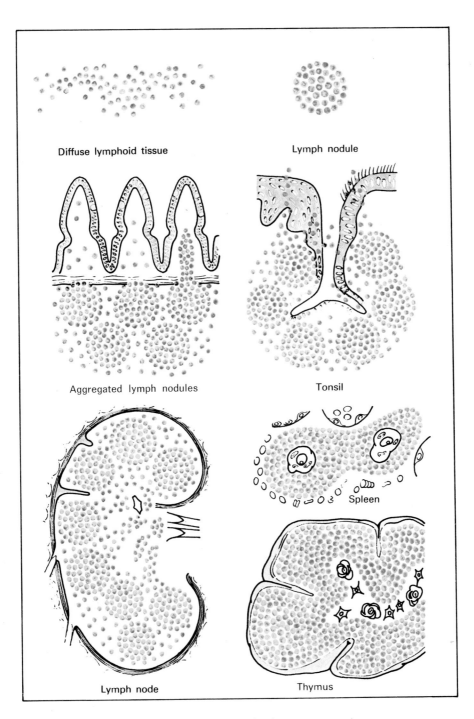

Fig. 10-39, a. Synopsis of lymphoid structures and organs.

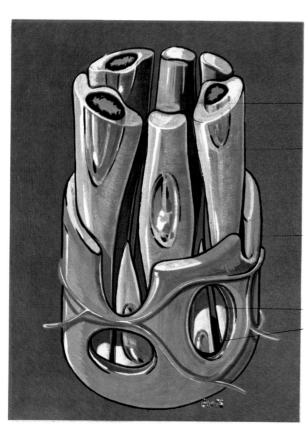

Fig. 10-40. A venous sinus of the spleen. Its wall consists of long fusiform endathelial can cells separated by spaces. The basement membrane has perforations.

- Fusiform endothelial cell
- Space through which the sinus communicates with the red pulp
- Reticular fiber
- Perforation of Basal membrane

11... Integument

The integument (Figs. 11-1, 11-2 and 11-3) is a multilayered organ which covers the body. It consists of *epidermis, dermis* and *tela subcutanea*. The integument (1) is a moisture barrier which prevents desiccation of the organism, (2) provides protection against mechanical injury and externally applied substances, (3) helps regu-

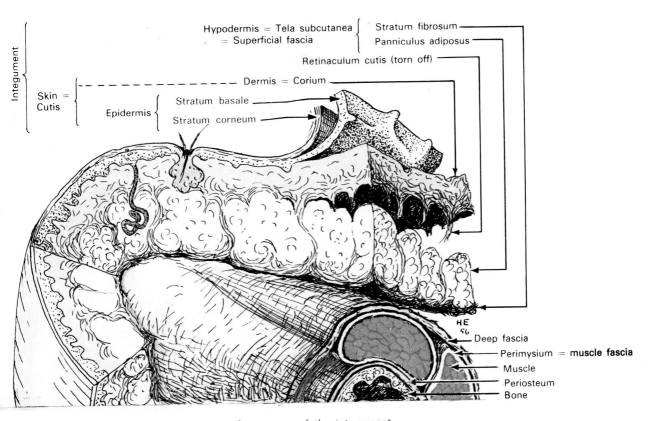

Fig. 11-1. Stereogram of the integument.

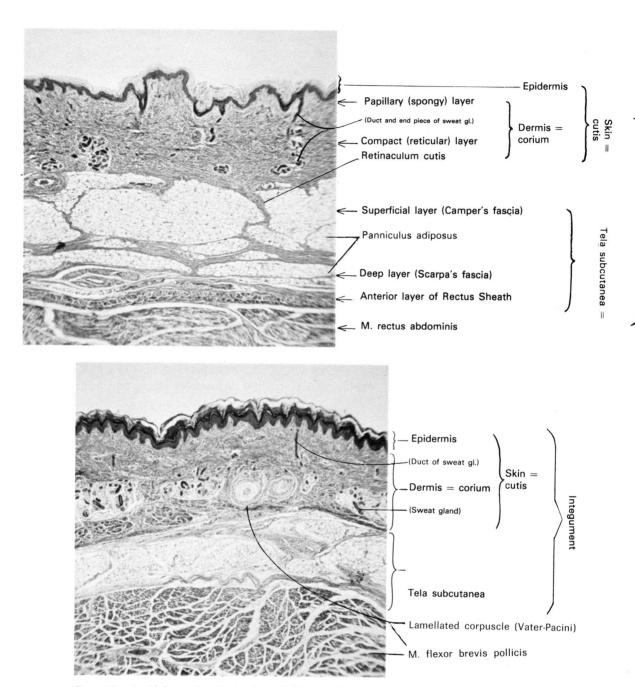

Fig. 11-2. A: Abdominal wall of infant, Goldner stain. 60 ×. B: Palmar integument of infant. 60 ×.

late body temperature, (4) excretes some waste products, (5) provides sensory perception, (6) is a site of vitamin D production and (7) protects the DNA of the organism from damage by ultraviolet radiation. The epidermis protects against penetration of hard objects, microorganisms and noxious substances. The dermis and tela subcutanea together act to (1) serve as shock absorbers, (2) distribute external pressures evenly and (3) adapt the body surface to the shapes of objects. Together with the epidermis which has adhesive qualities and provides friction, the dermis and tela subcutanea facilitate manual grasping and walking barefoot.

The cornified most external layer of the

epidermis (stratum corneum) provides an effective moisture barrier and prevents evaporation of body water. Severe burns which destroy this waterproof layer threaten the body with dehydration.

The integument helps to regulate body temperature by the following biologic mechanisms: insulation against cold is provided by hair, vasoconstriction of superficial capillaries conserves heat, vasodilatation of these vessels releases heat through the epidermis and evaporation of sweat lowers surface temperature.

A limited amount of excretory function is performed by the sudoriferous (sweat) glands. Sweat has slight bactericidal and fungistatic properties.

The integument functions as the largest and most important sensory organ, for it contains receptors for pain, temperature, touch, pressure, and two point discrimination.

Complexion is influenced by the presence or absence of *carotene* (a yellow pigment) and *melanin* (a dark-brown pigment) produced in the stellate *melanocytes* (see Chapter 3). Breakdown products of hemoglobin (hemes) may influence skin color when they circulate in the blood as a result of disturbed liver function (bilirubinemia). The skin becomes yellowish (jaundiced). Constriction and dilatation of superficial blood vessels produce pallor and blushing.

Skin

The skin (cutis) consists of an ectodermally derived stratified squamous epithelium, the *epidermis* and a dense connective tissue layer, the *dermis*, derived from mesoderm. The *retinacula cutis*, fibrous extensions of the dermis anchor the skin to the deep, fibrous layer of the *tela subcutanea*, a loose connective tissue layer. These retinacula traverse a more superficial fatty portion, the *panniculus adiposus*. Skin (epidermis + dermis) plus tela subcutanea constitute the integument (Figs. 11-1, 11-2 A,B and 11-3).

EPIDERMIS. The epidermis (Figs. 11-4, 11-9, 11;10, 11-11, 11-12, 11-13, 11-4 and 11-15) is a stratified, squamous, cornified epithelium whose *basal epithelial cells* (B) are anchored to the *basement membrane* (BM). In this stratum basale, cells adhere to one another along broad surfaces by

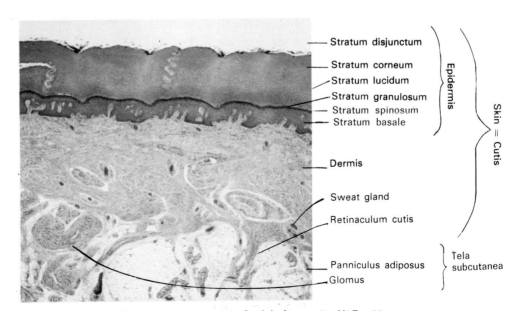

Fig. 11-3. Integument of adult finger tip, H&E. 56 ×.

desmosomes (*maculae adherentes*). In the next higher layer (stratum spinosum, S), desmosomes between neighboring cells appear to form "spines" which adhere to those of adjacent cells (formerly called "intercellular bridges"). *Tonofibrils* (which EM reveals are bundles of *tonofilaments*), occur randomly in the cytoplasm and form looping intracytoplasmic attachments with the desmosomes. They form bundles whose direction appears continuous with tonofibril bundles in the neighboring cells, creating a net-like effect throughout the stratum spinosum (Figs. 11-5 and 11-6).

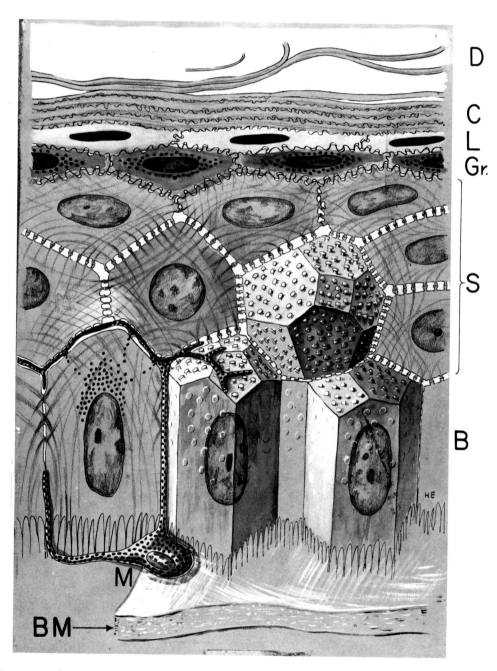

Fig. 11-4. The epidermis. BM = Basement membrane. B = Stratum basale. S = Stratum spinosum. Gr = Stratum granulosum. L = Stratum lucidum. C = Stratum corneum. D = Stratum disjunctum. M = Melanocyte.

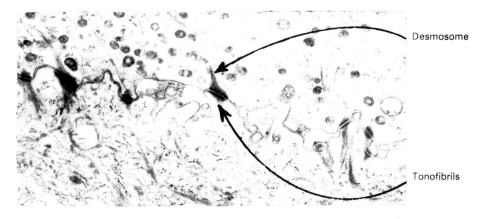

Fig. 11-5. Cellular junctions between hamster epidermal cells. Courtesy of Dr. T. Nakai, Chicago Medical school.

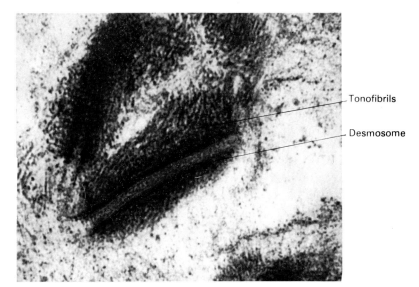

Fig. 11-6. Desmosome and tonofibrils in S. spinosum (monkey). Courtesy of Drs. Sun and White, University of Arkansas for Medical Sciences.

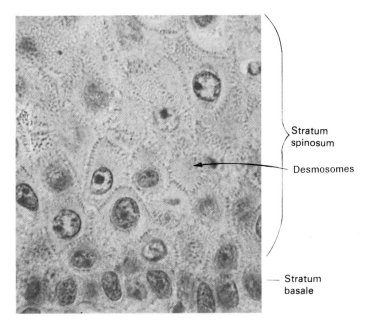

Fig. 11-7. Intercellular bridges in Stratum spinosum. 1900 ×.

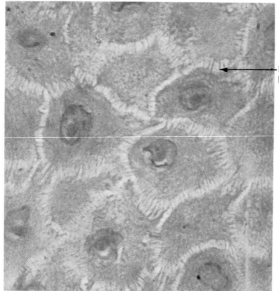

Fig. 11-8. Desmosomes (intercellular bridges) become very distinct in carcinoma of oral epithelium. H&E, 1900 ×.

Via light microscopy intercellular bridges occuring in normal stratified squamous epithelia (Fig. 11-7) become especially distinct in some forms of cancer (Fig. 11-8). Although cell multiplication normally is limited to the stratum basale, mitotic divisions also have been observed in the stratum spinosum. As a consequence of this growth process, cells are pushed toward the free surface so that the more superficial cells become flattened. As these flattened cells move farther away from the dermal capillaries, they recycle nucleic acids and synthesize lamellar granules and keratohyaline granules. The layer containing these strongly basophilic

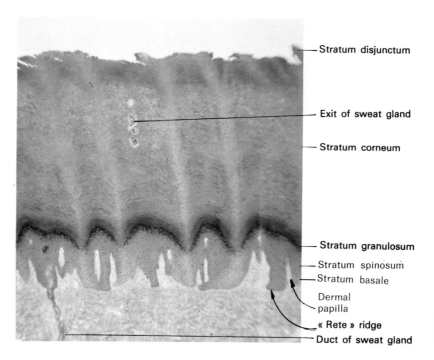

Fig. 11-9. Epidermis of finger tip, H&E. 700 ×.

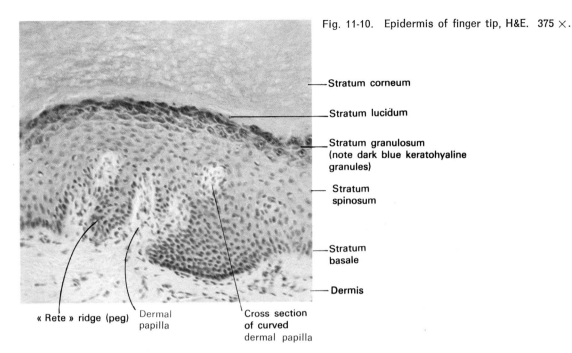

Fig. 11-10. Epidermis of finger tip, H&E. 375 ×.

- Stratum corneum
- Stratum lucidum
- Stratum granulosum (note dark blue keratohyaline granules)
- Stratum spinosum
- Stratum basale
- Dermis

« Rete » ridge (peg) Dermal papilla Cross section of curved dermal papilla

granules, is the *stratum granulosum* (Gr). The cells flatten even more during transformation into anucleate cornified cells. Sometimes an intermediate layer, the *stratum lucidum*, whose cells contain a highly refractive material, is interposed between the stratum granulosum and the stratum corneum. The stratum lucidum is limited to the thick epidermis of the palms and soles and occasionally in anal epidermis. It is not distinguishable in most other areas of the body. Cell death and shrinkage lead to further flattening and fusion with the wrinkled, dead cells of the *stratum corneum* (C). This layer consists mainly of *keratin* filaments

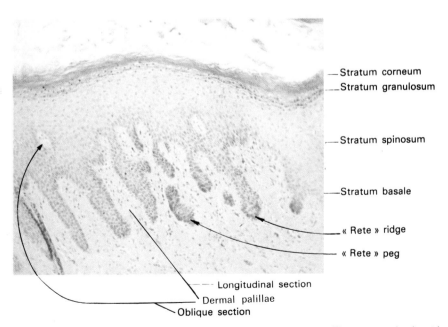

- Stratum corneum
- Stratum granulosum
- Stratum spinosum
- Stratum basale
- « Rete » ridge
- « Rete » peg

Longitudinal section
Dermal palillae
Oblique section

Fig. 11-11. Anal epidermis, H&E. 180 ×.

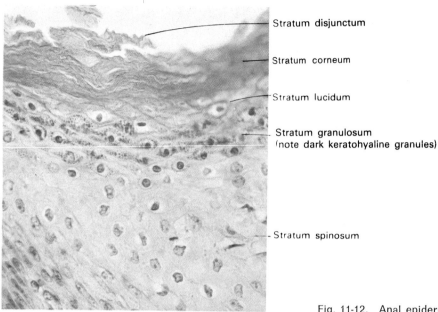

Fig. 11-12. Anal epidermis, H&E. 1000 ×.

derived from tonofilaments and an amorphous matrix. The surface layers of the stratum corneum become detached and form the *stratum disjunctum* (D).

Keratin is a fibrous, albuminoid protein. Because of its molecular structure it is relatively hard, elastic and resistant to friction. Keratin is synthetized primarily by stratified, squamous epithelia. The following is a list of locations where keratin exhibits increasing degrees of hardness: filiform papillae of tongue, stratum corneum of epidermis, hair, nails, antlers, claws and finally hoofs which consist of the hardest keratin. The hardness of tissue keratin is directly proportional to its sulfur content.

Branched, "dendritic" melanocytes (Fig.

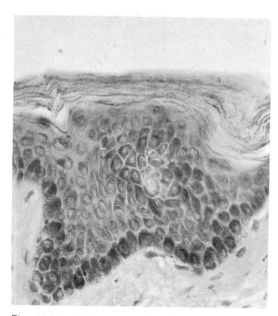

Fig. 11-13. Epidermis of axilla, pigmented, H&E. 750 ×.

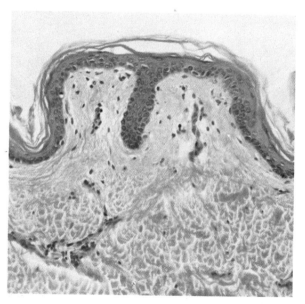

Fig. 11-14. Skin of face, H&E. 300 ×.

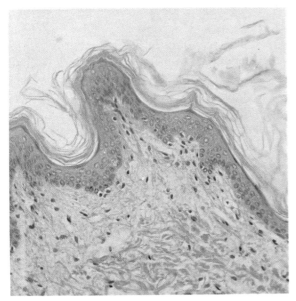

Fig. 11-15. Abdominal skin, H&E. 300 ×.

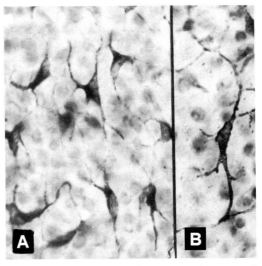

Fig. 11-16. Melanocytes from negro fetuses. A: 11 weeks; B: 12 weeks. (From skin spread flat on the slide Courtesy of Prof. A. Zimmermann, University of Illinois.)

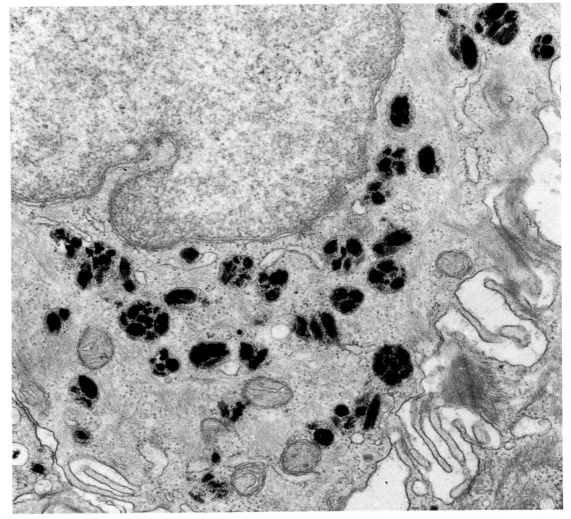

Fig. 11-17. Melanosomes (black) in an epidermal cell of the arm of a 90-year-old Caucasian man, biopsy. Courtesy of Prof. Klaus Wolf, Universität-Hautklinik, Vienna.

11-4, M and Fig. 11-16) provide pigment granules to cells of the stratum basale. When basal cells proliferate, they transport melanosomes (pigment granules) to epidermal cells of the higher layers (Figs. 11-13, 11-17, and 11-18). Figure 1-3 shows a melanocyte in its typical relation to the epidermis.

DERMIS. The dermis (Fig. 11-2, A) is a dense connective tissue layer that supports and nourishes the epidermis. It consists of a superficial, loose, *stratum papillare (papillary layer)*.

The papillary layer of the dermis forms projections, termed *dermal papillae* which are broad and low in some regions and slender and tall in others (Fig. 11-19). The papillae interdigitate with downward projections of the epidermis called *rete pegs* or *ridges* (Fig. 11-20, also seen in several previous pictures). The dermal papillae carry *intrapapillary capillary loops* (Fig.

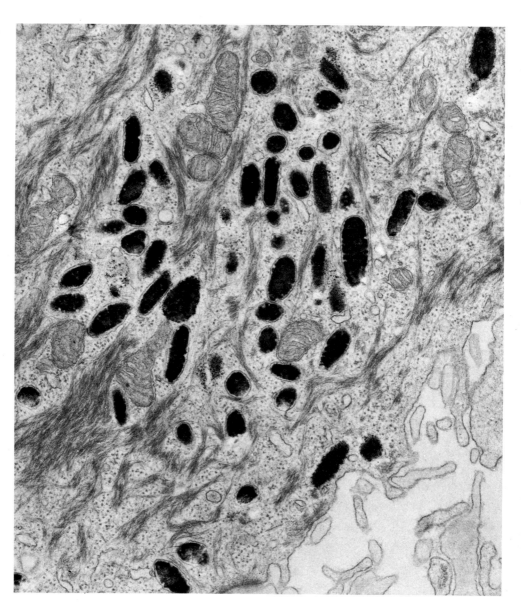

Fig. 11-18. Melanosomes and tonofibrils in epidermal cell of arm of 40-year-old negro. Biopsy. Courtesy of Dr. Klaus Wolf, Universitäts-Hautklinik, Vienna.

SKIN 247

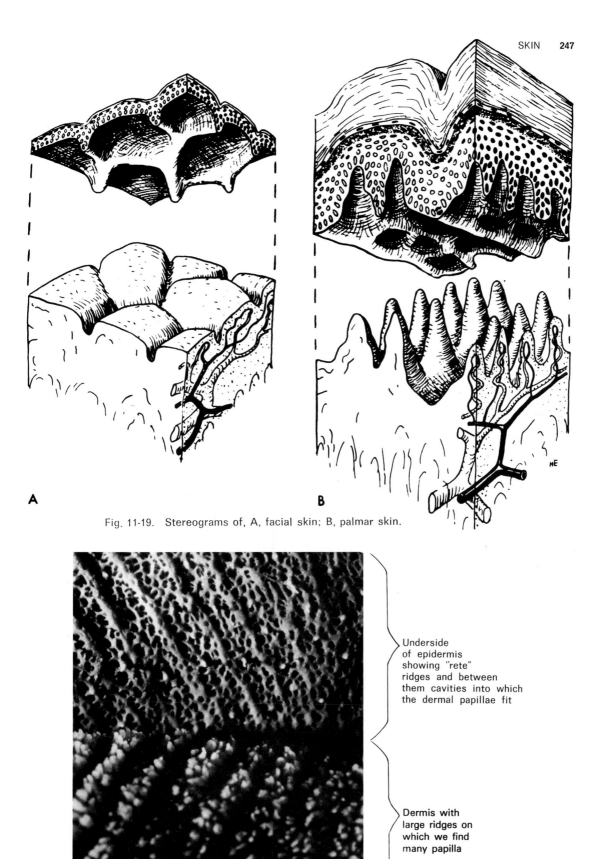

Fig. 11-19. Stereograms of, A, facial skin; B, palmar skin.

Underside of epidermis showing "rete" ridges and between them cavities into which the dermal papillae fit

Dermis with large ridges on which we find many papilla

Fig. 11-20. Epidermis (above) peeled off the dermis (below). Finger tip from an embalmed cadaver. The same effect can be achieved after boiling a beef tongue for 4 hours.

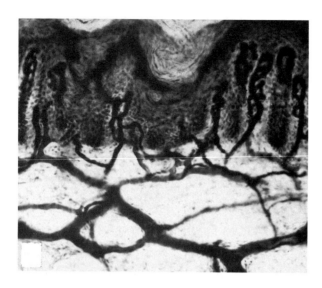

Fig. 11-21. Lead acetate injection of dermal arteries. (From A. C. Allen, The Skin, Mosby, St. Louis, 1954.)

Fig. 11-22. The deep surface of abdominal epidermis at various ages. (Reconstructions from serial sections by Dr. Allan A. Katzberg, School of Aviation Medicine, U.S.A.F.)

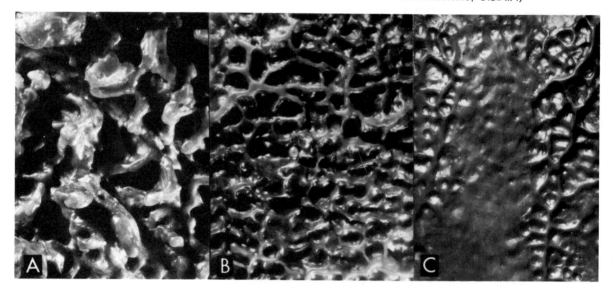

11-21) which nourish the epidermis. The dermal papillae also contain touch receptors. When the epidermis is stripped off the dermis (Fig. 11-20), one notices the papillae of the dermis (below) and their impressions into the deep surface of the epidermis (above). *Epidermal ridges* course between these impressions (see also Fig. 11-19). In the dermis some collagen fibers run parallel to the lines of tension in the skin, called *Langer's lines* by the surgeon. Surgical incisions made parallel to Langer's lines gape less and heal with less scarring than incisions made across them.

STRUCTURAL VARIATIONS OF THE SKIN. The structure of skin varies according to the pressure and friction to which it is exposed in any particular region. Facial epidermis is thin; the stratum lucidum and sometimes even the stratum corneum are absent (Figs. 11-14 and 11-19), A). In regions of this kind, thin rete ridges project downward between low, broad dermal papillae.

On the palms and soles (Figs. 11-9 and 11-19, B) the epidermis is thick, and a stratum lucidum is evident. Slender dermal papillae, arranged in irregular double or triple rows on top of dermal

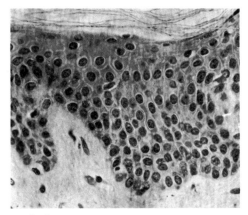

A Scalp

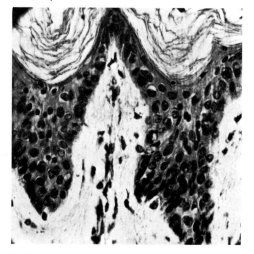

B Abdomen
Female, 37 years old.

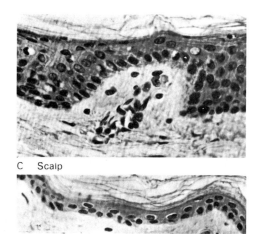

C Scalp

D Abdomen
Male, 91 years old.

Fig. 11-23. Age changes in the structure of the epidermis. In 37 year old woman, A, scalp; B, abdomen. In 91 year old man; C, scalp; D, abdomen.

ridges, fit into deep depressions of the epidermis and even produce superficial ridges. These superficial ridges form characteristic patterns responsible for finger prints, the study of which is called dermatoglyphics. These ridges can be seen in Figure 11-20.

Equally marked are variations in structure with age, as shown in Figure 11-22, A to C. In these pictures the deep surface of the epidermis is shown by means of wax-plate reconstructions. In an eight year old child (A) it has deep impressions for dermal papillae and large downward projections (rete ridges and pegs). At 55 years of age (B) the rete ridges are still impressive though much smaller and more regularly arranged. At 93 years of age (C) the deep surface of the epidermis has become almost flat. Figure 11-23, A and B, shows sections of epidermis of a 37 year old woman (A, from her scalp; B, from the abdomen). Figure 11-23, C and D, is from a 91 year old man, again representing scalp and abdominal epidermis. This flattening of the epidermis with advancing age is evident.

Tela Subcutanea

The *tela subcutanea (hypodermis, superficial fascia)* (Figs. 11-1, 11-2, A, B and 11-3) underlies the dermis. Fat, *panniculus adiposus*, is usually deposited in its superficial portion, as multicellular, rounded fat lobules. Its deep layer, the *stratum fibrosum*, unites with the dermis by broad sheets of dense, fibrous connective tissue, the *retinacula cutis*. Occasionally, the tela subcutanea may be divided into two layers as in the abdomen (the so-called Camper's and Scarpa's fasciae) (Fig. 11-2, A). The division of the panniculus adiposus (Fig. 11-24) by the flat retinacula cutis establishes a system of hydraulic pressure chambers, filled with semi-liquid fat, which act as shock absorbers. Figure 11-

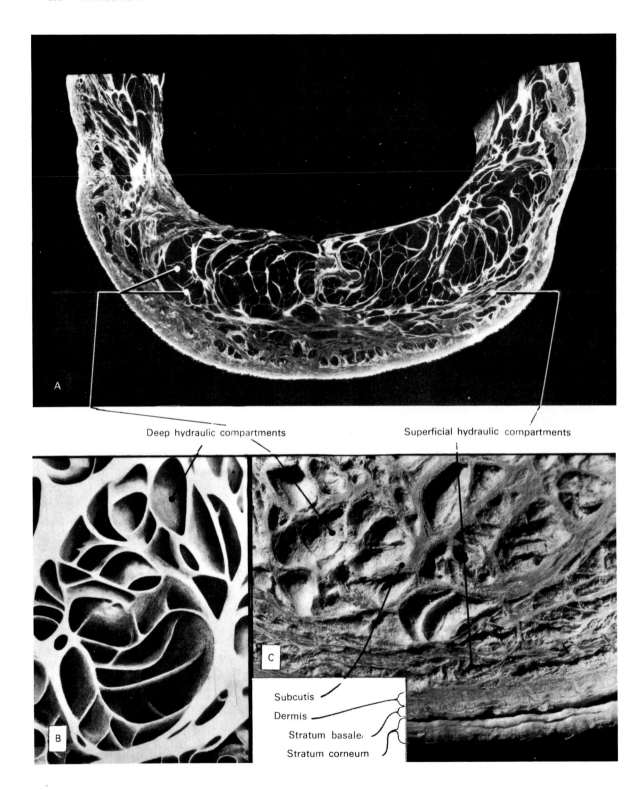

Fig. 11-24. Pressure chambers of the panniculus adiposus, in the integument of the heel. The walls between the compartments are retinacula cutis. (A and B from A. Blechschmidt: Morph. Jahrb. 73: 20-68, 1933.) C, Dry-preparation, with fat scraped out.

24, A, is a section through the integument of the heel. Figure 11-24, B, is a reconstruction of a few fatty pressure chambers of the heel. Figure 11-24, C, shows a dry preparation of the integument of the heel. The pressure compartments surrounded by thick connective tissue septa (retinacula cutis) can be visualized directly. Note the thickness of the stratum corneum.

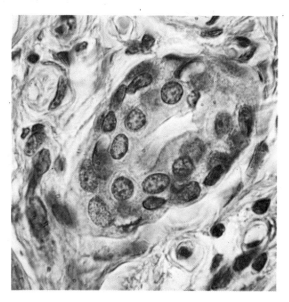

Fig. 11-26. Duct of sweat gland, Goldner stain. 1900 ×.

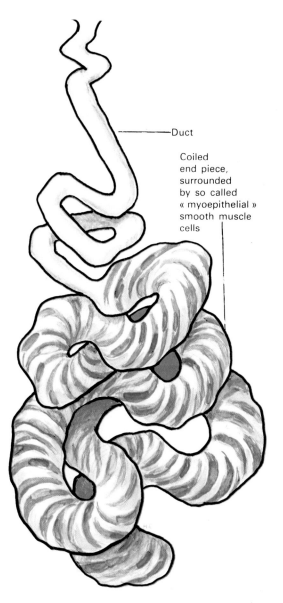

Fig. 11-25. A sweat gland surrounded by spiralling "myoepithelial" cells.

Appendages

SUDORIFEROUS (SWEAT) GLANDS. Sudoriferous glands (fig. 11-25) are simple, coiled tubular glands whose ducts pursue a cork-screw course and are lined with a double layer of cuboidal cells (Fig. 11-26). As it pursues its helical course through the epidermis, the duct of a sweat gland pierces the epidermis without retaining its own epithelial lining (Fig. 11-27, also Fig. 71-9). The coiled end pieces of sweat glands are located deep to the reticular layer of the dermis (Fig. 11-2) and consist of a single layer of secretory epithelial cells. Pressed between the epithelial cells and the basement membrane, are the *myoepithelial cells* which are arranged spirally around the glandular tubule (Figs. 11-25, 11-28, 11-29 and 11-31).

Myoepithelial cells containing myofilaments may be smooth muscle cells. Their contraction forces the secretory products from the end pieces into the duct.

Two kinds of sweat glands exist: merocrine and apocrine.

Merocrine sweat glands secrete a hypotonic fluid, which evaporates to produce

252 INTEGUMENT

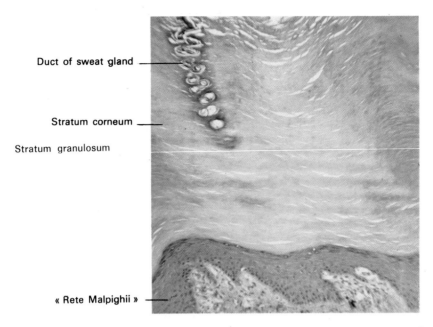

Fig. 11-27. Duct of sweat gland "piercing" the stratum corneum of palmar skin, H&E. 185 ×.

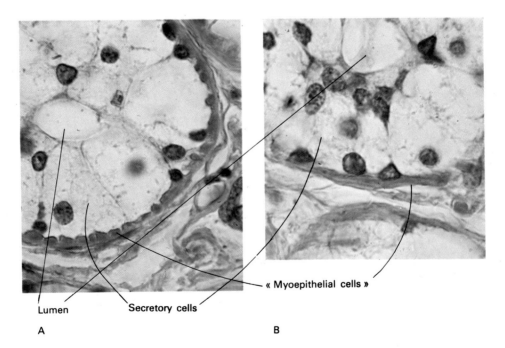

Fig. 11-28. Sections through eccrine sweat glands, Goldner stain. 1900 ×.

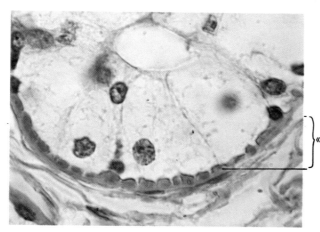

Fig. 11-28, C. Merocrine sweat gland with myoepithelial cells. Goldner stain, 1900 ×.

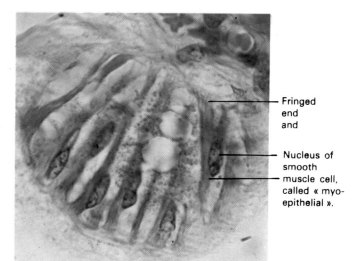

Fig. 11-29. Longitudinal sections through some socalled "myoepithelial" cells in the wall of an axillary merocrine (eccrine) sweat gland. Goldner stain, 1900 x.

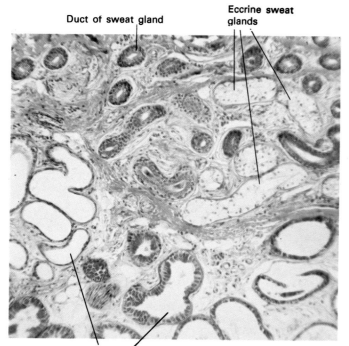

Fig. 11-30. The two kinds of sweat glands in the axilla, Goldner stain. 233 ×.

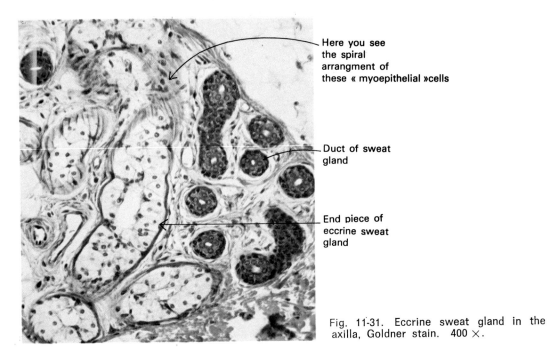

Fig. 11-31. Eccrine sweat gland in the axilla, Goldner stain. 400 ×.

cooling by a loss of body heat. The secretory cells have a clear, coarsely foamy appearance resembling mucous cells. Nuclei are round and may be located at any level in the cells (Figs. 11-31 and 11-28). Normally sweat contains about 0.3 per cent sodium chloride, 0.03 per cent urea and 0.07 per cent lactate.

Apocrine Sweat Glands. Darkly staining cells of apocrine sweat glands vary in shape from low cuboidal to high columnar

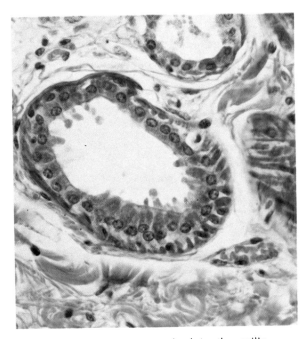

Fig. 11-32. Apocrine sweat gland in the axilla, Goldner stain. 833 ×.

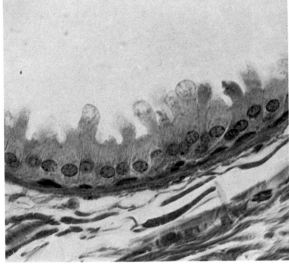

Fig. 11-33. Apocrine sweat secretion in a gland from the external auditory meatus. Goldner stain. 1900 ×.

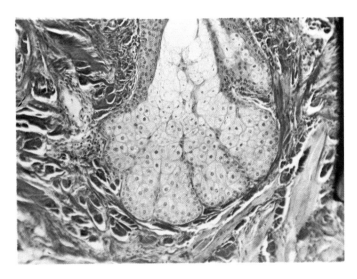

Fig. 11-34. Sebaceous gland from ala of nose, Goldner stain. 185 ×.

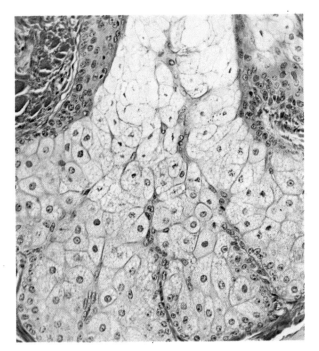

Fig. 11-35. Detail from Fig. 11-34. 375 ×.

during various phases of activity (Fig. 11-30, below) clearly evident in active tubules (Figs. 11-32 and 11-33). The glands are confined to the axilla, the areola of the mammary gland, the perianal and pubic regions, the linea alba and the nasal vestibule. In the external auditory meatus, modified apocrine sweat glands, e.g., the *ceruminous glands*, add their product to *sebum* of sebaceous glands to form *cerumen* (ear wax). Apocrine sweat glands nourish odor-producing bacteria. These organisms originally helped sexual attraction by creating a valued scent in ancient Greece, as recorded in Plutarch's biography of Alexander the Great. The odor is helpful to animals in recognizing each other. Deodorants destroy the bacteria without reducing secretion of sweat.

SEBACEOUS GLANDS. Sebaceous glands (Figs. 11-34; 11-35) belong to the holocrine type. Small cells at the periphery proliferate; and their descendants, closer to the lumen, accumulate oil in their cytoplasm. In the more central layers, more oil accumulates. The nuclei undergo necrosis, and the debris of the dead cells is extruded as sebum. Occasionally sebum

Fig. 11-36. Mechanical interaction of a collagen fiber and a sebaceous gland.

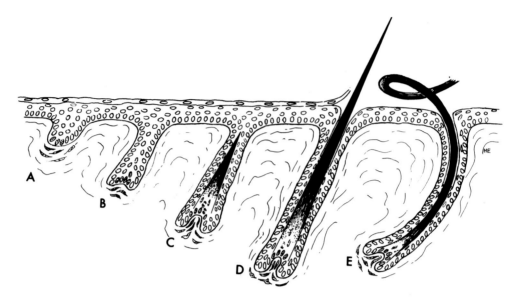

Fig. 11-37. Diagram of hair development.

hardens, mixes with keratin in the hair follicle, to form a *comedo* or "blackhead".

Inextensible collagen fibers in the dermis are wrapped around the turgescent and flexible sebaceous glands (Fig. 11-36) constituting a system of limited elasticity.

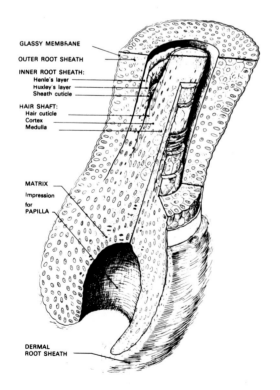

Fig. 11-38. A hair root.

HAIRS. Hairs are components of sensory organs (light touch receptors) which collectively act as heat insulators. The development of a hair in the fetus is illustrated in Figure 11-37. The *hair follicle* arises by an ingrowth of epidermis into the dermis (A). An accumulation of mesenchyme at the deep end of the follicle pushes into it (B) and becomes the hair papilla (C-E). The epidermal cells next to the papilla proliferate forming the matrix for a keratinizing cone (C) which pushes through the follicle (D). The first erupting hairs in the fetus break the periderm, the unicellular upper layer of the fetal epidermis. Curved follicles produce curly hair (E).

The root of a hair is shown stereographically in Figure 11-38 and in section in Figure 11-39 and Figure 11-40. The most important parts are the dermal hair papillae which nourish the matrix, i.e. that mass of epidermal cells which, by proliferation, produce hair shafts and push them upward. Keratinization of the shaft cells occurs at a higher level, but still deep in the hair follicle (Fig. 11-37). The layers shown in Figure 11-38 are seen only along a short stretch, a little above the matrix. In Figure 11-39 only those layers which can be clearly identified are labeled.

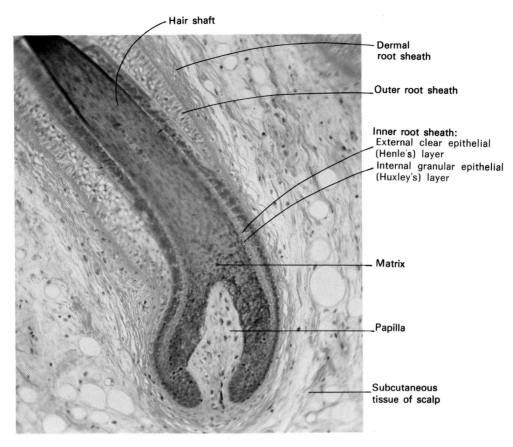

Fig. 11-39. A pigmented curved hair follicle producing a black curly hair. Galacyanin and Picro-fuchsin. 250 ×.

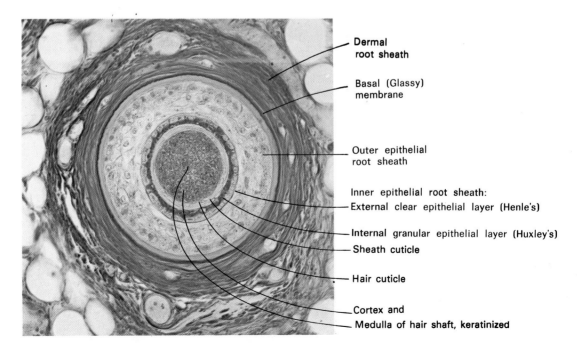

Fig. 11-40. Cross section of very moderately pigmented hair follicle of scalp, producing a blond hair. Azan, 416 ×.

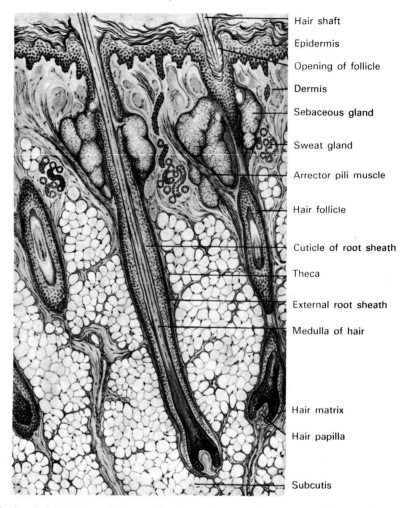

Fig. 11-41. Scalp of child. (From Gillison: Histology of the Body Tissues. Edinburgh, 1953.)

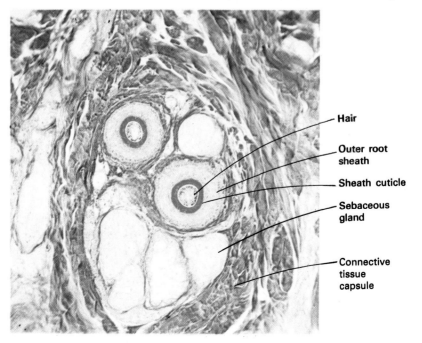

Fig. 11-42. A pilosebaceous organ from a tangential section of scalp. 185 ×.

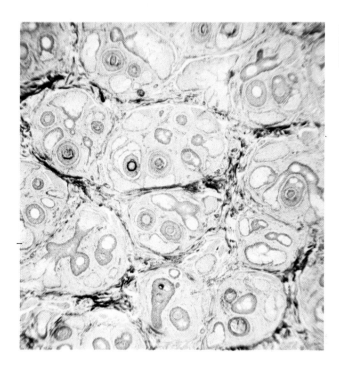

Fig. 11-43. Tangential section of scalp showing the individuality of pilosebaceous organs. Elastica stain, 60 ×.

Figure 11-41 demonstrates how terminal hairs are integrated into the integument of the scalp. Their follicles reach deeply into the subcutaneous tissue.

Each hair follicle is associated with sebaceous glands, but not all sebaceous glands are associated with hair follicles (Fig. 11-34). The combination is a *pilosebaceous organ*. Often, pilosebaceous organs possess connective tissue capsules (Fig. 11-42) which keep them separated from their neighbors (Fig. 11-43). Usually a hair is provided with an *arrector pili muscle*, a bundle of smooth myocytes inserted into the hair follicle along its deep flank and anchored in the superficial portion of the dermis in an oblique direction, so that its contraction causes the hair to "stand up" (Fig. 11-41).

The hairs of the newborn are delicate *lanugo hairs*. With advancing age, large *terminal hairs* developing in certain regions reach deeply into the tela subcutanea. They are classified according to anatomic location (Fig. 11-44). Lanugo hairs, present throughout life, are ubiquitous, growing in "bald" areas as well as among dense, terminal hair in both sexes (Fig. 11-45).

In normally "bald" areas, arrector pili muscles are often arranged in a flat basket-like pattern around and beneath a pilosebaceous organ (Fig. 11-46). Their con-

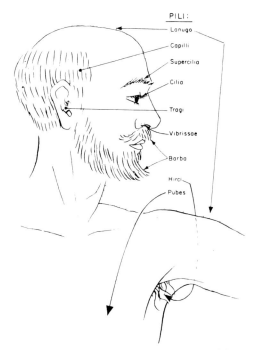

Fig. 11-44. Topographical nomenclature of hair.

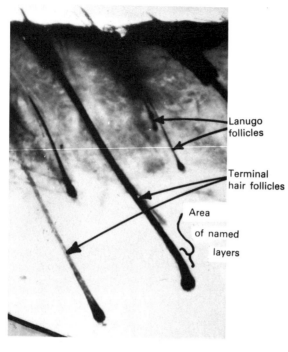

Fig. 11-45. Thick slice of scalp of woman.

traction produces "goose-bumps". Baldness of the scalp in adult males is due to a genetic factor, and not a disease, but a property which belongs to an individual's character. Eighty per cent of baldness is associated with a reduction of the thickness of the tela subcutanea of the scalp, due to absence or near absence of fat. Figure 11-47 shows the importance and necessity of a thick adipose layer for hair growth. All three specimens shown in that figure are from the middle parietal region, i.e., from the spot which, in some males, becomes bald first. A is from an adult woman; B from an adult man with "a good head of hair"; C from a bald man. Obviously, the tela subcutania of the bald man is too thin to accommodate large hair follicles; but many lanugo hairs are present. The panniculus adiposus in bald scalps may be totally absent, so that the dermis rests directly on the deep fibrous

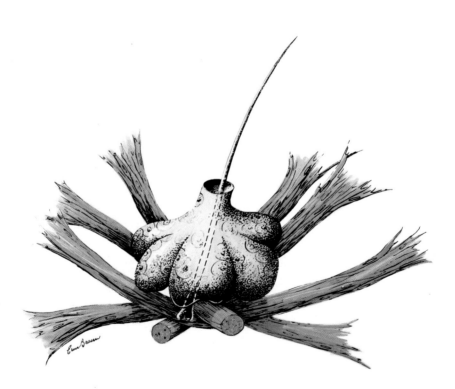

Fig. 11-46. An abdominal pilosebaceous organ with basket of smooth muscle fibers, 65-year-old man. Reconstruction from serial sections.

layer of the tela subcutanea. In such cases, the scalp is thinner than in the specimen in Figure 11-47, C.

The cycle of hair loss and replacement is illustrated in Figure 11-48. (A) At the end of a growth period (anagen), the hair matrix ceases to proliferate. During involution (catagen) the hair bulb changes shape; reflecting a temporary halt in the synthesis of the hair. (B) From the deep end of the shrunken follicle a new embryonic follicle grows downward. (C)

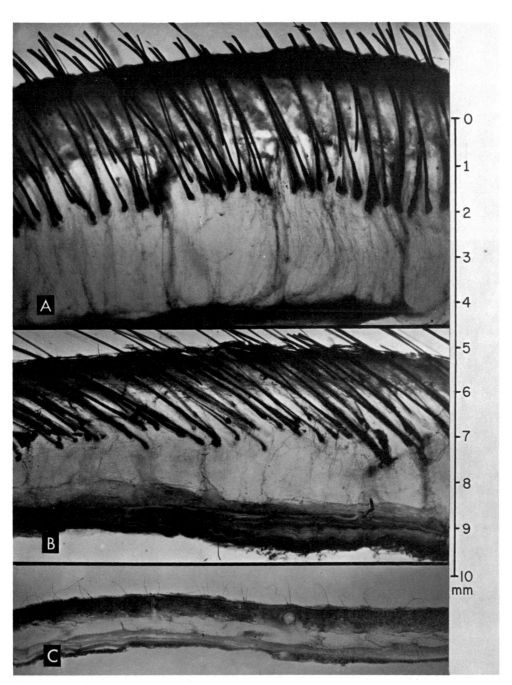

Fig. 11-47. Thick, unstained sections of parietal scalps. A, Of adult woman. B, Of adult man. C, Of bald man.

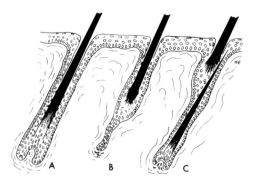

Fig. 11-48. Hair loss and replacement.

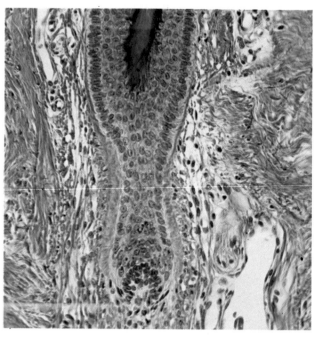

Fig. 11-49. Early telogen phase of hair replacement, in a tragus hair at the entrance to the external auditory meatus. 400 ×.

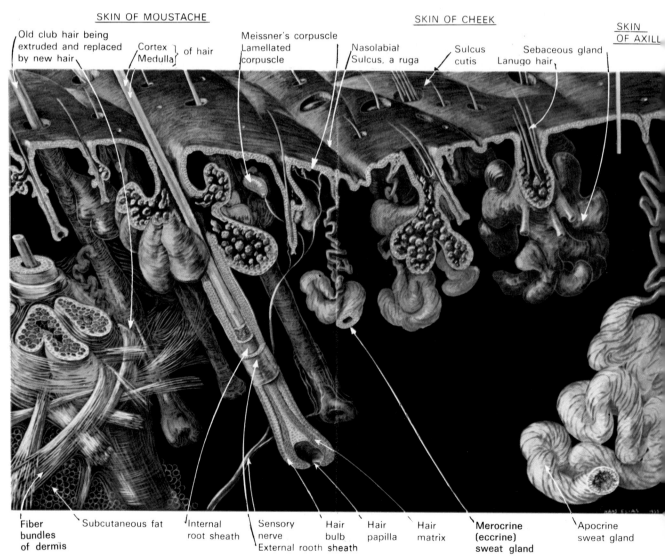

Fig. 11-50. Stereogram of facial integument. On the extreme right is an axillary sweat gland.

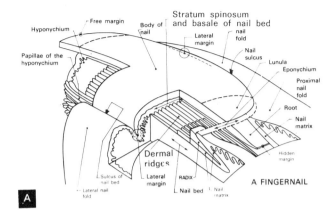

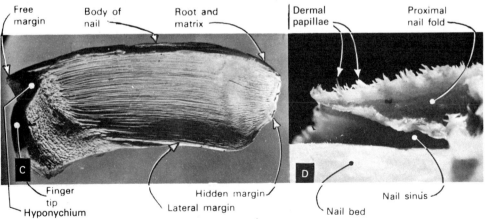

Fig. 11-51. A fingernail. A, Names of parts. B, Stereogram of fingernail. C, The epidermal portion of a nail stripped off and divided in half by a longitudinal cut seen from the left and from its deep aspect. D, Proximal part of the dermal component of the nail bed after removal of the epidermal component, longitudinally cut and seen from the left.

The new follicle produces a new hair which pushes a new opening for itself through the follicle while the old hair (telogen) falls out. Figure 11-49 shows the early telogen phase.

Figure 11-50 summarizes the essential points that have been discussed. It shows, among other things, various possible relationships between sebaceous glands and hairs. Since the major portion of that figure is a reconstruction of facial skin, it has some unusual features such as absence of arrector pili muscles and presence of a lamellated corpuscle at an unusually high level.

NAILS. The nail (Figs. 11-51, A and B) is a plate of hard stratum corneum, gliding distalward. It is produced by a *matrix* beneath the *proximal nail fold*. At this location the epidermis forms nail keratin without a stratum granulosum, preparatory to the production of the horny *nail plate*. The horny nail plate glides slowly over the nail bed directed by longitudinal dermal ridges. Strata granulosum and lucidum are absent in the nail bed. The epidermal part of the nail, with the *hyponychium* (Fig. 11-51, C) hemisected and seen from its deep aspect, shows the grooves produced by the dermal ridges and the pits (in the hyponychium) produced by dermal papillae. With special techniques, the entire epidermal part of the nail can be removed leaving the dermis. Such a preparation is shown in Figure 11-51, D. The thin skin adherent to the nail at the proximal end is called the *eponychium* or *cuticle*. The *lunula* is the visible distal extension of the matrix which is white because it consists of nucleated cells.

12...
Digestive Apparatus

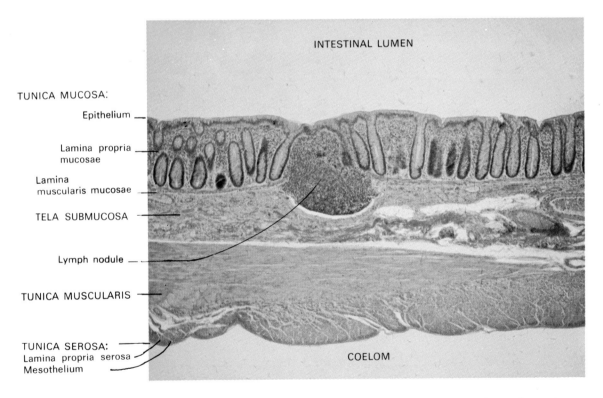

Fig. 12-1. The layers of the wall of the gastrointestinal tract, a complete mucous membrane (example: colon), H&E. 75 ×.

The organs for the intake, mastication, transport, digestion and absorption of food and drink are incorrectly called the gastrointestinal tract. Literally, this term pertains only to the stomach and intestines. To include the oral cavity, pharynx, esophagus, stomach, intestines, accessory digestive glands, and the anal canal, the term

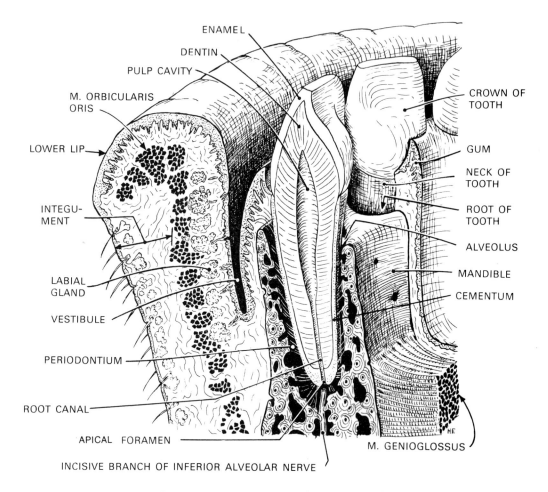

Fig. 12-2. Entrance to the alimentary canal.

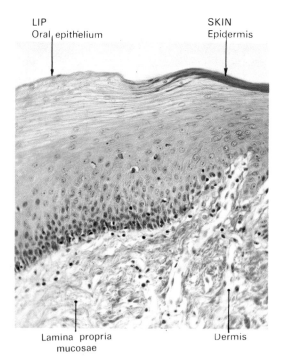

Fig. 12-3. Boundary between skin and oral mucous membrane at the lip. 390 ×.

preferred is the *digestive apparatus* or *digestive system*. The tubular structures have a wall organized into the layers exemplified by the wall of the colon (Fig. 12-1). From the inside out, these layers are:

Tunica Mucosa
  Epithelium
  Lamina propria mucosae
  Lamina muscularis mucosae
Tela Submucosa
Tunica Muscularis
Tela Subserosa
Tunica Serosa (Where in contact with coelum)
  Lamina propria subserosa
  Mesothelium

In the new nomenclature, the older

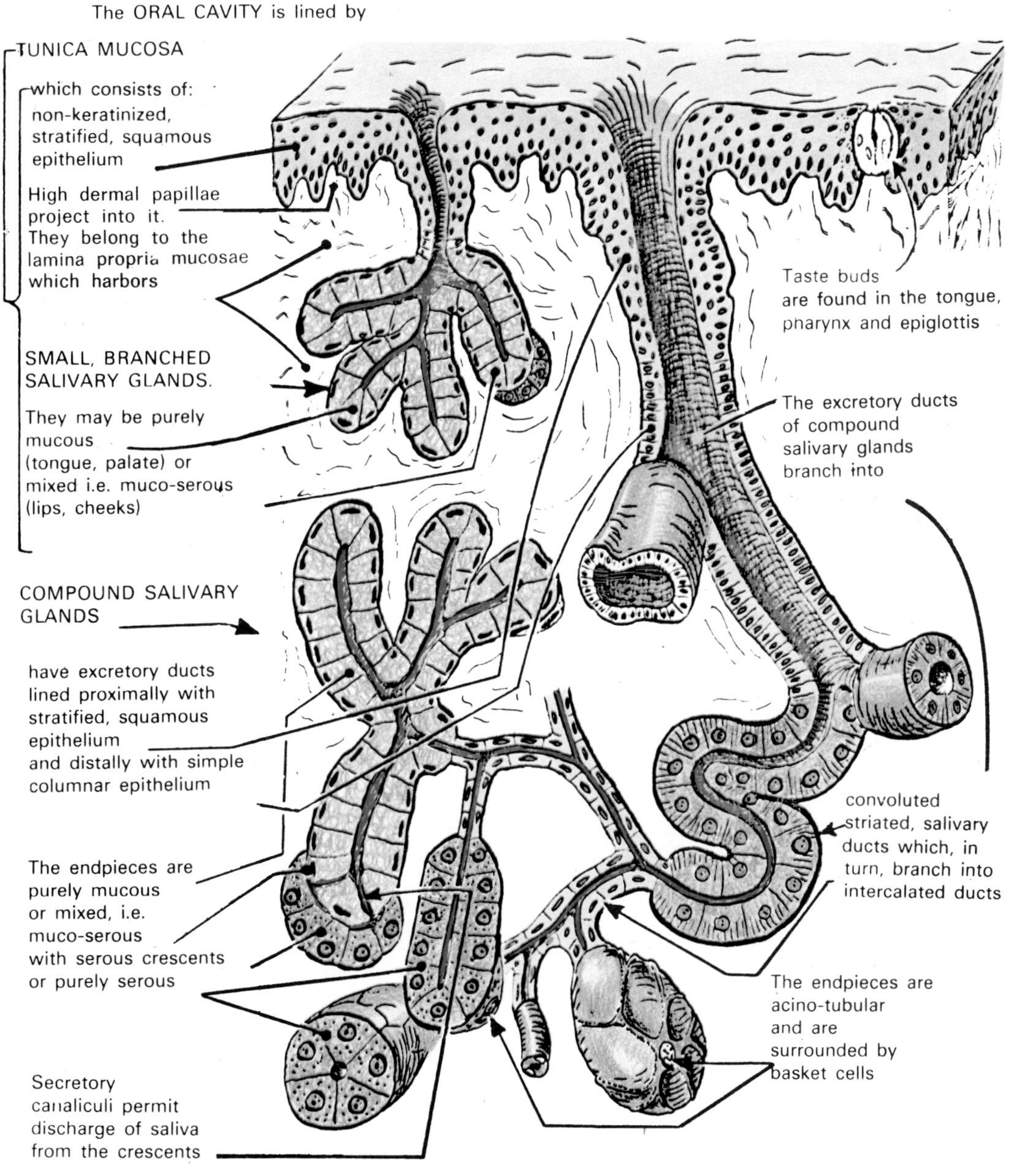

The ORAL CAVITY is lined by

TUNICA MUCOSA
which consists of:
non-keratinized, stratified, squamous epithelium

High dermal papillae project into it. They belong to the lamina propria mucosae which harbors

SMALL, BRANCHED SALIVARY GLANDS.

They may be purely mucous (tongue, palate) or mixed i.e. muco-serous (lips, cheeks)

COMPOUND SALIVARY GLANDS

have excretory ducts lined proximally with stratified, squamous epithelium and distally with simple columnar epithelium

The endpieces are purely mucous or mixed, i.e. muco-serous with serous crescents or purely serous

Secretory canaliculi permit discharge of saliva from the crescents

Taste buds are found in the tongue, pharynx and epiglottis

The excretory ducts of compound salivary glands branch into

convoluted striated, salivary ducts which, in turn, branch into intercalated ducts

The endpieces are acino-tubular and are surrounded by basket cells

Fig. 12-4. Structures connected with the oral cavity.

terms tunica submucosa and tunica subserosa have been changed to *tela submucosa* and *tela subserosa*. Tela is defined as any thin anatomical connective tissue layer resembling a web (Tela in Latin = web). Both submucosa and subserosa possess a network (web) of loose connective tissue housing plexuses of vessels and nerves.

Oral Cavity (Mouth)

The entrance to the digestive canal is guarded by the lips and teeth (Fig. 12-2).

Where the red portion of the lips begins, there is a transition from skin with its keratinized epithelium (*pars cutanea*) to a translucent stratified squamous epithelium which has no stratum corneum (*pars intermedia*). This epithelium is sufficiently transparent to permit the color of the blood in the stromal papillae to be seen. The red area of the lip is continuous with the oral mucosa (*pars mucosa*) (Fig. 12-3).

ORAL MUCOUS MEMBRANE. The *pars mucosa* of the oral cavity does not possess all the layers shown in Figure 12-1. It consists of a thick moist stratified squamous epithelium and a rather dense, but finely woven lamina propria mucosae. The superficial layer of stratified squamous epithelium of man (in most places) neither keratinizes nor develops mucous superficial cells; however in some places it may become keratinized. The lamina propria mucosae develops very tall, slender papillae which penetrate far into the epithelium without disturbing the smoothness of the epithelial surface (except on the dorsum of the tongue). These papillae contain blood capillary loops.

Salivary glands and taste buds are associated with the mucosa of the oral cavity.

SALIVARY GLANDS. Salivary glands (Fig.

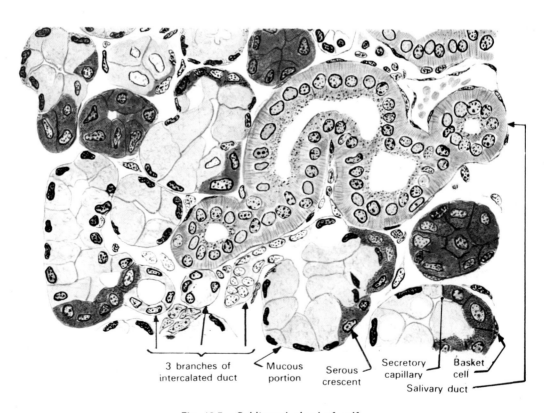

Fig. 12-5. Sublingual gland of calf.

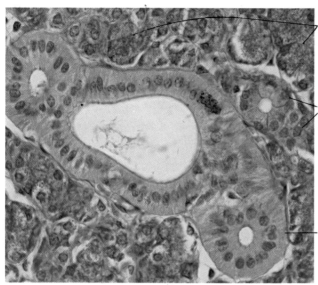

Fig. 12-6. Parotid gland, H&E. 750 ×.

Serous glandular end pieces

Intercalated ducts

Striated, salivary duct

12-4) are branched or compound tubuloacinar glands with three types of *acini* (secretory end pieces).

In *mucous acini*, the cytoplasm of the glandular cells is very light staining and foamy. The nuclei are compressed against the basal boundary of the cells.

Serous acini have dark-staining cells with spherical nuclei and an optically homogeneous cytoplasm which contains *zymogen granules*. The basophilia of the basal portion is caused by an abundant granular endoplasmic reticulum which synthesizes the protein zymogen granules. The protein product is combined with polysaccharide in the golgi complex as zymogen granules and stored in the apical portion (acidophilic) of the cell until secreted. EM reveals them as membrane-bound granules.

Mixed acini are basically mucous acini capped by serous cells aggregated in the form of *crescents (semiluna serosa)* (Figs. 12-4 and 12-5). Depending on whether serous or mucous cells prevail, mixed glands are called *seromucous* or *mucoserous*.

Branched smooth myocytes (sometimes called basket cells or myoepithelial cells) are found between the glandular epithelial cells and the basement membrane.

Small branched salivary glands are distributed as follows: (1) purely mucous glands, in the posterior portion of the tongue and in the palate; (2) mixed glands, in the lips, cheeks, pharynx, esophagus, trachea and bronchi; and (3) purely serous glands, deep to the vallate and foliate papillae. The mucous acini produce *mucus*, a slimy substance which lubricates the mouth and helps to form the bolus of food (N.B. mucous = adjective, mucus = noun). Serous cells produce a clear, aqueous liquid which contains *ptyalin*, a diastase capable of reducing starch and dextrin to maltose. The serous glands deep to the gustatory papillae not only provide an aqueous medium to put taste producing substances in solution, but also wash the taste buds clear to

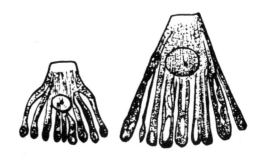

Fig. 12-7. Teased cells of salivary duct showing basal infoldings and mitochondria. (Pflüger, 1872.)

270 DIGESTIVE APPARATUS

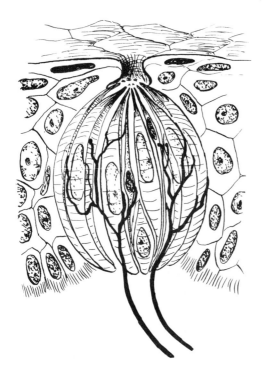

Fig. 12-8. Taste bud.

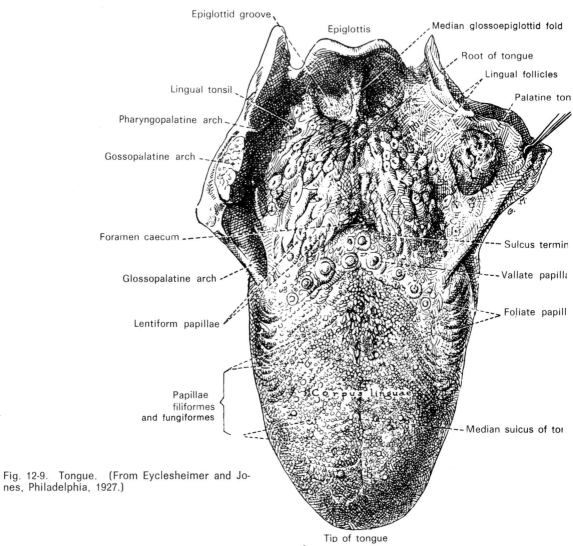

Fig. 12-9. Tongue. (From Eyclesheimer and Jones, Philadelphia, 1927.)

prepare them for another taste sensation.

Compound salivary glands (Figs. 12-4 and 12-5) are characterized by convoluted, *striated salivary ducts* and *intercalated ducts* (Fig. 12-6). The former are lined by a tall, simple cuboidal epithelium characterized by basal striations which are due to infoldings of the basilar surface of the cell membrane (Fig. 12-7). The high population of mitochondria causes the intense acidophilia of the basal half of the cell. The intercalated ducts are lined by squamous to a low, simple cuboidal epithelium. The *parotid glands* are said to be purely serous, but small nests of mucous and mucoserous acini are found in some human parotid glands. The *submandibular glands* are mixed, seromucous. The *sublingual glands* (Figs. 12-5) are mixed, mucoserous, but predominantly mucous glands.

The major salivary glands may be characterized as follows: The parotid gland (largest of the salivary glands) empties into the buccal cavity by a single main duct, the *parotid duct*, opposite the upper second molar tooth. Several interlobular ducts branch from the main duct and travel in the connective tissue septae of the lobules. Intralobular ducts leave the connective tissue septae and branch into the serous lobules. Simple columnar epithelium forms the intralobular ducts which lead directly to striated ducts. In comparing the parotid to other major salivary glands, it should be noted that the striated ducts are not as extensive as those of the submandibular gland, while those of the sublingual gland are minimal. In the parotid gland, the intercalated ducts are long and branching. (It is advisable for students to compare and contrast the parotid and the pancreas, because they have many features in common.)

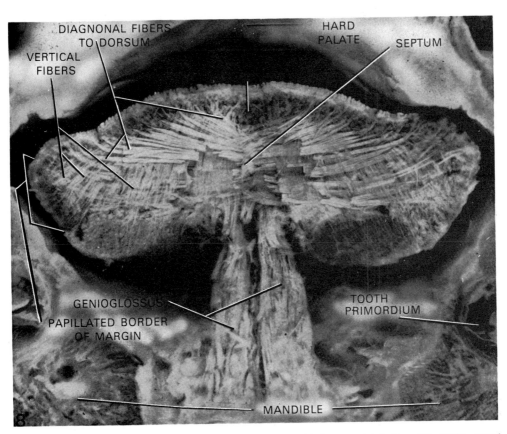

Fig. 12-10. Cross section of fetal tongue. (From Leon H. Strong: Anat. Rec. 126: 61-79, 1956.)

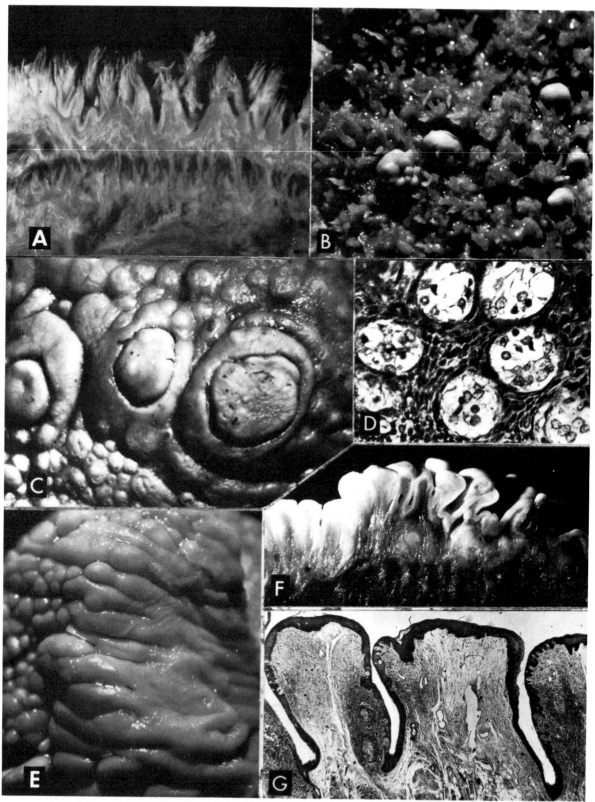

Fig. 12-11. Areas from the tongue of one 42 year old man. A, Filiform papillae seen in profile. B, Fungiform and filiform papillae (surface view). C, Vallate papillae (surface view). D, Tangential section of visceral epithelium of vallate papillae showing several taste buds in cross section. E, Foliate papilla (surface view). F, The same foliate papilla after slicing it in the direction transverse to the furrows. G, A small portion of the same specimen. The part shown in this photomicrogram is the place just beneath the letter D. Note taste buds at various places along the furrows, close to the top, and a microscopic tonsil in the left wall of the middle furrow.

The *submandibular gland* possesses a single duct which opens under the tongue on either side of the frenulum. The epithelium may be stratified or pseudostratified columnar. Interlobular ducts, recognized in connective tissue septae, have a simple columnar epithelium which continues into the intralobular ducts. The latter are not as numerous in tissue sections as one finds in the parotid. Long, branching striated ducts with low cuboidal to simple squamous epithelium lead to intercalated ducts.

The *sublingual glands* open by several ducts into the oral cavity next to the openings of the submandibular ducts. Interlobular ducts are present in sections, and intralobular ducts are numerous and very prominent. Striated ducts, while present, are few in number. Intercalated ducts are difficult to find. The acini are practically all mucous in character, but the secretory end pieces are capped by serous demilunes (*semiluna serosa*). Serous acini may be found occasionally. Adipocytes (fat cells) populate these glands as follows: the parotid has the most, while the sublingual has the least. The submandibular gland possesses a variable number between the other two.

TASTE BUDS. Taste buds (Figs. 12-4 and 12-8) are onion-shaped inserts of sensory epithelium in the stratified squamous epithelium. A taste bud consists of a single layer of tall, spindle-shaped cells which constitute a small area of *sensory epithelium*. Relatively thick, light staining *supporting cells* (which perhaps may function as taste cells) alternate with very slender dark staining *taste cells*. Each of these cells (except the few, basal replacement cells) ends distally in a slender tip containing numerous *taste hairs* (long microvilli and some microtubules). The taste hairs project into a tiny groove, the *taste pore*. The cells of this sensory epithelium are secondary sense cells (see Fig. 7-9, B), contacted by the ramified endings of the peripheal processes of unipolar, sensory

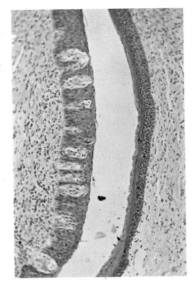

Fig. 12-12. Trench around vallate papilla with taste buds in visceral wall.

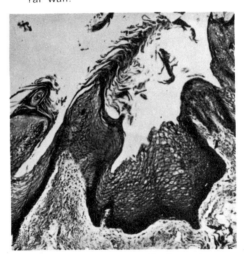

Fig. 12-13. A filiform papilla from the same specimen as shown in Fig. 12-11.

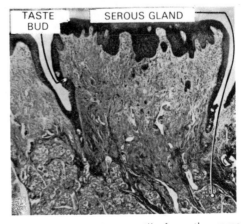

Fig. 12-14. A vallate papilla from the same specimen.

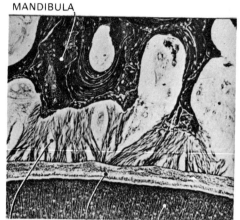

Fig. 12-15. Peripheral portion of the root of an incisor tooth including the periodontium and the mandibular bone. (Preparation by Dr. H. Sicher.)

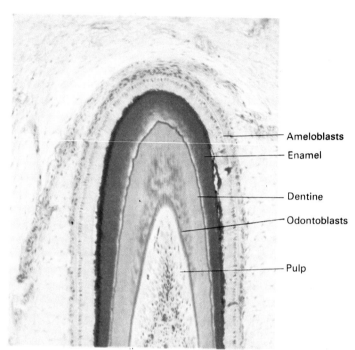

Fig. 12-16. Incisor tooth of a fetus, Azan. 185 ×.

nerve cells located in (a) the geniculate ganglion of the facial or seventh cranial nerve (for anterior part of tongue), (b) the petrous portion of the glossopharyngeal or ninth cranial nerve (for the posterior part of tongue) and (c) the nodose ganglion of the vagus or tenth cranial nerve (for pharynx and larynx). The axons of these nerve cells terminate in the nucleus of the *tractus solitarius*, a group of neurons in the brain stem involved in the perception of taste. Figure 12-11, D, shows a tangential section of the wall of a vallate papilla with six taste buds in cross section.

TONGUE. The tongue (Fig. 12-9) is formed by several skeletal muscle bundles, whose myocytes run somewhat perpendicularly to each other (Fig. 12-10), in three directions. The skeletal myocytes divide at their ends and insert into the lingual lamina propria mucosae just as facial muscles insert in the dermis (Fig. 6-11). The tongue, the only known structure that possesses branched skeletal myocytes, produces complicated movements which aid in chewing, swallowing and articulation.

The mucous membrane (tunica mucosa) of the tongue forms four types of papillae, each with a lamina propria mucosae core: (1) *filiform papillae*, which terminate in little bundles of delicate keratinized threads (Figs. 12-11, A and 12-13); (2) *fungiform papillae* (Fig. 12-11, B) which grow among the filiform papillae like mushrooms in a meadow (taste buds are located on the free surfaces of many fungiform papillae); (3) *vallate papillae* (Figs. 12-11, C and 12-14) surrounded by a sulcus (trench) and by a rim, have taste buds located in the visceral epithelium of the sulcus (Figs. 12-11, D, 12-12 and 12-19). The ducts of serous glands (von Ebner's) empty into the sulcus, washing away old gustatory material, so that new taste impressions can be received. (In a tangential section of the visceral epithelium of a vallate papillae (Fig 12-11, D), the taste buds are cut transversely); and (4) *foliate*

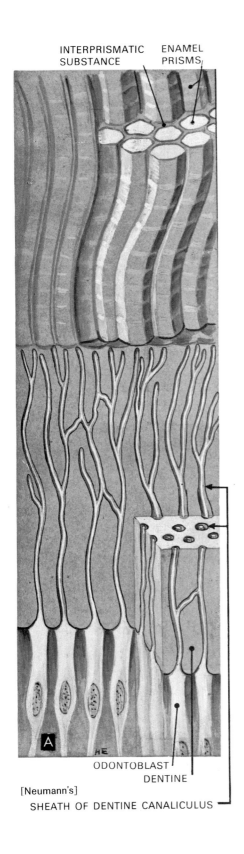

Fig. 12-17. Stereogram of the wall of a tooth.

papillae (Fig. 12-11, E), (not present in every adult) consisting of parallel ridges. In a section of foliate papillae cut transversely to the ridges (Fig. 12-11, F), one can notice that the ridges are separated by deep furrows. In a few adults, the epithelium of these furrows harbors many taste buds (Fig. 12-11, G).

Small subepithelial nodules of lymphatic tissue (*tonsils*) can occur in foliate papillae. An example of a microscopic tonsil is seen to the left of the central furrow in Figure 12-11, D. Ducts of serous glands (Fig. 12-11, G, lower left) empty into the grooves. (All the single illustrations in Figure 12-13 and 12-14 are from the tongue of one adult man.)

TEETH. Teeth are the evolutionary descendents of the placoid scales of sharks, remainders of the bony exoskeleton of the ostracoderms. The entire body of a typical shark is studded with these primitive teeth, the largest of which are found in the mouth. Only the oral teeth remain in mammals.

The human tooth is more elaborate in structure than the shark's placoid scales. The tooth, a very hard, hollow, thick walled organ consists of the layers shown in Figure 12-2. Human teeth are anchored in the maxilla and mandible by a *peridontal membrane* (Fig. 12-15) composed of collagen and elastic fibers to provide flexible suspension of the tooth in the alveolus. The peridontal membrane unites to the *cementum*, an especially hard kind of bone. The cementum, connected with the *dentine*, surrounds the pulp cavity (Fig. 12-16) containing areolar connective tissue, blood vessels and sensory nerve endings. The peripheral layer of pulp consists of columnar *odontoblasts* (Fig. 12-17) which produce and maintain the dentine. Processes extend from them into the *dentinal tubules* lined by a thin layer of connective tissue causing the dentine of the vicinity to appear darker than the other dentine. This delicate layer of connective tissue in the dentinal tubules is

276 DIGESTIVE APPARATUS

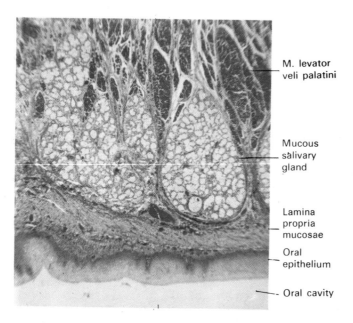

Fig. 12-18. Soft palate, 55 ×.

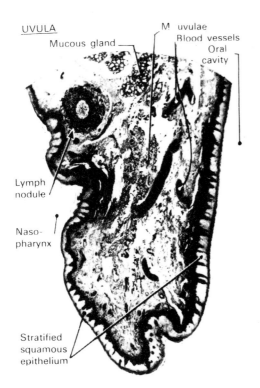

Fig. 12-19. Uvula.

Fig. 12-20. Sagittal section of posterior wall of oropharynx stained with iron hematoxylin, picrofuchsin and resorcin fuchsin to demonstrate the elastic layer.

The PHARYNX, in its upper portion, is lined with pseudostratified, ciliated epithelium, and the connective tissue that supports it is the lamina propria mucosae. But its lower portion is lined by stratified, squamous epithelium; and the connective tissue is divided into lamina propria mucosae and Tela submucosa

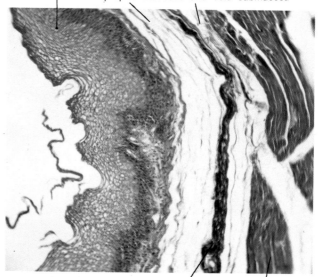

by a layer of elastic tissue which represents the upward continuation of the esophageal lamina muscularis mucosae. The tunica muscularis consists of skeletal muscle.

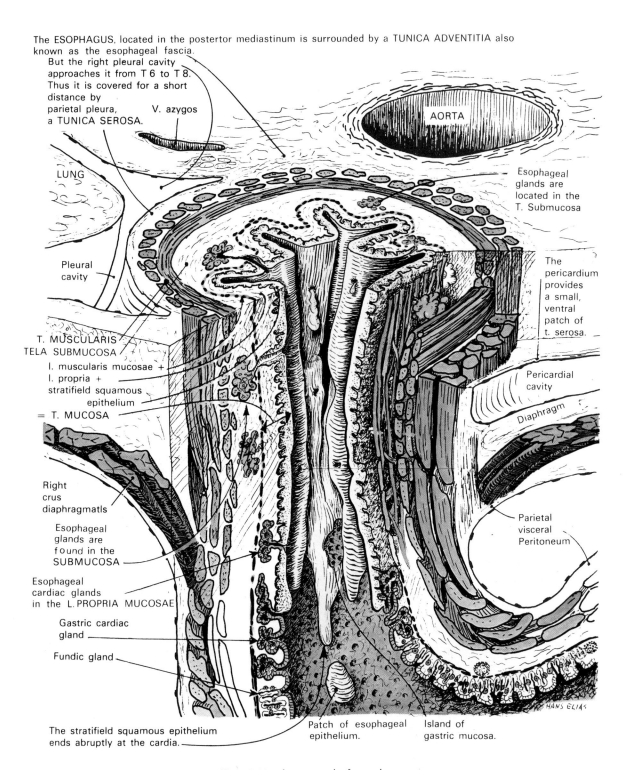

Fig. 12-21. Lower end of esophagus.

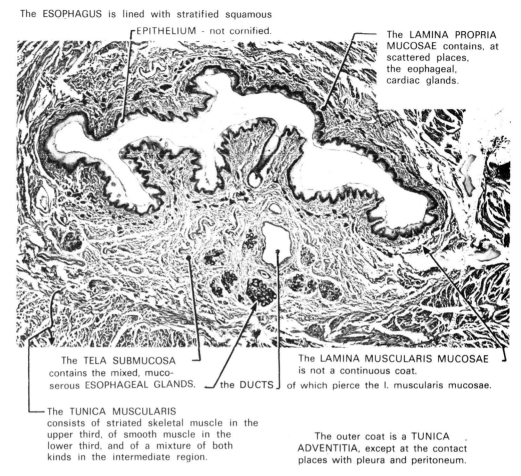

Fig. 12-22. Esophagus.

The ESOPHAGUS is lined with stratified squamous EPITHELIUM - not cornified.

The LAMINA PROPRIA MUCOSAE contains, at scattered places, the eophageal, cardiac glands.

The TELA SUBMUCOSA contains the mixed, muco-serous ESOPHAGEAL GLANDS, the DUCTS of which pierce the l. muscularis mucosae.

The LAMINA MUSCULARIS MUCOSAE is not a continuous coat.

The TUNICA MUSCULARIS consists of striated skeletal muscle in the upper third, of smooth muscle in the lower third, and of a mixture of both kinds in the intermediate region.

The outer coat is a TUNICA ADVENTITIA, except at the contact places with pleura and peritoneum.

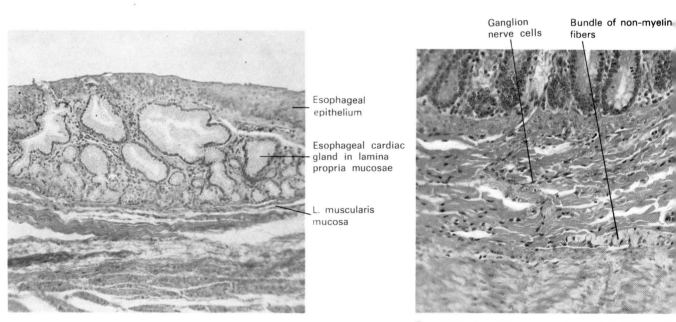

Fig. 12-23. Esophageal cardiac glands. H&E, 210×.

- Esophageal epithelium
- Esophageal cardiac gland in lamina propria mucosae
- L. muscularis mucosa

Fig. 12-23, A. Submucous plexus (of Meissner) in the ileum. H&E, 120×.

- Ganglion nerve cells
- Bundle of non-myelin fibers

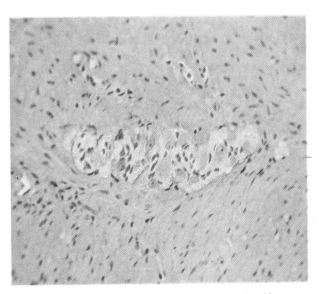

Fig. 12-24. A ganglion of the myenteric (Auerbach's) plexus of the colon, H&E. 375 ×.

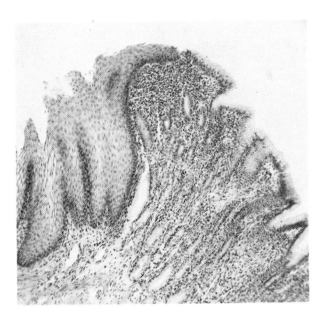

Fig. 12-25. Esophageal cardiac junction, H&E. 185 ×.

named the *lamella dentinalis* (Neumann's sheath) (Fig. 12-17).

Each tooth is divided into (1) a *root*, (2) a *neck* and (3) a *crown*. The *crown* of the tooth (*corona dentis*) is capped by *enamel* (Fig. 12-2) consisting of *enamel prisms* (Fig. 12-17), held together by *interprismatic substance*. Thick bundles of enamel prisms usually pursue a crooked and winding course with neighboring bundles running at angles to each other. Enamel, a cuticular product of the *ameloblasts,* is the hardest known substance of the body, consisting of 97 per cent inorganic material, chiefly fluor-apatite, hydroxy-apatite and some carbonates. The interprismatic substance, an extremely hard material of essentially the same minerals, contains slightly more organic matter which makes the enamel cap even tougher. Because ameloblasts disappear when the tooth erupts, enamel cannot be regenerated.

GUM. (Gingiva). Many collagen and elastic fibers (of the gingival lamina propria mucosae) continuous with the periodontal membrane, are inserted into the cementum, thus assisting in anchoring the tooth (Fig. 12-2). Many lymphocytes populate the lamina propria mucosae. The stratified squamous epithelium, frequently keratinized, normally has its visceral surface in contact with enamel; but in most persons it may become detached (Fig. 12-2) leaving a pocket in which tartar (dental calculus) can accumulate. .

LIPS AND CHEEKS. The lips and cheeks, with a core of skeletal muscle, are covered externally by skin and internally by oral mucosa containing mixed, mucoserous glands.

PALATE. The palate (hard and soft), a horizontal plate which separates the oral from the nasal cavity, is covered superiorly by pseudostratified, ciliated columnar epithelium (respiratory epithelium) (Figs. 13-1 and 13-2) and inferiorly by oral mucosa containing mucous glands (Fig. 12-18). The oral epithelium of the hard palate is

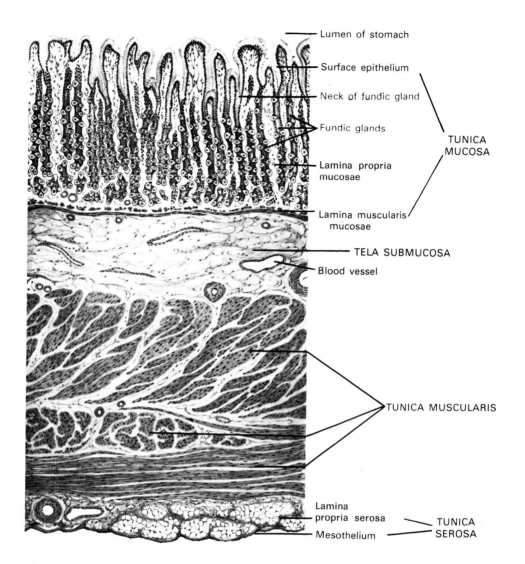

Fig. 12-26. Section through the wall of the body of the stomach. (From Gillison: Histology of the Body Tissues. E. & S. Livingstone, Ltd., Edinburgh, 1953.)

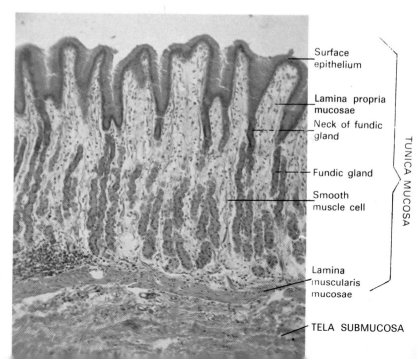

Fig. 12-27. The mucous membrane of the fundus of the stomach, H&E. 200 ×.

keratinized. The tunica mucosa of the hard palate adheres to the palatine bone in such a manner that the deep (superior) portion of its lamina propria mucosa blends with the periosteum. The core of the soft palate is formed by the skeletal muscles levator veli palatini and tensor veli palatini, intermingled with mucous salivary glands (Fig. 12-18).

UVULA. The uvula (Fig. 12-19) contains skeletal muscle. Distally, its core consists of areolar connective tissue. It is covered everywhere by typical oral mucosa.

PHARYNX. The pharynx (Fig. 12-20) is lined in its superior portions partly with pseudostratified and partly with stratified columnar epithelium; inferiorly, the epithelium is non-keratinized, stratified squamous. The connective tissue is divided into a lamina propria mucosae and a tela submucosa by a layer of elastic tissue which is continuous with the lamina muscularis mucosae of the esophagus. Mucous glands and taste buds are present. The tunica muscularis of the pharynx is formed in part by the stylopharyngeus and salpingopharyngeus muscles, but mainly by the three pharyngeal constrictors composed of skeletal muscle.

TONSILS. The tonsils are described with lympho-epithelial organs.

EPIGLOTTIS. The epiglottis has a core of elastic cartilage covered on both sides with oral mucosa containing a few taste buds, except for the dorso-caudal portion which is lined with respiratory mucosa.

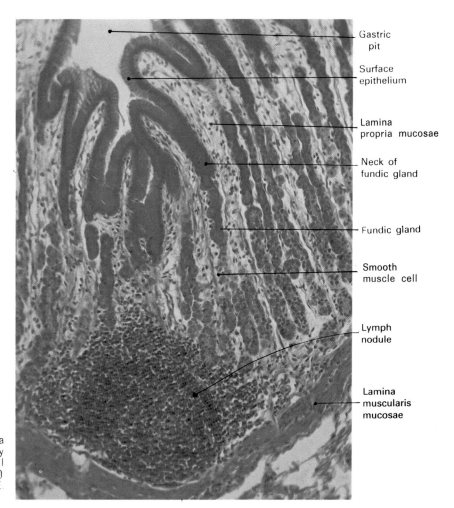

Fig. 12-28. The tunica mucosa of the body or fundus (the wall of both parts is alike) of the stomach, H&E 350 ×.

Esophagus

The esophagus (Figs. 12-21 and 12-22) possesses in its wall all the layers of a complete mucous membrane. The epithelium in man is non-keratinized stratified squamous (in many animals, however, it is keratinized). The lamina muscularis mucosae consists of longitudinally-oriented bundles of smooth myocytes, loosely arranged in the pattern of an irregular fence with large gaps. The tela submucosa consists of connective tissue which varies from loose to dense. The tunica muscularis possesses an inner circular and an outer longitudinal layer, with myocytes of the skeletal type in the upper third and of the smooth variety in the lower third. In the middle third, there is a mixture of both types, so that a gradual transition from skeletal to smooth muscle is noticed from superior to inferior. The outer coat is tunica adventitia, except for a few places where the esophagus comes close to the pleural cavity, so that the parietal pleura forms a patch of tunica serosa. This is explained in Figure 12-21.

Esophageal glands scattered as isolated, small clusters in the tela submucosa, are small, branched glands of the mixed type, containing purely mucous and purely serous acini as well as mucous acini with serous crescents. Occasionally, short striated ducts can be found in the glands. The excretory duct of each gland begins with simple cuboidal epithelium within the gland and terminates at the epithelial surface with a lining of stratified squamous epithelium. The duct pierces the lamina muscularis mucosae.

In the first part of the esophagus (at the level of the 5th tracheal cartilage) and at the terminal portion close to the cardia of the stomach, *esophageal cardiac glands* (Fig. 12-23) are found in the lamina propria mucosae. These are branched, tubuloalveolar glands lined by a mucous, simple columnar epithelium. These glands are similar in structure to the cardiac glands of the stomach. At the lower end of the esophagus some of them, located in the lamina propria mucosae of the esophagus, send their ducts downward to empty through gastric pits into the stomach (Fig. 12-21).

Enteric Plexuses

Two nerve plexuses pervade the entire alimentary canal from the esophagus to the rectum. (1) a delicate plexus of autonomic ganglia and nerve fiber bundles found deep in the submucosa is the *submucosal plexus* (of Meissner) (Fig. 12-23, A) and (2) a coarser plexus, the

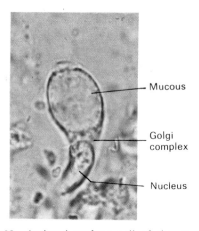

Fig. 12-29. Isolated surface cell of the stomach. Unstained, ordinary light, diaphragm narrowed.

myenteric plexus (of Auerbach), lodged between the circular and longitudinal layers of the tunica muscularis (Fig. 12-24 and 12-54). These plexuses contain the cell bodies of parasympathetic neurons, pre- and postganglionic para-sympathetic fibers, and postganglionic sympathetic fibers. Preganglionic fibers synapse with the neurons in the ganglia. Postganglionic fibers, both sympathetic and parasympathetic, innervate the effector organs, i.e., blood vessels, glands and musculature of viscera. There is increasing indirect evidence for the existence of "sensory" elements in the enteric plexuses. Among

ENTERIC PLEXUSES 283

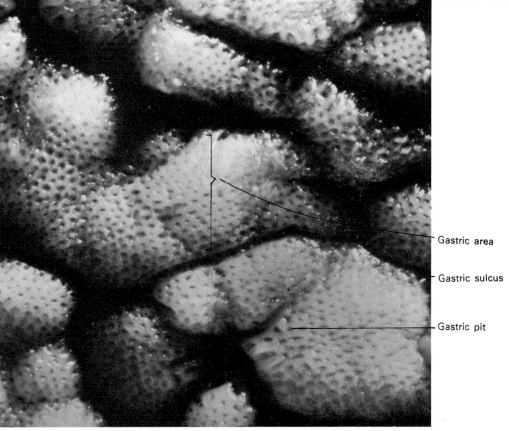

Fig. 12-30. Gastric areas and pits near cardia, surface view. 36 ×.

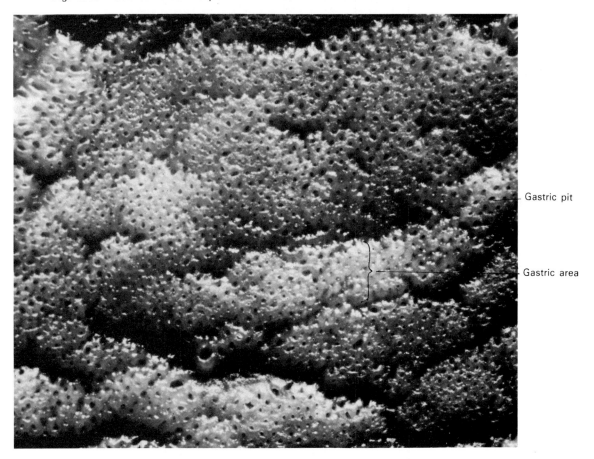

Fig. 12-31. Gastric areas and pits in fundus, surface view. 36 ×.

them must be pressure receptors which promote peristalsis, and chemoreceptors which promote glandular secretion as well as expulsion of bile from the gall bladder (see Fig. 12-65).

Esophageal-Cardiac Junction

Where the esophagus joins the stomach, the epithelium changes abruptly from stratified squamous to simple columnar. Though the epithelial change is abrupt (Fig. 12-25) the line of demarcation is not straight. Islands of gastric mucosa can be found in the distal end of the esophagus; and patches of esophageal epithelium may be present in the proximal portion of the stomach (Fig. 12-21). The outer layers of the esophageal wall in general are continuous with those of the stomach.

Stomach

Since the stomach (Fig. 12-26) is suspended in the peritoneal cavity, its outer lining is a serous membrane, the tunica serosa. This is followed by the thick tunica muscularis consisting of an outer longitudinal, a middle circular and an inner oblique layer between which lies the myenteric plexus. However, due to the rather complicated shape of the stomach, the words "longitudinal" "circular" and "oblique" are difficult to apply. The tela submucosa consists of loose connective tissue containing a submucosal plexus so

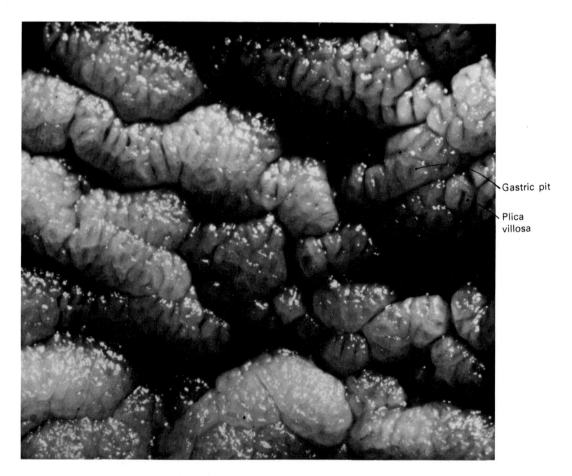

Fig. 12-32. Gastric areas and pits in the pyloric antrum. 36 ×.

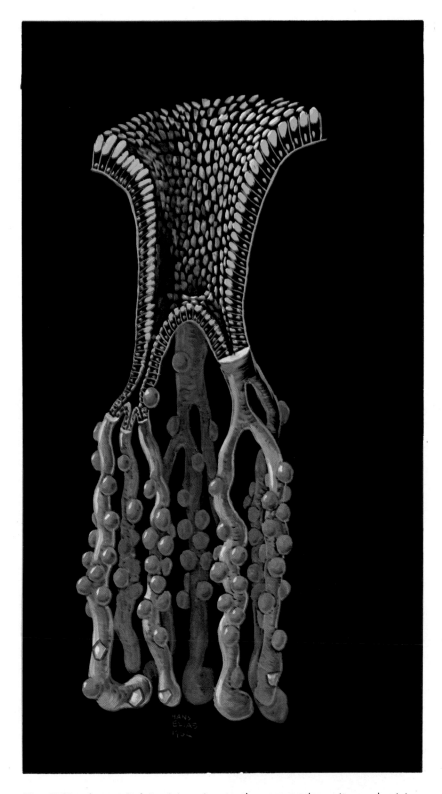

Fig. 12-23. A gastric pit giving rise to three secondary pits, each giving rise to three fundic glands. The hemispherical objects represent parietal cells; the stippled polygons, enteroendocrine cells. Note the mucous caps (white ovals) in the surface epithelium. (From Elias, der Magen, Ingelheim, 1964, by permission of Boehringer Sohn.) Source for Fig. 12-29 to 12-32 the same.

286 DIGESTIVE APPARATUS

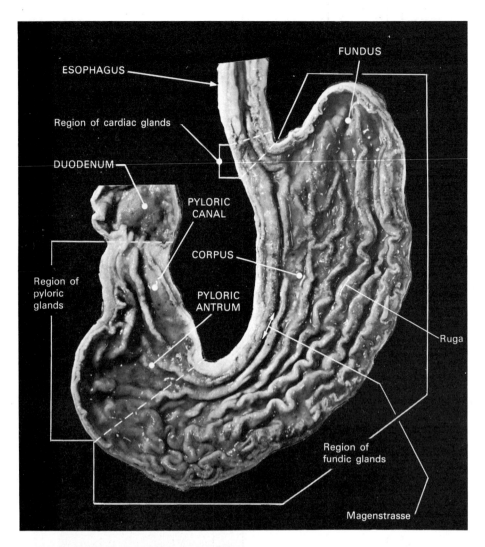

Fig. 12-34. Internal view of posterior half of stomach. (From Elias, der Magen, Ingelheim, 1964, by permission of Boehringer Sohn.)

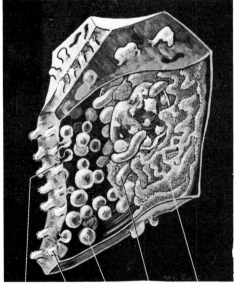

Fig. 12-35. A (left): Body principal exocrine (chief) cell. (From Elias, der Magen Ingelheim, 1964. With permission of Boehringer Sohn). B (facing page): Section of fundic gland. 1300 ×. Courtesy of Dr. D. H. Matulionis, University of Kentucky.

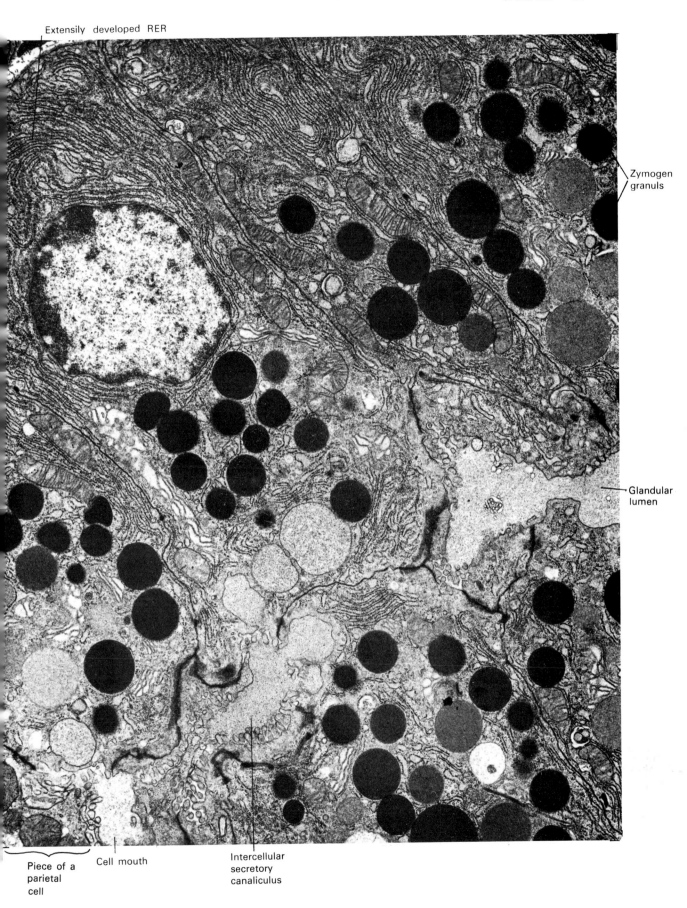

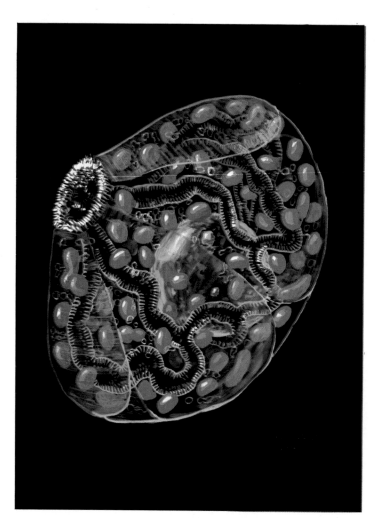

Fig. 12-36. A parietal cell showing intracellular secretory canaliculi, studded with microvilli and leading into a cell mouth. Intensely acidophil granules (red) have been identified as mitochondria. (From Elias, der Magen, Boehringer, Ingelheim, 1964.)

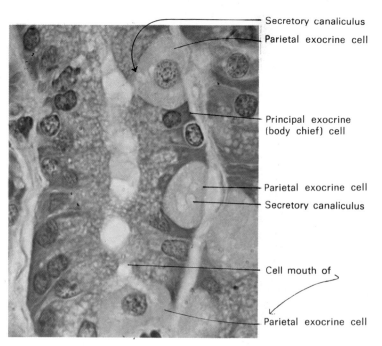

Fig. 12-37. Longitudinal, semithin (1 μm) section of a fundic gland, hematoxylin only. 1720 ×.

fine and delicate that it can rarely be seen. The tunica mucosa (Fig. 12-27) begins, externally, with the lamina muscularis mucosae from which smooth myocytes reach into the lamina propria mucosae. The lamina propria mucosae consists, as a rule, of a mixture of loose areolar

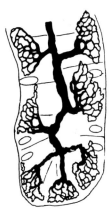

Fig. 12-38. Secretory caniculi in parietal exocrine cells. (Golgi, 1893).

and reticular connective tissue usually described as a diffuse, lymphoid tissue. However, of the free cells which populate the lamina propria mucosae, the lymphocytes are not the most numerous; instead there is a preponderance of plasma cells. Wandering eosinophils often can be encountered in this layer. A simple columnar, mucous epithelium covering the surface (Figs. 8-13 and 8-16) possesses cells which resemble goblet cells when isolated (Fig. 12-29); but within the epithelium, they are forced, by mutual compression, into a columnar shape. The mucoid substance secreted protects the stomach lining from autodigestion by its own enzymes. However, the *mucigen* of the surface cells is chemically different from mucigen of the goblet cells found elsewhere.

When viewing the inner surface of the stomach with a magnifier, one notices small depressions, the *gastric pits* (foveolae gastricae). They show regional differences: near the cardia (Fig. 12-30) in the fundus and body (corpus) (Fig. 12-31), the gastric pits appear as small, round openings. In the pyloric antrum and canal (Fig. 12-32), the pits form oblong slits. The pits are shallow in the cardia and deepest in the pyloric region. Each gastric pit branches into an average of three secondary pits (Fig. 12-33) from which the gastric glands arise. Further, an average of three glandular tubules arise from each secondary pit, so that approximately nine glands empty into one superficial, gastric pit (Fig. 12-33).

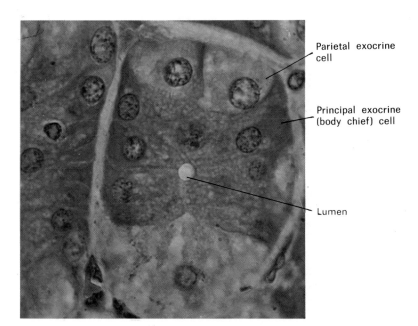

Fig. 12-39. Semithin transverse section of a fundic gland, hematoxylin. 1935 ×.

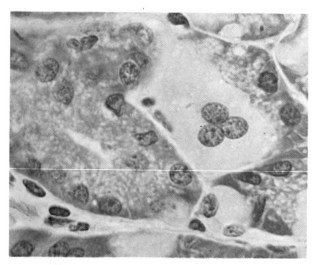

Fig. 12-40. A trinucleate parietal exocrine cell. 1800 ×.

GASTRIC GLANDS are confined to the tunica mucosa. Three types of gastric glands are characteristic for certain regions of the stomach. These regions are indicated in Figure 12-34.

Cardiac glands (Fig. 12-25) occupy a narrow zone at the cardia. Like the esophageal cardiac glands (Fig. 12-23) they are short, branched, tubulo-alveolar, mucous glands.

Fundic glands (Figures 12-27, 28, 33, 37, 38, 39, 40, 41, 42 and 43) occupy the body and fundus of the stomach. The long, thin fundic glands with narrow lumina are approximately one-fifth gastric pit and four-fifths gland. Their narrow necks are lined with low columnar or cuboidal mucous cells called *mucous neck cells*. The mucus of the mucous neck cells is acid mucopolysaccharide, whereas that of the surface cells is a neutral polysaccharide. As mucous neck cells proliferate, some daughter cells migrate towards the lumen of the stomach eventually replacing wornout surface cells. Other daughter cells migrate deep into the gastric gland and probably differentiate into *parietal* and *principal exocrine* (zymogenic) *cells*.

The fundic glands are lined by small, slightly basophilic principal exocrine cells which produce *pepsinogen*. These cells (Fig. 12-35, A and B) are basophilic because of the presence of a dense, granular endoplasmic reticulum in the basal part. Secretory zymogen droplets found in their cytoplasm empty into the glandular lumen

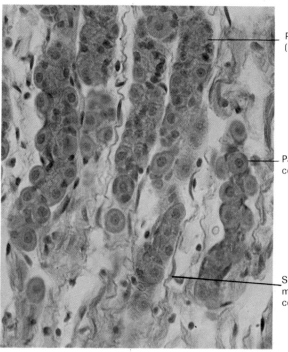

Fig. 12-41. Longitudinal section of fundic glands, H&E. 800 ×.

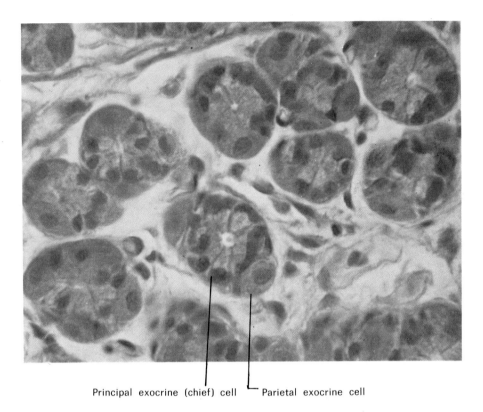

Principal exocrine (chief) cell — Parietal exocrine cell

Fig. 12-42. Cross sections of fundic glands, H&E. 1200 ×.

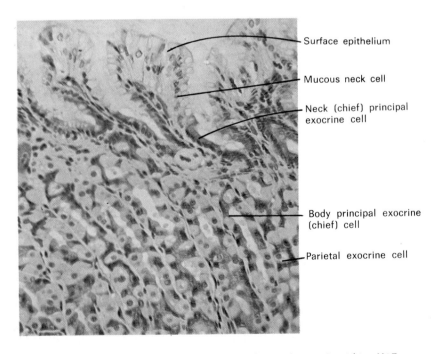

- Surface epithelium
- Mucous neck cell
- Neck (chief) principal exocrine cell
- Body principal exocrine (chief) cell
- Parietal exocrine cell

Fig. 12-43. Perpendicular section to the fundic surface. Semithin, H&E. 385 ×.

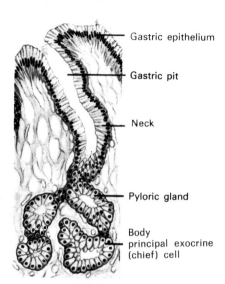

Fig. 12-44, A. Pyloric glands, (Gillison, 1953.)

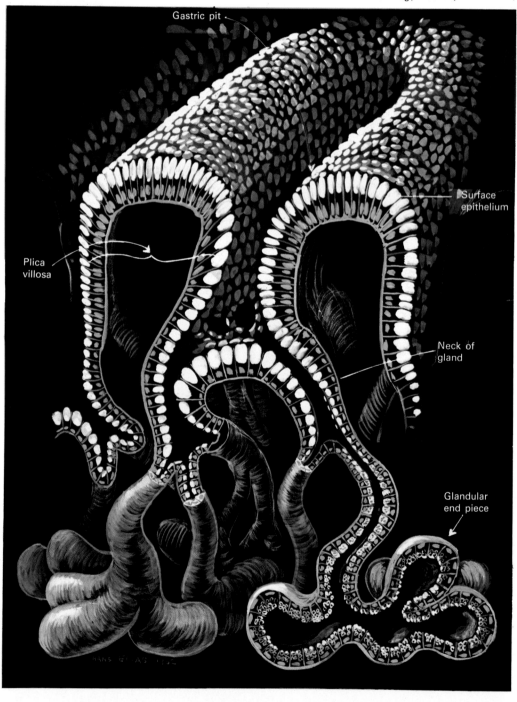

Fig. 12-44, B. A gastric pit in the pyloric antrum (from H. Elias der Magen 1965, Ingelheim, Courtesy of Boehringer Sohn.)

by discharging through the free cell surface. Along the luminal surface, the principal exocrine cells are studded with short microvilli which, in turn, bear delicate short threads, comparable to the microvillar antennulae of the gallbladder.

Wedged between the principal exocrine cells and often pushed toward the periphery away from the gland lumen are the large *parietal exocrine cells* (Fig. 12-36). Secretion from the most peripheral parietal cells can reach the glandular lumen through *intercellular canaliculi*. The parietal cells appear to be filled with strongly acidophilic granules. This acidophilia was difficult to reconcile with the known fact that these cells are instrumental in the production of *hydrochloric acid*. However,

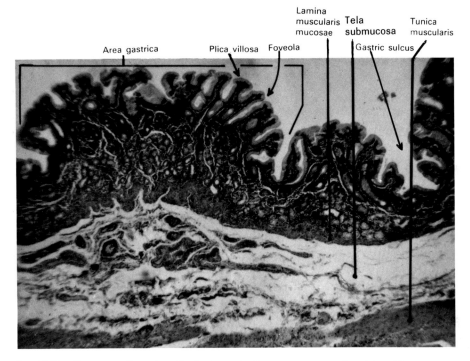

Fig. 12-45. Cross section of wall of stomach. Gastric areas and sulci. (From pyloric antrum.)

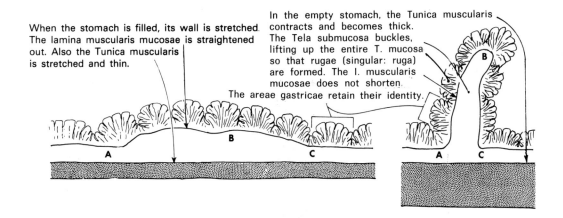

Fig. 12-46. Mechanism of adjustment of the stomach wall to capacity.

294 DIGESTIVE APPARATUS

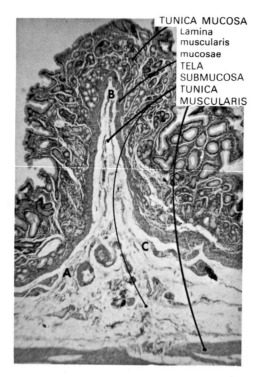

Fig. 12-47. A ruga from the pyloric antrum.

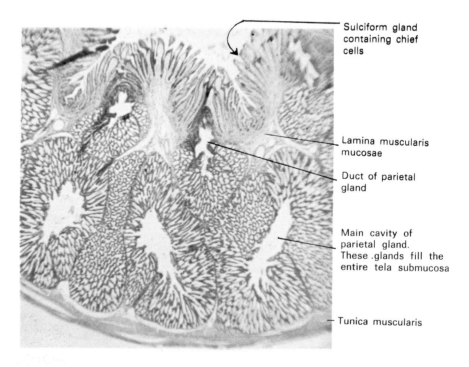

Fig. 12-48. Proventriculus of a chicken, H&E. 60 ×.

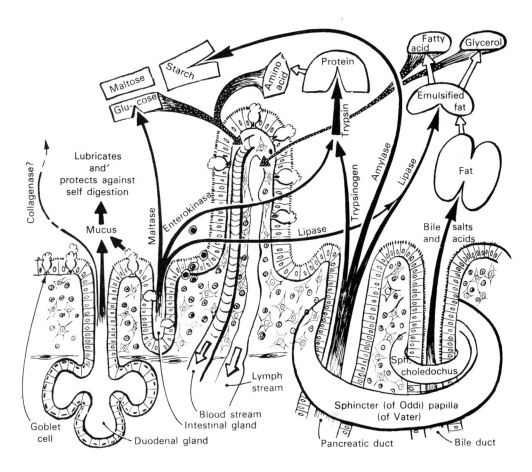

Fig. 12-49. Diagram of intestinal functions.

electron microscopy has shown clearly that the acidophilic granules are mitochondria. The hydrochloric acid produced at the cell surface may form as follows: (1) carbonic anhydrase forms carbonic acid from water and carbon dioxide, (2) the mixture of sodium chloride and carbon acid then yields hydrochloric acid and bicarbonate. In the human, the parietal cells are implicated in the production of *intrinsic factor* (of Castle) which is necessary for intestinal absorption of vitamin B12. That facet of the cell which faces the glandular lumen appears as a little mouth into which a network of *intracellular secretory canaliculi empty*. The canaliculi as well as the luminal surface are studded with microvilli which increase the surface area (Fig. 12-36). Secretory canaliculi can be visualized microscopically with the oil-immersion lens (Fig. 12-37) and can be demonstrated by silver impregnation techniques (Fig. 12-38). A few *gastric endocrine cells* occur deep in the gastric glands (Fig. 12-33, whitish stippled cells). The function of these cells will be considered with the intestine.

Pyloric glands (Figs. 12-44, A, and 12-44, B) occur in the pyloric antrum and in the pyloric canal as short, tortuous, branched, tubular mucous glands with fairly large lumina which empty into deep, wide pits. The pyloric glands are approximately one-half pit and one-half gland. In addition to mucus, the gastric endocrine cells of these glands produce the hormone *gastrin* when protein-rich food-stuffs contact the pyloric mucosa. After traveling in the bloodstream through the liver, heart,

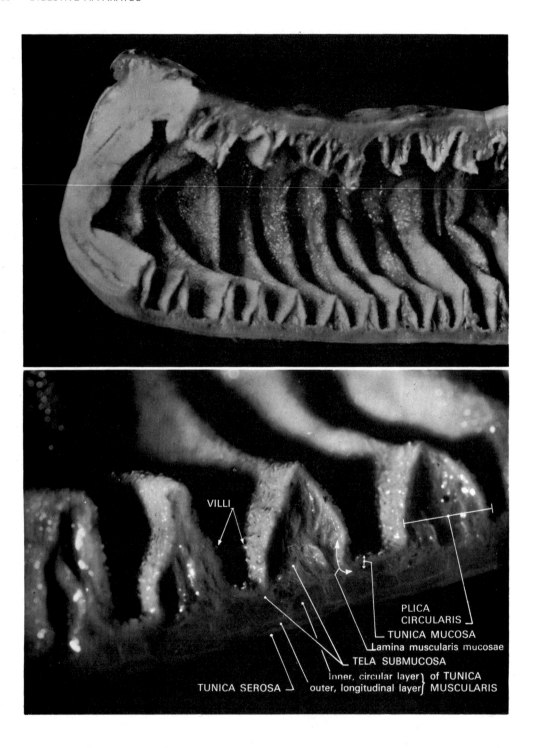

Fig. 12-50. Internal view of duodenum, surgical specimen. This piece of gut was filled with formol as soon as it was excised. As a consequence, the smooth muscle cells of the lamina muscularis mucosa contracted giving the circular plica a broader and lower shape than usual.

lung and aorta, the hormone returns to the stomach to stimulate the fundic glands to produce pepsinogen and hydrochloric acid. The pepsinogen is converted to pepsin in the presence of hydrochloric acid.

Small, permanent furrows (gastric sulci) in the tunica mucosa of the stomach (Figs. 12-30, 12-31 and 12-32) divide it into circumscribed, permanent hills, the *gastric areas* (Fig. 12-45).

When the stomach is full (Fig. 12-46, left) its internal surface, except for that in the pyloric region, appears smooth to the naked eye (though the gastric areas never disappear). When it is empty, the tunica muscularia contracts (Fig. 12-46, right). While the external surface remains smooth, internally, the tunica mucosa and tela submucosa are thrown into longitudinally oriented folds, the *gastric rugae* (Figs. 12-34, 12-46 and 12-47). Two of these longitudinal rugae near the minor curvature, touching at their tops, create a canal (the so-called "Magenstrasse" = stomach street) through which saliva and very small quantities of liquid may flow from the esophagus directly to the duodenum (Fig. 12-34).

PYLORUS. The pyloric sphincter is a local thickening of the circular layer of the tunica muscularis. Submucosal glands from the duodenum penetrate for a short distance into the pyloric canal.

The mammalian stomach, of which the human is a representative case, is not the only type of stomach found in vertebrates. A highly specialized form is the glandular stomach (proventriculus) of birds (Fig. 12-48), where all the principal exocrine cells are confined to the sulciform (slit-like) glands of the tunica mucosa and all the parietal exocrine cells are located in large submucosal, tubuloalveolar glands, the ducts of which are lined with mucous neck cells.

Even among mammals, and particularly in the ruminants, stomachs are highly specialized and very different from the hu-

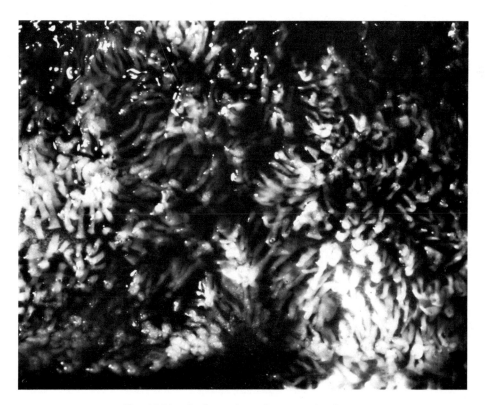

Fig. 12-51. Surface view of intestinal villi. 24 ×.

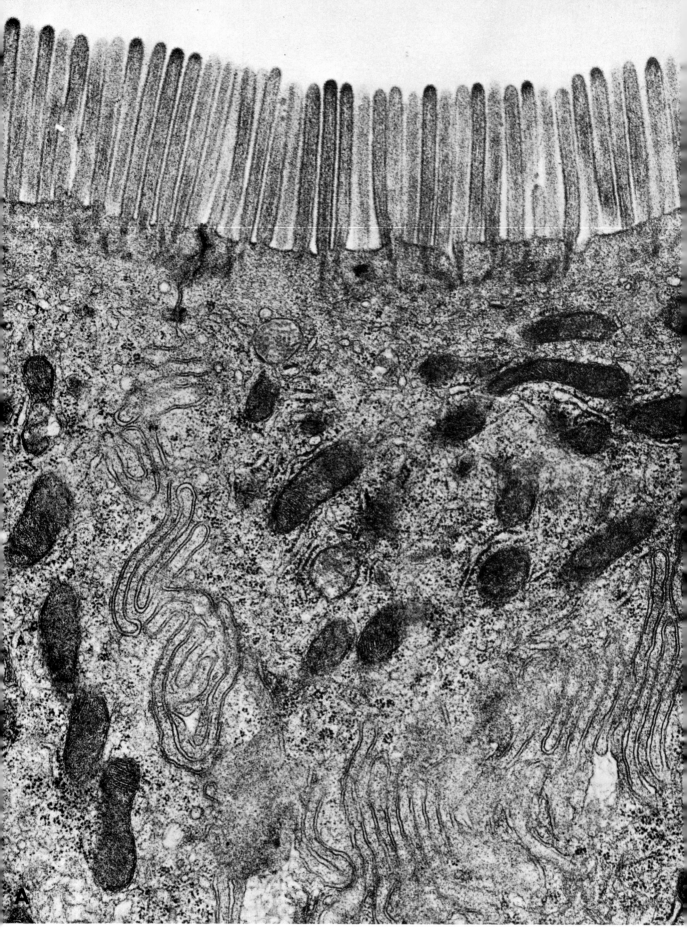

Fig. 12-52. A: Top of absorbing cell from small intestine. Picture by Prof. W. C. Forssman, Univ. of Heidelberg E. M.

man stomach (see textbooks of veterinary histology). The horse, the rat and Procavia have stomachs different from the human. In these three species, esophageal stratified squamous epithelium lines the proximal third of the body of the stomach leaving only the distal two thirds similar in structure to the human. Observations of this kind should alert experimental workers to the fact that animals are not just convenient replicas of man. Of importance is the transition from secretory epithelium of the stomach to absorptive epithelium of the small intestine.

Small Intestine

Digestion of food is completed in the small intestine, and foodstuffs as well as liquids are absorbed by its wall. The ma-

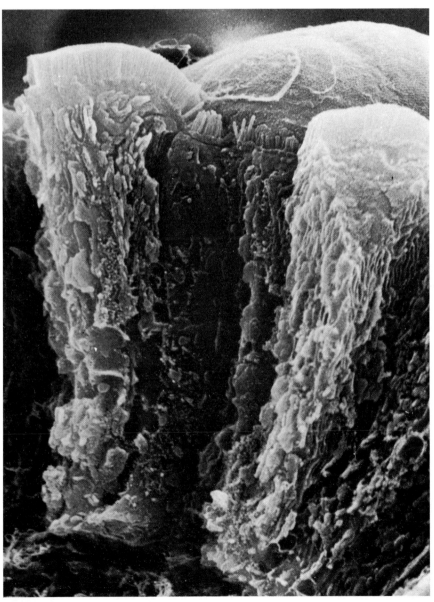

B

Fig. 12-52. B: Scanning EM of a columnar absorptive cells. Courtesy of Dr. J.-P. Revel, Cal. Tech.

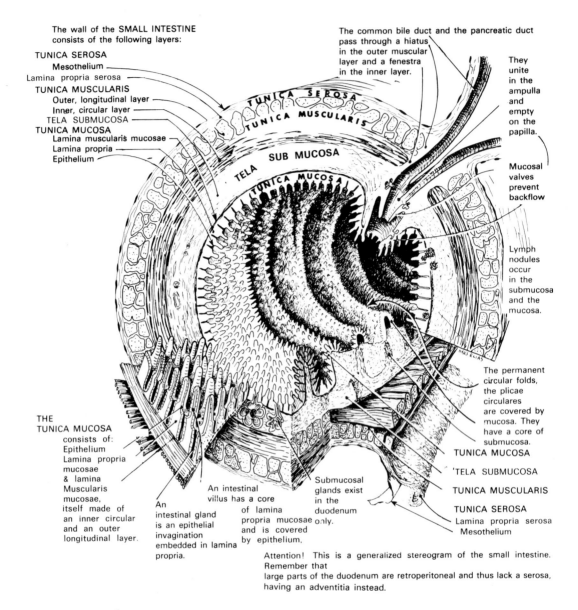

Fig. 12-53. The small intestine.

jor functions of the small intestine are summarized in Figure 12-49. Intestinal digestion is accomplished by secretions of the intestinal glands and the pancreas aided by components of bile produced in the liver. Duodenal glands and goblet cells are mainly protective in function, though the former may produce a weak, proteolytic enzyme.

Absorption occurs through cells of the surface epithelium.

The internal surface of the small intestine available for absorption is tremendously increased by: (1) circular folds (*plicae circulares*) (Figure 12-50), (2) villi (Fig. 12-51), and (3) microvilli (Fig. 12-52, A and B). The layers of the wall of the small intestine are shown in Figures 12-53, and 12-54 where their arrangement can be studied visually.

LAYERS OF THE WALL. The surface epithelium (Figs. 12-52, A and B, and 12-55) of the tunica mucosa is simple columnar with a "striated border" of microvilli.

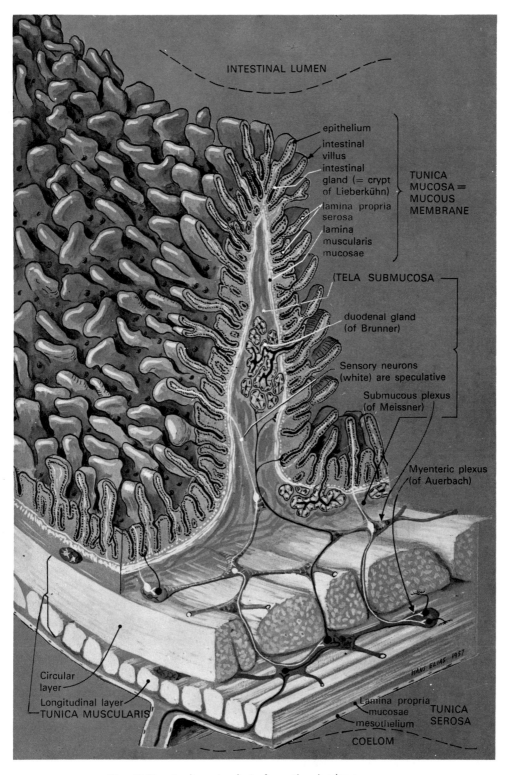

Fig. 12-54. A plica circularis from the duodenum.

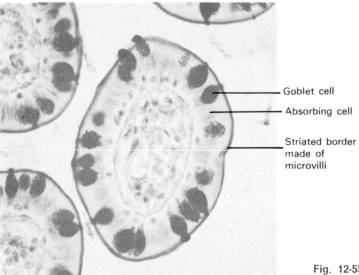

Fig. 12-55. Cross section through an intestinal villus. (Mucicarmin) 750 ×.

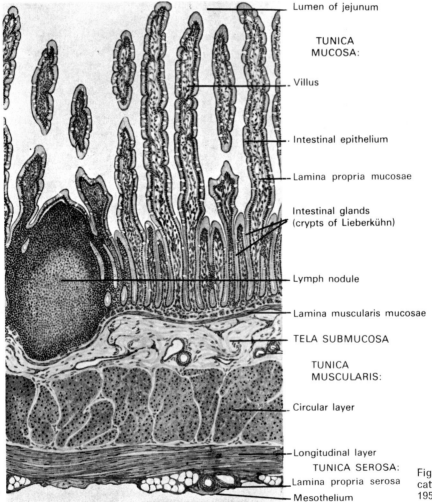

Fig. 12-56. Jejunum of cat. (From Gillison 1953.)

SMALL INTESTINE 303

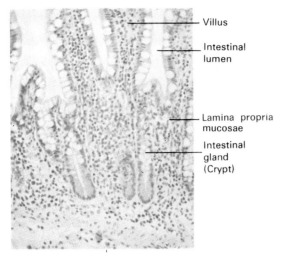

Fig. 12-57. Intestinal villi and crypts, H&E. 250×.

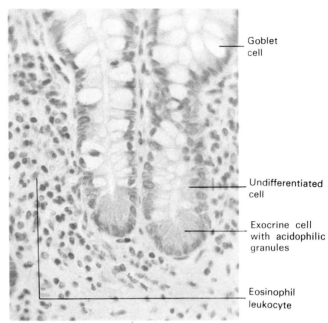

Fig. 12-58. Two intestinal glands (crypts of Lieberkühn), H&E. 675×.

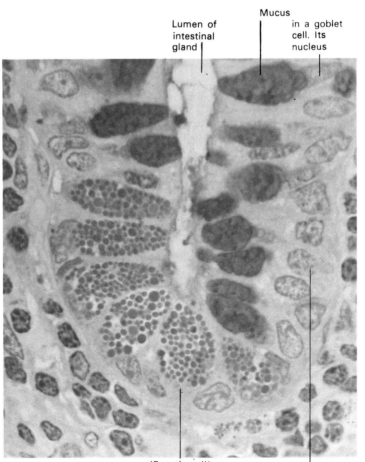

Fig. 12-59. Bottom of intestinal gland. Semithin section, H&E. 2500×.

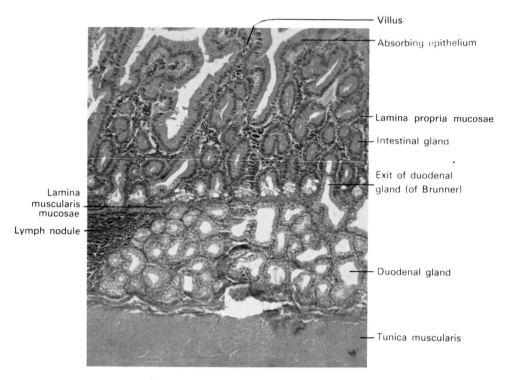

Fig. 12-60. Duodenum, H&E. 200 ×.

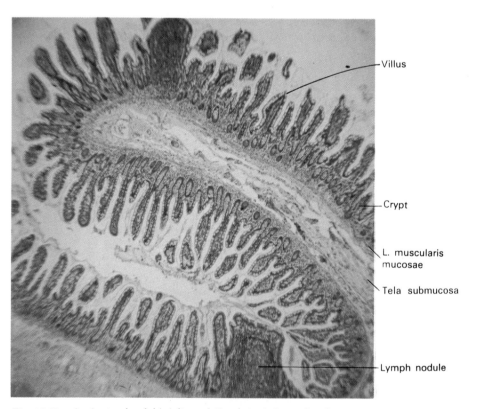

Fig. 12-61. A: A circular fold (plica of Kerckring) from distal jejunum, H&E. 75 ×.

Goblet cells found between the absorbing columnar cells can be identified best in sections through villi (Fig. 12-55).

The lamina propria mucosae is a mixture of some areolar and much reticular (lymphoid) connective tissue.

The lamina muscularis mucosae contains an inner circular and outer longitudinal layer, with small bundles of smooth myocytes extending into the lamina propria, even into the villi.

The tela submucosa consists of areolar connective tissue. The tunica muscularis possesses an inner circular and an outer longitudinal layer of smooth myocytes arranged in coarse bundles. The outermost layer is the tunica serosa, (the *visceral peritoneum*). A partial exception is the duodenum which is retroperitoneal. The anterior surface is covered with tunica serosa, while the posterior surface nests in a tunica adventitia. All of these features can be seen in Figures 12-53 to 12-56.

A delicate mucosal and a coarse submucosal plexus of lymph vessels (Fig. 10-11) is important in removing lipids and excess tissue fluids.

INTESTINAL GLANDS (CRYPTS OF LIEBERKUHN) extending from the surface epithelium into the lamina propria mucosae, are narrow, simple tubular glands (Figs. 12-

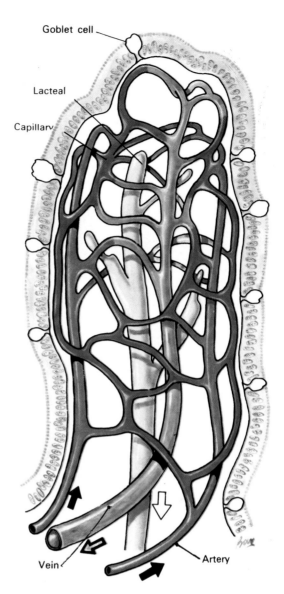

Fig. 12-62. Intestinal villi from duodenum.

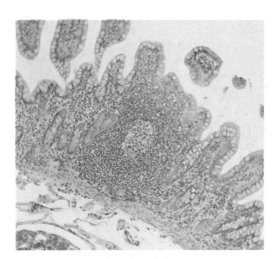

Fig. 12-61. B: A lymph nodule in the tunica mucosa of the distal jejunum, H&E. 200 ×.

57, and 12-58) lined by low, simple columnar epithelial cells interspersed with goblet cells. These glands open between adjacent villi. The epithelial cells in the crypts are lower than the more differentiated surface cells. Relatively undifferentiated cells replicate and subsequently migrate toward the tip of the villus, replacing old, worn-out cells which slough from the tip of the villus.

Exocrine cells with acidophilic granules (Figs. 12-58 and 12-59) reside in the bases of the intestinal glands. These cells (the former Paneth cells) reach their greatest number in the ileum and appendix, and occasionally, may occur in the large intestine. Though their precise role is not clear, there is increasing evidence that the acidophilic granules contain bacteriocidal enzymes (in rats). Earlier speculation that they contained a variety of digestive enzymes has not been substantiated. While other cells in the crypts appear to migrate, these can not divide or maneuver out of the crypts. They appear to differentiate from primitive cells which give rise to epithelial cells.

DUODENAL GLANDS (BRUNNER). *Duodenal submucosal* glands (Fig. 12-60) con-

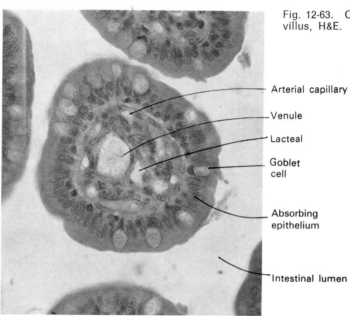

Fig. 12-63. Cross sections through an intestinal villus, H&E. 780 ×.

Fig. 12-64. Intestinal villi of rat; blood vessels injected with India ink, counter-stained with borax-carmine. 208 ×.

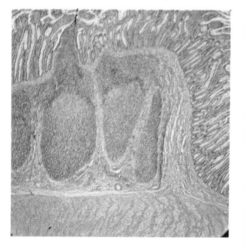

Fig. 12-64. A. Wall of ileum with aggregated lymph nodules. (Peyer's patch). H&E, 27 ×.

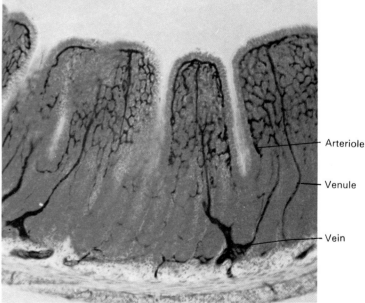

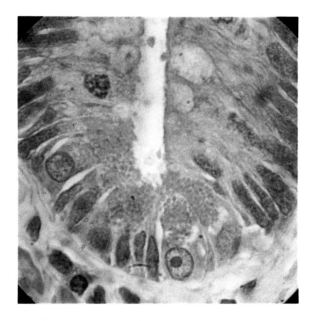

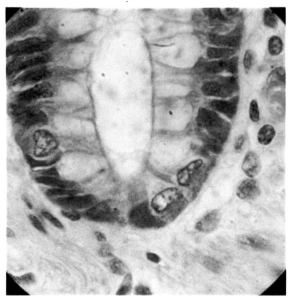

Fig. 12-65. Enteroendocrine cell (B) vs exocrine cell (A). Granules in exocrine cells (Paneth) empty into grand lumen. Granules of enteroendocrine cells, are delivered into capillaries next to base of cell.

fined to the duodenum are most numerous in its proximal portion, up to the duodenal papilla but may project in the pyloric canal. They become less numerous distally and are absent from the distal end of the duodenum, the jejunum and ileum. Duodenal glands are branched, alveolar, mucous glands located chiefly in the submucosa. Usually, they empty through intestinal glands, rarely through ducts of their own.

PLICAE CIRCULARES (KERCKRING). Plicae circulares (Figs. 12-50, 12-53, 12-54 and 12-61, A) are projections of the tela submucosa and therefore are covered with the complete tunica mucosa. The tunica muscularis does not participate in their formation. Unlike the gastric rugae, the plicae circulares of the small intestine are permanent structures which cannot be straightened out. In the proximal part of the duodenum, duodenal glands may extend into the core of the plicae. The circular folds are most prominent and numerous in the jejunum.

VILLI. Villi (Figs. 12-51, 12-54, 12-55, and 12-62) are projections, often cylindrical (Figs. 12-51, 12-55, 12-63), often broad (Figs. 12-54 and 12-64) and flat, covered with intestinal surface epithelium. Their vascular components, depicted in Figure 12-62, are embedded in a core of lamina propria macosae containing some smooth myocytes from the lamina muscularis mucosae. The network of capillaries with peripheral arteriolar supply and axial venous drainage is deployed just under the epithelium (Figs. 12-62 and 12-64). A branched lymph capillary, the *lacteal*, is located near the axis of a villus. The epithelium covering the villi glides continuously upward to replace the most apical cells which are shed. The epithelium (including goblet cells) is replenished by the vigorously proliferating cells of the intestinal glands. As these cells emerge from the crypts and slide onto the free surface, they undergo a profound change in structure and function. They acquire a tall columnar shape and develop a luminal striated border consisting of microvilli, thus changing their activity to absorption. The villi are tallest and most slender in the jejunum, shortest and broadest in the ileum.

NERVOUS PLEXUSES. The nervous plexuses pervading the entire digestive tract from the esophagus to the rectum are

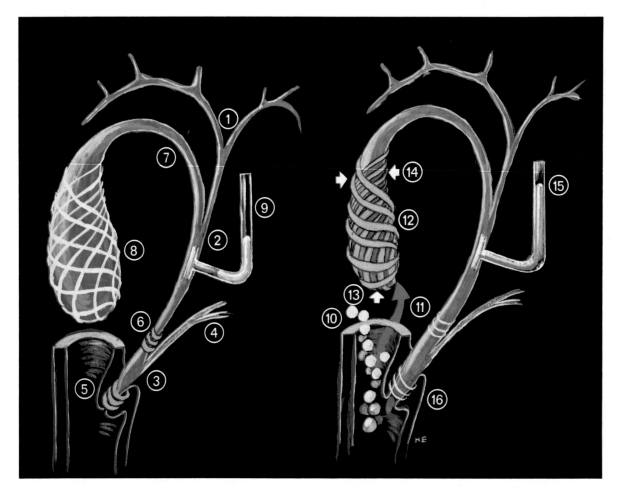

Fig. 12-66. Action of the musculature of the extrahepatic biliary passages. Bile, continuously produced in the liver, flows through the hepatic duct (1) into the common bile duct (2). In the duodenal papilla (3) it is mixed with pancreatic juice which arrives through the pancreatic duct (4). When the sphincter ampullae (5) and/or the sphincter choledochus (6) are closed, the bile backs up through the cystic duct (7) into the gallbladder (8). The sphincter of Oddi (5+6) is always closed when the bile pressure (9) is low. The arrival of fat (10) in the duodenum causes production of cholecystokinin (11) by the intestinal glands of the duodenum. Cholecystokinin is a hormone which elicits contraction of the gallbladder musculature (12). This elevates the fundus (13) and narrows the bladder (14). Thus, bile pressure (15) rises, and the sphincter of Oddi (16) opens.

shown in Figures 12-23, A, 12-24 and 12-54.

LYMPH NODULES. Isolated lymph nodules are found scattered in the lamina propria mucosae (Fig. 12-56, 60, 61, A&B). In the lower ileum, *aggregated lymph nodules* (formerly Peyer's patches), consisting of up to 70 nodules which impinge on the surface epithelium, usually are classified among the lympho-epithelial organs such as tonsils (Fig. 12-64, A). Penetration of the nodule bases through the lamina muscularis mucosae may leave isolated patches of smooth myocytes within the nodules. Aggregated lymph nodules, like the palatine tonsils, are more prominent in children. With aging, there is a concomitant regression and involution of lymphatic tissue, which correlates well with a decreased efficiency in the immune response.

HORMONES. The small intestine produces a variety of hormones or hormone-like substances in response to the presence of *chyme* in the intestine.

Previously, histologists recognized only

two types of enteroendocrine cells: *argentaffin cells*, which reduce silver directly (no pretreatment with extraneous reducing agents required), and *argyrophil cells* which require pretreatment with a reducing agent. The argentaffin cell was also called an *enterochromaffin cell* because the endocrine granules stained well with potassium bichromate, and other bichromate salts (Fig. 11-65). Intense studies in the last few years have made possible the identification of several additional entero-

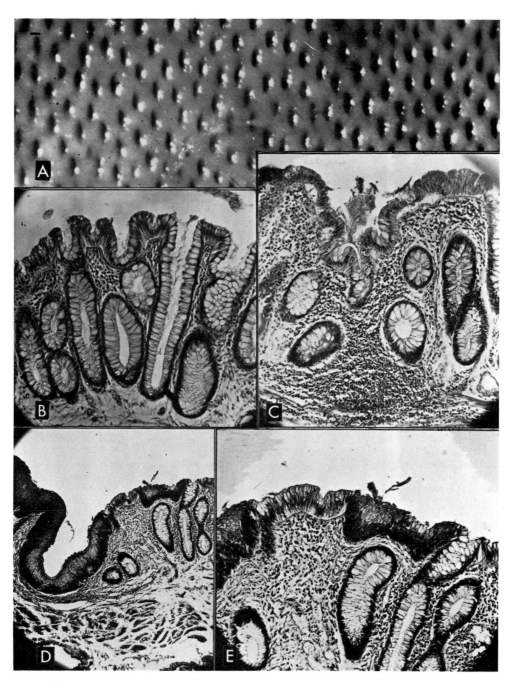

Fig. 12-67. The large intestine. A, Close-up view of inner lining of colon. B and C, Rectum. D and E, Upper-lower anal canal junction.

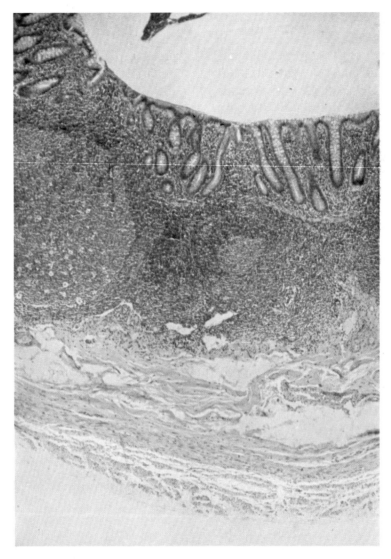

Fig. 12-68. A. Wall of vermiform appendix, H&E.

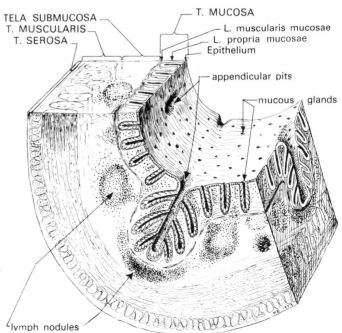

Fig. 12-68. B. Appendix vermiformis.

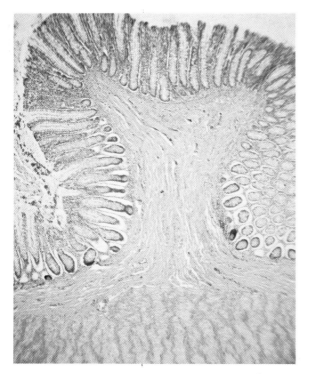

Fig. 12-69. Caecum with one plica semilunaris. H&E. 60 ×.

endocrine cells. Indeed, the gastrointestinal tract is now recognized as an important, diffuse endocrine organ. The hormones secreted by these cell types are primarily polypeptides. Accordingly, the secretory cells are designated as *APUD cells*, an acronym for "*Amine Precursor Uptake* and *Decarboxylation*". The APUD cells not only secrete a variety of hormones, but also can synthesize biogenic amines. Enteroendocrine cells, situated between the principal exocrine cells of gastric glands and between the secretory cells of intestinal glands, are recognized in light micrographs close to the lamina propria (Fig. 12-65).

Hormones produced in the stomach are: *serotonin* (Stimulates smooth muscle contractions), *endorphin* (has a morphine-like action), *histamine* (in phyloric glands, acting as a local stimulus on parietal cells to secrete HCl), *gastrin* (stimulates secretion of hydrochloric acid by the parietal cells), *somatostatin* (produced in the midzone of gastric glands, acts to inhibit release of growth hormone from the adenohypophysis), *glucagon* (acts as a biologic antagonist to insulin), *enteroglucagon* (in fundic glands, acts to reduce bowel motility thus permitting greater, time for digestion), and *vasoactive intestinal peptide* (in the stomach fundus; action like that of glucagon and secretin; also dilates small intestinal arteries, raises blood-glucose levels, and stimulates the secretion of the exocrine pancreas). The intestines secrete the following hormones:

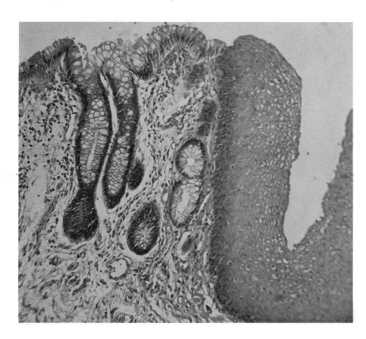

Fig. 12-70. Junction of simple columnar epithelium of the upper anal canal with the stratified squamous (mucosal) epithelium of the lower anal canal. H&E. 185 ×.

serotonin, endorphin, histamine, gastrin, gastroinhibitory polypeptide (inhibits secretion of gastric acid and pepsin, and stimulates insulin release), *somatostatin, motilin* (stimulates contraction of stomach smooth musculature), *glucagon, enteroglucagon, vasoactive intestinal peptide*, and *noradrenalin* (increases blood pressure).

Secretin is released into the blood

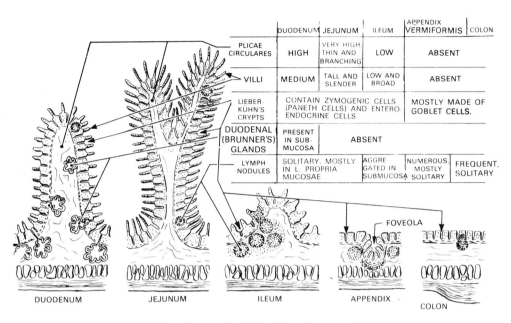

Fig. 12-71. Synopsis of the intestine.

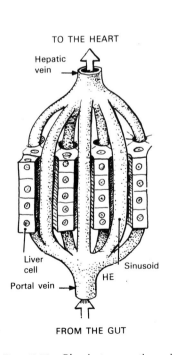

Fig. 12-72. Blood streams through the liver.

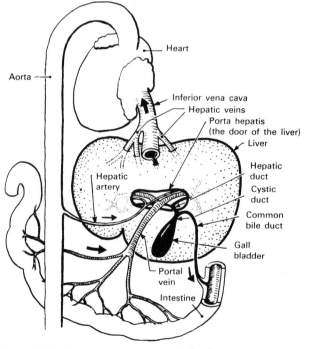

Fig. 12-73. Supply and drainage of the liver, seen from behind.

stream from the proximal portion of the small intestine. Secretin stimulates the pancreas to release large quantities of fluid containing bicarbonate ions. The bicarbonate reacts with hydrochloric acid in the chyme to yield sodium chloride and carbonic acid. The latter immediately dissociates into water and carbon dioxide.

Pancreozymin released by the proximal portion of the small intestine in response

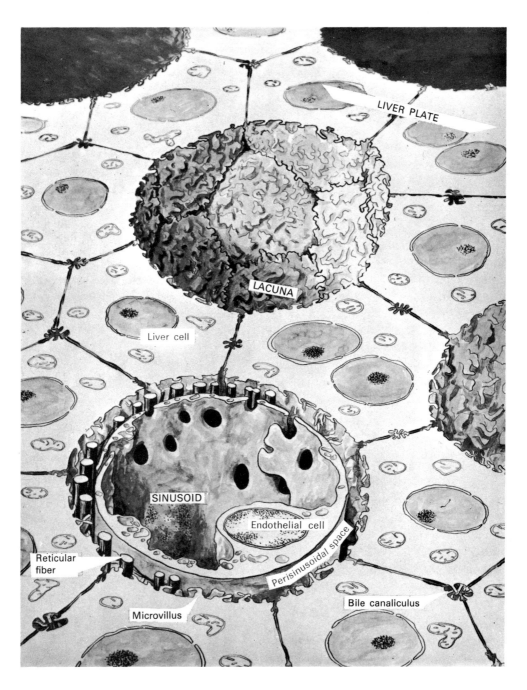

Fig. 12-74. Fine structure of the liver parenchyma. (From Elias: Das Leberparenchym. Nordmark-Werke, Hamburg, 1965.)

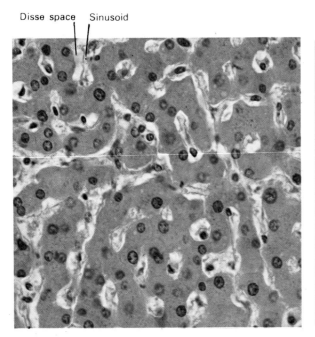

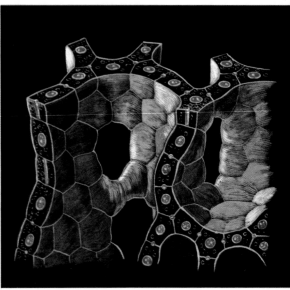

Fig. 12-75. Liver, autopsy (=necropsy) specimen.

Fig. 12-76. A liver plate.

to the presence of chyme, stimulates the pancreas to release large amounts of digestive enzyme.

Cholecystokinin released by the proximal portion of the small intestine in response to the presence of fats in the chyme, enters the bloodstream and causes contraction of smooth muscle in the wall of the gallbladder to release bile (study Figure 12-66). Bile squirts past the duodenal sphincter into the lumen of the small intestine with every passing wave of relaxation which precedes every wave of peristaltic contraction. The presence of fat in the chyme in the small intestine also causes the secretion of the hormone *enterogastrone* which inhibits the motility of the stomach.

Large Intestine

There are no villi in the large intestine, and hence the internal surface appears smooth. Regularly spaced pits, visible with a magnifying lens are the exit ducts of simple, tubular glands (Fig. 12-67, A). The surface epithelium and that lining the simple, tubular glands (crypts) is simple columnar with an abundance of goblet cells (Fig. 12-67, B and C).

VERMIFORM APPENDIX. In the vermiform appendix (Fig. 12-68, A and B) the intestinal glands are far apart from each other. In certain specimens, one finds large, oblong depressions, the appendicular pits, from which simple, tubular glands arise. The epithelium, lining lumen and the gland pits, consists of very small, ba-

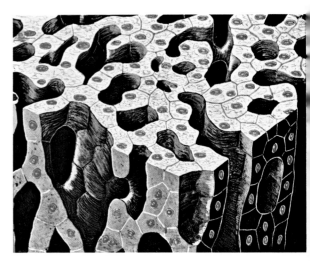

Fig. 12-77. The hepatic muralium.

Fig. 12-78. The network of sinusoids, suspended in the labyrinth of hepatic lacunae.

sophilic cuboidal cells. The tela submucosa contains a fair number of solitary lymph nodules but only during inflammation (appendicitis) do they approach the prominence of aggregated lymph nodes of the normal ileum.

CECUM, COLON AND RECTUM. Histologically, the large intestine is rather uniform. The crypts are evenly arranged, as is seen in an internal surface view (Fig. 12-67, A) and in sections (Figs. 12-67, B, 12-69 and 12-1). The lamina propria mucosae is rich in lymphoid tissue and solitary nodules (Fig. 12-1).

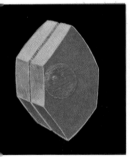
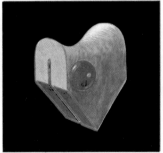
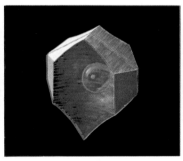
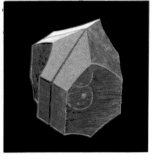

A Octahedron V ~ 12000 μm³ in the middle of a wall

B Pentahedron V ~ 9800 μm³ near a hole in a wall

C Decahedron V ~ 22000 μm³ connecting two corners

D Dodecahedron V ~ 35000 μm³ in a corner of the muralium

Fig. 12-79. The basic cell shapes of hepatocytes dependent on location in the muralium. Bile canaliculi lie between two or three liver cells (hepatocytes).

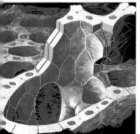

The basic cell shapes incorporated in the muralium simplex. Note also bile canaliculi.

TABLE 1	Approximate Volume (in μm³)	Area in Contact with Blood (in μm²)	Area in Contact with Other Liver Cells (in μm²)	Ratio of Area of Blood Contact / Area of Cell Contact
Type A Octahedron	12,300	1,344	1,536	0.875
Type B Pentahedron	9,840	1,344	968	1.4
Type D Decahedron	22,500	1,905	3,328	0.57
Type D Dodecahedron	34,850	2,022	3,584	0.56

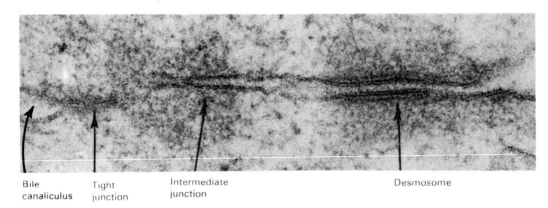

Bile canaliculus   Tight junction   Intermediate junction   Desmosome

Fig. 12-80. Connections between liver cells. (From C. Biava, Lab. Invest. *13*: 840-864, 1964.)

The longitudinal layer of the tunica muscularis forms the three equidistant bands, the *taenia coli*. Since these are shorter than the other coats of the colon, they cause periodic bulges of the colonic wall, the *haustra*. Folds between the haustra are the *plicae semilunares*. The plicae and haustra are in continuous motion. The serosa is incomplete, since large parts of the ascending and descending colon are retroperitoneal. Therefore, dorsally there are "bare areas" which are covered by an adventitia (Fig. 8-3).

At the junction of the superior and inferior parts of the anal canal, the simple columnar epithelium of the superior part changes abruptly to the stratified squamous epithelium of the inferior part. The margin is sometimes smooth as in Figure 12-70; and in other cases so intensely scalloped that in a median section of the junction (Fig. 12-67, D and E), patches of simple columnar epithelium seem to alternate with patches of stratified squamous epithelium. The internal anal sphincter is a thickened portion of the circular coat of the muscularis, and hence consists of smooth myocytes. The external anal sphincter is composed of skeletal muscle. Its myocytes are rarely packed in tight bundles and therefore exhibit the original shape of muscle cells (Fig. 6-14).

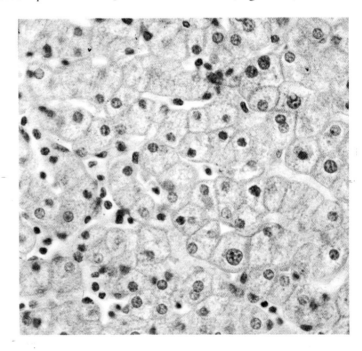

Fig. 12-81. Liver biopsy, H&E. 850 ×.

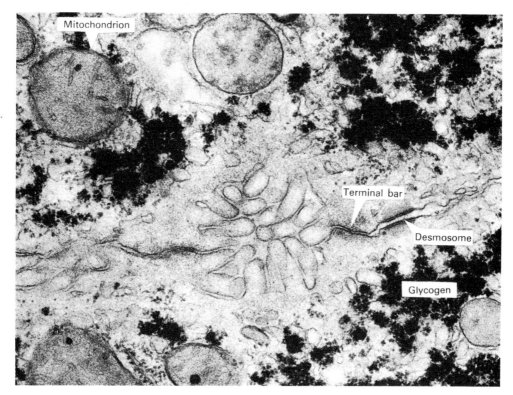

Fig. 12-82. Bile canaliculus delimited by two liver cells. (From C. Biava: Lab. Invest. 13: 840-864, 1964.)

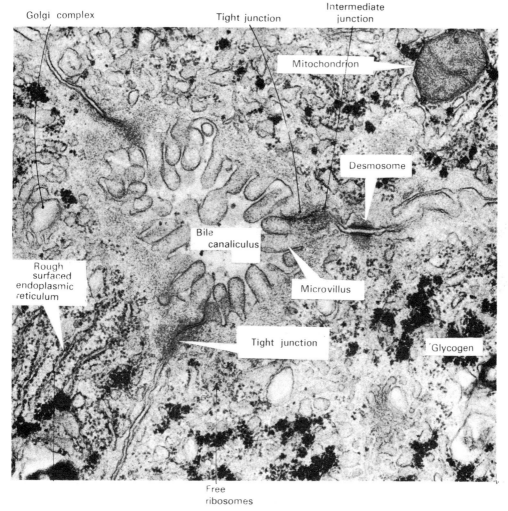

Fig. 12-83. Bile canaliculus delimited by three hepatocytes. (From C. Biava, Lab. Invest. *13*: 840-864, 1964.)

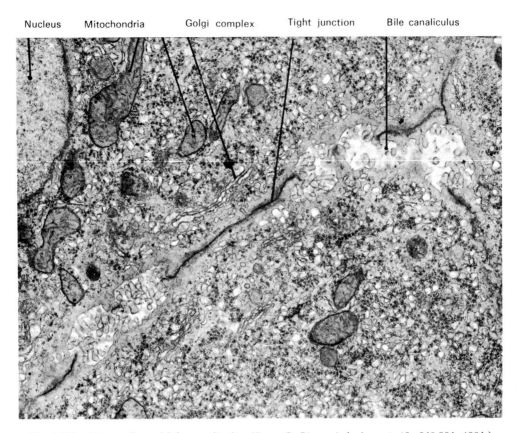

Nucleus Mitochondria Golgi complex Tight junction Bile canaliculus

Fig. 12-84. Liver cells and bile canaliculi. (From C. Biava: Lab. Invest. 13: 840-864, 1964.)

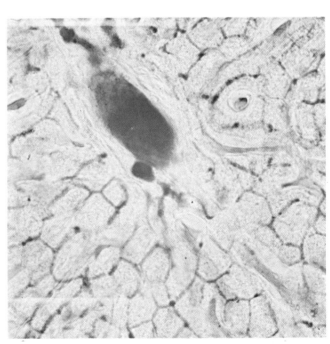

Fig. 12-85. Rabbit liver. The bile ducts and canaliculi are injected with Prussian blue suspended in gelatine, while the sinusoids are injected with carmine-gelatine. Meshes which surround a sinusoid are called vasozonal, those surrounding a cell are cytozonal. 850 x. The slide from which this photograph was taken in 1974 was prepared in Philip Stöhr's laboratory in Würzburg in 1887. It was a slide in the collection handed out to the students including Dr. Hans Elias' uncle, Dr. Siegfried Oppenheimer who gave him his collection.

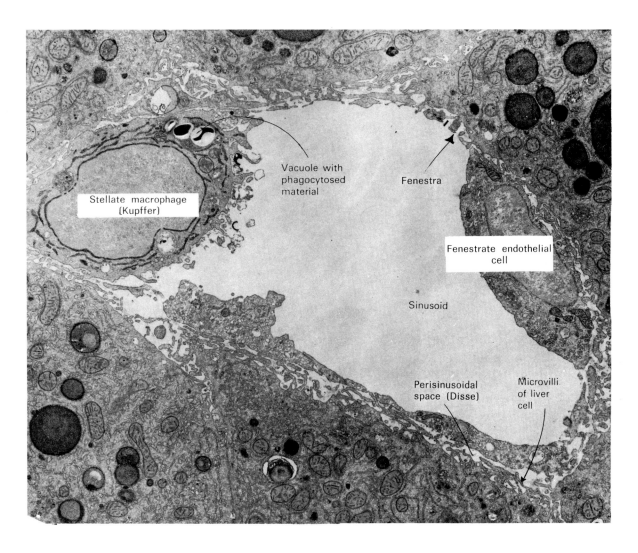

Fig. 12-86. A stellate macrophage and a fenestrated sinusoidal endothelial cell in the liver. The black reaction product indicates the presence of peroxydase. Courtesy of Drs. J.-J. Widmann and H. D. Fahim, Harvard Medical School.

Figure 12-71 gives a synopsis of the five major portions of the intestine.

Liver

The liver is a mass of endodermally derived cells around which blood streams from the intestine to the heart (Fig. 12-72 and 12-73).

The liver, although simple in structure, is perhaps the most versatile body organ. It is a: (1) storage site for food (glycogen, fat) and vitamins, (2) producer of various plasma proteins (albumin, fibrinogen, globulins, and prothrombin), (3) clearing filter for blood, (4) remover of wastes and (5) a digestive exocrine gland. The little evidence for the endogenous synthesis of heparin in the liver is questionable at this time.

The functions of storage, chemical synthesis, and excretion are performed by liver cells. By a poorly understood active transport mechanism, liver cells can move substances (such as bile dyes) against steep gradients of osmotic pressure. Liver cells also synthesize cholesterol, transform carotene into vitamin A, destroy sex hormones and detoxify drugs. Phagocytosis i performed by the *stellate macrophage cells*.

The liver is constructed of *hepatocytes* (liver epithelial cells), stellate macrophage cells (Kuffer cells), connective tissue (with very few perisinusoidal lipocytes, reticular and collagen fibers) (Fig. 12-74), and sinusoidal vessels.

The liver parenchyma is a cellular muralium simplex (Fig. 12-75), consisting of branched plates (*laminae hepatis*) one-cell thick (Fig. 12-76) which are interconnected like the walls in a building, leaving spaces, the *hepatic lacunae*, between them (Fig. 12-74 and 12-77).

Specialized capillaries, the *vas sinusoideum*, traverse the hepatic lacunae (Figs. 12-73, and 12-78).

The shape and size of each liver cell depends on its location in the muralium (Fig. 12-79, A-E). Number and chromosomal content (diploidy vs. polyploidy) of nuclei depend, in turn, on cell volume. Liver cell nuclei are spherical and possess a prominent nucleolus. Each hepatocyte is exposed to the hepatic lacunae by two or three of its facets (Figs. 12-75 through 12-79). The exposed hepatocyte surfaces are covered with short broad microvilli (Fig. 12-74). Hepatocytes adhere to one another by means of maculae adherentes (desmosomes), intermediate junctions and zonulae occludentes (tight junctions) (Fig. 12-80). The zonulae occludentes are found only along the bile canaliculi which communicate with the outer world. In comparing the liver parenchymal cells with a true epithelium, the minute portion of the cell surface turned toward the bile canaliculus is considered its apical surface; and the zonula occludens (tight junction) occupies the classical position of a junctional complex. During life, the hepatocytes usually bulge slightly into the lacunae toward which they are convex (Fig. 12-81). They are loaded with fairly large amounts of glycogen which appear as light spots in routinely treated slides (Fig. 12-81) and as black electron dense particles in electron micrographs (Figs. 12-82 and 12-83). After death, the liver cells become depleted rapidly of glycogen, so that they shrink and stain darker causing the hepatic lacunae to appear larger. This is why a necropsy liver (Fig. 12-75) looks so different from a biopsy liver (Fig. 12-81).

Bile canaliculi are formed by two or three grooves in the contact surfaces of

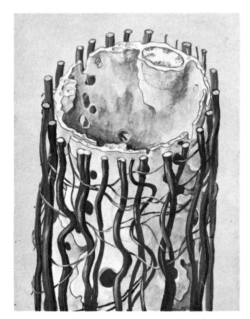

Fig. 12-87. A sinusoid supported by the network of reticular (argyrophil) fibers. (From Elias: Das Leberparenchym. Nordmark-Werke, Hamburg, 1965.)

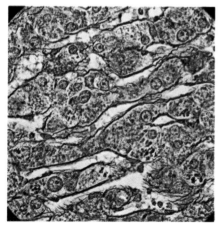

Fig. 12-88. Liver impregnated by the Wilder method. 625 ×.

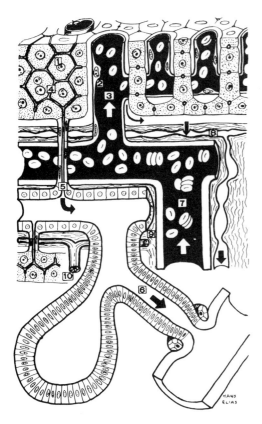

Fig. 12-89. Currents in the liver. 1, Liver cell. 2, Endothelial cell. 3, Sinusoid. 4, Bile canaliculus. 5, Ductule. 6, Common bile duct. 7, Portal vein. 8, Lymph vessel. 9, Periportal connective tissue. 10, Nerve. (From Popper and Schaffner: Liver. New York, McGraw-Hill. 1957).

two or three heaptocytes. The number of liver cells which participate in forming a canaliculus (two or three) depends on the geomtery of cell relations in the muralium (Figs. 12-74, 12-79, 12-82 and 12-83). Along the entire length of the vast network of bile canaliculi, hepatocytes are joined together by zonulae occludentes (Figs. 12-84). Cylindrical microvilli (Figs. 12-82 to 12-84) projecting into the lumina of bile canaliculi, serve to increaae the surface area.

It is possible for more than one bile canaliculus to form between two hepatocytes. The bile canaliculi form a continuous network pervading the entire liver (Fig. 12-85).

The mitochondria of liver cells (mostly spheroid, less frequently rod-shaped, occasionally cup- or disk-shaped; Figs. 12-83 and 1-9, A and B), are provided with a few small cristae mostly in the form of tubules. The granular endoplasmic reticulum consists of parallel, flat cisterns. Liver cells also contain agranular endoplasmic reticulum which may be continuous with the RER. The golgi complex is usually found not far from the bile canaliculi (Figs. 12-83 and 12-84).

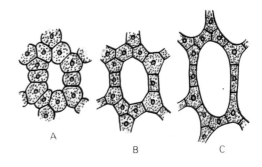

Fig. 12-90. Adaptability without disruption of the liver muralium to the degree of filling of the lacunae. A: Normal liver during life. B: Liver at autopsy. C: Congestion of the liver with blood due to heart failure.

Sinusoids of the liver are specialized capillaries lined by nonphagocytic endothelial cells. Phagocytic cells (Kupffer) are distinguishable by their stellate morphology and their content of endogenous peroxidase which can be blackened with diaminobenzidine (Fig. 12-86). Squamous, lobated endothelial cells are fenestrated, and overlap loosely without being attached firmly to one another (Figs. 12-74, 12-79 and 12-87). Some fibroblast-like lipocytes found in the *perisinusoidal space* (Space of Disse), store fat droplets which serve only as the solvent for vitamin A. These specialized lipocytes may be instrumental in fibrillogenesis. The entire sinusoid is supported, externally, by a network of reticular (argyrophilic) fibers (Figs. 12-87 and 12-88). Since the sinusoid with its surrounding fibrillar network is freely suspended in the lacuna, the latter is divided by the endothelial cells into the sinusoid and perisinusoidal space (of

Fig. 12-91. The continuous liver tissue and its arbitrary divisibility into hepatic or portal lobules. Only the latter are referred to as acini.

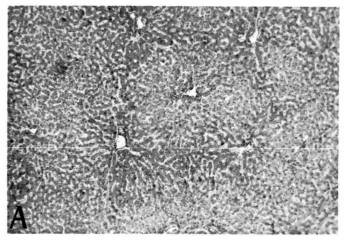

Imaginary outline of hepatic lobule — Central vein

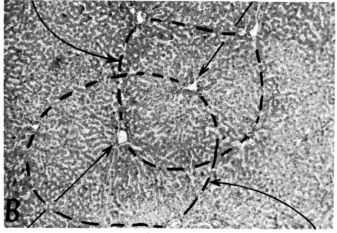

Portal canal — Imaginary outline of portal lobule (acinus)

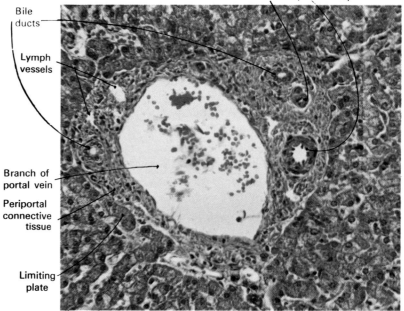

Fig. 12-92. A portal canal, Goldner stain. 400 ×.

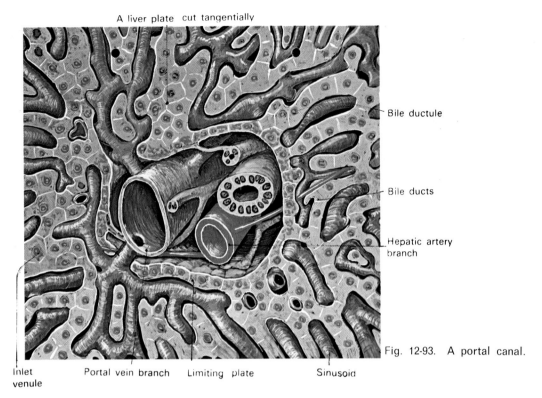

Fig. 12-93. A portal canal.

Disse), (Figs. 12-74, 12-75 and 12-86). During life (Fig. 12-81), the perisinusoidal space is so narrow as to be unresolvable with the light microscope in biopsy material. During the agonal phase, however, anoxia causes increased permeability of the sinusoidal wall to plasma, perhaps through enlargement of the pores and gaps in and between the endothelial cells. Thus, the perisinusoidal spaces become quite apparent after death (Fig. 12-75). Via the fenestrations in the endothelial cells and the gaps between them, the sinusoid openly communicates with the perisinusoidal space. Thus, hepatocytes come in direct contact with the plasma, while the reticular network prevents escape of blood corpuscles. Seepage of plasma through fenestrations and gaps in the sinusoidal wall results in the flow of liquid in the perisinusoidal space toward the portal canal. This fluid enters, by diffusion, into the interlobular lymph vessels.

The directions of flow of blood, tissue fluid, lymph and bile are shown in Figure 12-89.

The perisinusoidal space also communicates with clefts between liver cells, but never with bile canaliculi, for these are completely sealed off by zonulae occludentes. Only if the larger bile ducts are obstructed mechanically, as by a tumor or by gallstones, do the zonulae occludentes rupture so that bile spills into the perisinusoidal spaces and from there into the blood that flows in the sinusoids; in such cases bile also enters the lymph. Thus, obstructive jaundice is created.

The "plastic" hepatocytes can adapt their shapes to the volume of the lacunae, which depends on the volume of blood in the sinusoids and the volume of fluid in the perisinusoidal spaces (Fig. 12-90). The muralium of liver cells is continuous throughout the organ (Fig. 12-91), as is the labyrinth of lacunae, the network of sinusoids and that of the bile canaliculi.

The liver is surrounded by a perivascular fibrous capsule (*Glisson's*) (Fig. 8-7) covered with mesothelium wherever it faces the peritoneal cavity. Continuations of the capsule penetrate through the porta hepatis (Fig. 12-73) into the liver to provide support for blood vessels, ducts, lymph vessels and nerves. Against this periportal connective tissue (Figs. 12-92 and 12-93), the muralium is sharply set off by a single layer of liver cells, the *limiting plate*, itself a part of the muralium (Figs. 12-92, 12-93-and 12-94). If one compares the liver with a building, the liver plates correspond to internal walls with doors separating and connecting rooms; the limiting plate corresponds to the outer wall and to the walls of halls, corridors and staircases.

The inward extensions of the fibrous capsule (Fig. 12-95, to the right of 12) into the liver run in tunnels, called *portal*

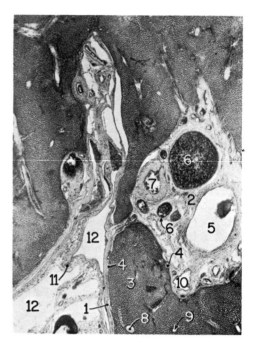

Fig. 12-95. Region of the ligamentum teres. To the left is the left lobe of the liver to the right the quadrate lobe. The arteries were injected with latex; therefore appear much thicker than during life. 1, Capsule (visceral peritoneum). 2, Portal canal. 3, Liver parenchyma. 4, Limiting plate. 5, Ramus quadratus venae portae. 6, Hepatic arterial branches inflated with latex. 7, Bile duct with epithelium lacerated. 8, Sublobular vein. 9, Central vein. 10, Lymph vessel. 11, Ligamentum teres hepatis. 12, Peritoneal cavity.

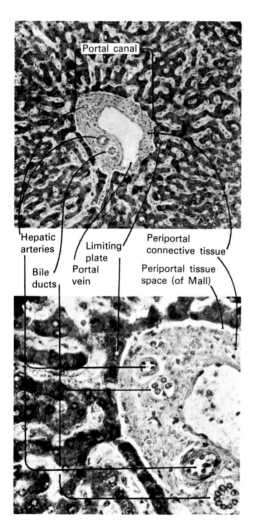

Fig. 12-94. A portal canal.

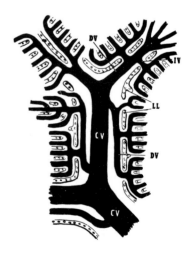

Fig. 12-96. Terminal portal vein branches.

LIVER 325

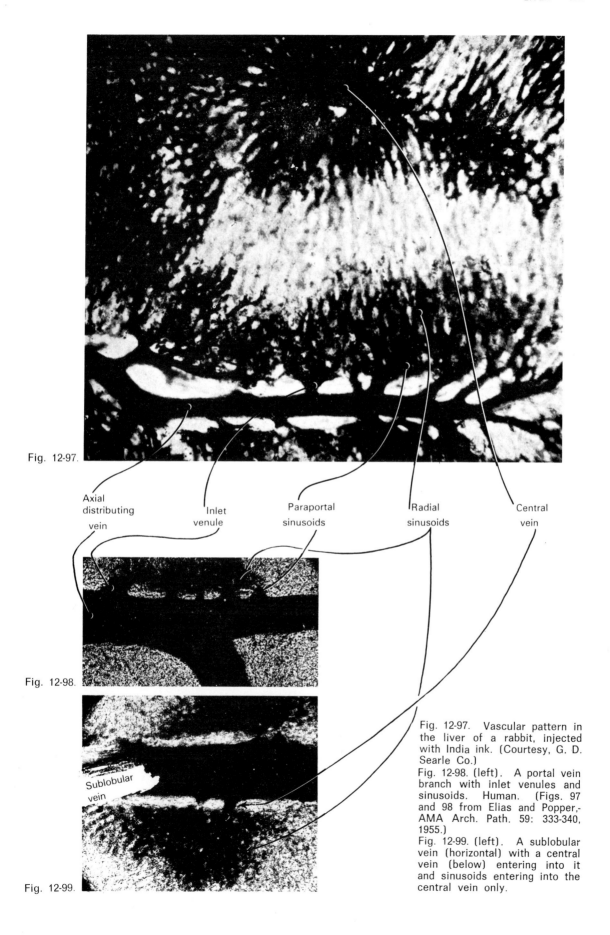

Fig. 12-97.

Axial distributing vein · Inlet venule · Paraportal sinusoids · Radial sinusoids · Central vein

Fig. 12-98.

Sublobular vein

Fig. 12-99.

Fig. 12-97. Vascular pattern in the liver of a rabbit, injected with India ink. (Courtesy, G. D. Searle Co.)

Fig. 12-98. (left). A portal vein branch with inlet venules and sinusoids. Human. (Figs. 97 and 98 from Elias and Popper,- AMA Arch. Path. 59: 333-340, 1955.)

Fig. 12-99. (left). A sublobular vein (horizontal) with a central vein (below) entering into it and sinusoids entering into the central vein only.

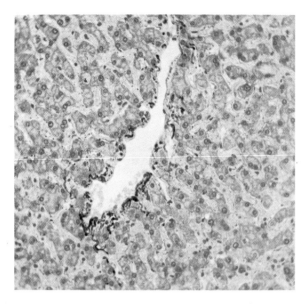

Fig. 12-100. A central vein with sinusoids entering into it, Azan. 375 ×.

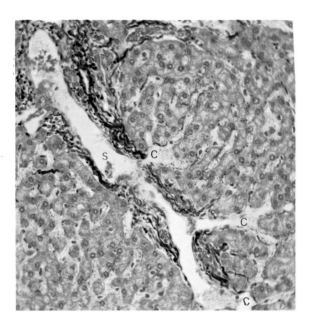

Fig. 12-101. Four central veins entering into a sublobular vein, Azan. 375 ×. Note that these tributaries to hepatic veins *are not* accompanied by ducts or arteries. C = Central veins; S = = Sublobular veins.

Lymph vessels and nerves also run in the portal canals.

The portal vein, upon entering the porta hepatis, divides into left and right trunks (fig. 12-73). The rami arising from the trunks enter into portal canals and divide with them. The mode of ramification is shown schematically in Figure 12-96. Conducting veins (CV) with diameters larger than 400 μm, give rise to distributing veins. From the walls of distributing veins, inlet venules (IV) arise which pierce the limiting plate (LL) beyond which they join paraportal sinusoids. The inlet venules supply the entire network of sinusoids (Fig. 12-97). Figure 12-97 shows inlet venules arising from an axial distributing vein in a rabbit; the same arrangement from a human liver is shown in Figure 12-98.

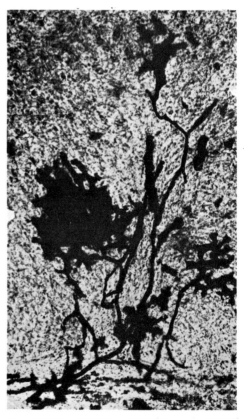

Fig. 12-102. Hepatic arterioles and arterial capillaries injected with India ink. They open into sinusoids (large, black spots). (From Elias and Petty: Anat. Rec. 116: 9-18, 1953.)

canals (Figs. 12-92 and 12-93), surrounded by the limiting plate. Portal canals contain branches of the *portal vein*, the *hepatic artery* and tributaries of the *hepatic duct*. These three tube systems constitute the *"portal triad"*.

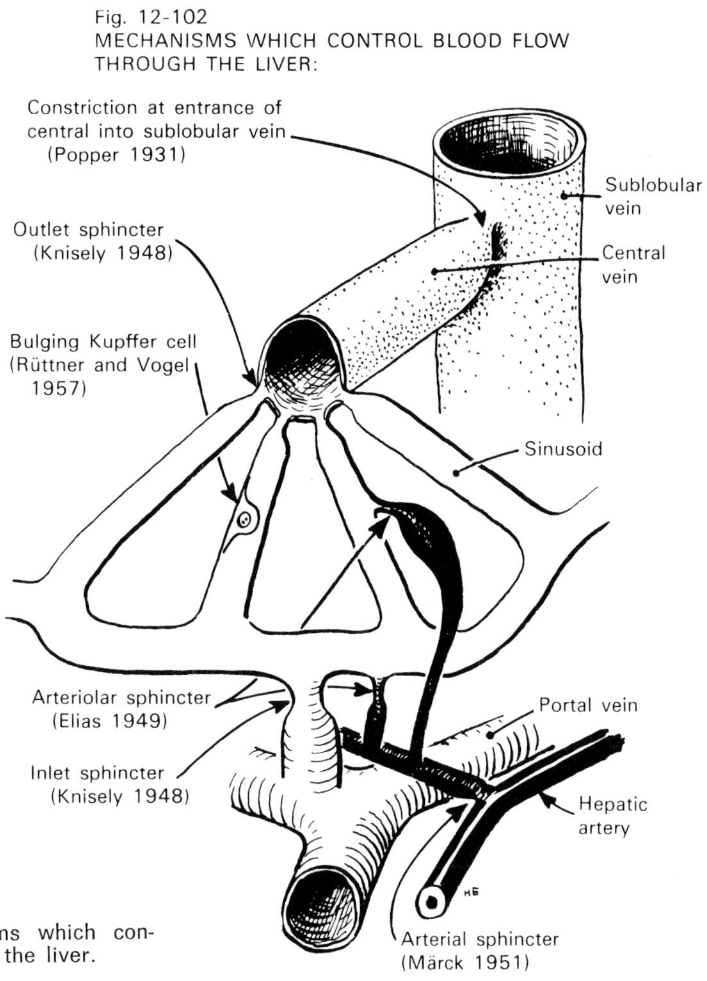

Fig. 12-102
MECHANISMS WHICH CONTROL BLOOD FLOW THROUGH THE LIVER:

Fig. 12-103. Mechanisms which control blood flow through the liver.

Sinusoids converge toward the smallest tributaries of the hepatic veins, called *central veins*. In man all the sinusoids enter into central veins as seen in Figures 12-99, 12-100 and 12-101. (In many kinds of animals the arrangement is different.) Central veins enter *sublobular veins* (Figs. 12-99 and 12-101) which empty through collecting veins of ever increasing caliber. The largest veins are *hepatic veins* which join the inferior vena cava (Fig. 12-73).

Hepatic arteries accompany branches of the portal vein in the portal canals. One to five arteries may run in one portal canal (Figs. 12-92 and 12-93), forming a plexus which surrounds the branch of the portal vein. Certain arterial branches supply the walls of bile ducts and other structures in the portal canal, and drain into inlet ven-

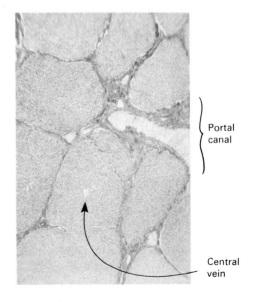

Fig. 12-104. Pig liver, Azan. 45 ×.

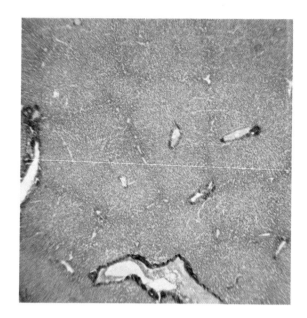

Fig. 12-105. Human liver, Azan. 45 ×.

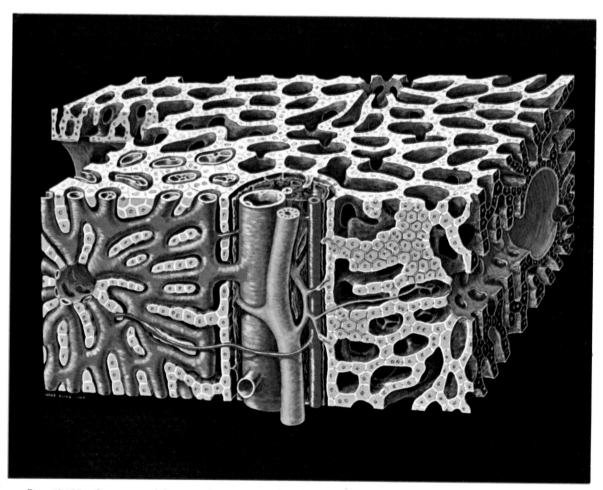

Fig. 12-106. Summary of liver architecture. Pink: The muralium composed of one-cell-thick plates made of liver cells (hepatocytes). Green: Bile canaliculi, bile ductules (cholangioles) and bile ducts. Violet: Portal vein branches and sinusoids. Red: Branch of hepatic artery and arterioles. Blue: Hepatic veins including three central veins and one sublobular vein. Sinusoids are fenestrated. Yellow: Lymph vessels.

Fig. 12-107. Vessels and ducts in a dissected human liver seen from behind and below. Colors as in Fig. 12-105. (From Elias and Sherrick, Morphology of the Liver, Academic Press, N. Y., 1969.)

ules. Most of the hepatic arterial blood, however, enters directly into sinusoids at all lobular levels (Fig. 12-102). At points of branching, the hepatic arteries are provided with sphincters. Sphincters also seem to exist around portal inlet venules and around outlets of the sinusoids.

Blood flow through the liver is controlled at various levels by specific mechanisms as summarized in Figure 12-103.

Lymph and bile flow in a direction opposite to that of the blood flow (Fig. 12-89).

With normal blood pressure, many sinusoids run radially toward the *central vein*, producing a pattern, which appears to divide the liver (faintly) into *hepatic lobules*, each with a *central vein* in its geometric center (Fig. 12-91, B). However, when the blood pressure gradient between portal and hepatic veins drops or is reversed (experimentally or pathologically), the sinusoids radiate from portal canals, and *portal lobules* appear (Fig. 12-91, B). Portal lobules are sometimes called *acini*. The liver shown in Figure 12-91 is from an autopsy specimen where no blood pressure exists. One can divide it, arbitrarily, into portal or hepatic lobules (Fig. 12-91, B). However, all these divisions are artificial, for in reality the entire liver is one uninterrupted continuum.

The hepatic lobule may be divided into three different areas or zones. The liver parenchyma nearest the portal canal is the portal or paraportal zone. The area around the central vein is in the central or paracentral zone. Between the portal (sometimes called peripheral) and central zones is the intermediate or midzonal region. The different zones have different structural and functional characteristics. For example, after complete glycogen depletion caused by fasting, feeding

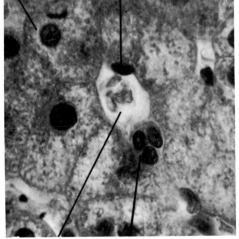

Fig. 12-108. Sinusoid and intralobular ductule (surgical biopsy specimen).

330 DIGESTIVE APPARATUS

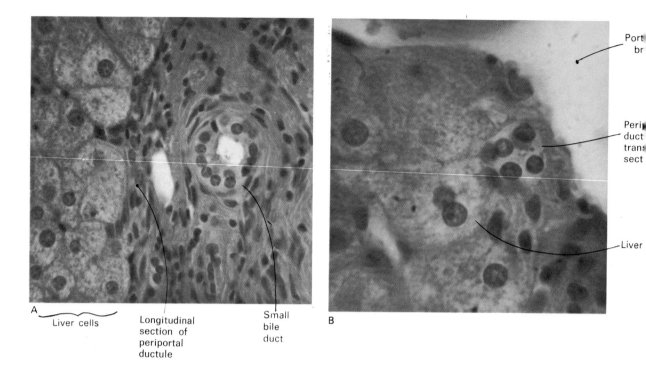

Fig. 12-109. A and B. Marginal area of a portal canal. Biopsy, H&E. (From Elias. Die Gallenwege, Boehringer, Ingelheim, 1970.)

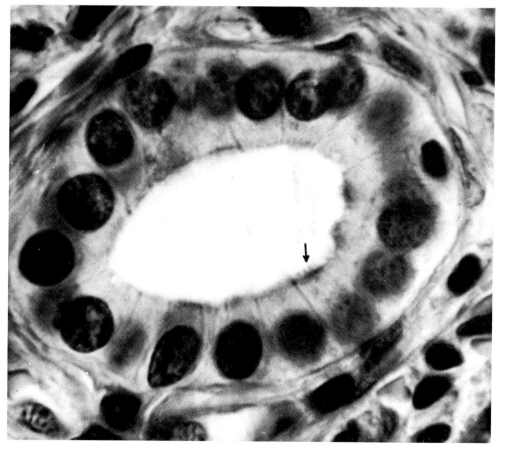

Fig. 12-110. A small bile duct showing microvilli (arrow). Biopsy. 2000 ×.

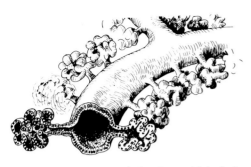

Fig. 12-111. Stereogram of the duct which drains the quadrate lobe and the inferior and intermediate portions of the left lobe of the liver. At the upper left, the duct which drains the superior left segment enters the larger bile duct.

causes the deposition of glycogen in the periportal zone first and the paracentral zone last.

The liver of adult pigs (Fig. 12-104) is divided into distinct lobules by connective tissue septa which connect portal canals. Some investigators consider this lobulation as a degenerative change, because the liver of newborn pigs is not lobulated and resembles the human liver (Fig. 12-105), the parenchyma of which is a continuum. Different kinds of lobulation occur in various animals such as the polar bear, the camel, the raccoon and others.

When liver parenchyma is lost, as by surgical removal, the remaining hepatocytes enter into DNA synthesis and subsequent mitosis with the result that the original weight of the liver is completely and accurately restored (regeneration).

Figure 12-106 summarizes the architecture of the human liver. It shows, among other things, that portal canals run almost perpendicular to the hepatic veins. This interdigitation exists on the microscopic as well as on the macroscopic level (Fig. 12-107). Except in the region of the hilus, arteries and ducts accompany the branches of the portal vein.

DUCTS. *Biliary passages* begin with the

Opening of mucous gland

Fig. 12-112. Internal surface of common bile duct. 30 ×.

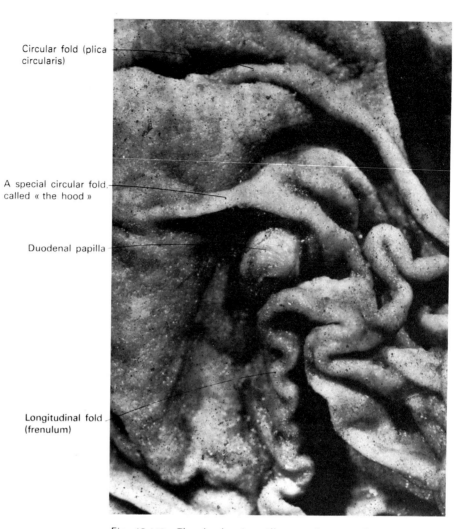

Fig. 12-113. The duodenal papilla seen from inside the intestine.

bile canaliculi. The bile canaliculi form networks of polygonal meshes between the liver cells (Figs. 12-79, E and 12-85) and empty through intraparenchymal ductules (Fig. 12-108) and paraportal ductules (*cholangioles, "canals of Hering"*) into *bile ducts* of ever increasing caliber lined by simple cuboidal to columnar epithelium (Figs. 12-93, 12-106 and 12-109). The cross-sections of the smallest ductules (Figs. 12-108 and 12-109, A and B) are smaller than that of a single hepatocyte. The epithelium of the entire system of bile ductules and ducts is provided with widely spaced microvilli visible even with the light microscope (Fig. 12-110). The ducts form networks in the portal canals (figs. 12-93 and 12-106) which converge into extrahepatic bile ducts (Fig. 12-111) and the latter into the *hepatic duct* (Fig. 12-73). This duct branches into the *cystic duct* and the *common bile duct (ductus choledochus)*. These extrahepatic bile ducts are lined by a tall, simple columnar epithelium with microvilli. Around the smaller extrahepatic bile ducts, mucous and mixed glands are arranged in two opposite rows (Fig. 12-111). The tunica mucosa of the common bile duct forms low folds and shallow circular depressions (Fig. 12-112). Small, mucous glands located mainly in the tela submucosa empty into these depressions. The tela submucosa and tunica muscularis of the common bile duct are

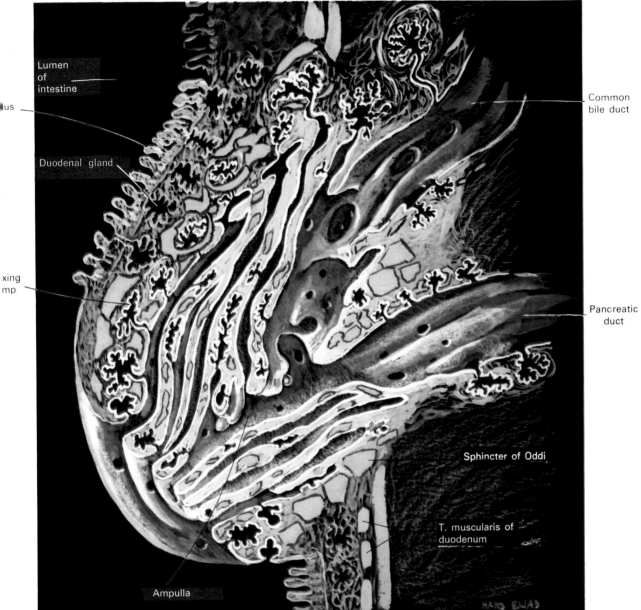

Fig. 12-114. Reconstruction of the duodenal ampulla and papilla. (From Elias, Die Gallenwege, Boehringer, Ingelheim, 1970.)

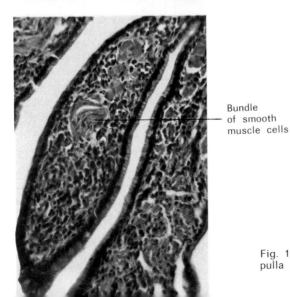

Fig. 12-115. Musculature in a valve of the duodenal ampulla (same source), gallocyanin and picrofuchsin.

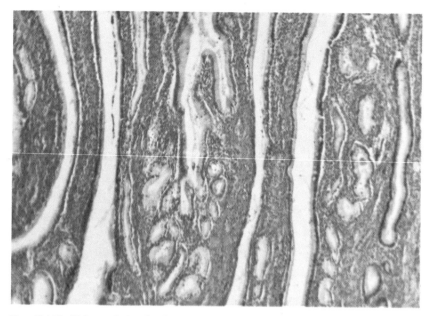

Fig. 12-116. Valves of the duodenal ampulla containing branched mucous glands (same source), gallocyanin and picrofuchsin.

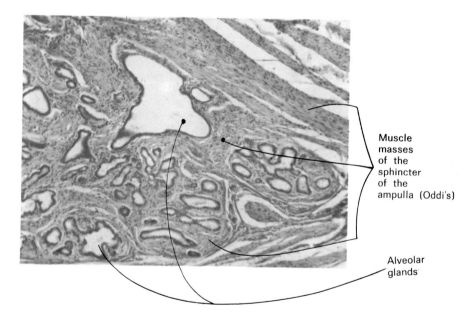

Fig. 12-117. Alveolar glands surrounded by strong smooth muscle bundles in the wall of the duodenal papilla (same source), gallocyanin and picrofuchsin.

Fig. 12-118. Internal view of the gall bladder. (From Elias, Die Gallenwege, Boehringer, Ingelheim, 1970.)

interwoven. The relative amount of smooth myocytes increases toward the end of the common bile duct, where it forms the *sphincter of the ampulla (Oddi)*. In all infants and in many adults, the end of the common bile duct unites with the end of the main pancreatic duct in the *hepato-pancreatic ampulla (Vater)*, located in the wall of the duodenum. A portion of intestinal wall containing the ampulla protrudes into

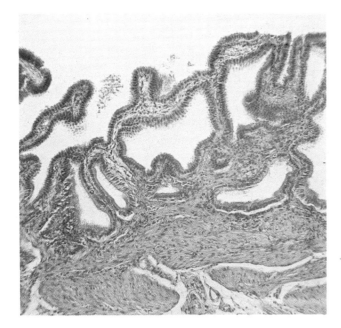

Fig. 12-119. Wall of the gall bladder, hematoxylin and picrofuchsin. 200 ×.

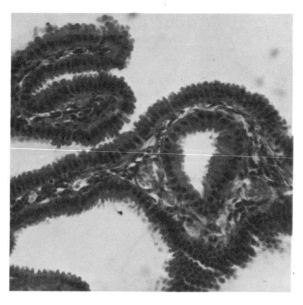

Fig. 12-120. Ridges of the gall bladder, hematoxylin and picrofuchsin. 620 ×.

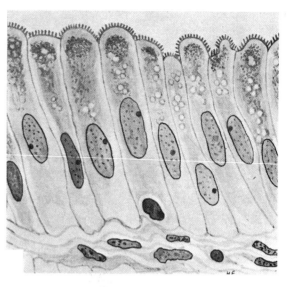

Fig. 12-121. Epithelium of the gall bladder drawn from slides prepared by Drs. Eglitis and Hayes, Ohio State University.

the gut forming the *duodenal papilla* (Figs. 12-113 and 12-114). The mucosal folds of the common bile duct gradually increase in height and complexity so that, in the intramural portion of the duct and in the ampulla, they enclose branched diverticula (Fig. 12-114) which prevent flow of chyme into the ducts. In addition, these mucosal folds contain bundles of smooth myocytes which serve to mix bile and pancreatic juice (Fig. 12-115). Also branched mucous glands are found in these valves (Fig. 12-116). In the outer wall of the ampulla large alveolar glands are found surrounded by thick bundles of smooth myocytes which are constituents of the sphincter of the ampulla (Fig. 12-117). Alternating contraction and relaxation of these muscles could expell and re-aspire liquid from and into these alveoli aiding the mixing process.

The Gallbladder

On its external anterior surface, the gallbladder is covered with an adventitia that is fused with the capsule of the liver. On its posterior surface and at the apex, its covering is a tunica serosa, the visceral peritoneum.

The internal surface of the gallbladder shows connecting ridges and folds (Fig. 12-118). In an oblique section (Fig. 12-119), connections between ridges can be seen. The lamina muscularis mucosae and hence the tela submucosa are absent. The very thick tunica muscularis consists of three spiralling layers. The folds of the

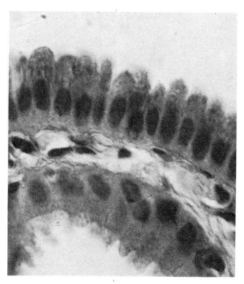

Fig. 12-122. Epithelium of the gall bladder, hematoxylin and picrofuchsin. 1900 ×.

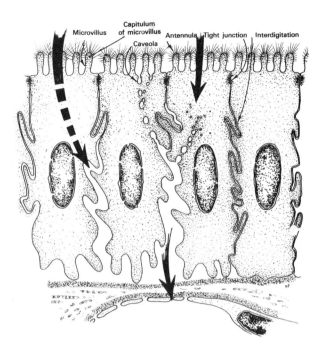

Fig. 12-123. Diagram of gallbladder epithelium based on the work by Kaye, Lane, Wheeler and Whitlock. At rest: right. During water reabsorption: left. (Anat. Rec. 151: 369, 1965.)

Fig. 12-124. Internal view of cystic duct. 100 ×. (From Elias and Sherrick, Morphology of the Liver, Academic Press, N. Y., 1969.)

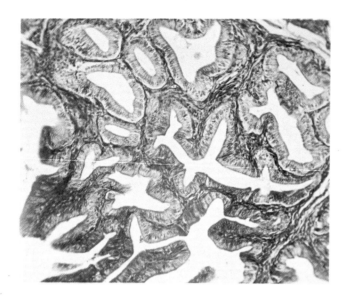

Fig. 12-125. The mucosal muralium of the cystic duct, Azan. 200 ×.

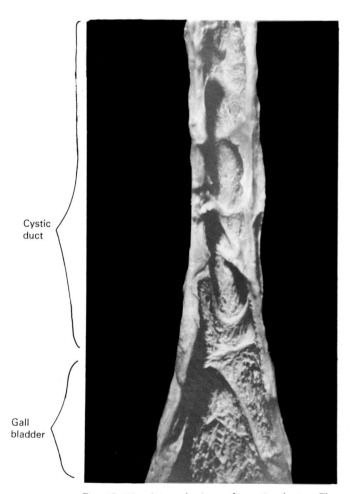

Fig. 12-126. Internal view of cystic duct. The large, internal ridges are parts of the spiral valve. 3 ×.

gallbladder are supported by a highly vascularized, areolar lamina propria mucosae covered by a tall simple columnar epithelium specialized for water reabsorption (Figs. 12-119, 12-120 and 12-123). These cells (Fig. 12-121) have typically dome-shaped apices covered with approximately 140 deeply staining microvilli on each cell luminal surface. The microvilli are studded with thin filamentous processes, the antennulae of the microvilli. The luminal margins of the cells are connected tightly by junctional complexes. Laterally, the cells interdigitate with one another. The cytoplasm of the columnar epithelial cells contains granules and bubbles of characteristic sizes at different levels. During rest, the intercellular spaces are narrow and interdigitations are tight throughout the height of the epithelium. This condition is shown in Figure 12-123 at the right. During active water re-absorption (Fig. 12-123), the intercellular spaces and interdigitations open up (left and middle). Water must pass through the distal cell surfaces, since the zonulae occludentes prevent intercellular passage. Entrance into the cells occurs either by a diffusion-like process (upper arrow in Fig. 12-123) or by micropinocytosis. The water is then released into the intercellular spaces. Three possible, hypothetical

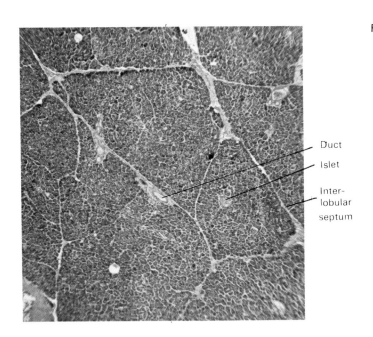

Fig. 12-127. Pancreas, H&E. 55 ×.

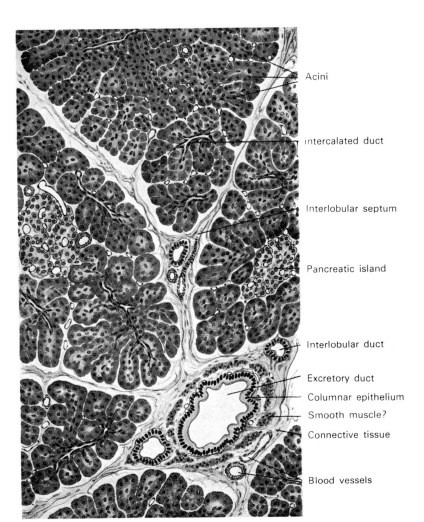

Fig. 12-128. Pancreas. (From Gillison; Histology of the Body Tissues. Edinburgh, 1953.)

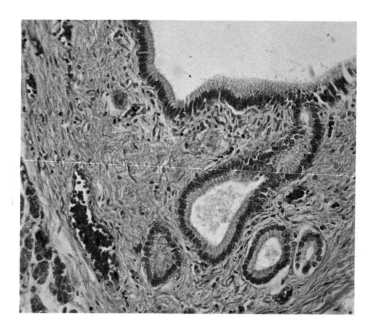

Fig. 12-129. Small and large interlobular ducts in the pancreas, 420 ×.

mechanisms for releasing water are illustrated in Figure 12-123. Finally, the water must pass through two basement membranes and endothelial fenestrations into a capillary (bottom of Fig. 12-123).

In the gallbladder, bile is stored and concentrated five to ten-fold. When bile is needed in the intestine for the emulsification of fat, cholecystokinin is released into the bloodstream by the enteroendocrine cells in the intestinal crypts. The sphincter of the ampulla relaxes and the tunica muscularis of the gallbladder contracts, lifting its fundus so that, as through a syphon, bile will flow freely. Figure 12-66, summarizes the mechanisms of bile storage and release. Deficient bile production and/or release results in poor absorption of fats and of the fat-soluble vitamins, A, D, E and K.

The tunica mucosa of the cystic duct resembles that of the gallbladder, but the

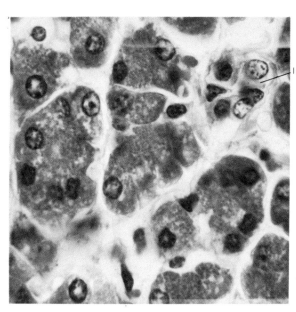

Fig. 12-130. Pancreas, showing an intercalated duct. 1900 ×.

THE GALLBLADDER 341

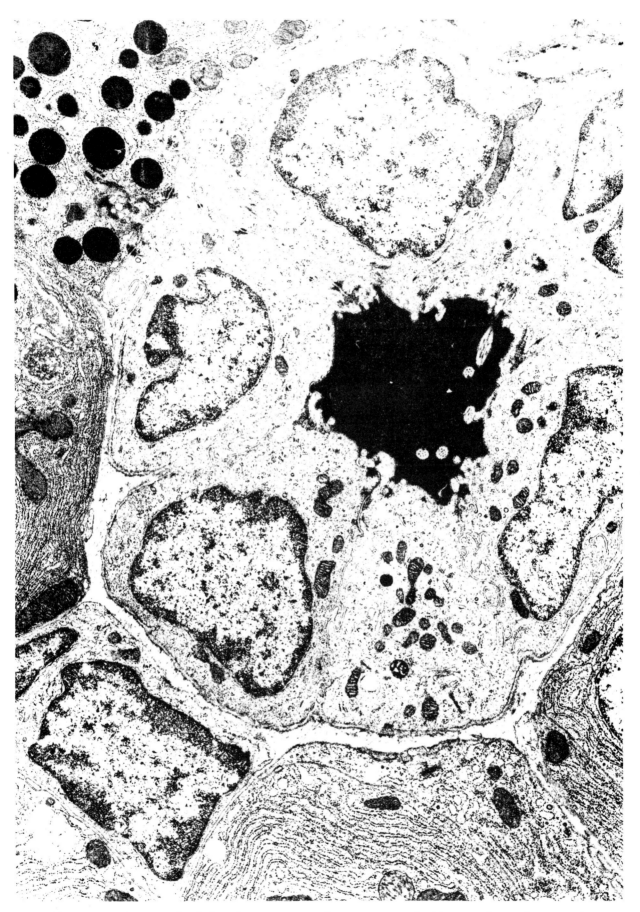

Fig. 12-131. An intercalated duct from pancreas. Lumen contrasted with peroxydase. Courtesy of Prof. Horst Kern, University of Heidelberg. 11200 ×.

folds are higher and the diverticula between them more extensive. This creates a muralium mucosum (Figs. 12-124 and 12-125). The epithelium of the cystic duct resembles that of the gallbladder. The muscularis is poorly developed; but high folds of mucous membrane, collectively called the *spiral folds of the cystic duct* (Heister's valves), project into the duct lumen (Figs. 12-124 and 12-126).

Pancreas

The pancreas (Figs. 12-127 and 12-128) is a compound gland with serous acini and islands of endocrine tissue. Quite aptly this organ is called, in some countries, the "abdominal salivary gland".

EXOCRINE PORTION. The excretory ducts of the pancreas which empty onto the simple columnar epithelium of the duodenum are lined by simple columnar epithelium near the exit and by simple cuboidal epithelium distally. As small ducts gradually converge, the epithelium becomes taller, and in the large ducts it is columnar (Fig. 12-129). The ducts branch repeatedly into the smallest, *intercalated ducts* (Fig. 12-130), which are lined by pale-staining, simple, cuboidal epithelium. Interestingly enough some of their cells are ciliated (Fig. 12-131), reminiscent of their common endodermal heritage with respiratory epithelium. The intercalated ducts penetrate the acini as noted by the presence of *centroacinar cells* within the acini (Figs. 12-132 and 12-134). The pyramidal pancreatic exocrine cells which compose the acini and tubules have spherical nuclei. The basophilia of the basal portion of the acinar cells is attributed to the presence of granular endoplasmic reticulum arranged in the form of concentric, perforated lamellae (Figs. 12-134 and 1-10, B). The apical portion of these exocrine cells contains highly refractile, acidophilic *zymogen granules* (Fig. 12-134). Intra- and intercellular canaliculi (Figs. 12-133

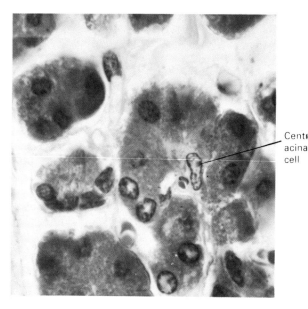

Fig. 12-132. A pancreatic acinus with a centroacinar cell. 1900 ×.

and 12-134) connect with the acinar lumen. The pancreatic cells produce amylase, lipase, trypsinogen, chymotrypsinogen, ribonuclease, deoxyribonuclease and sodium bicarbonate in response to *pancreozymin* and *secretin*.

ENDOCRINE PORTION. The *pancreatic islets (of Langerhans)* (Figs. 12-135 and 12-136) are connected with the ducts and acini of the exocrine pancreas by narrow bridges reflecting their common ancestry

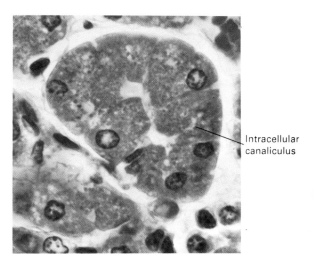

Fig. 12-133. A pancreatic acinus. 1900 ×.

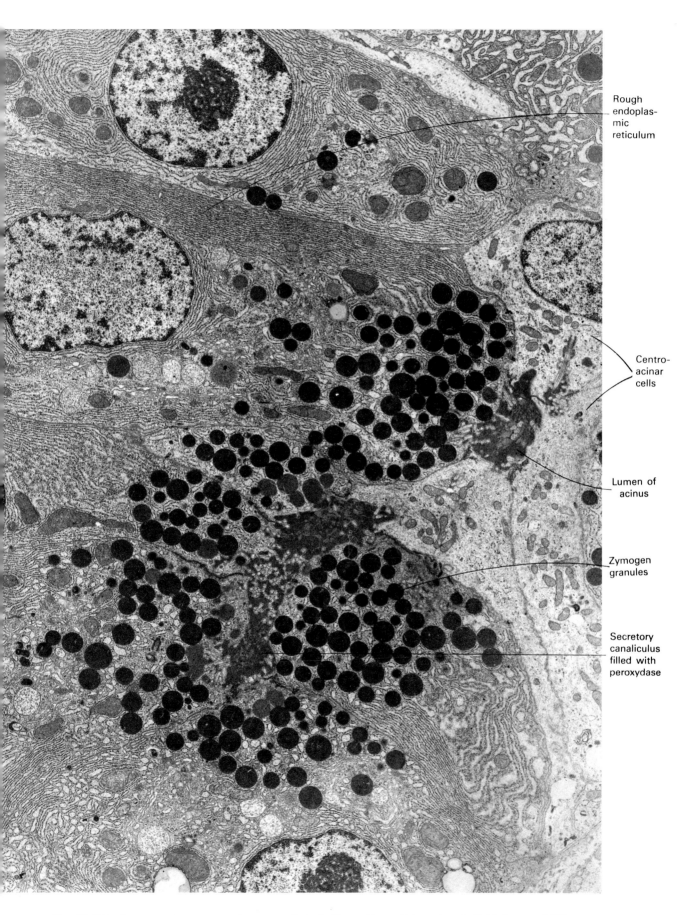

Fig. 12-134. A pancreatic acinus. Courtesy of Prof. Horst Kern, University of Heidelberg. 6425 ×.

Fig. 12-135. A pancreatic islet (Langerhans). 375 ×.

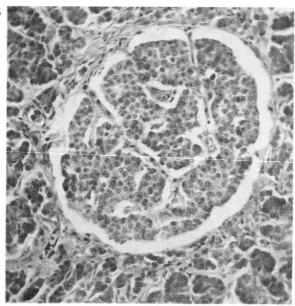

from duodenal endoderm (Fig. 12-137). Such an island is a richly vascularized little muralium duplex (Fig. 12-135) with a sphincter controlling the blood supply to each individual islet.

It is virtually impossible to distinguish different cell types in routine H&E preparations. However, with special staining techniques or electron microscopy, at least four different types of cells can be identified: (1) the *alpha endocrine cells* which produce *glucagon* (hyperglycemic factor), (2) the more numerous *beta endocrine cells* which produce *insulin* (hypoglycemic factor), (3) *delta endocrine cells* (Fig. 12-136) and (4) *undifferentiated chromophobic islet cells*. The *delta or definitive islet cells* (D or A Cell) resemble the alpha endocrine cells. Evidence is mounting from immunocytochemical studies,

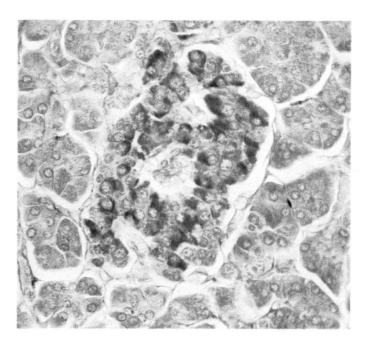

Fig. 12-136. A pancreatic islet. Beta cells stained blue with aldehyde fuchsin, photographed from a slide prepared by Dr. Anna-Mary Carpenter, University of Minnesota. 955 ×.

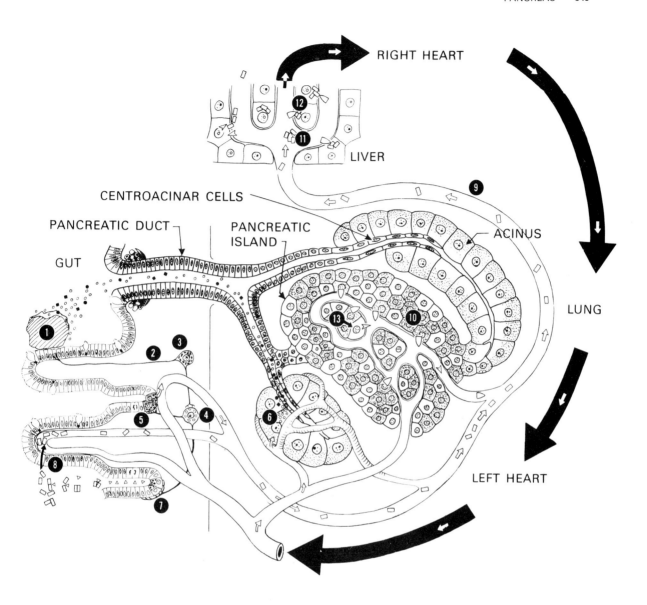

Fig. 12-137. Diagram of pancreatic and related functions.

that these cells produce two important hormones: gastrin and somatostatin. The *undifferentiated islet cells* are chromophobic, that is, they fail to stain with routine stains. They may be reserve cells or terminal alpha endocrine cells.

Some functional relationships between the intestine, the pancreas and liver are illustrated in Figure 12-137: the presence of a bit of food (1) in the gut is perceived by hypothetical chemoreceptors (2) from the submucosal plexus (3). The neurons of the latter signal the "sensation" to motor cells in the myenteric plexus (4) which in turn stimulate the intestinal mucosal cells (5) to produce the hormones pancreozymin and secretin (white arrows). These hormones enter the blood stream, travel through the portal vein (9), through the liver, the right side of the heart, the lungs, the left side of the heart, the aorta, and finally through the celiac and pancreaticoduodenal arteries into the pancreas, where they stimulate the exocrine cells (6) to produce sodium bicarbonate and enzymes, shown as black diamonds, black dots and white circles, which can act upon the food (1). Myenteric motor nerve cells also signal the epithelial cells and exocrine cells with acidophilic granules (Paneth) of the intestinal glands (7) to secrete peptidases and maltase (little white triangles). The latter splits the disaccharide maltose (double rectangles) into the monosaccharide glucose (single rectangles). Glucose is absorbed by the intestinal villi (8) to enter the blood stream. Through the portal vein (9) the glucose is delivered to the liver. Insulin (white boats), produced by the beta cells (10) of the pancreatic islands, also enters the liver through the portal vein and causes glucose to be synthesized to glycogen (groups of three rectangles) and to be stored in the liver cell (12). Insulin also causes glycogen storage in adipose tissue and muscle fibers. Glycogen is remobilized into glucose by the action of glucagon (darts) produced by the alpha cells (13) of the pancreatic islands.

13...
Respiratory apparatus

Respiration occuring in every cell of the body requires intake of oxygen and elimination of carbon dioxide and water. Oxygen is delivered to the cells, and carbon dioxide and water are removed via the bloodstream. Exchange of these substances occurs in the lungs, two sac-like thoracic organs opened to the environmental air by a system of passages. *Conducting passages* and *respiratory spaces* must be distinguished. All conducting passages from the nose to the bronchioles (except the terminal bronchioles) are lined by pseudostratified to simple cuboidal epithelium with goblet cells (respiratory epithelium) (Fig. 13-1). The epithelial cells bear vibratile cilia, and microvilli. Some cells carry only microvilli; but most cells bear cilia and microvilli (many of which are branched) (Figs. 13-2; 13-3 and 13-4). The stroke of the cilia is directed toward the pharynx. Thus, a mucous film containing impurities is moved from the nasal passages posteriorly into the pharynx and from the lungs and trachea superiorly into the pharynx, from where these materials are directed into the esophagus.

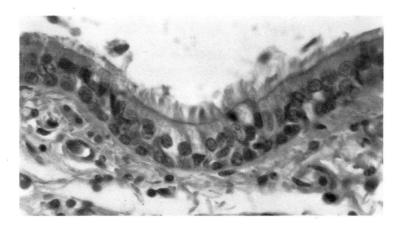

Fig. 13-1. Respiratory epithelium from bronchus, H&E.

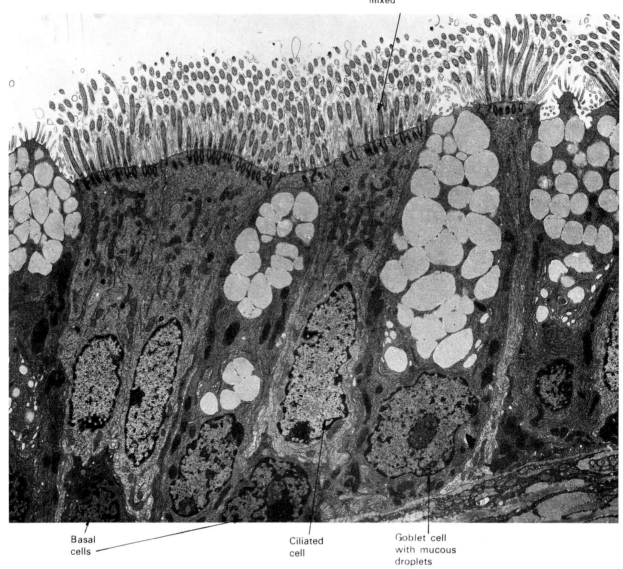

Fig. 13-2. Respiratory epithelium of the nose. 5200 ×. Courtesy of Dr. D. H. Matulionis, University of Kentucky.

Nose

The nose is the normal gateway to the respiratory system. The internal aspect of the lateral wall of the external nasal pyramid is shown in Figure 13-5. If one cuts the external nasal pyramid along the dotted line in Figure 13-6, A, the cut-off piece, B, will contain parts of the septal and alar cartilages; and a section such as shown in Figure 13-7 results. The external nasal pyramid is supported by bone (at a higher level) and by hyaline cartilages. Externally and in the region of the vestibule, the cartilages are covered by integument. Bulk is created by fat in the deeper layers, by interlacing skeletal myocytes and by a very thick fibrous dermis in which hair follicles (vibrissae), sweat glands and sebaceous glands occur. A few large apocrine sweat glands are usually present.

Internally, the nasal cavity is divided into a left and a right half by the nasal septum (Fig. 13-8). Curved bony processes projecting from the lateral walls into the nasal cavity, are the superior, middle and inferior conchae. The bone of the conchae is of the very delicate *spongiosum* variety (Fig. 13-11). The spaces bounded by them are the superior, middle and inferior meatus through which air streams on the way from the nostrils to the choanae and into the nasopharynx to be partially freed from dust particles by the cilia of the respiratory epithelium and to be warmed by the venous sinuses. Open, though narrow, passages connect the nasal cavity with the paranasal sinuses and air cells. The conchae and walls of the nose are covered by respiratory mucosa (Figs. 13-8, 13-9 and 13-11). The paranasal sinuses likewise are lined with a respiratory mucosa, but very reduced in thickness (Fig. 13-11, left).

The nose of primates is much smaller and simpler in structure than that of most mammals, possessing highly convoluted conchae and a more extensive olfactory epithelium.

The cause of the reduction in size and functional importance of the primate nose is the forward transposition of the eyes (see Fig. 16-15, C).

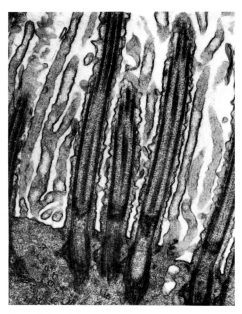

Fig. 13-3. Cilia and branched microvilli of nasal, respiratory epithelium. Courtesy of Dr. D. H. Matulionis, University of Kentucky. 3233 ×.

Above the limen (Fig. 13-5), the nasal cavity is lined by *respiratory mucosa* (Fig. 13-9) which contains seromucous glands and a plexus of large veins and venous sinuses resembling the erectile tissue found in the genitalia. The "erection" of this tissue, i.e., filling of the venous sinuses, occurs in response to histamine and impedes passage of air during colds and "hay fever". The *respiratory epithelium* is pseudostratified, ciliated, columnar with goblet cells of the same kind as shown in Figure 13-2.

The *olfactory mucosa* is less vascular than the respiratory mucosa; it contains branched, serous olfactory glands (of Bowman). Its pseudostratified epithelium (Fig. 13-10) consists of *sustentacular* (supporting) cells (B, C, D) and a special kind of primary sense cell, the olfactory neurosensory epithelial cell. The dendrites of the latter are non-typical cilia floating in a layer of serous fluid probably excreted by

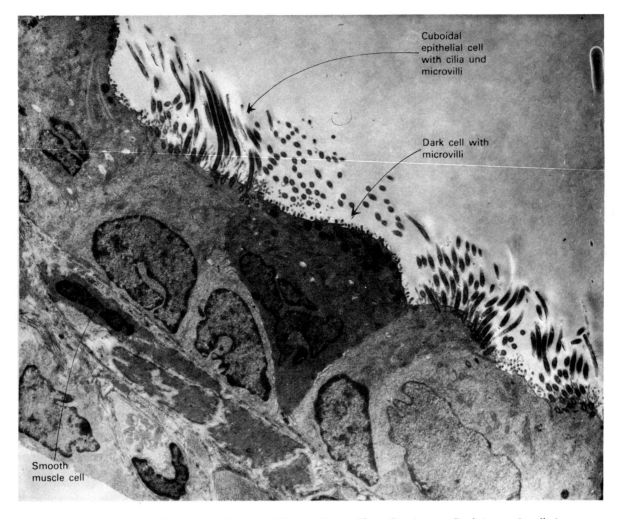

Fig. 13-4. Wall of small bronchus of rat. 4000 ×. (From Elias, Die Lunge, Boehringer, Ingelheim, 1973.)

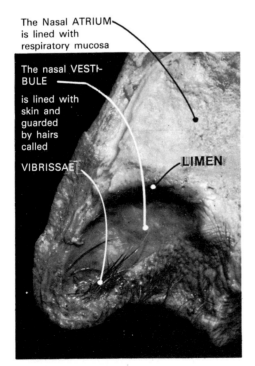

Fig. 13-5. Lateral wall of external, nasal pyramid, seen from inside.

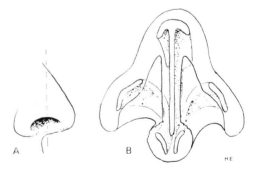

Fig. 13-6. Diagram to show the plane of cutting for Figure 13-7.

the olfactory glands. At their bases, the olfactory cilia have the typical 9 + 2 structure of kinocilia; but toward their apices, all but two of the filaments are lost so that they are very thin and flexible. The axons of the olfactory cells, constituting the first cranial nerve, end in vast arborizations in the glomeruli of the olfactory bulb (Fig. 7-16, center), where they form interlacing synapses.

provided with respiratory and stratified squamous epithelium. The lamina propria mucosae contains mucoserous glands. These are most prominent in the ventricular folds, the so-called "false vocal cords".

At the level of the *glottis* (the true vocal folds and the opening between them), the laryngeal mucosa is supported by the vocalis muscle which is covered by a wedge of dense, elastic connective tissue, known as

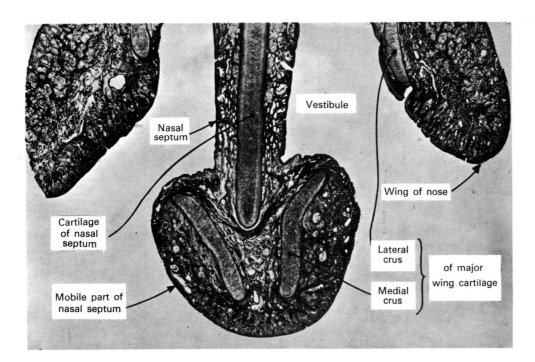

Fig. 13-7. Frontal section of external nasal pyramid

Larynx

The larynx (Figs. 13-12, 13-13 and 13-14) is supported by a group of cartilages. The thyroid, cricoid and arytenoid cartilages are of the hyaline kind; the corniculate, cuneiform and epiglottal cartilages as well as the vocal process and the apex of the arytenoid cartilage are of the elastic variety. These cartilages are connected by skeletal muscles. Externally, the larynx is covered by a tunica adventitia; internally it is lined by a complex mucous membrane

the *vocal ligament*. Stratified squamous epithelium covers the vocal ligament; whereas respiratory epithelium lines some other spaces of the larynx (Figs. 13-12 and 13-13).

Minute, arcuate muscle fibers swing out from the vocalis muscle into the vocal ligament. According to L.H. Strong, these so-called "pitch fibers" which make up the modulatory fascicles determine, by differential contractions, the length of the vibrating segment of the glottis.

Air forced out of the lungs causes the

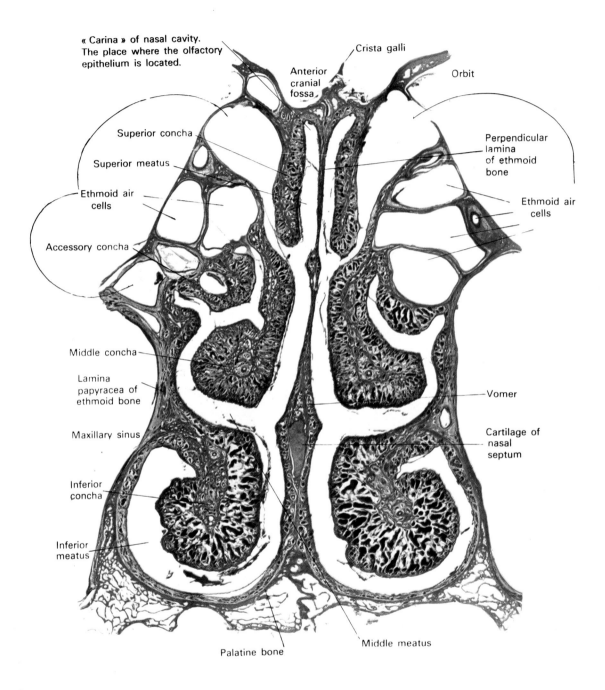

Fig. 13-8. Frontal section of nose at a deep location.

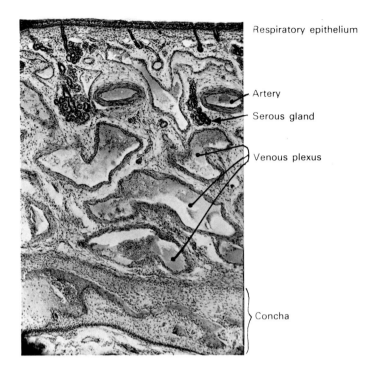

Fig. 13-9. Middle concha of calf.

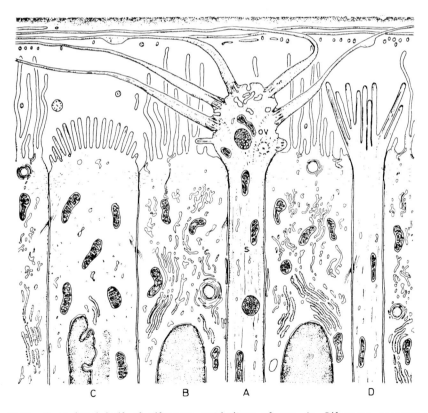

Fig. 13-10. Superficial half of olfactory epithelium of cat. A: Olfactory, sensory cell process bearing specialized cilia. At their base, these cilia have the typical structure of kinocilia. Distally, they taper. Their thin distal portions contain only two single filaments. These thin ends float in the surface film which is secreted by olfactory glands (Bowman's). B, C, D: Different types of supporting epithelial cells. (From K. H. Andres, Der Olfaktorische Saum der Katze, Z. Zellforsch, 96: 250-275, 1969.)

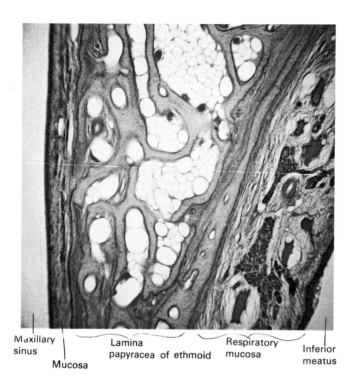

Fig. 13-11. Lateral wall of nose with a narrow portion of the maxillary sinus, H&E. 65×.

Maxillary sinus | Mucosa | Lamina papyracea of ethmoid | Respiratory mucosa | Inferior meatus

vocal ligament to vibrate with a specific number of cycles per second which results in a definite pitch. Altering the length of the vibrating segment modulates the pitch. Others believe that the degree of contraction of the entire vocalis muscle determines pitch. Small tonsils (Fig. 13-14) are found in the wall of the laryngeal ventricle.

Trachea

The larynx is continuous with the trachea (Fig. 13-15, A), a tube held open by incomplete rings of hyaline cartilage in its wall (Fig. 13-15, B). The posteriorly located openings in the rings (Fig. 13-15, C) are bridged by bundles of smooth myocytes, the *trachealis muscle*. Internally, the trachea is lined by a mucous membrane which bears typical respiratory epithelium (Figs. 13-15, C, and 13-16). The tela submucosa harbors mucoserous glands (Figs. 13-15, C, 13-16 and 13-17).

Since the trachea (Fig. 13-15, A) does

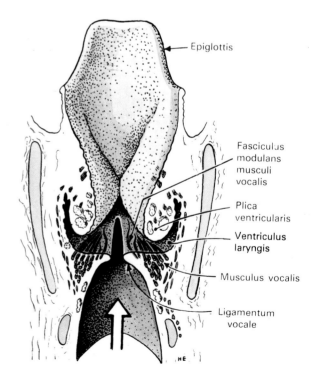

Fig. 13-12. Anterior half of larynx seen from behind.

Fig. 13-13. Frontal section of larynx. (From a preparation by Dr. Leon H. Strong, Chicago Medical School.)

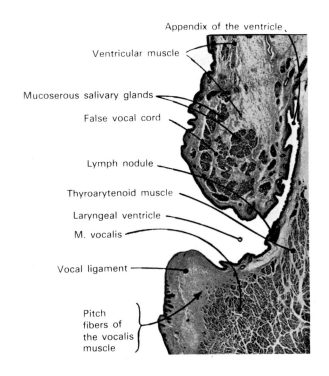

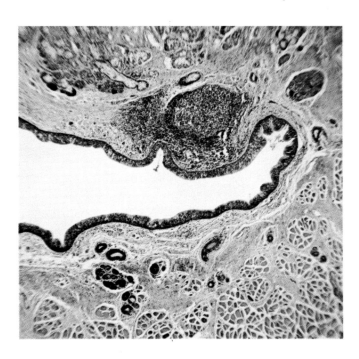

Fig. 13-14. The laryngeal ventricle with a little tonsil, Hansen's iron-hematoxylin and eosin.

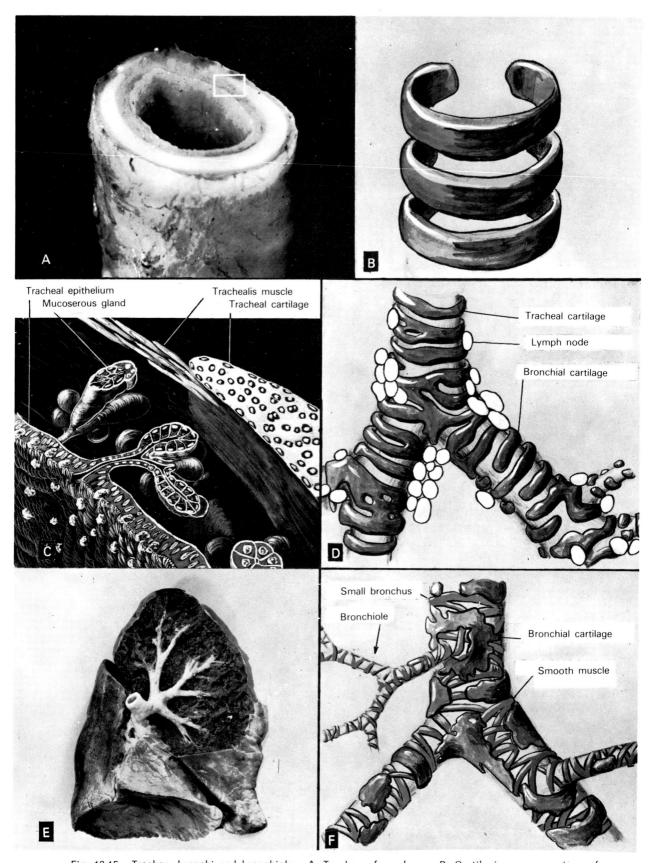

Fig. 13-15. Trachea, bronchi and bronchiole. A, Trachea of newborn. B, Cartilaginous, open rings of trachea. C, Stereogram of wall of trachea. D, Cartilages and lymph nodes at the bifurcation of the trachea. E, Lung of 13 year old boy. F, Small bronchi and bronchioles. (From Film strip by Elias, courtesy of U.S.P.H.S.)

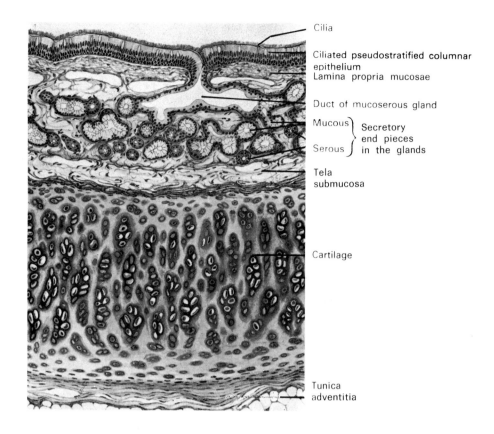

Fig. 13-16. Wall of trachea. (From Gillison: Histology of the Body Tissues. Livingstone, Edinburgh, 1953.)

not touch any body cavity, its outer coat is an adventitia and, therefore, a sharp external outline is lacking.

Bronchial Passages

The *bronchi* and *bronchioles* resemble the trachea in histological structure (Fig. 13-18). Near the bifurcation of the trachea (Fig. 13-15, D) the cartilages become complicated in shape. As the conducting air passages branch (Fig. 13-15, D, E, and F) their caliber decreases, the amount of cartilage in their walls is reduced (Fig. 13-15, D and F), and cartilage rings give way to irregular plates (Fig. 13-15, F). Passages supported by cartilage are called bronchi while those having no such support are known as bronchioles. During expiration, the mucous membrane of the bronchi is thrown into longitudinal folds (Fig. 13-19, A).

Nonciliated cells occurring in the epithelium of the bronchioles, called bronchiolar exocrine cells (clara cells) are presumably secretory. Ultrastructurally these cells display features characteristic of secretory cells, i.e., rER, large golgi complex and apical secretory granules that are revealed by protein stains. In humans, the cells appear sparingly in the bronchioles and terminal bronchioles. However, in smokers the number increases dramatically.

There are 16 steps of branching of the conducting airways in adult man, resulting in approximately 33,000 terminal bronchioles in each lung. The *terminal bronchioles* are lined by simple cuboidal epithelium, with some ciliated epithelial patches. Nonciliated cells occurring in the epithelium of the bronchioles, called bronchiolar

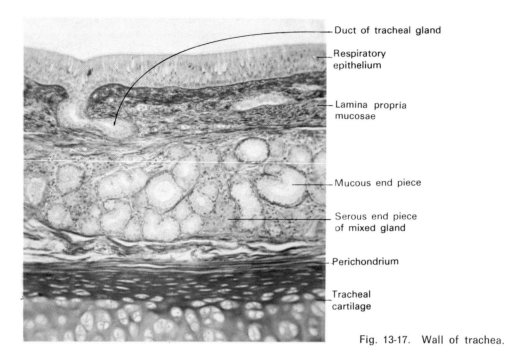

Fig. 13-17. Wall of trachea.

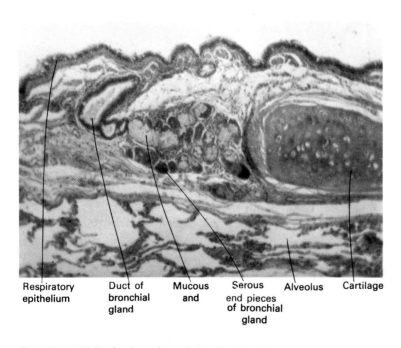

Fig. 13-18. Wall of a bronchus of the fourth order. H&E.

Fig. 13-19. Some aspects of the lung. A, C, G and H, From boy, 13 years old. B and D, From dog. A, B and C, Wet specimens. H, Dry preparation (Semper method).

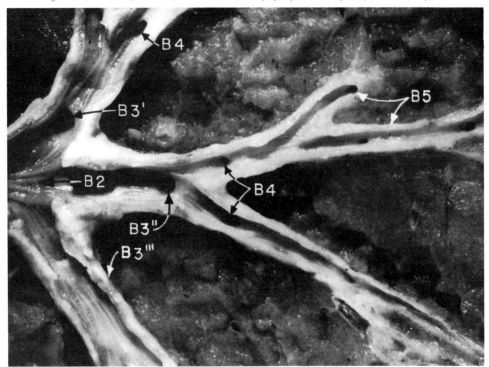

B = Bronchus
B2 = Bronchus of the second order: Bronchus lobaris superior sinister
B3 = Bronchus of third order: B3' = B. segmentalis apicoposterior
B3'' = B. segmentalis anterior
B3''' = B. segmentalis linguaris
B4 = Bronchi of fourth order
B5 = Bronchi of fifth order

Fig. 13-19, A. Bronchi of 13 year old boy.

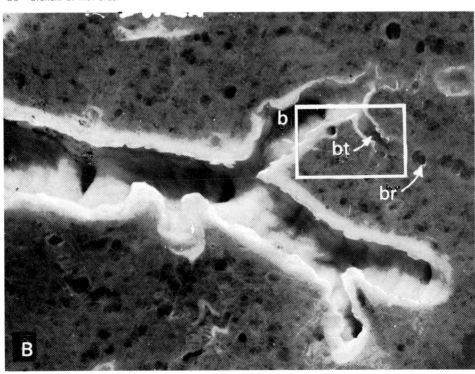

b = Bronchiolus
bt = Bronchiolus terminalis
br = Bronchiolus respiratorius

13-19, B. Bronchi of dog.

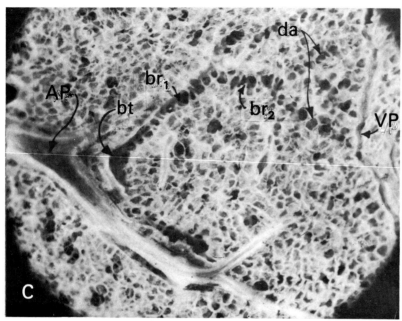

13-19, C. Terminal and respiratory bronchioles, alveolar ducts and blood vessels of boy, 13 yrs.

br_1 = Bronchiolus respiratorius of first order
br_2 = Bronchiolus respiratorius of second order
bt = Bronchiolus terminalis
da = Ductulus alveolaris
AP = Ramus arteriae pulmonalis
VP = Radix venae pulmonalis

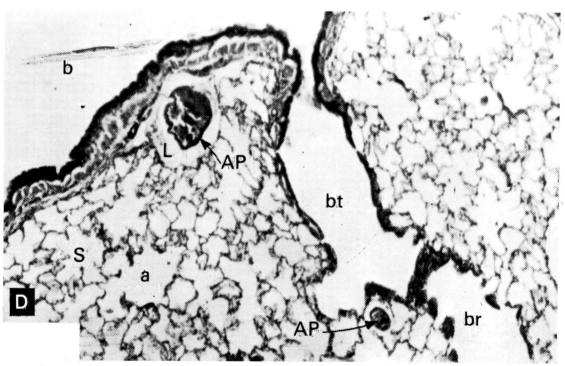

b = Bronchiolus
br = Bronchiolus respiratorius
bt = Bronchiolus terminalis
a = Atrium
S = Sacculus alveolaris
L = Vas lymphaticum
AP = Ramus arteriae pulmonalis

13-19, D. Detail from lung of dog (rectangle in 19, B).

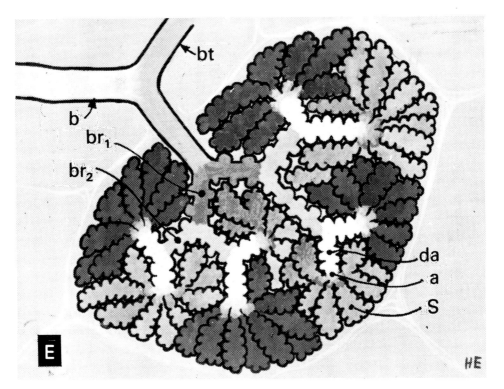

b = Bronchiolus
br₁ = Bronchiolus respiratorius of first order
br₂ = Bronchiolus respiratorius of second order
bt = Bronchiolus terminalis
da = Ductulus alveolaris
a = Atrium
S = Sacculus alveolaris

13-19, E. Diagram of lobule.

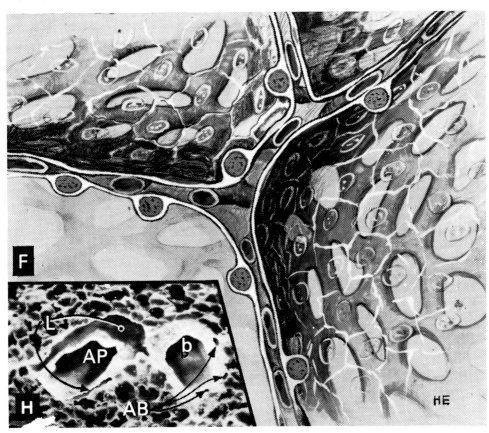

b = Bronchiolus
L = Vas lymphaticum
AP = Ramus arteriae pulmonalis
AB = Ramus arteriae bronchialis

13-19, F. Diagram of interalveolar septum. H Detail from a dry preparation of lung of boy, 13 yrs.

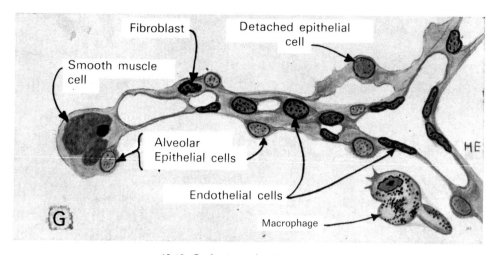

13-19, G. An interalveolar septum.

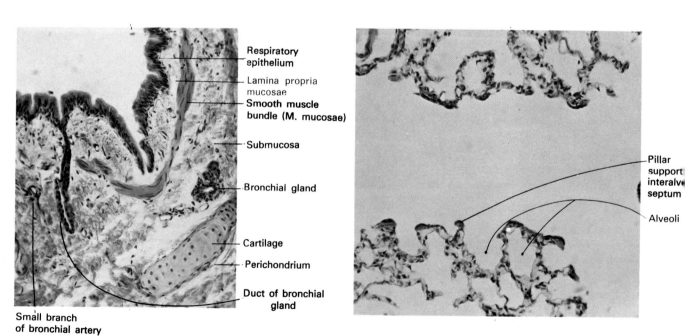

Fig. 13-20. Small bronchus, H&E. 100 ×.

Fig. 13-21. Branching respiratory bronchiole. 100 ×.

exocrine cells ("clara cells") are presumably secretory. Ultrastructurally these cells display features characteristic of secretory cells, i.e., rER, large golgi complex and apical secretory granules that are revealed by protein stains. In humans, the cells appear sparingly in the bronchioles and terminal bronchioles. However, in smokers the number increases dramatically.

All bronchial passages are surrounded by spiralling and circular bundles of smooth musculature, corresponding to a lamina muscularis mucosae (Fig. 13-15, F). Small mucoserous bronchial glands located in the lamina propria mucosae (Fig. 13-18) and in the tela submucosa (Fig. 13-20) become less numerous toward the periphery. The walls of the bronchi are supplied by branches of the bronchial artery, small vessels which course the lamina propria mucosae of the bronchi. Branches of the pulmonary artery accompany the bronchi (Fig. 13-19, C, D and H). Lymph nodes (Fig. 13-15, D) and solitary lymph nodules are often found alongside the bronchi. Lymph drainage occurs mainly through lymph vessels in the adventitia of

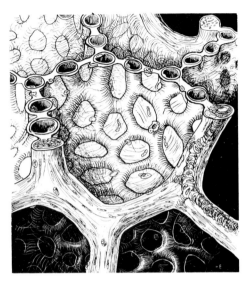

Fig. 13-22. View from an alveolar sac into some alveoli.

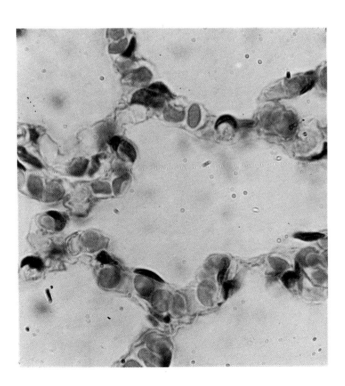

Fig. 13-23. Interalveolar septa, H&E. 1250 ×.

pulmonary arterial branches (Fig. 13-19, D, H). Tributaries of pulmonary veins run far from the bronchial tree.

The terminal bronchioles branch into three generations of *respiratory bronchioles* (Fig. 13-19, C and D), passages lined by simple cuboidal epithelium, with alveoli incorporated into their walls (Fig. 13-21). From the last respiratory bronchiole, of which we have about 260,000 in each lung, there arise three generations of *alveolar ducts*, passages whose entire walls are lined by alveoli. Each terminal alveolar duct (two million per lung) ends in a space called the *atrium* from which two *alveolar sacs* arise. These are blindly end-

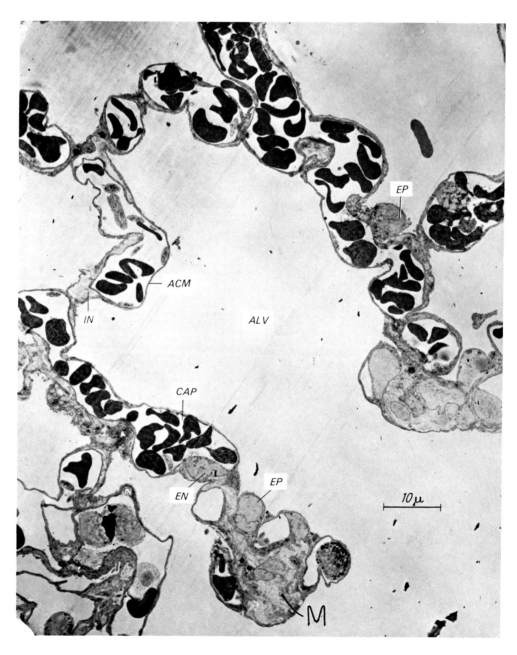

Fig. 13-24. Interalveolar septa. ACM = Air-capillary membrane. ALV = Alveolus. CAP = Capillary. EN = Endothelium. EP = Epithelial cell. IN = Interstitial tissue. M = Muscular trabecula. (From Weibel: Morphometry of the Human Lung. Springer, Berlin-Heidelberg, 1963.)

ing spaces lined everywhere by alveoli. All together, there are 23 generations of branching, about four million alveolar sacs; and since there are about 30 alveoli around an alveolar sac, each adult human lung contains approximately 150 million alveoli. Those arising from respiratory bronchioles and alveolar ducts are included in that number.

All passages and spaces which arise from a terminal bronchiole constitute a *lobule* (Fig. 13-19, E).

Alveoli

Gas exchange occurs through the walls of the alveoli. An *alveolus* (Fig. 13-22) is a small bag, open on the side which faces the respiratory bronchiole, the alveolar duct or the alveolar sac (Figs. 13-21 and 13-24). In the terminal portions of the

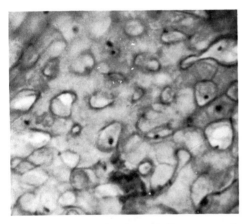

Fig. 13-25. Interalveolar capillary network. (From Weibel: Morphometry of the Human Lung. Springer, Berlin-Heidelberg, 1963.)

lung (alveolar ducts and alveolar sacs), neighboring passages are so close together (Fig. 13-21) that the alveoli which arise from them come into direct contact. The thin walls between the alveoli are the *interalveolar septa* (Figs. 13-19, F and G and 13-22 to 13-24). Along their free margins these septa may be supported by trabeculae of smooth muscle bundles (Figs. 13-19,

G, 13-21, 13-22 and 13-24, M) and at other places by pillars of collagen fibers. These muscular trabeculae form polygon nets (Fig. 13-22), whose meshes are the openings to the alveoli. They are related to the alveoli as door frames to rooms. Arterioles and venules may travel along some muscular trabeculae (Fig. 13-22, right). Alveoli may communicate with each other through the *interalveolar pores* (of Kohn) which are approximately 10 μm in diameter.

The interalveolar septum is, in essence, a dense net of wide capillaries (Figs. 13-19, F, 13-22, 13-23, 13-24 and 13-25). Between the capillaries there are interstitial areas which contain a few fibroblasts elastic, reticular and collagen fibers.

The fine structure of the interalveolar septa is illustrated diagrammatically in Figure 13-26, stereographically in Figure 13-27 and documented in electron micrographs, Figures 13-28 and 13-29. As these figures attest, the core of the interalveolar septum is a network of capillaries, supported by delicate connective tissue. The septa are covered by an extremely thin, simple squamous alveolar epithelium, so thin that it cannot be seen with the light microscope (Fig. 13-23). The respiratory epithelial cells (sometimes called Type I pneumonocytes) are visible only where their nuclei are located. The basement membrane of this epithelium and the basement membrane of the capillaries may fuse in a few places (Figs. 13-26, 13-27, 13-29). The respiratory epithelium also covers the fibrous and muscular pillars. As a rule, the air-blood barrier consists of respiratory epithelium, two fused basement membranes, often with some tissue fluid between, and capillary endothelium. Its mean total thickness is approximately 0.5 μm. Toward the alveolar lumen, the respiratory epithelium is covered by *surfactant*, a phospholipid material rich in lecithin. A surfactant is a surface tension reducing material essential for normal respiration. If the surface

RESPIRATORY APPARATUS

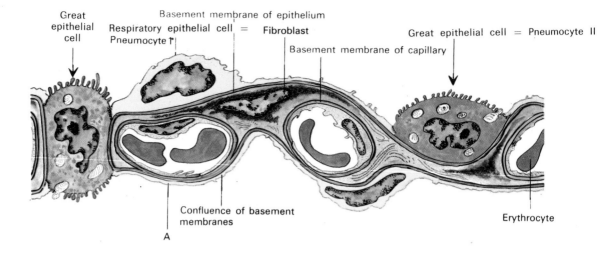

Fig. 13-26. Diagram of the fine structure of an interalveolar septum. (From Elias, Die Lunge, Boehringer, Ingelheim, 1973)

Fig. 13-27. Stereogram of the fine structure of an interalveolar septum. To understand this picture it must be compared with Fig. 13-26. (From Elias, Die Lunge, Boehringer, Ingelheim, 1973.)

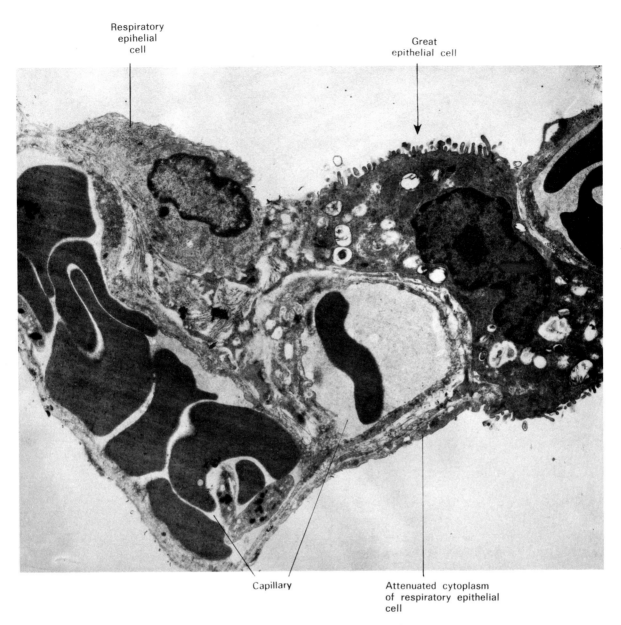

Fig. 13-28. A portion of an interalveolar septum of the rat (From Elias, Die Lunge, Boehringer, Ingelheim, 1973.)

tension in the alveolous equalled that of water, normal, quiet inspiration and expiration could not be accomplished with the thoracic musculature.

Surfactant is produced by the *great epithelial cells* which are interspersed among the Type I pneumonocytes. The great epithelial cell (Type II pneumonocyte, great alveolar cell, septal cell) is irregularly cuboidal in shape. The characteristic feature of the great epithelial cell is the presence of many multivesicular and multilamellar bodies of cytosomes and microvilli on its alveolar surface. These cells may be incorporated into the alveolar epithelium, or they pierce the entire interalveolar septum touching the lumina of two adjacent alveoli (Figs. 13-26, 13-27 and 13-28).

Oxygen must traverse the following structures on its way to hemoglobin: surfactant, a thin respiratory epithelial cell and its basement membrane; the basement

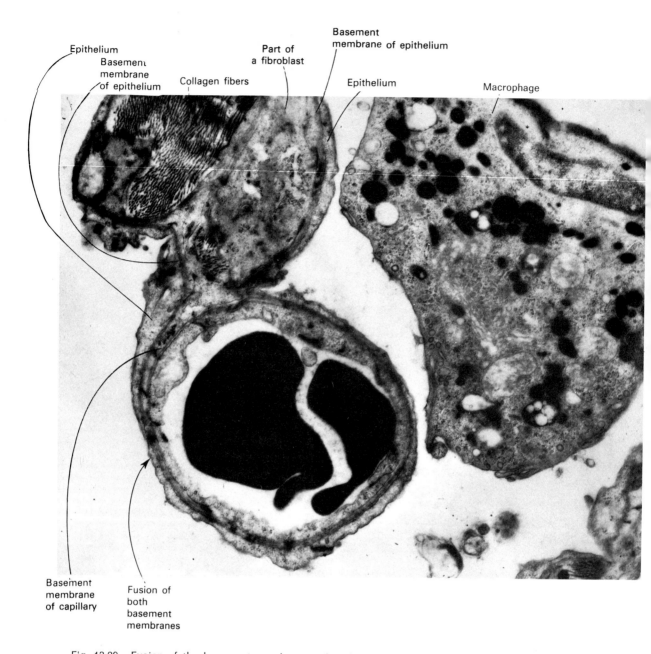

Fig. 13-29. Fusion of the basement membranes of endothelium and epithelium (rat) (left) and dust cell (right). (From Elias, Die Lunge, Boehringer, Ingelheim, 1973.)

membrane of the endothelium, a thin endothelial cell, the plasma of the blood and the plasmalemma of the erythrocyte. Only at a few places (A in Fig. 13-26, also Figs. 13-27, right and 13-29, bottom) is the barrier thinner.

In Figures 13-19, D, F and G, 13-23, 13-24, 13-26 and 13-28, note that the interalveolar septum is exposed to air on both sides. The total internal respiratory surface of both adult lungs is estimated to amount to approximately 75 m^2 (about 790 square feet).

Alveolar macrophages (dust cells) (Fig. 13-29) are free macrophages often found in the alveoli. They ingest dust particles with many pseudopodial processes and possess a variety of cytoplasmic inclusions.

Dwellers of industrial urban areas often

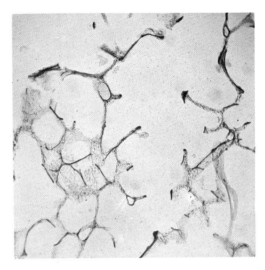

Fig. 13-30. Elastic fibers in the lung. Resorcinfuchsin. 375 ×.

show deposits of carbon in the connective tissue along the bronchi, especially in lymphoid structures, a condition known as *anthracosis*. Externally their lungs appear gray in contrast to the bright pink lungs of country residents.

Respiratory Mechanics

Ventilation of the lungs is accomplished by rhythmic expansion and contraction of the thorax. In this process, the caliber of the bronchi changes relatively little owing to their cartilaginous support, but their angles of branching change. The respiratory portions, however, expand and collapse.

An important constituent of the interalveolar connective tissue are elastic fibers (Fig. 13-30). Elastic fibers are also numerous in the visceral pleura (not stained in Fig. 13-31).

Expiration occurs almost automatically because of the presence of so much elastic tissue. Inspiration must be forcefully accomplished by skeletal muscles which expand the thorax, creating a negative pressure in the pleural cavity. This counteracts the contraction of the elastic fibers. After death, when surfactant production has ceased, the surface tension of the alveolar liquid film plus the contraction of the elastic tissue tends to cause collapse of the lungs against the rigidity of the chest wall; but a negative pressure remains for a while in the pleural cavity which keeps the alveoli open.

Pleura

The visceral pleura, a typical serous

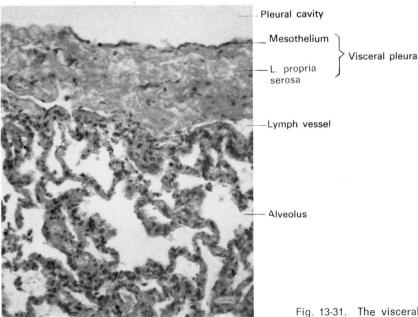

Fig. 13-31. The visceral pleura, H&E. 375 ×.

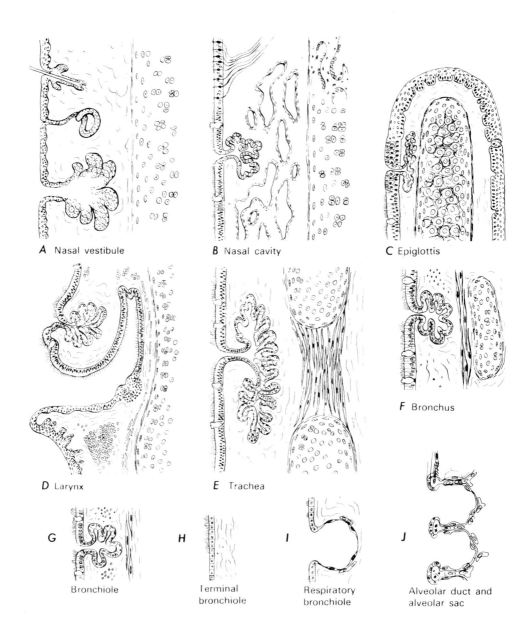

Fig. 13-32. Synopsis of the respiratory system.

membrane (Fig. 13-31), covers the lung. Beneath its mesothelium is a connective tissue layer rich in elastic fibers (not stained in Fig. 13-31). The elastic connective tissue is continuous with that of the interlobular and the interalveolar septa.

Evolution of the Lung

Figure 13-33 gives a synopsis of the evolution of the vertebrate lung, as follows:

A, The lungs have evolved from the swim bladder of fishes: among Ganoidea,

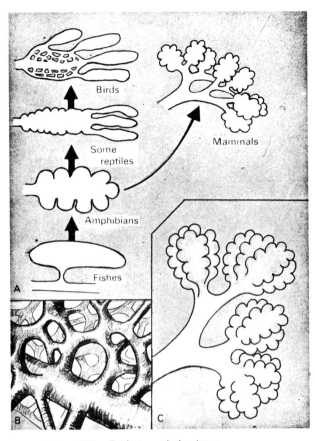

Fig. 13-33. Evolution of the Lung.

the swim bladder acquired an auxiliary respiratory function by developing vascular septa. The amphibian lung is a single, though large, alveolar sac. In some reptiles (snakes, chameleons), air sacs are appended caudally to the lung. In birds, the highest stage of perfection is reached; their lung is a network of anastomosing tubes without dead ends, connected with air sacs (Figs. 13-33, A, above left and 13-33, B). The mammalian lung evolved as an aberrant side branch. Each alveolar sac is homologous to an entire amphibian lung.

B, Fresh air streams freely through the air capillaries of the birds' lungs.

C, The mammalian lung has dead ends. Therefore, pure fresh air rarely reaches the respiratory tissue (Fig. 13-33, C).

NEURO-ENDOCRINE (APUD-TYPE) CELLS OF THE LUNG.

Cells with neuroendocrine (NE) characteristics have been identified in the pulmonary epithelium, in much the same way that these cell types populate the gastrointestinal tract. Electron microscopic, histochemical studies and immuno-reactivity show that these cells have features similar to the APUD cells of the GI tract. Groups of these cells occur as innervated neuroepithelial bodies (NEB). Both the NE and NEB are particularly prominent in the fetus and in the neonate, but also occur in the adult. The dense core granules of these cells are spherical in shape, and measure about 120 nm in diameter. The NE cells are scattered throughout the entire length of the tracheobronchial tree, while the NEB appear to be confined to the epithelium of intrapulmonary airways, particularly at the bifurcations of the bronchi. While the precise function of the NE cells is not known, current data implies that these cells are related to lung development and to neonatal adaptation.

14...
Urinary Organs

Kidney

The two kidneys remove water and waste, particularly the end products of protein metabolism, from the blood. Frequently, substances in the kidneys are actively transported against osmotic pressure gradients by a process not fully understood.

LOBES. Each kidney consists of two grossly visible structural components, when the kidney is sectioned: (a) a smooth *medulla* which exhibits parallel streaks, and (b) the mottled *renal cortex*. The medulla forms individual units, the *renal pyramids*, which are surrounded by cup shaped portions of the cortex, the *renal lobes*. The pyramids are separated from one another by compressed interconnected masses of cortex, the *renal columns* which form cortical septae (Figs. 14-4 and 14-5).

The kidney of warm-blooded animals (Fig. 14-1) consists of several individual lobes which may remain independent of each other, or fuse partially or completely. Figure 14-1 shows a progressive series of fusion of lobes. On one extreme are the birds and Cetacea (Fig. 14-1, A) with hundreds of lobes (most of which are completely separated; on the other extreme are the unilobular kidneys of such diversified mammals such as horses, rodents, rabbits, etc. (Fig. 14-1, E). Fusion of lobes begins with the cortical portions and then by the medullary parts. In the human newborn, lobes are usually demarcated by deep furrows (Figs. 14-2, 14-3). During childhood, fusion of the cortical portions usually causes the external surface of the kidney to become smooth, while the medullary *pyramids* remain separated (Fig. 14-1, C). Not infrequently, complete fusion of the cortical portion occurs very early as in the case of the six month old infant (Fig. 14-4). In other cases the interlobar furrows though less sharp than in childhood, persist into old age. Figure 14-5 illustrates the human kidney in greater detail.

GROSS ANATOMY OF THE ADULT KIDNEY.

The entire kidney may be compared to a very thick-walled spill-proof cup with introverted marqins (Figs. 14-4, 14-5). In the kidney the space (as in the cup), called the *renal sinus* is filled with fat in which the *renal pelvis* and its *calyces* are embedded. The entrance to the sinus is the *hilus*. The entire kidney is invested with a fibrous capsule surrounded by perirenal fat which is continuous with the fat of the renal sinus. The renal pelvis is a dilatation of the ureter which branches into three to five

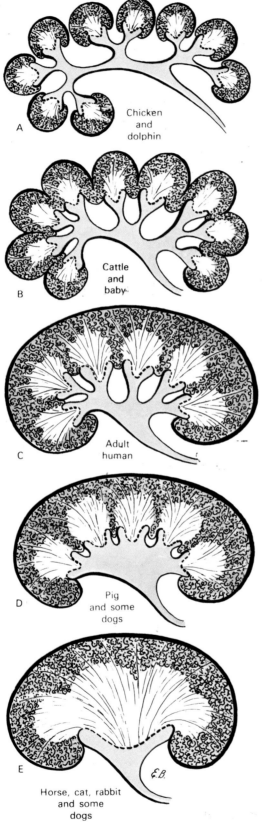

Fig. 14-1. Lobes of the kidney and their fusion.

chambers named the *major calcyes*, which in turn branch into two or three *minor calcyes*. The medulla, divided into pyramids faces the renal sinus, and the apex of each pyramid which abuts into a minor calyx is termed a *renal papilla* (Fig. 14-6). The papilla indents the distal wall of the minor calyx so that it is surrounded by the cup-like face of the minor calyx (Fig. 14-7). The cup of the calyx appears to end blindly, but in fact, long straight tubules project form the minor calcyes into the medullary tissue and penetrate into the cortex. Bundles of straiqht tubules recognized in the cortex are known as *medullary rays*.

The renal cortex may be divided into a peripheral *zona externa* (Fig. 14-8) and a juxtamedullary *zona interna*. The cortex consists of *renal corpuscles* and *convoluted tubules*, sometimes collectively referred to as the *cortical labyrinth* (Fig. 14-9). Renal corpuscles are absent from the outer cortex to a depth of 1 to 2 mm.

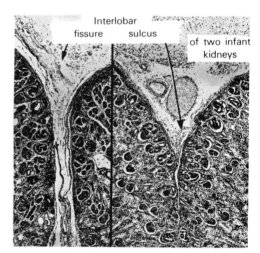

Fig. 14-2. Interlobar fissure and septum in two infants.

Medullary rays beqin with a broad base near the medulla (Fig. 14-9) and become thinner toward the capsule (Fig. 14-8).

THE NEPHRON. The structural and functional unit of the kidney is the nephron (Figs. 14-10, 14-11), a very complicated

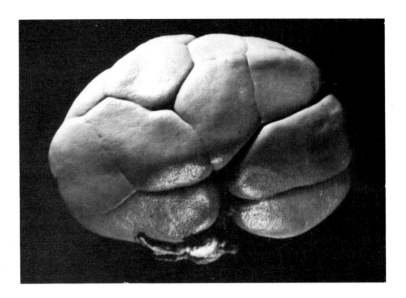

Fig. 14-3. Kidney of 2-year-old girl. The interlobar fissures are beginning to disappear. Various depths of furrows and degrees of fusion can be seen. (From Elias, Steinhausen and Pfeiffer, Die Niere, Boehringer, Ingelheim, in press.)

convoluted tube, which is conveniently simplified in these diagrams. This tube has a blind end, enlarged into a ball which is pushed in as shown in Figure. 14-13. This first part of the nephron is the *glomerular capsule* into which a tuft of vascular channels, *the glomerulus*, fits. This tuft of specialized capillaries is supplied by the *afferent glomerular arteriole* (red arrow, Fig. 14-12) and drained by the *efferent glomerular arteriole* (blue arrow). Glomerulus plus capsule form the *renal corpuscle* (Fig. 14-12). The region where the afferent and efferent arterioles feed and drain the glomerulus is the *vascular pole*. Exiting the capsule on the opposite side is the *proximal convoluted tubule* of the nephron, an area designated as the *tubular pole* of the renal corpuscle (Fig. 14-13). As indicated in Figure 14-12, the glomerular capsule is closed by epithelium having an internal (visceral) and an external (parietal) layer. Between these two layers is the *lumen of the capsule* (Fig. 14-12). The capillaries of the glomerulus together with the internal epithelium forms a unique ultrafiltration mechanism. The liquid in the capsular lumen (yellow in Fig. 14-13) is primary urine which flows into the rather long proximal convoluted tubule. The next segment of the nephron is the *ansa nephroni* (loop of the nephron), which consists of a *descending limb*, a *thin limb* and an *ascending limb* (Fig. 14-10). The nephron terminates at the point where the distal convoluted tubule empties into a renal collecting tubule. Before reaching this connection, however, the distal convoluted tubule passes close to the vascular pole where the tubular epithelium forms the *macula densa* (see below). Several nephrons may empty into one collecting tubule. Both proximal and distal convolutions are highly contorted, much more so than they appear in the simplified Figures 14-10 and 14-11. The distal convoluted tubule returns toward the renal corpuscle of its origin and passes between the afferent and the efferent arteriole, which are arranged in the form of the letter V. The distal convoluted tubule passes through this V in a direction perpendicular to both vessels (Fig. 14-10, A). At this precise place, an accumulation of nuclei in the epithelium of the distal convoluted tubule facing the afferent glomerular arteriole, aggregate as the *macula densa* epithelial cells. Hypothetical chemoreceptors in the macula densa "sense" sodium concentration in the distal tubule and regulate the glomerular filtration rate.

The vascularization of the nephron and the direction of blood flow are shown in

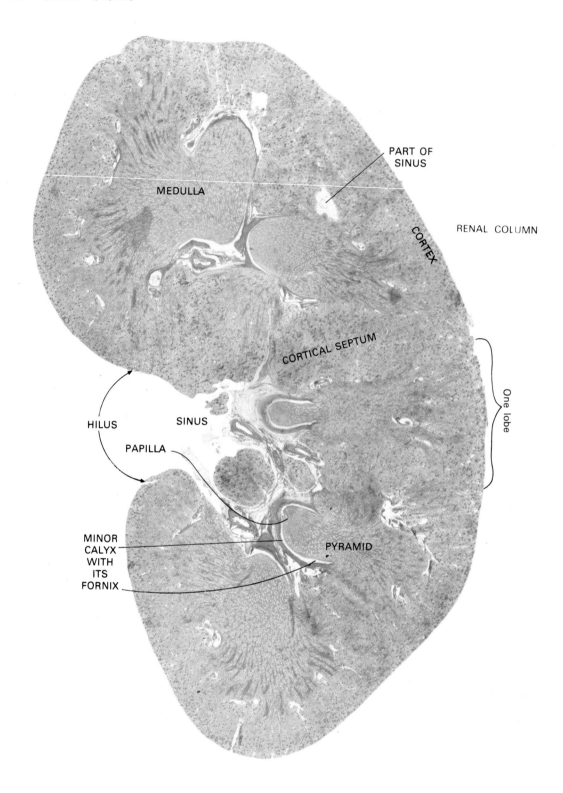

Fig. 14-4. Kidney of 6-month-old boy, H&E. 46×.

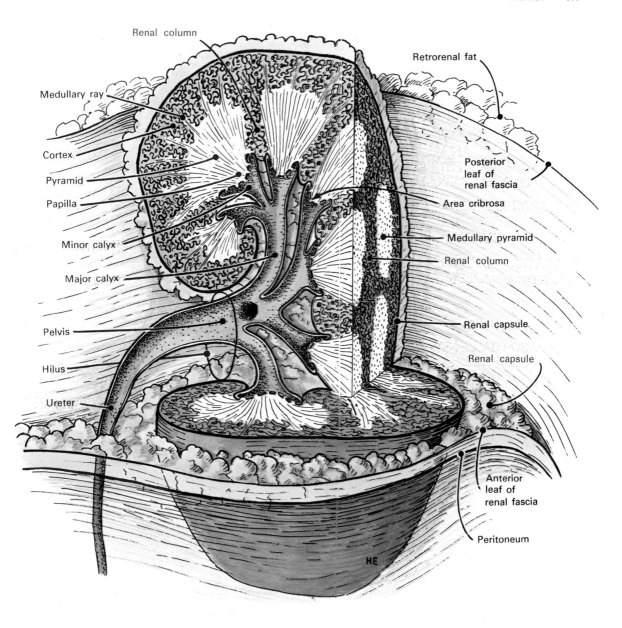

Fig. 14-5. Adult, human kidney.

Figure 14-11. By reference to Figure 14-10, the names of the vessels shown in Figure 14-11 can be identified.

Nephric loops reach various depths according to the location of their renal corpuscles (Fig. 14-14). Those from the most external nephrons are located entirely within medullary rays and do not reach the medulla proper (Fig. 14-15). On the other extreme are the loops from juxtamedullary nephrons (including those which arise in renal columns) which extend as far as the renal papillae (Fig. 14-16). The thin seqements of the loops vary in their location within the loops according to the location of the nephron (Fig. 14-14).

LOBULES. A lobule of the kidney is a small portion of a lobe, with each medullary ray forming the core of a lobule (Fig. 14-17). A lobule consists of all the nephrons which drain into the collecting tubules of one medullary ray. All the neph-

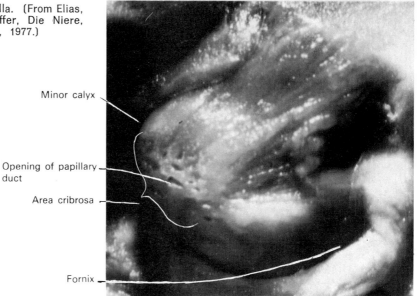

Fig. 14-6. A renal papilla. (From Elias, Steinhausen and Pfeiffer, Die Niere, Boehringer, Ingelheim, 1977.)

ric loops of one lobule are located in the same medullary ray. The arteries and veins which supply the cortex are located between lobules. The *afferent arterioles* which supply the glomeruli arise from *interlobular arteries* (Fig. 14-17).

The juxtamedullary cortex, however, is not organized into lobules, because its nephrons reach the pyramids directly (not through medullary rays).

BLOOD VESSELS. *Interlobar arteries* supply *arcuate arteries* from which *interlobular arteries* arise. The latter give rise to the *afferent arterioles* (vasa afferentia) which supply the glomeruli (Figs. 14-11 and 14-17, 14-18). The *efferent arterioles* (vasa efferentia) branch into capillary networks

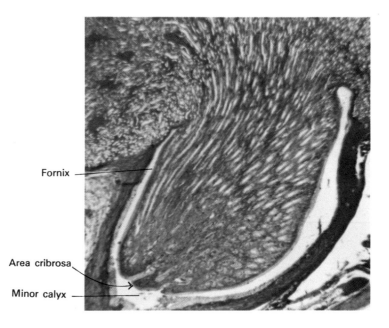

Fig. 14-7. Renal papilla and minor calyx with fornix. 6-month-old infant, Goldner stain. (Same case as Fig. 14-4.)

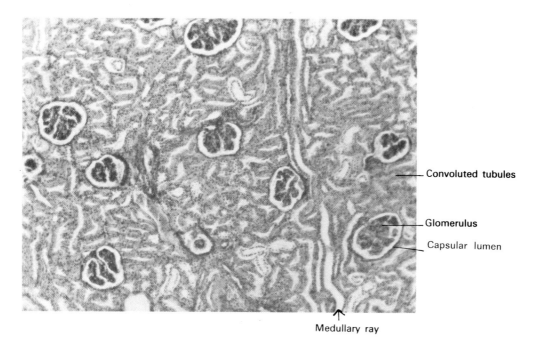

Fig. 14-8. Intermediate renal cortex, 13 mm from capsule, H&H. (Same source.)

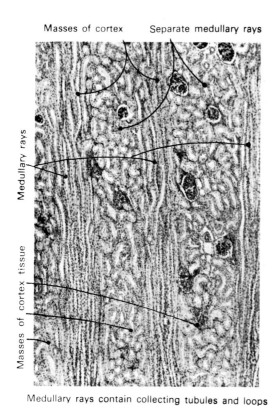

Fig. 14-8. B: Section of renal cortex perpendicular to the capsule, passing longitudinally through three medullary rays (6 month old infant).

(Fig. 14-11) which drain through interlobular, arcuate and interlobar veins.

The efferent arterioles of peripheral glomeruli (zona externa).(Fig. 14-19, above) send capillaries to the cortex; juxtamedullary glomeruli (zona interna) (Fig. 14-19, below) send capillaries to their own zona and the medulla.

"Straight" arterioles (arterial vasal recta), which arise from juxtamedullary efferent arterioles descend into the medulla. By a hair-pin turn they become the venous vasa recta (Fiq. 14-16). The vasa recta are united into bundles, The *fasciculus vascularis* (Figs. 14-20 and 14-21). Together with the nepric loops and with the collecting tubules, they form a complicated counter current system thought to be instrumental in the formation and concentration of urine. The venous vasa recta empty into interlobular veins.

The efferent arterioles of peripheral glomeruli in rats run straight to the surface without prior branching. There they form "welling points", branching into 3-6 arterial capillaries which branch into peritubular plexuses.

The Juxtaglomerular Complex

At the vascular pole of the renal corpuscle, modified smooth muscle cells in the wall of the afferent arteriole (Fig. 14-22) contain (1) more RER than regular smooth muscle cells, (2) a well-developed golgi complex and (3) membrane-bound secretory granules. These specialized smooth muscle cells are a part of the *juxtaglomerular (JG) complex*. The JG cells are baroreceptors which produce *renin* in response to a drop in blood pressure in the afferent arteriole. Renin is a substance which indirectly increases systemic blood pressure by converting angiotensinogen to angiotensin I which is converted to angiotensin II in the lung. Angiotensin II is a vasoconstrictor which also acts as a trop(h)ic hormone, stimulating the release of aldosterone from the zona glomerulosa of the adrenal cortex. Aldosterone causes an increased resorption of sodium ions in the distal convoluted tubule which, in turn, elevates the osmolarity of the blood, thus promoting increased water resorption. This produces an increase in blood volume with a consequent rise in blood pressure.

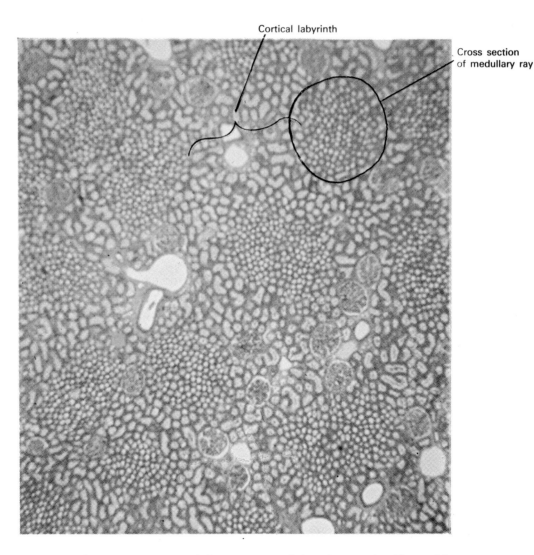

Fig. 14-9. Section through juxtamedullary cortex parallel to the capsule. The medullary rays are cut transversely. They are surrounded by cortical labyrinth, Goldner stain. 80 ×.

KIDNEY 381

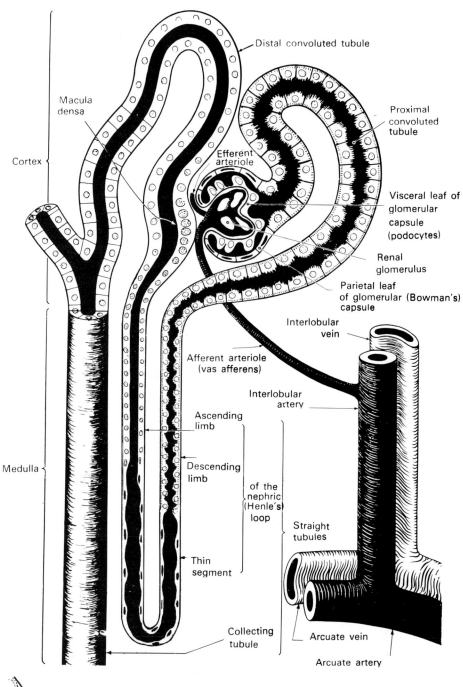

Fig. 14-10. Diagram of a nepluron.

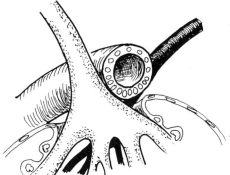

Fig. 14-10, A. At the lower left: Passage of the distal convoluted tubule forming the macula densa between the afferent and efferent arteriole.

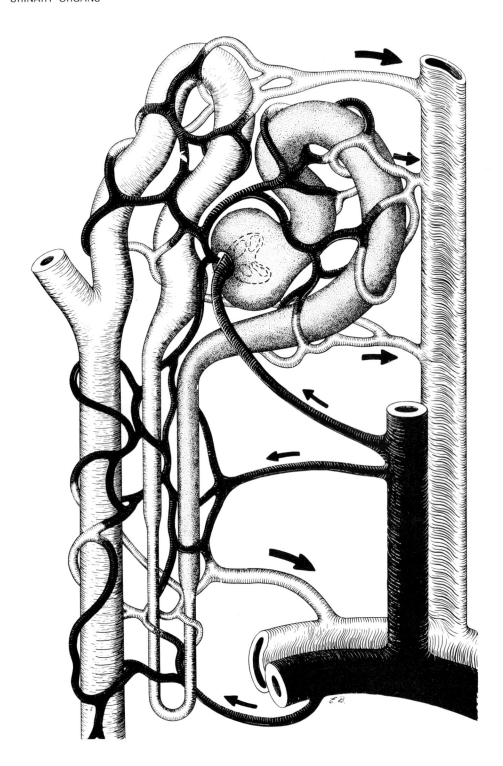

Fig. 14-11. Vascularization of the nephron.

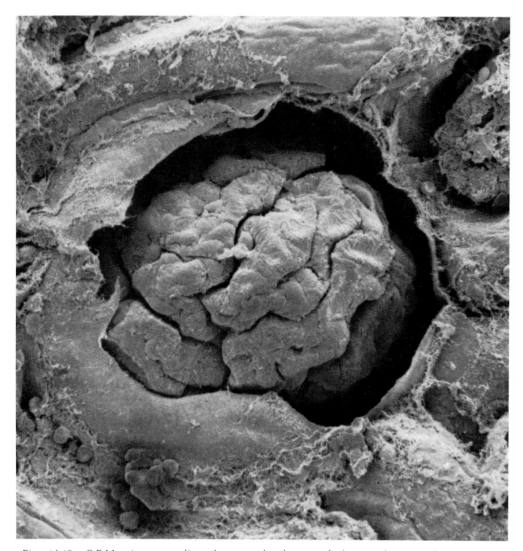

Fig. 14-12. S.E.M. view revealing the capsular lumen of the renal corpuscle. (1040 ×). (Courtesy Dr. Walter S. Tyler, University of California, Davis).

The optically clear epithelial cells of the *macula densa* of the distal convoluted tubule are juxtaposed to the juxtaglomerular endocrine cells of the afferent arteriole. Surrounding the macula densa and the juxtaglomerular cells is an island of extraglomerular, perivascular cells (Polkissen or polar cushion cells). The *juxtaglomerular complex* (JGC) (Figs. 14-22 and 14-23) consists of (1) the macula densa, (2) the juxtaglomerular cells, and some authors include, (3) the islets of perivascular mesangial cells (See also Fig. 14-18).

THE RENAL CORPUSCLE. The renal corpuscle (Figs. 14-13, 14-24) consists of the *glomerular capsule* and of the *renal glomerulus*, a tuft of blood channels which arises through a distribution chamber from the afferent arteriole. The glomerular blood channels branch and anastomose within an almost continuous mass of specialized endothelial cells. Unlike capillaries (Fig. 14-25, left), all the blood channels of one glomerulus are surrounded by one common, highly folded basement membrane (Fig. 14-25, right). Occasionally, between the blood channels *mesangial* cells are inserted. The endothelium together with mesangium (to be defined below) forms a branched, tunneled sheet covered by the basement membrane, seen in the thin sections of Figures 14-26 and 14-27, B. The

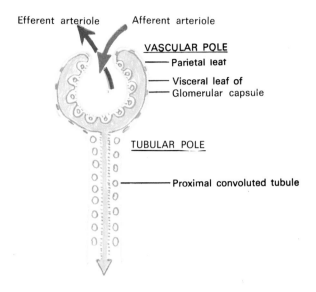

Fig. 14-13. Diagram of the renal corpuscle.

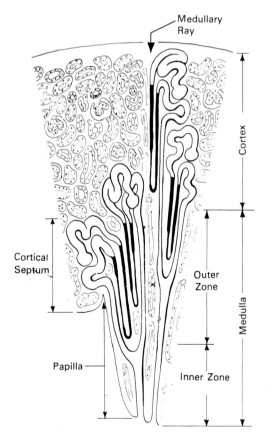

Fig. 14-14. Location of nephric loops and of the thin segments as dependent on the location of their renal corpuscles.

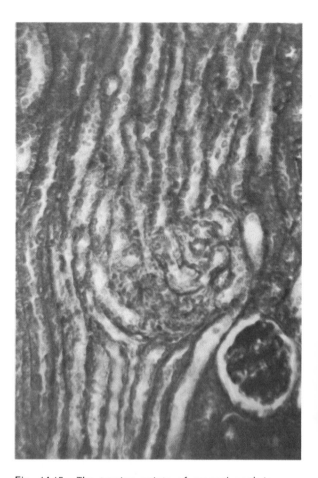

Fig. 14-15. The turning points of several nephric loops from external nephra within a medullary ray. Note that the so-called hairpin turn is wide leaving several tubules between the descending and ascending limbs. Thus, if counter-current plays a role in urine concentration, there is no direct exchange between parts of the *same* nephron. From 6-month-old infant, Goldner stain.

three-dimensional organization of this highly vascularized, branched sheet, called the vascular lamina of the glomerulus, is seen in the reconstruction from 35 serial sections shown in Figure 14-27 A. The electron micrograph of Figure 14-28, A and B, shows the uniqueness of the vascular lamina of the glomerulus even more convincingly. The glomerular blood channels are not permanent, but are temporary tunnels whose positions frequently shift, as can be seen in transilluminated kidneys of living amphibians. Shifting has not been confirmed for the mammalian kidney, because it may be a much slower process. Most importantly, the glomeruli

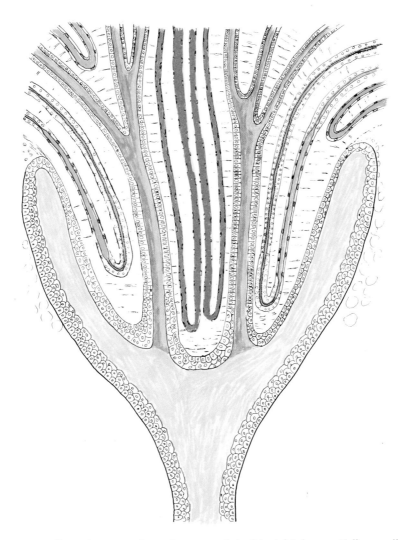

Fig. 14-16. Renal papilla and minor calyx. Green: nephric (Henle's) loops. Yellow: collecting tubules, papillary ducts and pelvis. Red: Vasa recta.

in higher forms usually are located too deep in the cortex to be observable during life. Blood coursing through is collected in the center into the collecting chamber of the efferent arteriole before exiting (Figs. 14-24, 14-29). Connective tissue cells which penetrate into the vascular lamina of the glomerulus constitute the *mesangium*. Mesangial cells could be instrumental in bringing about the shifting of blood channels, if such shifting actually occurs. As connective tissue derivatives, the mesangial cells provide structural support to the glomerular vessels. In addition they are actively phagocytic and assist in maintaining the integrity of the glomerular basement membrane. The mesangium also contains some strands of electron-dense material similar in appearance and continuous with the basement membrane (Fig. 14-26, A).

The glomerular basement membrane supports specialized epithelial cells known as *podocytes* (Figs. 14-28, A; 14-32 and 14-33) which make up the *visceral leaf of the glomerular capsule*.

The podocytes have long, branched, cytotrabeculae (Figs. 14-32) 14-33 and 14-34) whose terminal branches, the *cyto-*

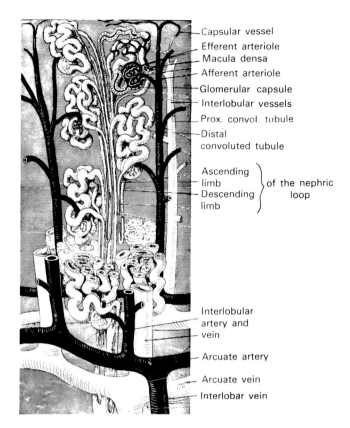

Fig. 14-17. A lobule of the kidney.

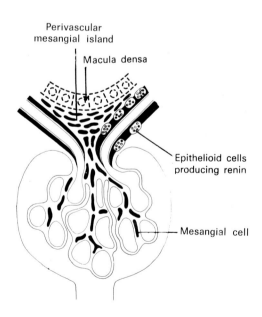

Fig. 14-18. Kriz's concept of the juxtaglomerular complex (courtesy of Prof. W. Kriz, University of Heidelberg).

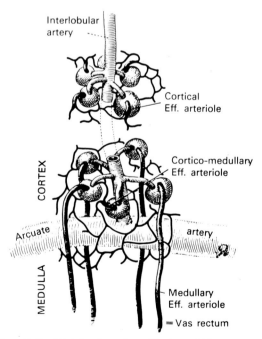

Fig. 14-19. Distribution of capillaries which arise from arteriolae efferentes. (From Edwards: Anat. Rec. 125: 521-530, 1956, courtesy of Wistar Institute.)

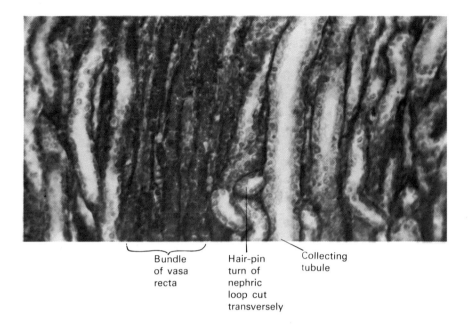

Fig. 14-20. Longitudinal section through "straight" tubules and blood vessels (vasa recta) in the renal medulla of a 6-month-old infant. (From Elias, Steinhausen, Pfeiffer, Die Niere, Boehringer, Ingelheim.)

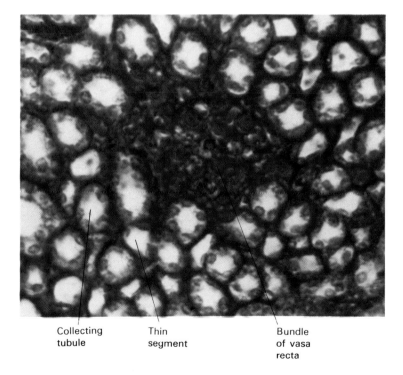

Fig. 14-21. Cross section through tubules and blood vessels in the renal medulla. (Same source as Fig. 14-18).

podia, interdigitate with each other. The slits between the cytopodia are covered by thin diaphragms or slit membranes (Figs. 14-35 and 14-36), organized like-zippers. They possess "slits" just large enough for the passage of molecules known to be filtered.

The cell body of the podocyte (Fig. 14-33) is elevated above the level of the cytopodia which contacts the basement membrane. Many of these cytopodia are covered by parts of the podocyte body or by its flat extensions, called *podocytic membranes*. Thus, a *subpodocytic lacuna* is

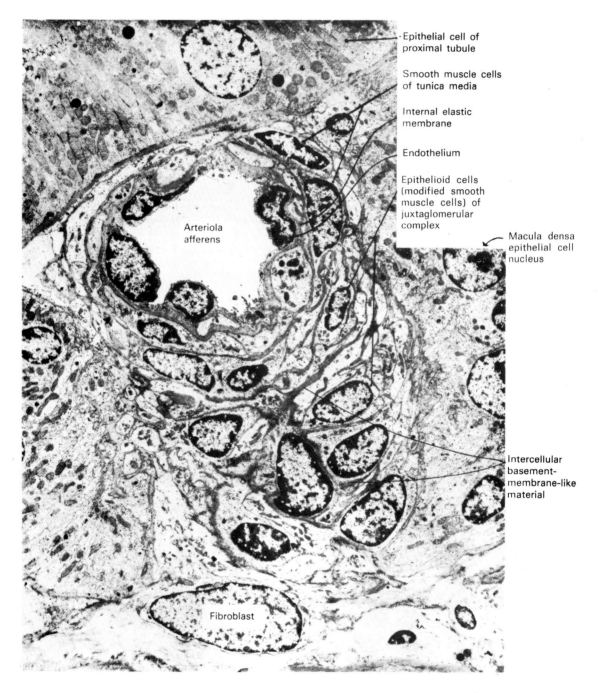

Fig. 14-22. Juxtaglomerular apparatus of rat. Courtesy of Dr. A. Schechter, Massachussetts General Hospital.

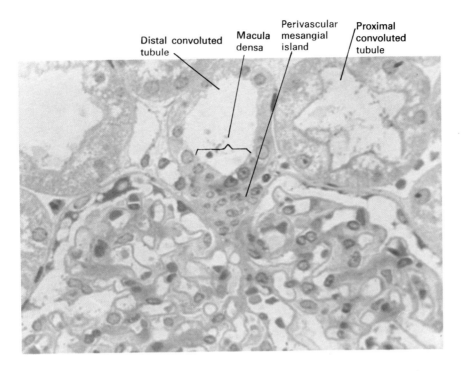

Fig. 14-23. Juxtaglomerular apparatus, semithin section, toluidine blue. Photographed from a slide prepared in the laboratory of Prof. A. Bohle, Tübingen. (From Elias, Steinhausen, Die Niere, Boehringer, Ingelheim.)

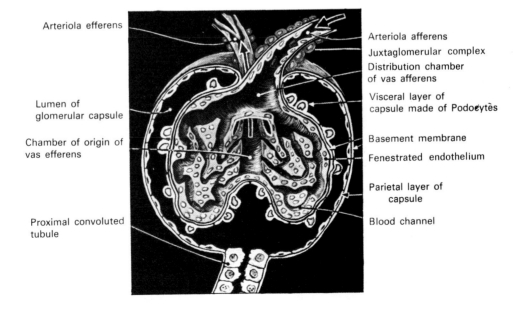

Fig. 14-24. Stereogram of renal corpuscle. (From Elias: The Renal Glomerulus by Light and Electron Microscopy. Res. Serv. Med. 46: 1-28, 1956.)

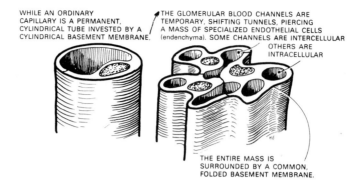

Fig. 14-25. Schematic comparison between a typical capillary and the glomerular blood channels.

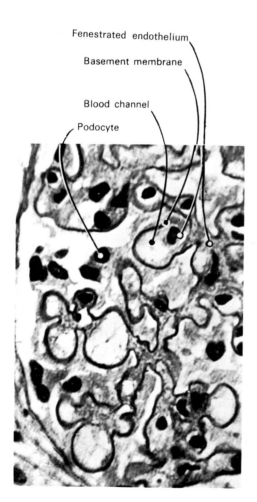

Fig. 14-26. A small portion of a glomerulus. Borst's hot formol-hematoxylin method, 2 μ.

gs. 14-37, 14-38 and 14-39). Occasionally, the "slits" communicate openly with the capsular lumen as seen in Figure 14-28, A and in Figure 14-40, right. The question is still debated whether such open communication exists during life.

An extremely thin, *fenestrated endothelium* is interposed between a blood channel and the basement membrane. The fenestrations of this attenuated endothelium can be seen in a tangential section (Fig. 14-39). The basement membrane between the podocytes and endothelial cells is the only continuous structure in the wall between these cells.

The total cross-sectional area of the afferent arterioles is greater than that of the efferent arterioles. As a result, the blood pressure is high in the glomerular blood channels. The hydrostatic pressure difference is approximately 19 mm of mercury between afferent and efferent arterioles. Blood plasma may contact the basement membrane by passing through the fenestrations in the endothelium.

Molecules with a molecular weight of 60,000 or more are retained in the blood plasma. Molecules with a molecular weight of 5,000 pass easily through the basement membrane and subsequently through the cytopodial slits. Even serum albumin molecules (molecular weight 50,000) may pass if these oblong particles are directed perpendicularly to the cytopodial slits, the main filtration barrier (Fig. 14-36).

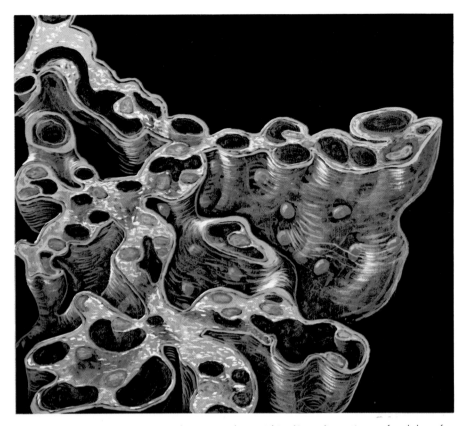

Fig. 14-27. A: Reconstruction from serial semithin (2 μm) sections of a lobe of a glomerulus. (From Elias, El Glomerulo Renal, Wander, Mexico City, 1970).

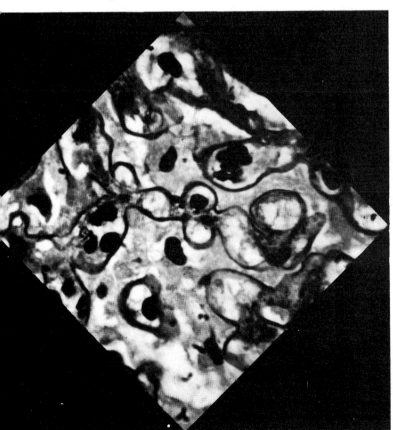

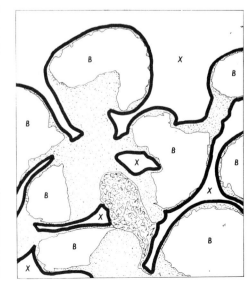

Fig. 14-27. B: The upper section of the series from which the model Fig. 14-27, A was prepared. Method of Borst: Fresh specimen boiled for 10 seconds in formol. Paraffin section, 2 μm thin. Delafield's hematoxylin. This technique, also used for Fig. 14-26, permits deep blue-violet staining of the basement membrane. (Same source as 14-27, A).

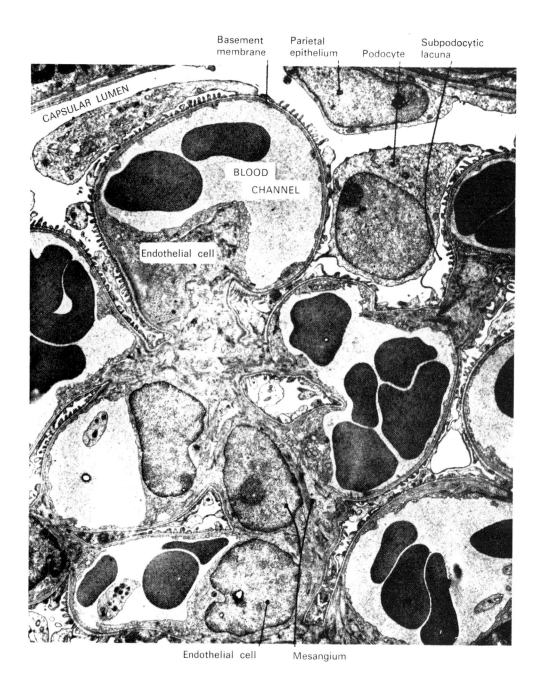

Fig. 14-28. A. A portion of a renal corpuscle of a mouse. (From J. Rhodin: Atlas of Ultrastructure. W. B. Saunders Co., Philadelphia, 1963.)

Fig. 14-28. B. A tracing of the same electron microgram designed to clarify Fig. 14-28, A. X = Extraglomerular area. B = Blood channels. Black = Basement membrane. Stippled = endenchyma. Shaded = mesangium.

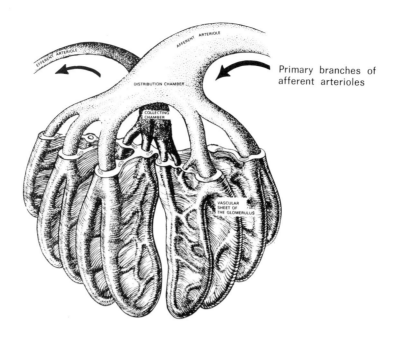

Fig. 14-29. Stereogram of glomerulus. (From Elias: Anat. Anz. 104: 26-36, 1957.)

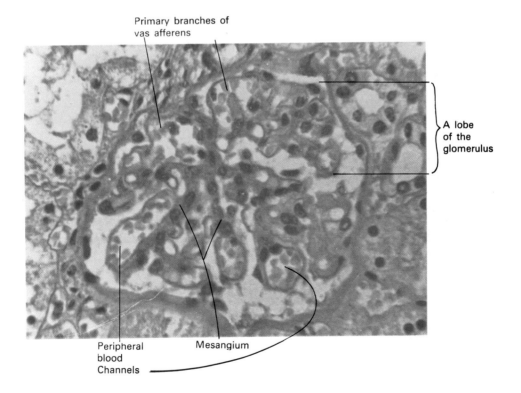

Fig. 14-30. Renal corpuscle of a 7-year-old boy. Death was by drowning. This glomerulus shows a highly developed mesangium. (From Elias, El Glomerulo Renal, Wander, Mexico City, 1970.)

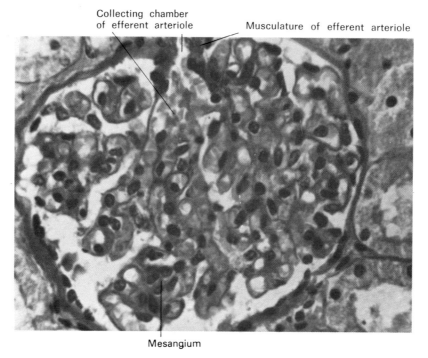

Fig. 14-31. Another corpuscle from the same kidney as Fig. 14-28, but with poorly developed mesangium. These two glomeruli were located next to each other. (Same source.)

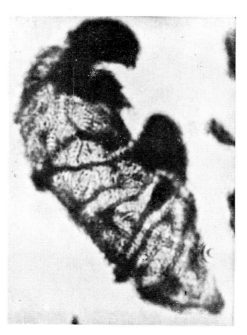

Fig. 14-32. Podocytes stained with molybdic hemotoxylin. (From Kulenkampff: Ztschr. Anat. 117: 520, 1954.)

It is not clear how the filtrate passes from the subpodocytic lacunae to the capsular lumen. The glomerular filtrate has essentially the same composition as blood plasma minus the materials of high molecular weight. Various questions and possible interpretations are illustrated in Figures 14-41 and 14-42.

THE PROXIMAL CONVOLUTED TUBULE. The proximal convoluted tubule, whose lumen is continuous at the tubular pole with that of the capsular lumen (Figs. 14-13, 14-24, 14-42, 14-43), consists of low columnar or cuboidal epithelial cells with scalloped outlines and brush borders (Figs. 14-44; 14-45). Rod-shaped mitochondria, in the striated basal cytoplasm, impart an acidophilia to these cells. The tubule is enclosed in a basement membrane continuous with that of the capsular parietal epithelium. The basal cell surface shows deep infoldings (Fig. 14-45). The brush border consists of microvilli (Fig. 14-45) between which are found small pinocytotic depressions, or caveolae. "The brush border" contains alkaline phospha-

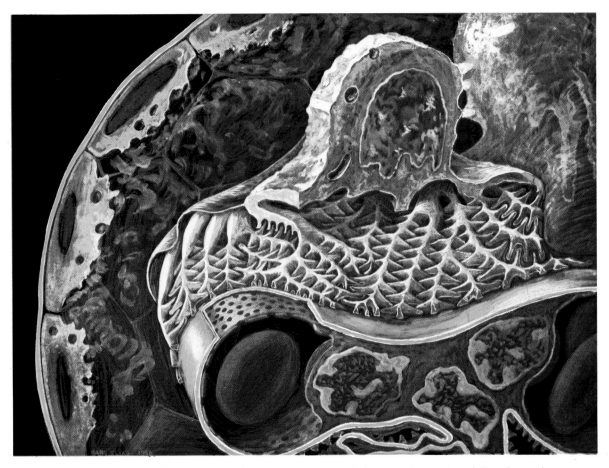

Fig. 14-33. Stereogram of renal corpuscle, two podocytes and the vascular lamina of the glomerulus. (From Elias, El Glomerulo Renal, Wander, Mexico City, 1970.)

tase, oxidative enzymes and mucopolysaccharides filling the spaces between the microvilli.

The main function of the proximal convoluted tubules is re-absorption of up to 90 per cent of the filtered water, electrolytes, glucose, most of the amino acids and small proteins in an isosmotic solution. The proximal convoluted tubular epithelium transports, among other waste products, urea and uric acid from the blood into the urine; and it secretes creatinine.

THE NEPHRIC LOOP (Ansa). The epithelium of the proximal convoluted tubule continues with that of the *descending limb of the nephric loop*. Its epithelium maintains the character of the proximal tubule but is slightly lower (Figs. 14-46 and 14-47). The descending limb continues into the *thin segment*, where the epithelium changes to simple squamous. In the *ascending limb* of the loop, the epithelium is low cuboidal. The thin segment of the loop contains microvilli and is freely permeable to water and sodium. The thicker ascending portion of the loop (Fig. 14-48) contains sparse, widely spaced, short microvilli. The epithelium is permeable to water and pumps sodium from the tubular lumen into the surrounding stroma. This process increases osmotic pressure in the stroma with the result that water is drawn out of the thin seqment by osmosis, and sodium enters the descending limb by diffusion.

THE DISTAL CONVOLUTED TUBULE. The epithelium becomes cuboidal in the distal convoluted tubule. In both the ascending

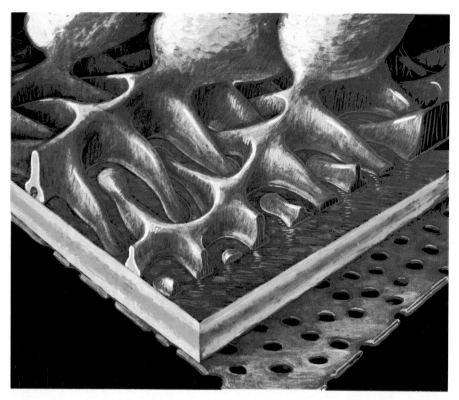

Fig. 14-34. The glomerular ultrafilter. Pink: fenestrated endothelial lamina. Green: basement membrane. Yellow: processes of a podocyte. (Same source as Fig. 14-31, with permission of C. H. Boehringer Sohn, Ingelheim.)

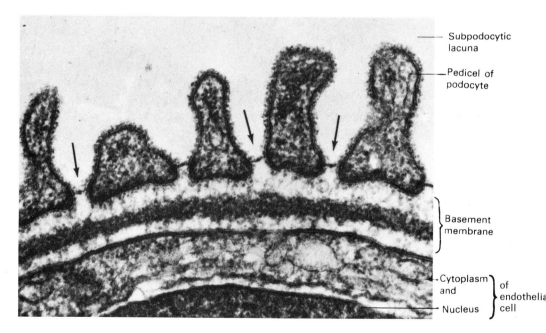

Fig. 14-35. The glomerular ultrafilter. Arrows indicate central ridge of slit diaphragm. (From Rodewald and M. J. Karnovsky, J. Cell Biol. 60: 423-433, 1974. With permission of Rockefeller U. Press.)

limb of the nephric loop and the distal convoluted tubule, the optically clear epithelial cells have only a few, scattered microvilli. The cells of the distal convoluted tubule actively pump certain ions, e.g., potassium, sodium, ammonia, out of the lumen. Aldosterone, a hormone of the adrenal cortex, is necessary for the reabsorption of sodium by the distal convoluted tubule. The distal convoluted tubule is the site of acidification of the filtrate. Antidiuretic hormone (ADH from the neurohypophysis or posterior lobe of the pituitary) increases the permeability of the distal convoluted tubule to water.

THE COLLECTING TUBULES. The collecting tubules, lined by clear cuboidal epithelium (Fig. 14-48), converge into *papillary ducts* which are lined by simple columnar epithelium (Fiq. 14-16). Fluid entering the collecting tubule is hypotonic, but as it passes through the tubule, water is drawn out into the hypertonic intestitium. The fluid that ultimately leaves the collecting tubules is therefore concentrated and is called *urine*. The collecting tubules are permeable to water under the influence of ADH. The concentration of ADH depends on a balance between the outer world and the internal medium. Abundant water in-

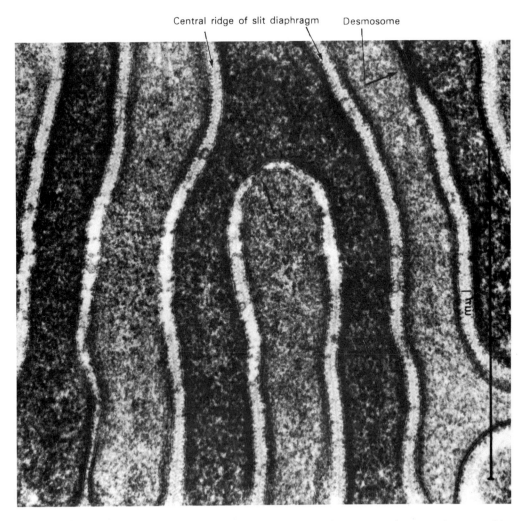

Fig. 14-36. Tangental section of foot processes, fixed with tannic acid. Note the zipper-like slit diaphragm. (From Rodewald and M. J. Karnovsky, J. Cell. Biol. 60: 423-433, 1974. With permission of Rockefeller U. Press.)

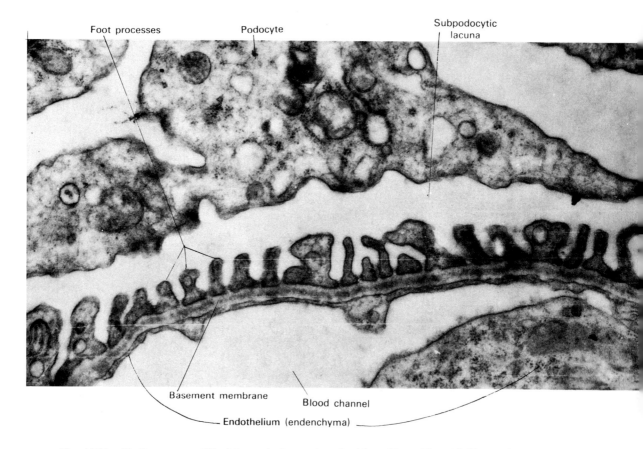

Fig. 14-37. Medium power EM picture of glomerulus of rabbit. (From Elias, El Glomerulo Renal, Mexico City, with permission of C. H. Boehringer, Ingelheim.)

Fig. 14-38. Small portion of glomerulus of rabbit to show pinocytosis, podocytic membrane and subpodocytic lacuna. (From Elias, Allara, Elias, and Murthy: Z. Mikro. Anat. Forsch. 72: 344-365, 1965.)

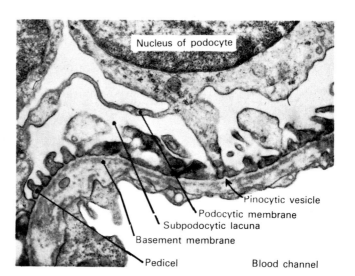

take will decrease ADH production by the hypothalamus; thirst and external aridity will increase ADH production. Loss of ADH production results in poor concentration of the fluid in the lumen of the collecting tubule. The clinical syndrome is called *diabetes insipidus*.

THE RENAL PAPILLA. The renal papillae, i.e., those portions of the medullary pyramids which project into a minor calyx (Figs. 14-4, 14-6, 14-7 and 14-16), contain papillary ducts and the lower portions of a few nephric loops which arise from juxtamedullary nephrons. The papillae also contain the lowest portions of the straight blood vessels (*vasa recta*). Both the neph-

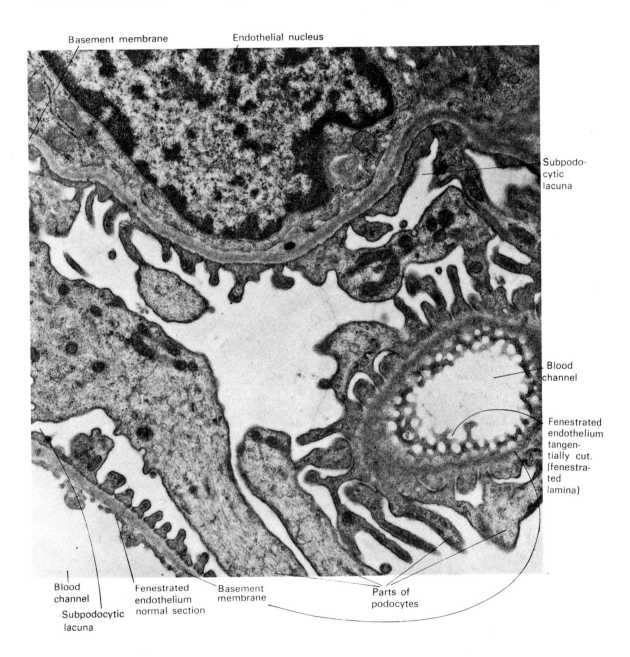

Fig. 14-39. Part of a glomerulus of a rat. Courtesy of Drs. Forssmann and Brühl, University of Heidelberg.

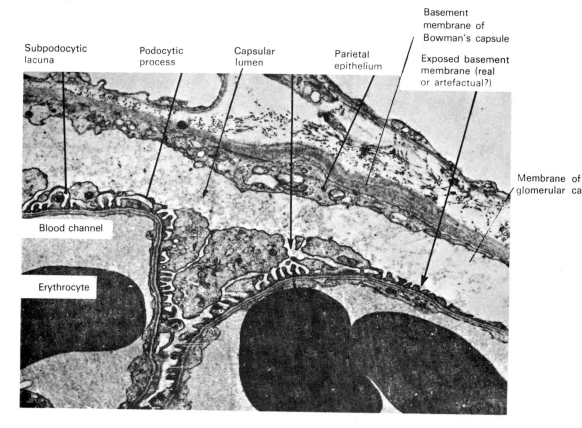

Fig. 14-40. External part of renal corpuscle of rabbit.

ric loops and the vasa recta are forced to make hairpin turns in the papilla. All these tubular structures are embedded in the papillary interstitial connective tissue. The latter consists mainly of interstitial fluid, but suspended in it are the interstitial cells.

When the urea concentration in the fornix of the minor calyx is high, the osmotic pressure in the papillary interstitium is increased; this, in turn, leads to higher concentration of urine. Indirectly then, high protein consumption imparts a certain resistance against dehydration.

The renal papilla is covered by transitional epithelium (Figs. 14-16 and 14-49), which enters into the papillary ducts for a short distance;(Figs. 14-16 and 14-49).

The papillary ducts empty into minor calyces covered by transitional epithelium.

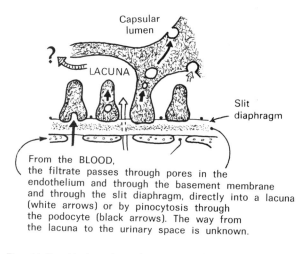

Fig. 14-41. Various hypothetical possibilities concerning the glomerular filtration mechanisms.

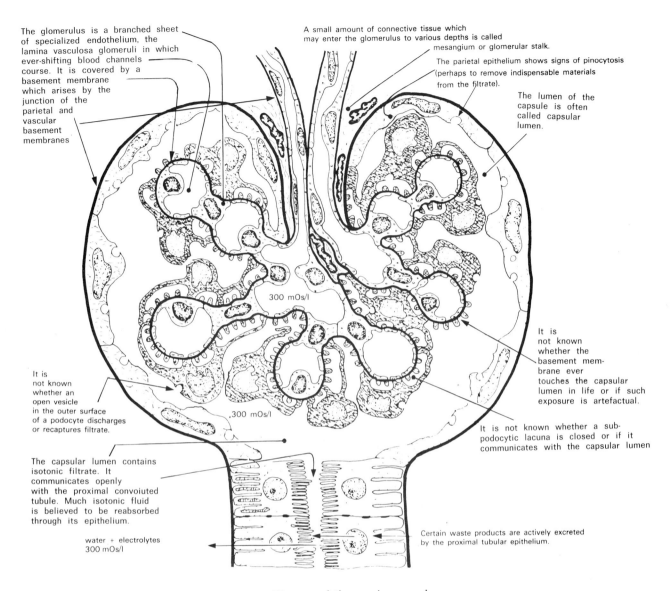

Fig. 14-42. Diagram of the renal corpuscle.

The surface of the renal papilla, being perforated by the foramina of the papillary ducts, is called the *area cribrosa* (sieve-like area, Fig. 14-6).

Urinary Passages

The major and minor calyces (Figs 14-5 and 14-49) the *renal pelvis*, the *ureter* (Fig. 14-50) and the *urinary bladder* (Figs. 14-51 to 14-54) are lined by transitional epithelium. Blood capillaries touch and indent its basal layer. The muscularis, thin in the pelvis and calyces, increases in thickness caudalward. All muscle bundles are continuous with each other; by around the calyces, they form sphincters.

In the wall of the ureter, a lamina propria mucosae and tela submucosa are virtually absent (Fig. 14-50). The very strong tunica muscularis is practially in contact with the epithelium. There seems to be an internal circular layer of smooth myocytes, a middle longitudinal layer and an external circular layer, but in reality all three

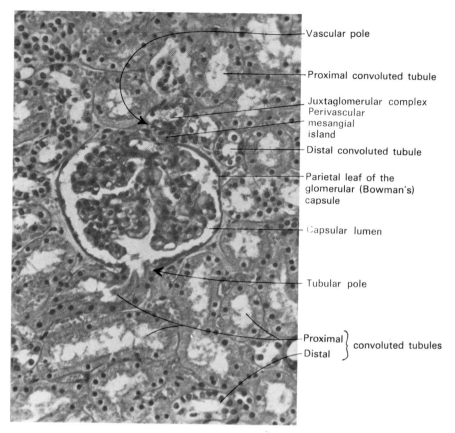

Fig. 14-43. Kidney cortex. (From Elias, El Glomerulo Renal, Mexico City, 1970.) With permission of C. H. Boehringer Sohn, Ingelheim.)

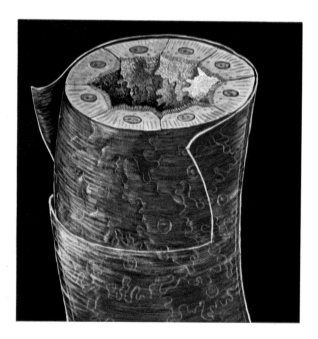

Fig. 14-44. Proximal convoluted tubule.

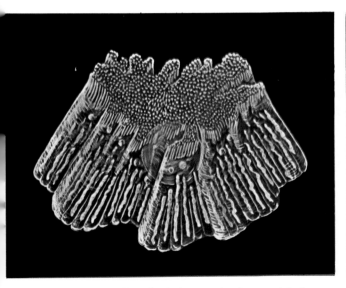

Fig. 14-45. A: A cell of the proximal convoluted tubula.

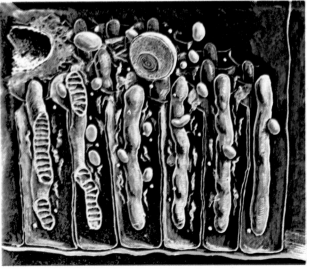

Fig. 14-45. B: Basal portion of a cell of the proximal convoluted tubule.

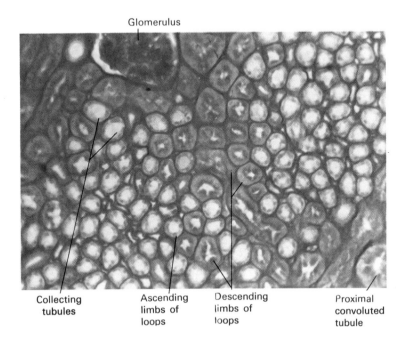

Fig. 14-46. Cross section of a medullary ray. The descending limbs of the nephron (Henle's) loops are grouped together in the middle. They are surrounded by ascending limbs. (From Elias, Steinhausen and Pfeiffer, Die Niere, Boehringer, Ingelheim).

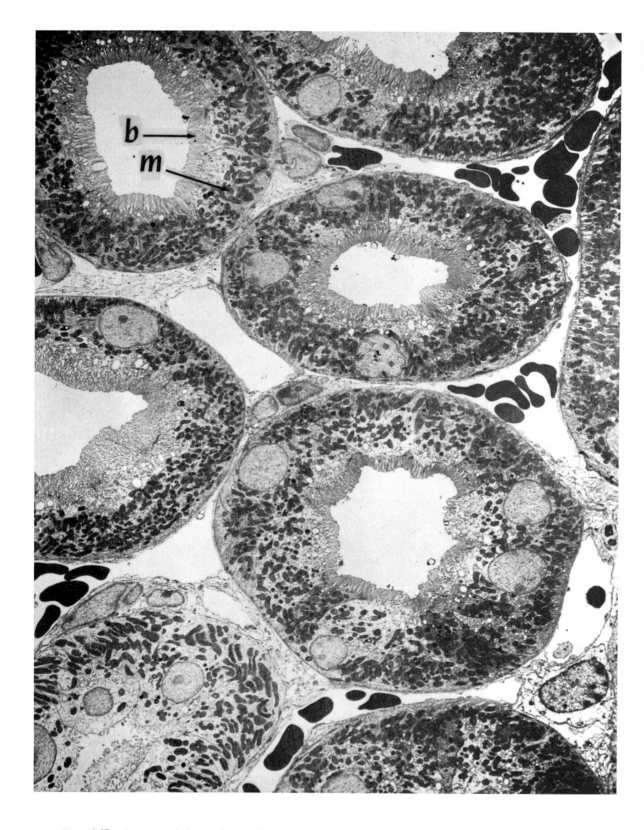

Fig. 14-47. A group of descending limbs of nephric loops and one ascending limb (lower left) in a medullary ray of a mouse. Courtesy of Dr. Johannes Rhodin, University of Michigan. b: Brush border; m: mitochondrion.

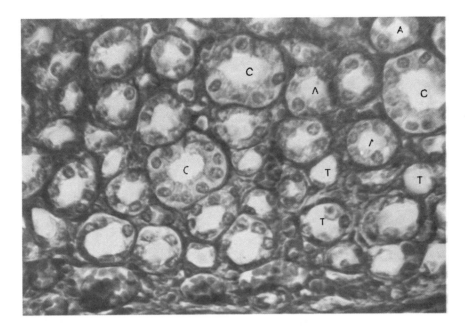

Fig. 14-48. Cross section of kidney medulla. A: Ascending limbs. C: Collecting tubules. T: Thin segments. (From Elias, Steinhausen and Pfeiffer, Die Niere, Boehringer, Ingelheim.)

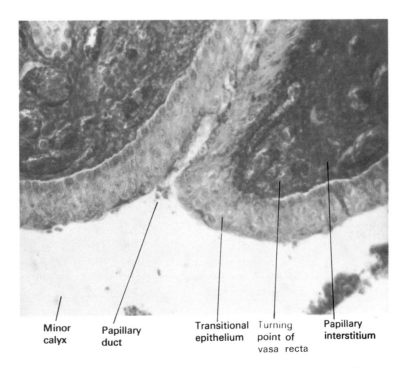

Fig. 14-49. Area cribrosa of a 6-month-old infant. (From Elias, Steinhausen, and Pfeiffer, Die Niere, Boehringer, Ingelheim.)

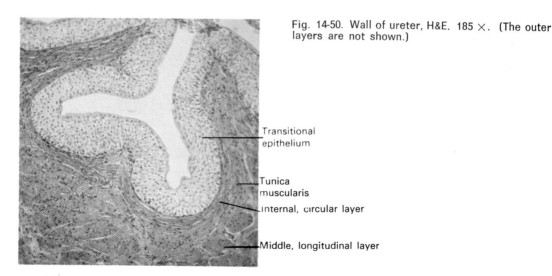

Fig. 14-50. Wall of ureter, H&E. 185 ×. (The outer layers are not shown.)

Transitional epithelium
Tunica muscularis
Internal, circular layer
Middle, longitudinal layer

layers are continuous with each other. The inner circular myocyte bundles swing outward and downward to form the longitudinal bundles which, in turn, swing further out returning to a circular course. The outer layer of this retropertioneal tube is, of course, an adventitia.

The wall of the urinary bladder (Figs. 14-51, 14-52 and 14-53) is, in many human cases, a complete mucous membrane where a distinct 2-layered lamina muscularis mucosae establishes the boundary between the lamina propria mucosae and the tela submucosa. Deep to the latter, follows a very powerful tunica muscularis which consists of several layers of obliquely oriented smooth myocytes. Above and behind the outer layer is a serous membrane, the parietal peritoneum. Anteriorly, laterally and inferiorly it is a tunica adventitia.

Pressure receptors (one of them a tactile corpuscle, seen in the tunica submucosa, Fig. 14-51) increase, by reflex action, the contraction of the external sphincter vesicae and help retain urine, until a degree of pressure is reached that requires emptying. At that moment, the musculature of the bladder wall contracts, while the external sphincter relaxes. Such baroreceptors as tactile corpuscles, belonging to the general afferent somatic system, bring the bladder pressure to consciousness and cause the organism to seek out a place to urinate.

The structure of the transitional epithelium of the bladder changes according to the state of filling of the organ (Fig. 14-54).

The female urethra is lined by patches of transitional, pseudostratified and stratified epithelium. Small mucous qlands are present. The deep layer of its lamina propria contains an erectile plexus of veins. The thick muscularis has an inner longitudinal and outer circular layer.

The male urethra is desribed with the male genital organs.

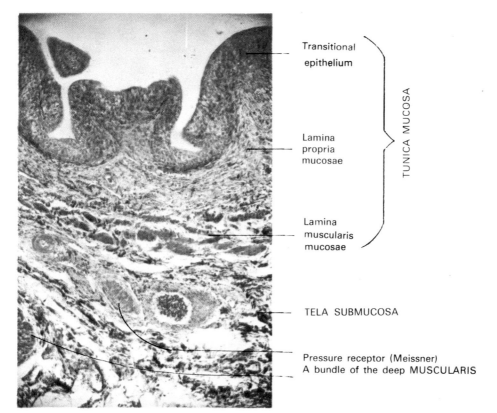

14-51.

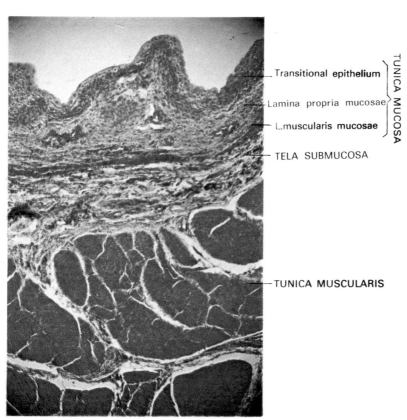

14-52.

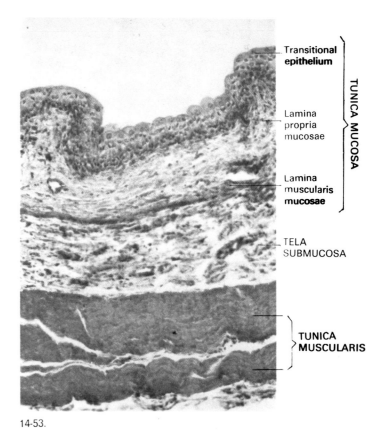

Figs. 14-51, 14-52 and 14-53. Sections of three human urinary bladders to show that in man, contrary to traditional belief, the wall of the bladder possesses all the layers of a complete mucous membrane.

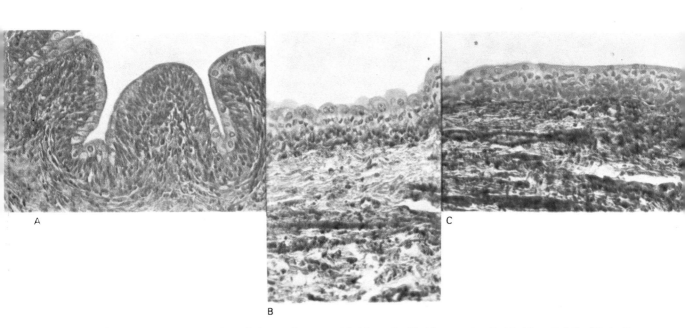

Fig. 14-54. Transitional epithelium of urinary bladder. A: Bladder empty; B: bladder slightly filled; C: bladder full, Azan. 750 ×.

15...
Endocrine system

Unlike exocrine glands which discharge their products through ducts, the *endocrine glands* secrete specialized products, called *hormones*, into the blood stream. In the older literature, endocrine glands were commonly referred to as ductless glands or glands of internal secretion, terms which are appreciated for their descriptive sense. Hormones do not act locally, but are dis-

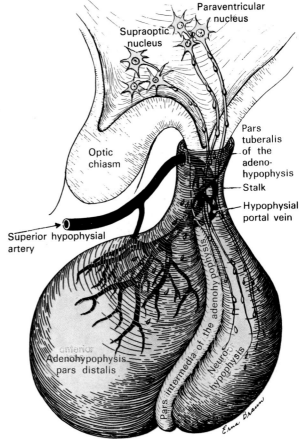

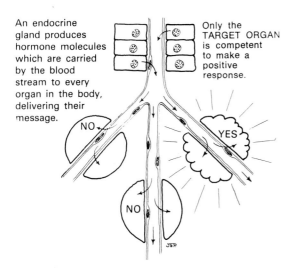

Fig. 15-1. Hormones affect specific target organs.

Fig. 15-2. The hypophysis cerebri.

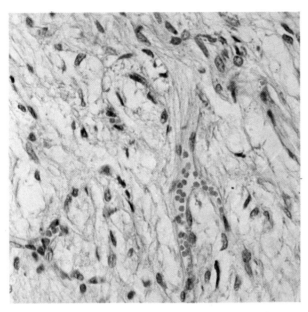

Fig. 15-3. Neurohypophysis (posterior lobe of the hypophysis cerebri), H&E. 650 ×.

tributed by the blood stream throughout the body to act at specific sites (target organs) where minute quantities elicit a response in specific cells. Target cells have specific receptors with the competence to stimulate a response to a particular hormone (Fig. 15-1).

The endocrine glands include the *hypophysis cerebri* (pituitary gland), *thyroid glands, parathyroid glands, suprarenal glands,* and *epiphysis cerebri* (pineal gland). Other organs discussed elsewhere possess endocrine functions: *placenta* (cf. textbook of embryology), *pancreatic islets* (Chapter 12) and the *enteroendocrine cells* of the digestive tract (Chapter 12). Evidence also exists that the thymus, spleen and liver may possess endocrine functions, but these usually are not classified as endocrine glands. The testes and ovaries, in addition to producing products required for procretion, secrete important hormone products (Chapter 18).

Hypophysis Cerebri

The *hypophysis cerebri* (pituitary gland) (Fig. 15-2) is suspended from the base of the brain and housed in a depression of the sphenoid bone, the *sella turcica*. The hypophysis cerebri, commonly referred to as the "master gland", directs much of the bodily activities, including other endocrine organs through its tropic (trophic) hormones. Two major portions can be distinguished:

1. *Neurohypophysis* (posterior lobe), originates from the base of the diencephalon, as an outgrowth of neural ectoderm.

2. *Adenohypophysis* (anterior lobe) composed of three areas: (1) *pars distalis*, (2) *pars intermedia* and (3) *pars tuberalis*. The entire lobe develops from an evagination of ectodermal epithelium in the roof of the primitive oral cavity (*stomadeum*) into the sella turcica.

NEUROHYPOPHYSIS (Posterior Lobe). The *neurohypophysis*, characterized by the structure of brain tissue, remains connected to the floor of the diencephalon via the *infundibular stalk*. The cells which compose the delicate stroma of glial elements are called *central gliocytes* (pituicytes) (Fig. 15-3). Axons of neuron cell bodies located in the *hypothalamus*, pass through the stalk of the infundibulum and maneuver through the central gliocytes. These axons, constituting the *hypothalamo-hypophyseal tract*, do not appear to synapse with other cells, but terminate as end-arborizations in the walls of blood capillar-

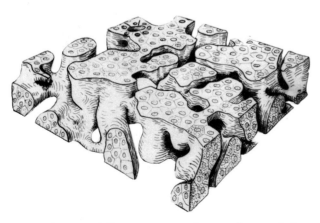

Fig. 15-4. Reconstruction of parenchyma of the adenohypophysis (anterior lobe of hypophysis cerebri.)

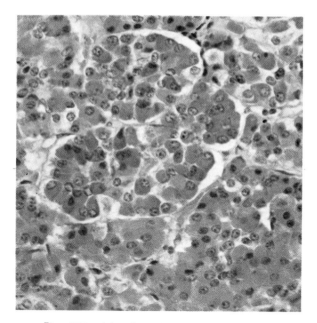

Fig. 15-5. Adenohypophysis. (Anterior lobe of pituitary). H&E. 650 ×.

ies where they discharge their hormones.

Thus, the neurohypophysis is not, per se, an endocrine gland, but a relay station in which hormones synthesized elsewhere are stored and released into the blood stream.

Neuron cell bodies, in the *paraventricular* and *supraoptic nuclei* of the hypothalamus, produce hormones which are transported within their unmyelinated axons through the hypothalamo-hypophyseal tract into the neurohypophysis. There the *neurosecretory substance* accumulates in the terminal expansions of the axons where they may be seen in many longitudinal and cross sections of the nerve fibers. When the neurosecretory substance becomes particularly prominent, the resulting mass is referred to as an *accumulation of neurosecretory corpuscles* (Herring body). This secretory material is discharged into the blood stream as required. Two hormones are stored and released in the neurohypophysis: (1) *oxytocin*, which stimulates myometrial contraction and ejection of milk from lactating mammary glands and (2) *vasopressin* or *antidiuretic* hormone (ADH), which inhibits diuresis by causing water reabsorption from distal convoluted and collecting tubules of the kidney and raises blood pressure by causing contraction of smooth myocytes in the blood vessels.

ADENOHYPOPHYSIS (Anterior lobe). *Pars distalis*. The major portion of the adenohypophysis consists of the pars distalis, a continuum of crooked cords and plates of varying thickness (Fig. 15-4), supported by reticular fibers and tunneled by a network of sinusoids. Some of these vessels arise from the hypophyseal portal veins which collect blood from the hypophyseal stalk. A few of the parenchymal cells are small, stain very lightly and form little clusters in the centers of the cords. These are the *chromophobe endocrine cells* (gamma cells) considered to be either reserve cells or degranulated chromophil cells (Fig. 15-5). There are investigators who maintain that these cells produce the hormone ACTH. Surrounding the chromophobe endocrine cells and lying adjacent to the sinusoids are the *chromophil endocrine cells*, more than half of which are *acidophil endocrine cells* (alpha cells) and the rest are *basophil endocrine cells* (beta cells).

The acidophil endocrine cells can be distinguished further (via electron microscopy) into: (1) *somatotrop(h)ic endocrine cells* and (2) *mammotrop(h)ic endocrine cells*. The somatotrop(h)ic endocrine cells contain electron dense, membrane bound granules (diameter 300 to 350 nm) and a relatively large golgi complex. These granules are the source of *somatotrop(h)in (STH)* or *growth hormone*, which stimulates the epiphyseal plate in growing long bones. Metabolism of proteins, fats and carbohydrates is regulated by the activity of somatotrop(h)in. Activity of other hormones is enhanced by this hormone. The mammotrop(h)ic endocrine cells resemble the somatotrop(h)ic endocrine cells, but their granules are larger (600 to 900 nm in diameter). The hormone *prolactin (LTH)* or *mammotrop(h)in*, produced by these cells stimulates lactation in the breast.

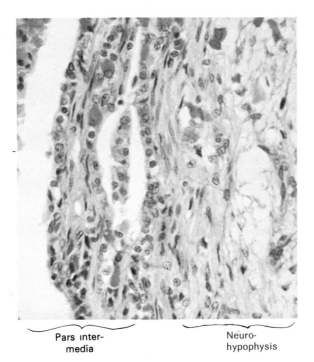

Fig. 15-6, A. Relationship of the pars intermedia (of the adenolypophysis) to the neurohypophysis. H&E. 650 ×.

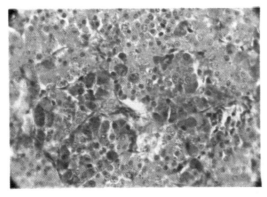

Fig. 15-6, B. Adenohypophysis of the hypophysis cerebri.

The *basophil endocrine cells* are subdivided into three basic cell types: (1) *thyroptrop(h)ic endocrine cells*, (2) *corticotrop(h)ic endocrine cells* and (3) *gonadotrop(h)ic endocrine cells*. The thyrotrop(h)ic cells produce *thyroid stimulating hormone (TSH)*, which stimulates the secretory activity of the thyroid follicular endocrine cells. *Corticotrop(h)ic cells* produce *adrenocorticotrop(h)ic hormone (ACTH)* which stimulates the suprarenal cortex to secrete glucocorticoids. Two subtypes of gonadotrop(h)ic endocrine cells can be distinguished by electron microscopy, based on granule size and location in the cytoplasm: (a) a cell with granules 250 nm in diameter (in the male), secretes *interstitial cell stimulating hormone (ICSH)* which causes the testicular interstitial endocrine cells to secrete testosterone, or in the female, produces *leuteinizing hormone (LH)* which initiates the secretion of progesterone in the ovary, and (b) a cell with abundant granules (200 nm in diameter) secretes *follicle stimulating hormone (FSH)* which in the female stimulates growth of the ovarian follicle and in the male acts on the seminiferous tubules to produce spermatozoa.

Pars intermedia. The pars intermedia, which is poorly developed in man, consists of a few irregular cell cords and follicles composed mainly of basophilic endocrine cells (Fig. 15-6, A). Some follicles are lined with ciliated cuboidal cells and many contain colloid. It is not rare to encounter similar follicles which migrate to the pars distalis. A few cell cords and scattered cell groups extend into the pars nervosa for a short distance. The hormone of

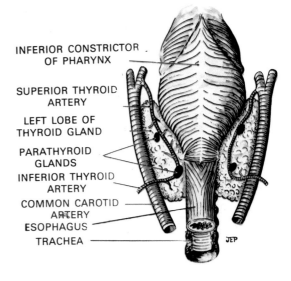

Fig. 15-7. Thyroid and parathyroid glands, posterior view. The isthmus is not visible from this aspect.

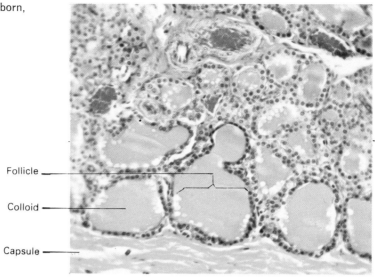

Fig. 15-8. Thyroid gland of new-born, Goldner stain. 350 ×.

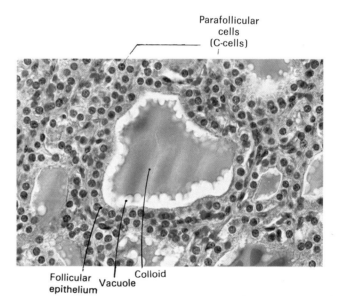

Fig. 15-9. A thyroid follicle, Goldner stain. 650 ×.

the pars intermedia, *melanocyte stimulating hormone* (*MSH* or *intermedin*), causes expansion of pigment cells when injected into amphibians, reptiles and fishes. In mammals, MSH causes increased pigmentation, probably by stimulating synthesis of melanin.

Pars tuberalis. The pars tuberalis, a superior extension of the pars distalis, embracing the anterior and lateral aspects of the infundibulum is composed primarily of chromophobe cells.

HYPOTHALAMIC CONTROL OF THE ADENOHYPOPHYSIS. The adenohypophysis is under the nervous and hormonal control of the hypothalamus. Neurosecretory cells in the hypothalamus produce specific *releasing* (RF) or *inhibiting* (IF) *factors*. Examples of releasing factors are: TSH^{RF}, FSH^{RF}, LH^{RF}, STH^{RF} and $ACTH^{RF}$. An example of an inhibiting factor is LTH^{IF}. The following is an illustration of how the system works. As the blood level of estrogen rises, neurosecretory cells of the hypothalamus secrete LH^{RF} which is released into the *hypophyseal portal circulation*. LH^{RF} is delivered to the pars distalis where it stimulates the release of LH by

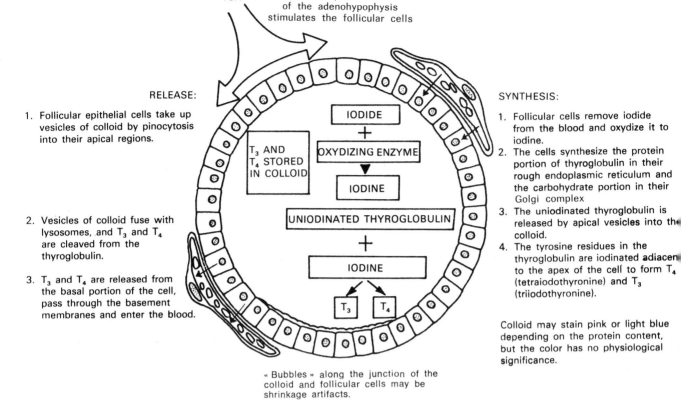

Fig. 15-10. Diagram of the activities of a thyroid follicle.

the gonadotrop(h)ic endocrine cells. The hypophyseal portal system is, therefore, an important vascular connection between hypothalamus and hypophysis. Superior hypophyseal arteries form a capillary plexus around the hypothalamic neurosecretory cells. Veins drain into the pars distalis where they re-branch into another capillary-sinusoidal plexus. The hypophyseal portal system is the means by which RF or IF reaches the pars distalis from the hypothalamus.

Thyroid Gland

The thyroid gland (Fig. 15-7) is located on the anterior and lateral sides of the trachea just inferior to the thyroid cartilage and larynx. It consists of two lobes connected across the midline by an isthmus (Fig. 8-35). The hormones produced by the thyroid are T_4 (*thyroxine* or *tetraiodothyronine*) and T_3 (*triiodothyronine*). T_4 and T_3 are the hormones which control the rate of metabolism in the body. The thyroid also produces *thyrocalcitonin*, a hypocalcemic factor which is antagonistic to parathyroid hormone. Thyrocalcitonin decreases bone resorption and thereby decreases blood calcium.

The fibrous capsule sends delicate septa into the gland dividing it into a series of connected, rather indistinct "lobules". The functional unit is the *follicle*, a closed bag lined by a single layer of normally low cuboidal *follicular endocrine cells* supported by a basement membrane. The follicular cells surround a central cavity containing a thick albuminous mass called *colloid* (Figs. 15-8 and 15-8). Areolar connective tissue interwoven with a reticular network separates adjacent follicles and supports the rich capillary bed.

Thyroid follicles are of many sizes, and the larger ones are often irregularly shaped (Fig. 15-8). Different functional states are manifested by the shapes of the cells and condition of the colloid (Figs. 15-10 and 15-12). In the newborn, the follicles are very active, and many bubbles (vacuoles) may be seen at the periphery of the colloid (Fig. 15-9); in adults follicles are less active, thus more are in the storage phase (Fig. 15-16, left). The epithelium is low. Hard colloid, often showing parallel artifactual cracks, is encountered. Liquid vacuoles are rare.

Thyroglobulin is synthesized by the follicular endocrine cells and secreted into the follicle. Thyroglobulin is a large glycoprotein which is stored in the colloid. The protein portion of the thyroglobulin is synthesized in the rough endoplasmic reticulum of the epithelial cells, while the carbohydrate portion is synthesized in the golgi complex. Apical vesicles release newly manufactured thyroglobulin into

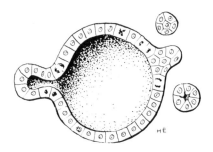

Fig. 15-11. Follicular budding.

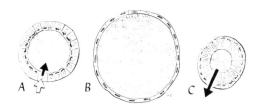

Fig. 15-12. A, secretion; B, storage; C, absorpion (hypothetical).

Fig. 15-13. Hyperactive follicle.

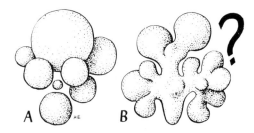

Fig. 15-15. Conflicting concepts of thyroid structure. A, Isolated follicles; B, follicles united into a lobule.

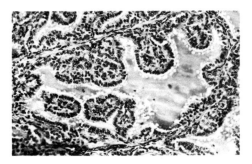

Fig. 15-14. Thyroid hyperplasia. (Preparation by Dr. P. Szanto Cook Country Hosp.)

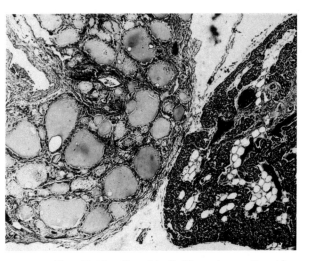

Fig. 15-16. Thyroid (left) and parathyroid (right), H&E. 80 ×.

the colloid. Iodide taken up by the epithelial cells is oxidized to iodine. Tyrosine residues in thyroglobulin are iodinated in the colloid adjacent to the apex of the cell, and linked together to form tetraiodothyronine (T_4) and triiodothyronine (T_3). Under TSH stimulation, the follicular endocrine cells engulf colloid by endocytosis; and the colloid-containing vesicles assimilated into the apical region of the cells, fuse with lysosomes. T_3 and T_4 are cleaved enzymatically from the thyroglobulin and released from the basal portion of the cell, to pass through basememt membrane and sinusoidal endothelium into the blood.

Multiplication of thyroid follicles occurs by budding (Fig. 15-11). When increased demands are made for thyroid hormone production, as in severely cold environments, TSH induces the epithelium to undergo hyperplasia and hypertrophy (Figs. 15-13 and 15-14). The result is increased *basal metabolism*. If an infant fails to produce enought thyroid hormone, *cretinism* results; if the secretion is inadequate in the adult, *myxedema* develops. A goiter is an enlarged thyroid gland. Since 3 dimensional reconstructions of thyroid follicles have not yet been made, we do not know exactly whether each follicle is isolated, or if connections exist between them (Fig. 15-15).

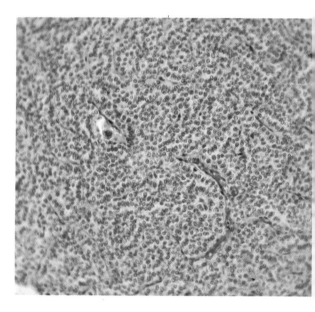

Fig. 15-17. Parathyroid of new-born, Azan 350 ×.

Parafollicular endocrine cells (C cells) reside between follicular cells, but fail to protrude into the colloid filled lumen. The follicle basement membrane retains these cells close to the follicular endocrine cells. Other parafollicular endocrine cell aggregate in the loose connective tissue interstices between follicles. These specialized cells secrete the hormone thyrocalcitonin

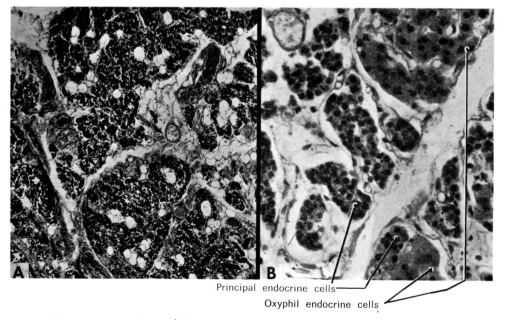

Principal endocrine cells
Oxyphil endocrine cells

Fig. 15-18. Parathyroid of 36 year old man. A, Low power; B, high power.

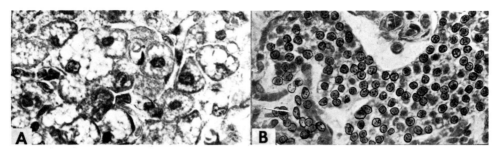

Fig. 15-19. A and B, two areas from one parathyroid, in a woman 28 years old to show structural variability.

which lowers serum calcium by supressing bone resorption (Fig. 15-9).

The Parathyroid Gland

There are usually four parathyroid glands (Figs. 15-7 and 15-16, right), two on the posterior surface of each lobe of the thyroid just external to its capsule. They are tiny, oblong, flattened, brownish bodies about 6 mm in length.

Loosely organized connective tissue septa and trabeculae from the delicate capsule invade the gland to serve as passageways for the blood vessels. They also divide the parenchyma into irregularly connected masses of cells (Figs. 15-16, right to 15-20), occasionally containing small lumina lined with columnar cells (Fig. 15-21). The parenchyma is composed of two kinds of cells: (1) principal endocrine cells and (2) oxyphilic endocrine cells (Figs. 15-20 to 15-22). The term "oxyphilic", synonymous with "acidophilic" has little to do with oxygen. Usually the *principal endocrine cells* are very small and have little cytoplasm; thus their large nuclei appear close together, resembling lymphocytes. In some cases (Fig. 15-19, A) they are large

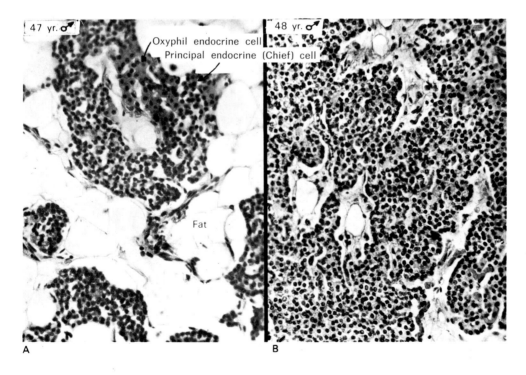

Fig. 15-20, A. Parathyroid of man, 47 years old. Fig. 15-20, B. Parathyroid of man, 48 years old.

and contain many vacuoles. Large capillaries emanating from septal vessels course between the cords and masses of cells. A reticular fiber network accompanying the septa and surrounding the blood vessels provide stromal support to the cells.

At the end of the first decade of life, the *oxyphilic endocrine cells* appear. Larger than most principal cells, they have an acidophilic cytoplasm attributed to an abundance of mitochondria (Fig. 15-23). Oxyphil endocrine cells tend to aggregate in small clumps near the capsule. Although they increase in number with advancing age, their function is unknown. The principal endocrine cells produce *parathyroid hormone (hypercalcemic factor)* which participates in the regulation of calcium metabolism. Parathyroid hormone increases the resorption of bone (releases calcium) and increases intestinal absorption of calcium.

If excess parathyroid hormone is secreted, the calcium level in blood elevates at the expense of the calcium content of the bones. When the hormone is deficient, the blood calcium level falls, and the patient suffers from extreme nervous excitability and tetany. Thus, the parathyroids are essential to life.

With advancing age, an increasing number of unusually large adipocytes invade the connective tissue septa, a process

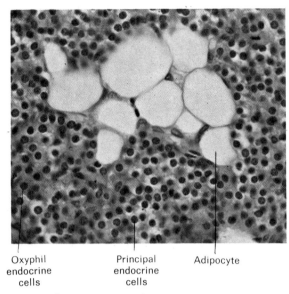

Fig. 15-22. Parathyroid of adult.

called involution. This results in a division of the parenchyma into thinner cords and masses, which contributes to the extreme structural variation (Figs. 15-18 to 15-20, B). Under low power the parenchyma of older adults may be confused with bone marrow because of the presence of very large fat cells (Fig. 15-22). The fibro-fatty stroma is almost characteristic of this gland, particularly in the adult.

Suprarenal Glands

The two suprarenal glands (adrenal glands) located at the superior poles of the kidneys are enclosed in renal fascia and embedded in perirenal fat.

Each consists of an inner medulla of neural crest origin (ectoderm) and an outer cortex which arises from peritoneal mesothelium (mesoderm). Its complicated blood supply and drainage is illustrated in Figure 15-24. The gland is surrounded by a capsule composed of collagen and reticular fibers.

CORTEX. The cortex may be divided into a *zona glomerulosa, zona fasciculata* and *zona reticularis* (Fig. 15-25 and 15-26).

Stroma. From a capsule, consisting mainly of fibrous connective tissue, mem-

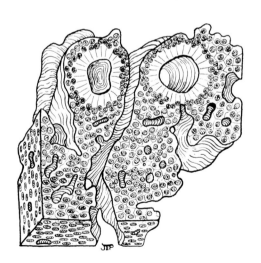

Fig. 15-21. Cell groups and follicles in the parathyroid, reconstruction.

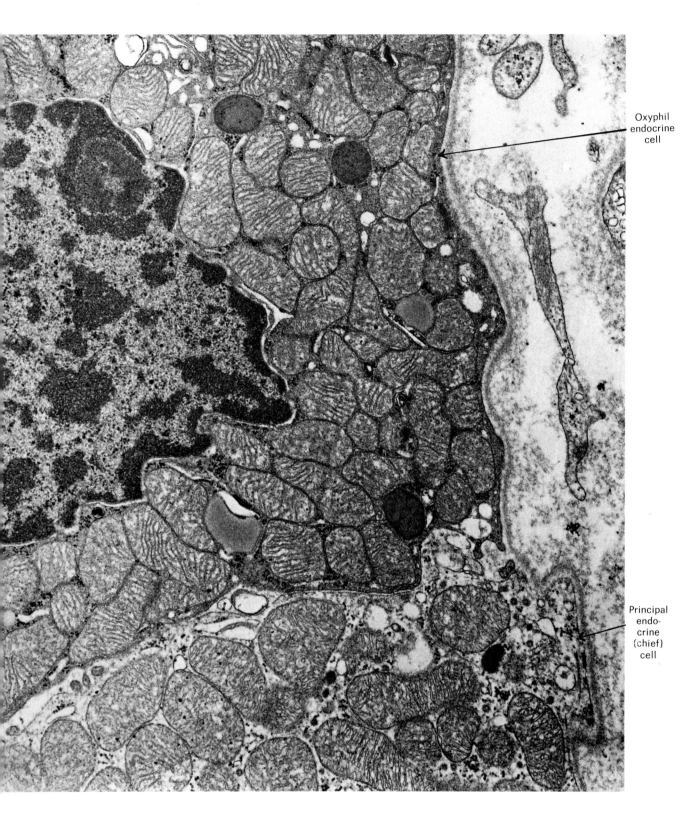

Fig. 15-23. An oxyphil endocrine cell (above) and a principal endocrine cell (below) of human parathyroid. (Courtesy of Drs. C. N. Sun and H. J. White, University of Arkansas for Medical Sciences).

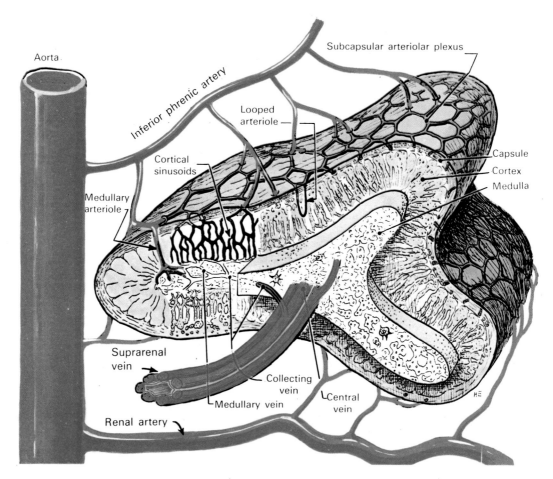

Fig. 15-24. Vascularization of the right suprarenal gland.

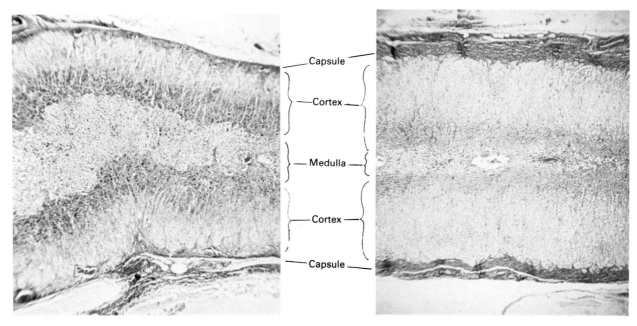

Fig. 15-25. Two suprarenal glands of adults, Goldner stain. 55 ×.

branous septae (Fig. 15-27) supported by networks of reticular fibers project into the zona glomerulosa and divide it into cell masses of different sizes and shapes (Fig. 15-28). The structural variations of the normal human adrenal cortex are so great that many pictures are needed to convey an approximate idea of this complexity. Up to 5 structural types (Fig.

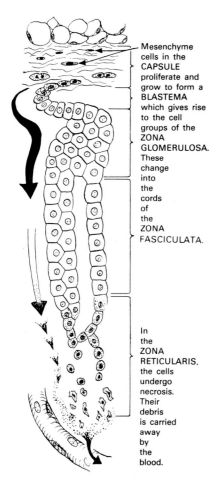

Fig. 15-26. Life history of adrenocortical cells.

15-28) may occur side by side in one normal gland, but in systemic diseases the picture becomes more uniform. At the interface of the zona fasciculata and zona glomerulosa, the membranous septa subdivide, as fingers of a glove from the palm and surround the individual cords of cells (Fig. 15-29).

Vasculature. Sinusoidal capillaries, composed of fenestrated endothelial cells, arise from an arterial plexus located at the junction of the capsule and zona glomerulosa, and form vascular basket-works in the connective tissue spaces surrounding the cell groups (Fig. 15-27). At the junction with the zona fasciculata these sinusoids aim toward the medulla, paralleling the cords. The vessels are so numerous that adjacent endothelial walls practically touch each other. They form an almost complete vascular sheath for the individual cords (Fig. 15-29) so that every cell contacts one or more capillaries.

Blastema and Zona Glomerulosa. A blastema of proliferating mesenchymal cells located in the capsule continuously gives rise to new cell groups of the zona glomerulosa (Figs. 15-26 and 15-28). The architecture of the zona glomerulosa, as mentioned above, undergoes far reaching variations. Eight structural types have been found in healthy adults (Figs. 15-28, 15-30, 15-32 and 15-33), but not more than five in any one individual. The infant suprarenal cortex shows greater uniformity (Fig. 15-34) and consists of very small cells. One should be aware of the existence of many structural types in normal individuals. Sixteen additional types have been found in patients suffering from diseases. The zona glomerulosa secretes the *mineralocorticoids* (example, aldosterone)

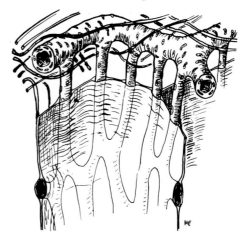

Fig. 15-27. Arterial plexus in the fibrous capsule and capillaries in the membranous septa.

which prevent loss of sodium in the urine, and also maintains a proper balance between sodium and potassium concentrations in the body.

Zona Fasciculata. The zona fasciculata has a more uniform architecture fashioned by simple branching, connected cords and short plates. The large cells of the zona fasciculata contain an abudance of cytoplasmic lipid droplets which is removed during routine slide preparation, giving the cells a spongy or foamy appearance.

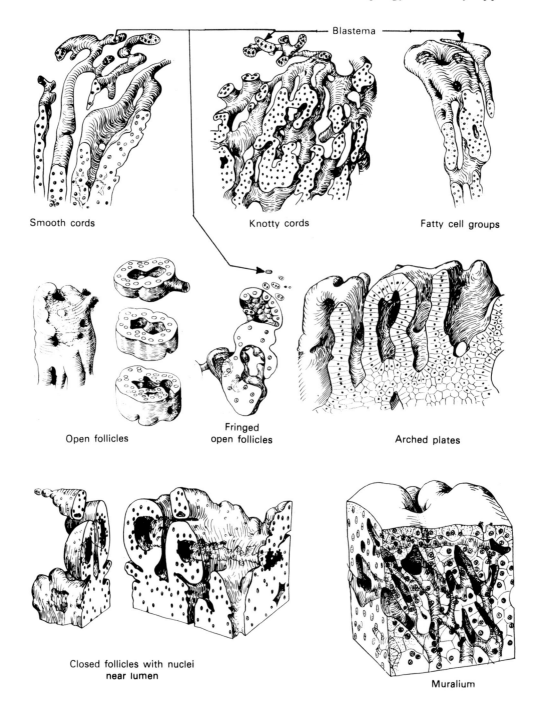

Fig. 15-28. Various structural types of normal human adrenal cortices. Reconstructions from serial sections. (From Elias and Pauly, Endocrinology *58*: 714-789, 1946.)

The cells of this zone are responsible for the production of the *glucocorticoids* (example, hydrocortisone) which regulate the conversion of proteins to carbohydrate (diabetogenic effect) and depress the lymphoid system. Production sites for the various steroids of the suprarenal cortex have been localized in the rat but may not be the same in man.

Zona Reticularis. The zona reticularis (Fig. 15-30) has thin, connected cords and isolated necrotic cells usually embedded in a dense but delicate connective tissue.

Fig. 15-29. Stereogram of zona fasciculata.

This is the zone where most of the cells degenerate and die. In males and females it is known to produce some *sex hormones* (androgen and minute quantities of estrogen and progesterone), particularly under pathologic conditions. The life history of adrenocortical cells from their origin in the blastema to their death in the zona reticularis is outlined in Figure 15-26. Cell migrations, so pronounced in man, monkey, dog and hoofed animals, plays a very minor role in rodents. In all species some mitotic figures may be found in all zones.

In wasting diseases, the suprarenal cortex undergoes severe atrophy (Fig. 15-35).

MEDULLA. The morphology of the suprarenal medulla is variable. In some cases it has the structure of a cellular muralium (Fig. 15-36); in others it consists of ill-defined cell groups (Fig. 15-37, A and B). Two types of medullary endocrine cells (of neural crest origin) can be identified in the medulla: (1) a *clear endocrine cell* or *epinephrine cell* and (2) a *dense endocrine cell* or *norepinephrine cell*. The secretory granules are about the same size, but those of the dense endocrine cell appear more electron dense (via electron mi-

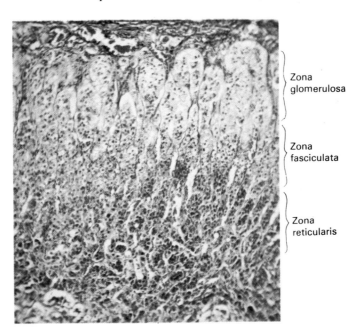

Fig. 15-30. Suprarenal cortex, type of arched plates, Goldner Stain. 175 ×.

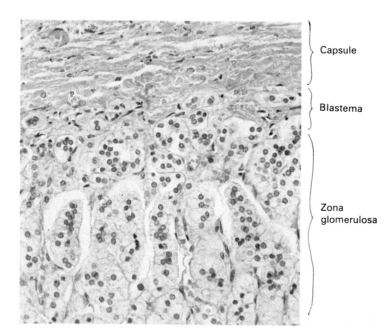

Fig. 15-31. Suprarenal capsule, blastema and zona glomerulosa, H&E. 350 ×.

croscopy). The norepinephrine cells give strong reactions with silver and iodide stains, and autoflourescence in these is also stronger.

The medulla is tunneled by a network of venous sinusoids (Fig. 15-36) supplied by (1) medullary arteries that pass directly from the subcapsular arterial plexus into the medulla (without sending branches to the cortex), (2) by sinusoids of the suprarenal cortex, and (3) occasionally by a fairly large central artery (3 per cent of human cases). Thus, arterial blood is mixed in the sinusoids with venous blood. Drainage of the entire suprarenal gland proceeds through veins having bundles of longitudinally oriented smooth myocytes in their walls (Fig. 15-38). The longitudinal musculature is very prominent in the central vein of the suprarenal gland medulla and also in the suprarenal vein outside the gland. The medullary endocrine cells contain small granules that stain brown with chromium salts (chromaffin reaction) due to the high content of catecholamines. The catecholamines augment the action of

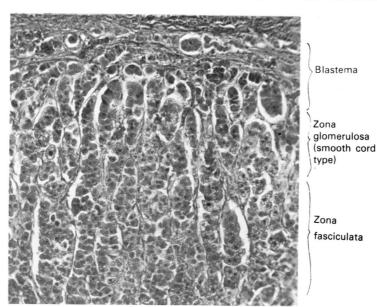

Fig. 15-32. Suprarenal cortex with active blastema. Lipoid depleted type of smooth cords, Goldner stain. 245 ×.

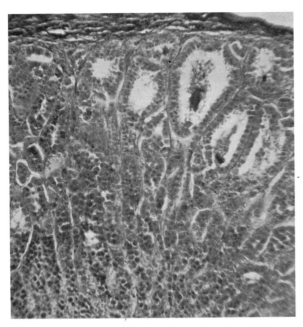

Fig. 15-33. Zona glomerulosa with closed follicles. Goldner stain, 260 ×.

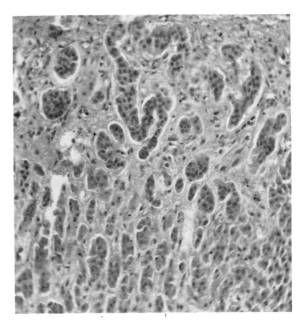

Fig. 15-35. Severe atrophy of the suprarenal cortex in a terminal cancer patient, H&E. 385 ×.

the sympathetic nervous system by increasing heart rate, raising blood pressure and generally preparing the organism for "flight or fight".

Small portions of medullary tissue, the *paraganglia*, are found outside the suprarenal gland along the sympathetic chain. Together with the medulla, these constitute the *chromaffin system*, so called because of its affinity to chromium salts.

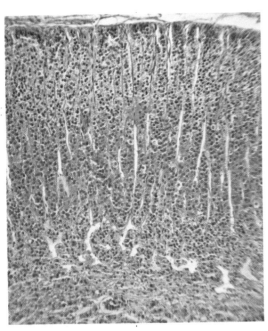

Fig. 15-34. Suprarenal cortex of infant, H&E. 175 ×.

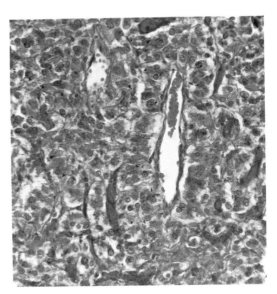

Fig. 15-36. Muralium type of suprarenal medulla, Masson stain. 450 ×.

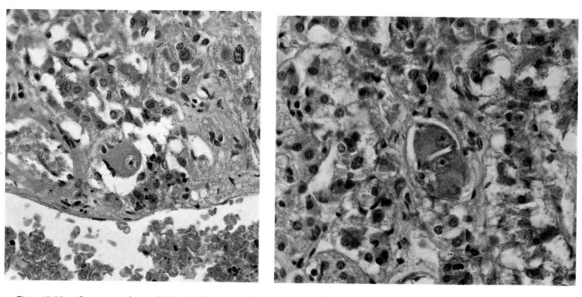

Fig. 15-37. Suprarenal medulla. Cells irregularly arranged. Note ganglion cells. H&E. A: 700 ×; B: 725 ×.

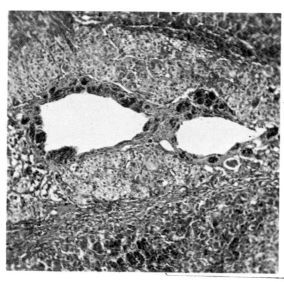

Longitudinal muscle bundles in the wall of a tributary to the central vein of the suprarenal medulla gland

Medulla

Fig. 15-38. Suprarenal medulla with vein, Goldner stain. 180 ×.

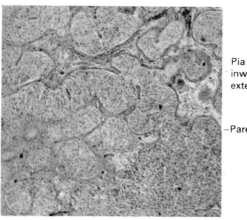

Pia mater, inward extension

Parenchyma

Fig. 15-39. Epiphysis cerebri (pineal). Azan stain. 106 ×.

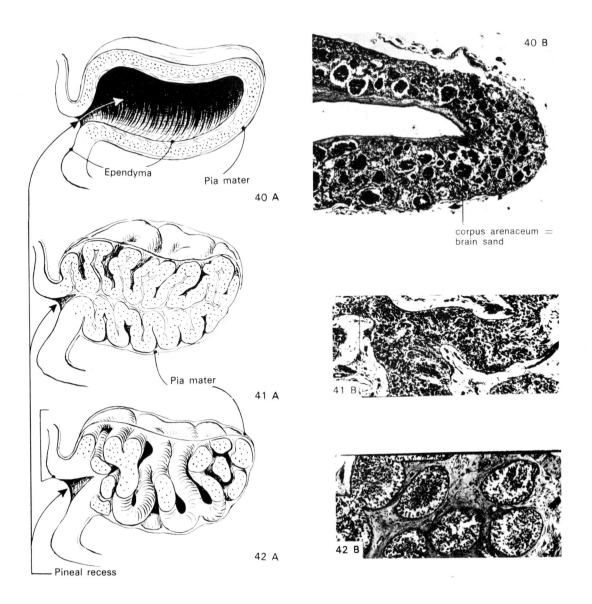

Figs. 15-40, 15-41 and 15-42. Three structural types encounted in the epiphysis cerebri.

The medulla is innervated by preganglionic fibers of the sympathetic nervous system and contains sympathetic postganglionic nerve cell bodies (Fig. 15-37).

Epiphysis Cerebri (Pineal body)

The epiphysis cerebri (pineal body) is a small pine-cone shaped diverticulum from the superior, caudal wall of the diencephalon that may contain a cavity continuous with the third ventricle. A fibrous capsule derived from the pia mater sends numerous vascularized septa into the body of the organ (Fig. 15-39). In rare cases the epiphysis cerebri remains a simple sac (Fig. 15-40, A and B), but it is usually thrown into folds to form crooked plates (Fig. 15-39 and 15-41). Sometimes these folds are broken up into cords (Fig. 15-42, A and B). The parenchyma of the epiphysis may: (1) retain the histologic structure of the embryonic neural tube

(Fig. 15-44, A), (2) resemble brain tissue (Fig. 15-44, F) or (3) be composed of cells which have processes with club-shaped ends in contact with blood vessels (Fig. 15-43). Neuroglia cells support the parenchyma and may be intermingled with pial, areolar connective tissue in the septa.

A clue to the identification of the epiphysis cerebri is the presence of *corpora arenacea* (brain sand) (Fig. 15-40, 8 and 15-44, F). These are mulberry-shaped concretions, usually staining dark blue with routine H and E stains. Although found in infancy, the corpora arenacea increase with age. Corpora arenacea occasionally are found in the environment of the epiphysis, i.e., in the habenula, posterior commissure and pretectal area. In CAT-scans the epiphysis serves as a major landmark because of its radio-opacity. The epiphysis secretes *melatonin* which causes congregation of melanin granules, a blanching effect. The gland also has an antigonadotrop(h)ic effect. Administration of melatonin delays sexual development; whereas pinealectomy soon after birth results in precocious sexual development.

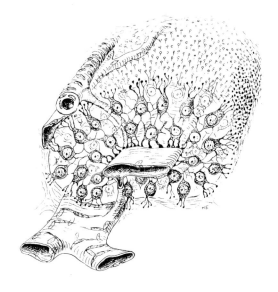

Fig. 15-43. Del Rio-Hortega's concept of epiphyseal structure.

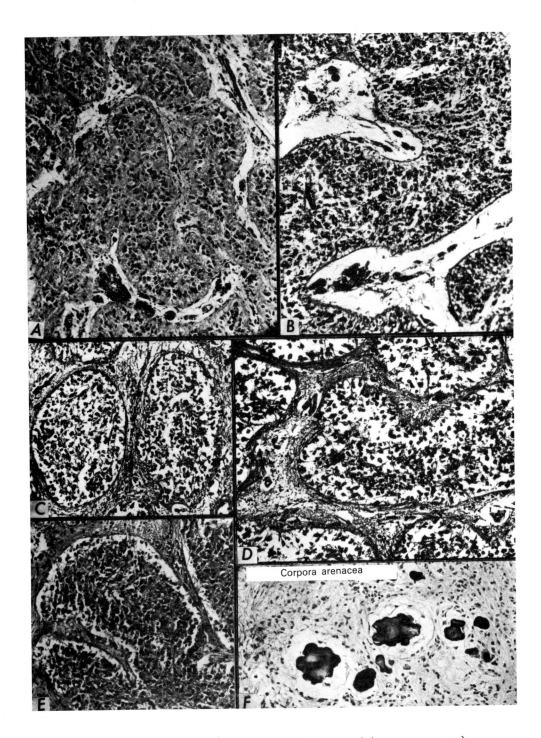

Fig. 15-44. Various types of epiphyses. F: Brain sand (corpora arenacea).

16...
The eye

Major Divisions

The eye, a spheroid camera obsucra, is constructed of three concentric coats: (1) the outer *tunica fibrosa bulbi* (Fig. 16-1), (2) the middle *tunica vasculosa bulbi (uvea)* (Fig. 16-2) and (3) the inner *tunica interna (sensoria) bulbi* or *retina* (Fig. 16-3). Figures 16-1, 16-2 and 16-3 demonstrate the relationships of the three coats, and their named subdivisions. The iris (Fig. 16-4) divides the space between the *cornea* and the *lens* into *anterior* and *posterior chambers*, which contain aqueous humor and communicate with each other through the *pupil*.

The eye, as an optical instrument, has four functional components: (1) a protective coat, (2) a nourishing coat, (3) a dioptric system and (4) a layer of sensory perception and integration.

Tunica Fibrosa Bulbi

Protective Coat - *Sclera*

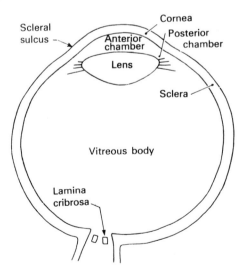

Fig. 16-1. Tunica fibrosa bulbi.

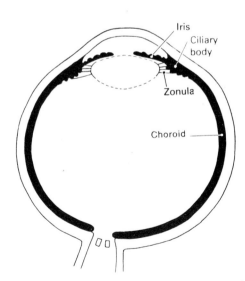

Fig. 16-2. Tunica vasculosa bulbi. (Uvea) and tunica fibrosa.

432 THE EYE

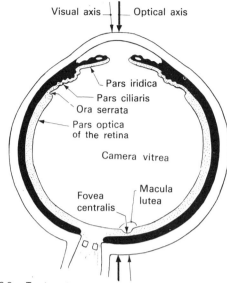

Fig. 16-3. Tunica interna (sensoria) bulbi (stippled), tunica vasculosa bulbi, (black) and tunica fibrosa (white).

The sclera, the posterior five-sixths of the tunica fibrosa bulbi, is a hard shell of dense fibrous connective tissue that protects and supports the eyeball. A narrow, pigmented zone near its internal surface is called the *lamina fusca* (Fig. 16-11). *Melanocytes* populating this area produce the pigment melanin. The sclera is continuous with the *dura mater encephali*. The region where the optic nerve fibers exit from the retina is the sieve-like *area cribrosa* (Fig. 16-1). The *substantia propria* of the cornea, which together with the *substantia propria* of the sclera constitutes the bulk of the tunica fibrosa bulbi, will be described with the dioptric apparatus.

Tunica Vasculosa Bulbi (Uvea)

Nourishing Coat - Choroid

The *choroid* (Figs. 16-2, 16-11, 16-13 D, E), forming the posterior five-sixths of the tunica vasculosa bulbi, is practically co-extensive with the sclera and continuous with the pia mater encephali surrounding the optic nerve. The choroid may be divided into four layers, externally to internally: (1) *lamina suprachoroidae*, (2) *substantia propria* (a vascular layer), (3)

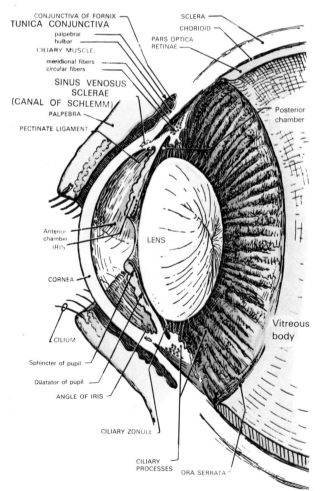

Fig. 16-4. The dioptric apparatus of the eye.

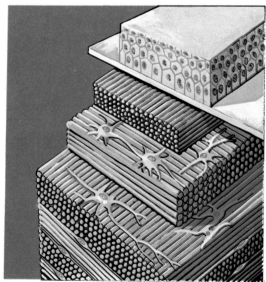

Fig. 16-5. Diagramatic recontruction of the cornea.

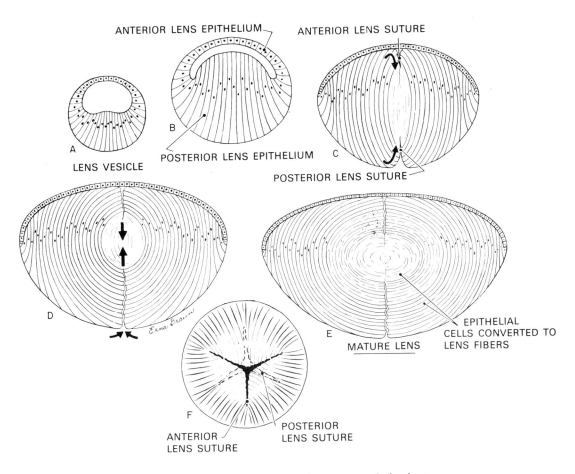

Fig. 16-6. Development and structure of the lens.

lamina choroidocapillaris (capillary layer), and (4) the *complexus basalis* (formerly the lamina vitrea or glassy membrane of Bruch) (Fig. 16-11). The essential feature of the choroid is that it possesses a large number of choroidal melanocytes that characterize it as a pigment layer. An abundance of loose connective tissue with blood vessels provides nourishment for the outer portions of the retina, while the internal layers of the retina are vascularized by the retinal vessels (Fig. 16-13, F).

Dioptric Apparatus

The dioptric apparatus of the eye, shown in Figure 16-4, consists of the cornea (the anterior sixth of the tunica fibrosa), the aqueous humor within the anterior eye chamber, the lens and the vitreous body.

CORNEA. The transparent cornea (Fig. 16-10, A) is covered externally by a stratified squamous, non-keratinized *anterior epithelium* with its thick basement membrane, the *anterior limiting lamina* (*Bowman's* membrane). The core of the cornea is the dense, fibrous *substantia propria* (Fig. 16-10, A) made of strata of parallel fibers running at various angles to each other in discrete layers and enclosing between them flattened, stellate fibroblasts (Fig. 16-5). Posterior to the substantia propria lies the elastic *posterior limiting lamina* (of *Descemet*) and finally a layer of simple cuboidal cells, the *posterior epithelium*.

The cornea is continuous with the sclera. Deep to the *sulcus sclerae* (Fig. 16-

Fig. 16-7. Anterior lens epithelium in cataract due to tuberculous iridocyclitis. (From a slide in the collection of the Armed Forces Institute of Pathology.)

1) is the *scleral venous sinus* (the circular canal of Schlemm) (Fig. 16-4). The inner surface of the cornea is interconnected with the iris and ciliary body by the *pectinate* ligaments. Minute openings between the fibers of this ligament are known as the *iridocorneal angular spaces* (of Fontana). Through these openings the anterior chamber communicates with the scleral venous sinus. Aqueous humor is a clear liquid produced by the *ciliary processes*. It flows across the posterior chamber through the pupil into the anterior chamber, drains through the iridocorneal spaces into the scleral venous sinus and from there into the anterior ciliary veins. This flow pattern controls intraocular pressure and when defective, results in *glaucoma*.

AQUEOUS HUMOR. The aqueous humor contained in the anterior chamber of the eye provides the second refractive medium in the dioptric apparatus, followed by the lens.

LENS. The biconvex lens (Fig. 16-4), bounded by two paraboloid surfaces, develops from epidermal epithelium which is cuboidal anteriorly and columnar posteriorly. The posterior columnar cells are extremely long, curved prisms, called *lens fibers*. Their arrangement is best understood when the development of the lens is considered, as shown in Figure 16-6. Wrinkling of the anterior lens epithelium, as seen in Figure 16-7, or calcium deposition within the lens, results in *cataract* formation.

SUSPENSORY LIGAMENT. The elastic lens suspended at the ciliary body by fibers of the zonula ciliaris (non-extensible suspensory ligament composed of delicate collagen fibers) (Fig. 16-4), is stretched so that it assumes a slightly flattened shape. In this position it has, in the normal eye, the correct curvature for focusing the entire dioptric system at infinity. (It is *far accommodated*.) (Fig. 169, left).

CILIARY BODY. The ciliary body a portion of the tunica vasculosa (Figs. 16-2 and 16-4), consists of radially arranged folds, the *ciliary processes* (Fig. 16-10, B). It is covered on its internal surface by an extension of the retina, the pars ciliaris retinae. The retina, in this area, is composed of two layers: a pigmented epithelium (external leaf) and non-pigmented epithelium (internal leaf) (Fig. 16-10, D).

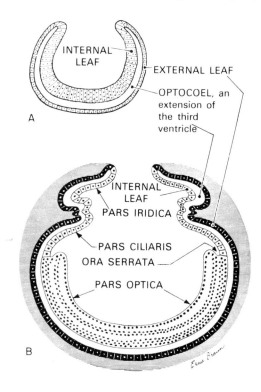

Fig. 16-8. Leaves of the retina. A, In an early embryo; B, in a later embryo.

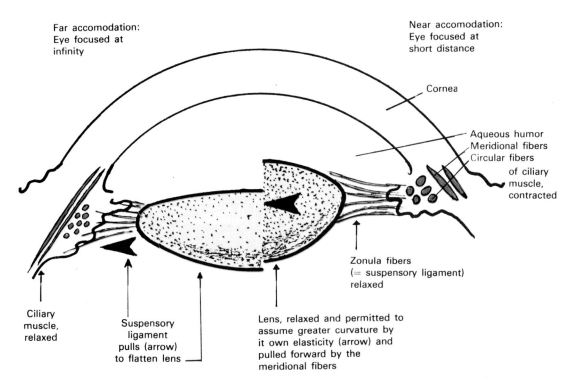

Fig. 16-9. Mechanism of near accomodation. The lens is elastic and tends to increase its curvature (arrow at right). At rest (left half of picture) the lens is stretched flat by the zonula fibers (green). This traction which counteracts the elasticity of the lens is indicated by the arrow on the left side. The ciliary muscle (brown) is relaxed, the lens focused at infinity. Near accomodation is accomplished by contraction of the circular fibers of the ciliary muscle (right, bright red) which permits relaxation of the elastic lens whereby it increases its curvature and thus decreases its focal distance. Contraction of the meridional fibers of the ciliary muscle (bright red, extreme right of picture) moves the lens forward, increasing its near focusing, and brings it into a region in which the eyeball is less wide, contributing again to increased curvature.

Three of the ciliary processes are cut coronally in Figure 16-10, B. The ciliary body contains the *ciliary muscle* (Figs. 16-4, 16-9, and 16-10, B) with meridional and circular smooth myocytes. Contraction of this smooth muscle permits relaxation of the lens. The inherent elasticity of the lens thus can become effective, and the loss of tension exerted by the suspensory ligament allows the lens to reestablish its own spheroid shape. This produces an increased curvature on both the anterior and posterior surfaces and a decrease in focal distance (*near accommodation*). This mechanism is illustrated in Figure 16-9. In late middle age, the lens gradually loses its elastic power to contract and *presbyopia* results.

VITREOUS BODY. The transparent gel-like vitreous body fills the space posterior to the lens and ciliary body. It consists of a very loose network of widely scattered connective tissue cells and very delicate fibers, the *vitreous stroma*. The meshes within this network are filled with a watery liquid, the *vitreous humor*. The vitreous body, the most posterior component of the dioptric apparatus of the eye, resembles embryonic connective tissue similar to that of the umbilical cord.

IRIS. The iris is a circular diaphragm with a circular aperture, the *pupil*. In the irideal stroma immediately posterior to the epithelium of the anterior surface of the iris, are found chromatophores which contain dark brown pigment (melanophores). In a blue iris, guanin containing chromatophores (*guanophores*) in the form of crystal plates are located in front of the melanophores. Blue color is created when a colorless, moderately translucent substance is located before a black back-

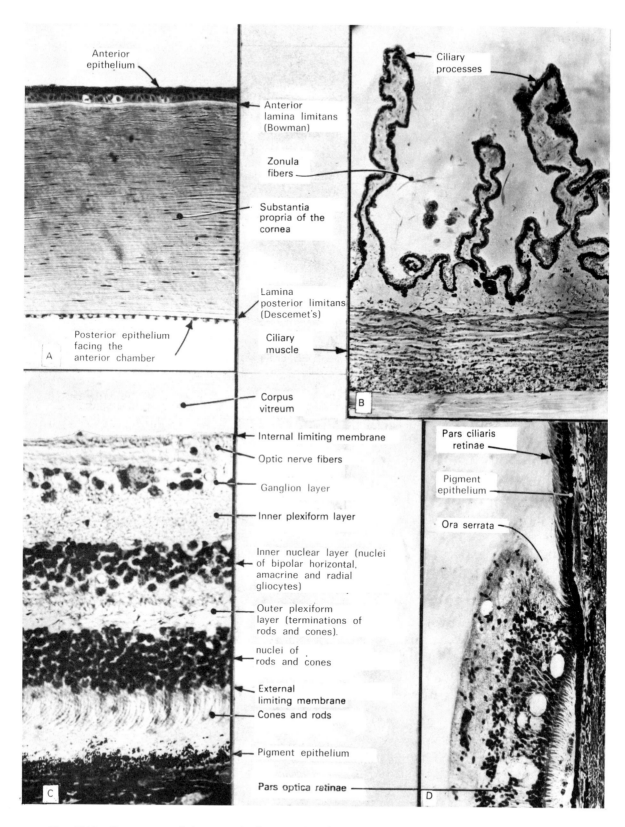

Fig. 16.10. Some parts of the eye. A, Cornea. B, Ciliary processes, tangentially cut. C, Layers of the retina. D, Ora serrata. (From slides in the collection of the Armed Forces Institute of Pathology.)

ground, just as the sky appears blue because tiny colorless particles float in the air in front of the black void of the universe. The blue color appears only when colorless particles are illuminated. The blue color in the feathers of many birds is due to this same phenomenon of the "semiopaque" medium. If a yellow pigment (lipochrome) appears in cells (lipophores or xanthophores) in front of the guanophores, eyes may have a greenish tint.

Contraction of the *sphincter pupillae*, a layer of smooth myocytes encircling the pupil, constricts the pupil. Contraction of the *dilator pupillae* smooth myocytes, arranged radially around the pupil dilates the pupil. The *sphincter pupillae* is innervated by parasympathetic fibers; whereas the *dilator pupillae* is innervated by sympathetic fibers. The "drops" that an ophthalmologist puts in a patient's eyes to dilate the pupils, contain epinephrine, a sympathetico-mimetic agent (acts like the sympathetic nervous system). Atropine, the poison of night shade plants is occasionally used for pupilary dilatation if a long effect is desired. The sphincter and dilator muscles of the iris control the size of the pupil and thus the amount of light which enters the dioptric apparatus. In dim light the pupil increases in diameter by contraction of the dilator pupillae, in bright light the pupil decreases in diameter by contraction of the sphincter pupillae. The latter also increases the depth of focus.

Retina

The retina lining the inner surface of the tunica vasculosa is divided into three portions, as illustrated in Figure 16-3. As an extension of the diencephalon, the retina (after invagination of the optic vesicle to form the optic cup, Fig. 16-8, A) is composed of an internal and an external leaf which enclose the intraretinal space, itself an extension of the third ventricle. The external leaf is a pigmented epithelium,

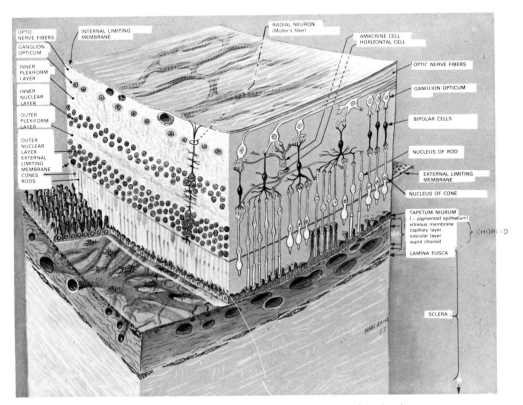

Fig. 16-11. Stereogram of the retina, choroid and sclera.

the stratum pigmentosum. The internal leaf contains no pigment. The subdivision of the retina into three parts is explained in Figure 16-8, B. While the stratum pigmentosum is a simple epithelium throughout its extent, the internal leaf shows regional differences. In the *iridial portion of the retina* as well as in its *ciliary portion* (Figs. 16-8, B and 16-10, B and D), the

extension of the brain, is the light-receiving and transmitting apparatus. It is constructed of three layers of nerve cells as illustrated on the right side of Figure 16-11 and supported by specialized neuroglia cells, the *radial glial cells* (Müller's fibers). The layers of the retina through which light passes to reach the rods and cones are: *internal limiting layer* of glial cells,

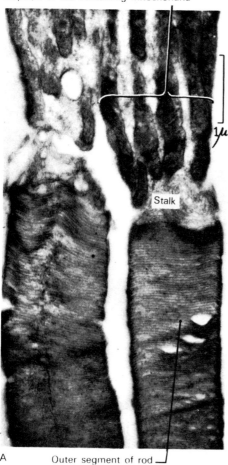

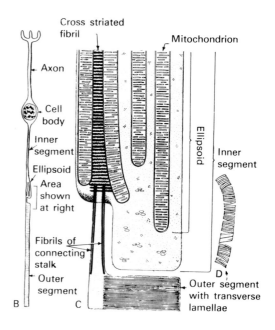

Fig. 16-12. Ultrastructure of outer segment of rod. A and C: from Sjöstrand, J. Cell Comp. Phys. *42*: 15-44, 1953. B: General diagram. D: Drawing by Franz Schultze, 1872.

optic nerve fiber layer, ganglion cell layer[1], *internal plexiform layer, internal nuclear layer, external plexiform layer, external nuclear layer, external limiting layer* and the photosensory *neuroepithelial layer of rods* and *cones*.

The rods and cones are the light-receiving elements, whose specialized dendrites, called *external segments*, stick out through the external limiting membrane. Since the intraretinal space is embryologically a part of the third ventricle, the cells which line it, i.e., the rods and cones and the stratum pigmentosum, could be considered to be specialized ependymal cells.

The rod (*epitheliocytus bacillifer*) is a primary neurosensory cell, i.e., a neuron whose dendrite receives a stimulus and whose axon transmits the impulse to an-

internal leaf is formed of a simple cuboidal to low columnar epithelium. The *optical part of the retina* begins abrupty at the *ora serrata* (Figs. 16-3, 16-4, 16-8, B and 16-10, D).

The *optical portion of the retina*; an

other neuron. The dendrite of a rod consists of the parts shown in Figure 16-12: the external segment is, along most of its length, located on the outer side of the external limiting layer. Beyond that layer it becomes a little enlarged into the highly refractive "ellipsoid" which contains many mitochondria. A thin connecting stalk joins the ellipsoid to the external segment. Via electron microscopy, the connecting stalk has the structure of a specialized cilium, with nine peripheral pairs of microtubules; but lacks the central pair of microtubules seen in kinocilia. Electron microscopy (Fig. 16-12, A and C) has confirmed the lamellated structure of the external segment demonstrated in 1872 by Schultze (Fig. 16-12, D). The rods are highly sensitive, and their external segment can be covered by pigment of the *stratum pigmentosum* for protection against intense illumination (*light adaptation*). In dim light, the pigment retracts (*dark adaptation*). Both conditions are shown in Figure 16-11 at the right. Rod vision is based on the decomposition, by light, of *rhodopsin* (*visual purple*) contained in the external segment. It is important to distinguish between the two words "accommodation" (which means focusing through the lens) and "adaptation" (change of retinal sensitivity).

The retinal cones (*epitheliocytus conifer*) (Fig. 16-11) are neurosensory color receptors. The external segment of a cone is shorter than that of a rod, and is divided into a very thick ellipsoid and a small conical apical portion (Fig. 16-11, right). Many rods, (but only one or a few cones), synapse with one *bipolar cell* (Fig. 16-11); and one to several bipolar cells synapse with one multipolar neuron ganglion cell layer (Fig. 16-11). These large neurons project their axons through the optic nerve, optic chiasm and optic tract into the brain (lateral geniculate body with collaterals to the superior colliculus and to the pretectal region).

Horizontal neuronal cells and *amacrine neuronal cells* of the internal nuclear layer (Fig. 16-11) may serve to create a unified image out of numerous, individual spots of light. The exact position of each retinal cell type and its perikaryon, as well as their synaptic relationships, should be carefully studied in Figure 16-11.

These cell types are best demonstrated by special techniques. Routine slides present a drastically simplified picture showing 10 layers as seen on the left side of Figure 16-11 and in Figure 16-10, C.

Rods predominate in the peripheral, regions of the retina, cones in the central. At the end of the visual axis (Fig. 16-3) the retina has a yellow spot the *macula lutea* which is almost coextensive with a shallow depression, the *fovea centralis* (Fig. 16-13, B and C). At this place, only cones are found, each connected individually with one bipolar cell which, in turn, synapses individually with one multipolar neuron of the ganglion cell layer. This one-to-one synaptic conduction assures great acuity of vision and is expressed in the brain by the fact that the tiny fovea centralis is projected upon a large part of the visual cortex, giving the foveal image prominence in consciousness. At the fovea centralis, bipolar and multipolar neurons of the ganglion cell layer are bent to the side. It is this arrangement that produces the shallow depression. While in certain

---

[1] "Ganglion cell layer" is a misnomer, for this extensive layer of large multipolar neurons is neither organized nor functions like any ganglion. Functionally, it corresponds to the so-called "sensory nuclei" of the gray matter in the central nervous system. Like the latter, this group of nerve cells receives impulses from an afferent neuron and conveys them by means of its axons to the thalamus. The part of the thalamus involved is the lateral geniculate body. While the impulse to other sensory "nuclei" is transmitted by the first-order neuron. In the retina the decision of whether the bipolar cells are to be considered first or second-order neurons is a matter of taste. At any rate, ganglion cell layer fibers go to the thalamus. On the way to the cortex, there is only one more neuron (thalamo-cortical, in this special case called geniculo-calcarine).

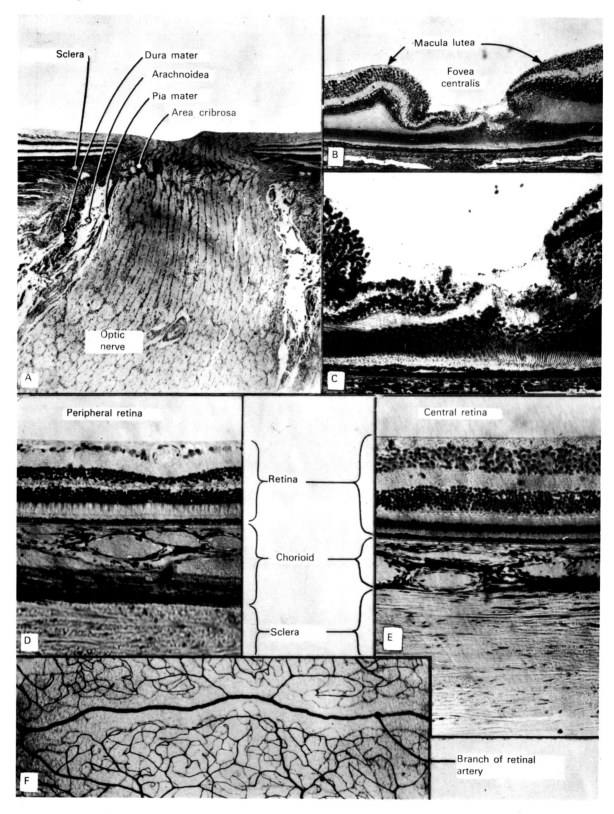

Fig. 16-13. The retina. A, Blind spot. B and C, Fovea centralis. D, Tunics of the eye near the equator. E, Tunics of the eye near the end of the optical axis. F, Retinal vessels injected with India ink, from dog. (A-E from slides in the collection of the Armed Forces Institute of Pathology.)

animals (particularly birds) the fovea centralis is constructed so that its borders act as a magnifier, in man (Fig. 16-13, B and C) its shape is irregular, making it less efficient.

The yellow color of the macula lutea is due to the actual presence of a yellow pigment. Whether this pigment is localized in the vitreous body or permeates the entire remaining retina, is not clear. It acts as a yellow filter and, like a yellow filter used in photography, increases visibility at great distances.

Persons who maintain that impulses from the periphery to the cerebral cortex "always" involve a chain of three neurons, deny that the rods and cones are neurons. If they were, the visual chain pathway would include four neurons. Rods and cones then would be comparable to hair cells of the inner ear which are specialized epithelial cells (see Chapter 17). In fact, by position, the rods and cones occupy the place of ependymal cells, and thus they may be considered specialized ependymal cells.

The predominance of cones at the end of the visual axis and of rods nearer the ora serrata is expressed in regional variations of thickness of the various retinal layers, as is seen in Figure 16-13, D and E.

Where the optic nerve fibers (axons of the multipolar neurons of the ganglion cell layer) leave the retina, at the *blind spot (optic disk, optic papilla)*, all three tunics are interrupted (Fig. 16-13, A). Here the sclera, with perforations called the *area cribrosa*, is devoid of the choroid and the retina. The blind spot can be demonstrated when looking at the fly in Figure 16-14 with the right eye only. At a distance of approximately 16 inches (40 cm), the frog will disappear from sight; its image will have fallen on the blind spot.

Stereoscopic Vision

The great convexity of the cornea, the high refractive indices of the dioptric media and the concavity of the retina make it possible for an object lying slightly behind the equator of the eye to be faintly perceived (Fig. 16-15, A). In animals with eyes placed laterally, panoramic vision with an anterior and posterior field of binocular vision is thus facilitated (Fig. 16-15, B). In primates (Fig. 16-15, C), the eyes are placed in a forward position. Nearby objects are seen under slightly different angles (parallax), because both eyes are distant from each other. This arrangement helps in depth perception or stereoscopic vision. Proprioceptive impulses from the extrinsic eye muscles and, perhaps, perception of intraocular pressure produced by accommodation help in estimating distances.

Lids and Conjunctiva

The lids (*palpebrae*) (Fig. 16-16) are folds of skin containing externally the orbicularis oculi muscle and internally a hard plate of fibrous tissue, the *tar-*

Fig. 16-14. Demonstration of the blind spot of the right eye.

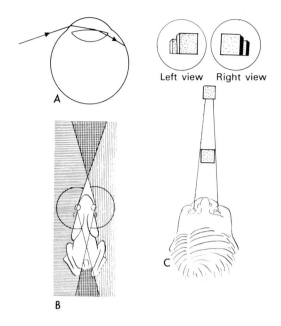

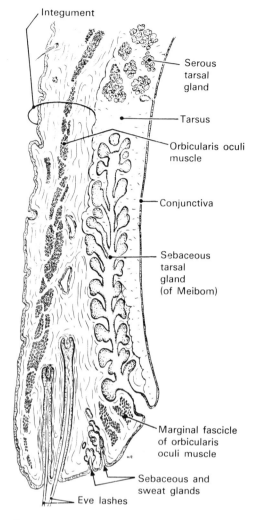

Fig. 16-15. Panoramic versus stereoscopic vision.

sal plate, in which are embedded the branched, sebaceous, *tarsal glands (of Meibom)*. Along the edges are hairs, called *eyelashes (cilium)*, as well as some ciliary sweat glands (Moll) and sebaceous glands (Zeis). Internally, the lid is lined with an epithelium two cells thick, the *palpebral conjunctiva*, which is reflected upon the sclera as the *bulbar conjunctiva* and is continuous with the corneal epithelium. Serous glands are found in the tarsus.

Fig. 16-16. Sagittal section of upper lid.

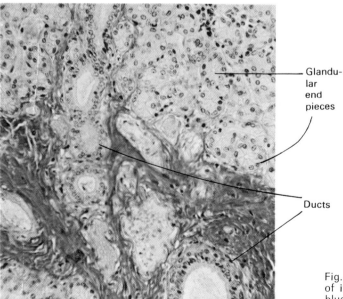

Fig. 16-17. Lacrimal gland and some of its ducts. The stroma is stained blue, Azan. 375 ×.

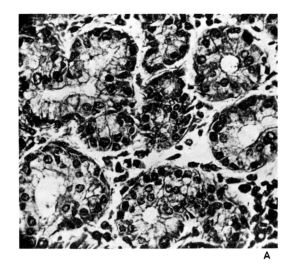

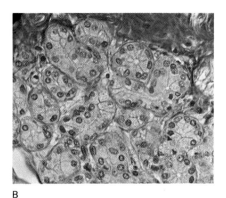

Fig. 16-18. Lacrimal gland. A: H&E, 700 ×; B: Azan, 500 ×.

Attached to the tarsus is the smooth tarsal muscle (of Muller).

Lacrimal Glands

Several compound, tubulo-alveolar, serous *lacrimal glands*, empty by separate ducts into the *conjunctival sac*. The glandular end-pieces may have the morphology of serous glands (Fig. 16-17), or they may be foamy; and in some of the end-pieces one finds two layers of cells (Fig. 16-18). The ducts (Fig. 16-17 and Fig. 16-19) are lined by simple cuboidal to pseudostratified or even stratified columnar epithelium.

The Neurea Theory

From a functional point of view it seems paradoxical that light must pass several thick layers of neural tissue before it hits the outer segments of the rods and cones; for while the light passes through this tissue it undergoes diffusion. Ernst Haeckel presented a plausible theory to explain the inversion of the retina. He speculated that one of the earliest ancestors of the chordates was a little animal, called the Neurea (Fig. 16-20, A) which had an open neural plate during its adult life. It may have looked very similar to a frog neurula. The neural plate of all vertebrate embryos has two lateral enlargements (optic placodes) in the diencephalon region. Light falls directly (white arrows) upon these optic placodes (Fig. 16-20, A and B). In this condition, the development of light receptors on the exposed side of the optic placode would be convenient. The eyes of the cephalopods develop in a different

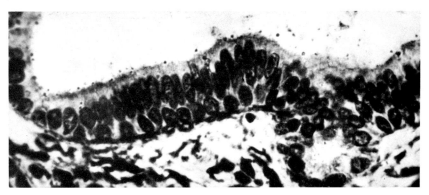

Fig. 16-19. Epithelium of duct of lacrimal gland, H&E. 1000 ×. On the left side, the epithelium is simple cuboidal, in the middle it is stratified columnar.

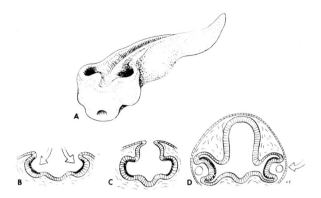

Fig. 16-20. Haeckel's neurea theory to explain the paradoxical phenomenon that the light must pass a thick sheet of brain tissue before it hits the rods and cones.

manner; thus their retinas retain the original position. In vertebrates, the neural plate becomes closed, and, with it, the optic placode becomes entrapped inside the body (Fig. 16-20, C). During final development of the eye, the optic vesicle becomes secondarily invaginated to form the optic cup. The retina which once was the optic placode, however, retains its original orientation. Fortunately, the deficiency of vision which would be caused by our inverted retina has been overcome by the fovea centralis. The fovea permits sharp and distinct vision in a very small area only; and this may be the cause of our habit to think analytically, i.e., to pay attention to one little thing at a time.

17...
The ear

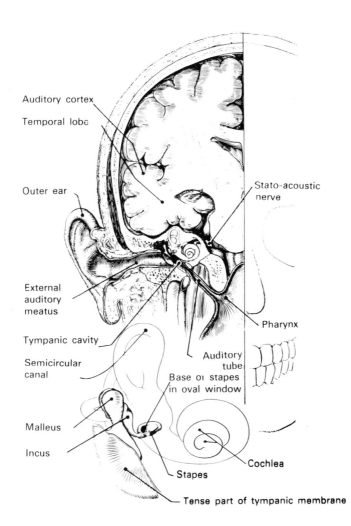

Fig. 17-1. Topography of the ear. (Drawing by B. Melloni, courtesy of Abbott Laboratories.)

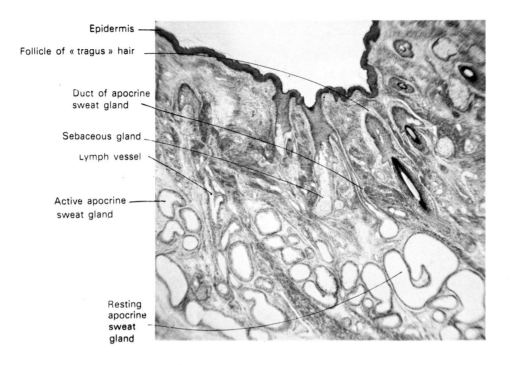

Fig. 17-2. Lining of external auditory meatus, Goldner stain. 65 ×.

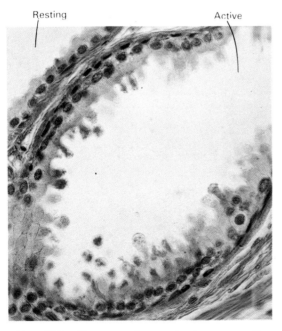

Fig. 17-3. Two ceruminous glands, one resting (upper left) and the other active (middle of picture). Apocrine secretion is evident. Goldner stain, 785 ×.

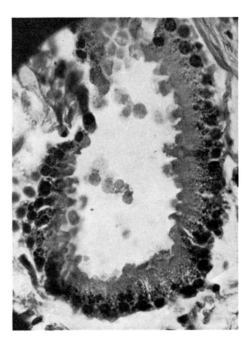

Fig. 17-4. An active ceruminous gland showing pigment in the basal portions of the epithelial cells. 780 ×.

Sheltered in the petrous portion of the temporal bone is the statoacoustic organ (Fig. 17-1).

Outer Ear

AURICLE. The auricle (external ear) consists of elastic cartilage covered by integument.

EXTERNAL AUDITORY MEATUS. The external -auditory meatus (Figs. 17-1, 17-2) is supported externally by elastic cartilage and at a deeper location, lies within a bony canal. It is lined by integument which contains sebaceous glands and modified, apocrine sweat glands, called *ceruminous glands* (Figs. 17-3 and 17-4). Both types of glands contribute to the formation of cerumen (ear wax). The pigment seen in Figure 17-4, is responsible for its brown color.

Middle Ear

The *auditory tube* (Fig. 17-1), lined inferiorly by pseudostratified, ciliated, columnar epithelium (with occasional patches of stratified, squamous, non-keratinizing epithelium), contains seromucous glands and lymph nodules (and, in children, small tonsils) in its mucosa. Superiorly, the epithelium is formed of simple columnar cells. The *tympanic cavity* including the auditory ossicles) and the mastoid air cells are lined by simple cuboidal epithelium supported by an extremely thin lamina propria which is practically identical with the periosteum. The *tympanic membrane*, which consists of radial and circular collagen and elastic fibers, is covered laterally by thin skin but medially by simple cuboidal epithelium. The plate of the stapes is fastened in the oval window by dense connective tissue.

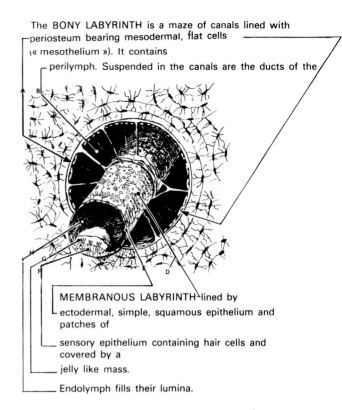

Fig. 17-5. Suspension of the membranous labyrinth in the bony labyrinth.

Inner Ear

MEMBRANOUS AND OSSEOUS LABYRINTH. - The inner ear is a complex system of canals and ducts. The ducts form the *membranous labyrinth*, which is suspended in a system of canals in bone, the *osseous labyrinth* (Fig. 17-5).

The membranous labyrinth arises as an invagination of the embryonic epidermis; hence, the space within it, filled with *endolymph*, is a captured portion of outer world. Its lining, therefore, is a true epithelium. The flat cells which line the osseous labyrinth are sometimes, misleadingly, termed "mesothelium". The osseous labyrinth contains *perilymph*.

The structural plan presented in Figure 17-5 undergoes local variations. Figure 17-6 shows the outlines of the bony labyrinth

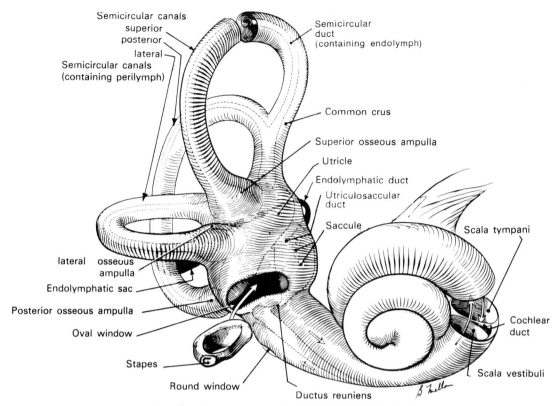

Fig. 17-6. Bony labyrinth. (Drawing by B. Melloni, courtesy of Abbott Laboratories.)

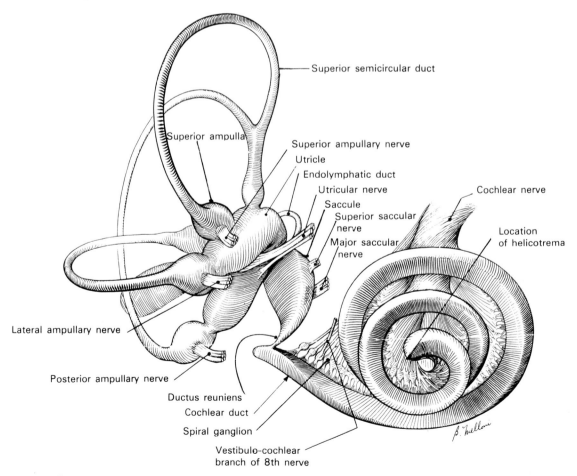

Fig. 17-7. Membranous labyrinth. (Drawing by B. Melloni, courtesy of Abbott Laboratories.)

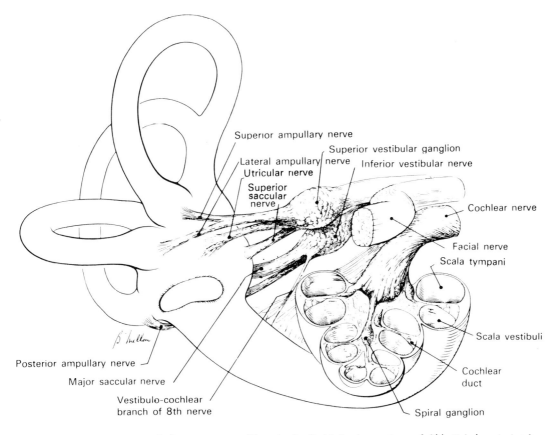

Fig. 17-8. Innervation of the inner ear. (Drawing by B. Melloni, courtesy of Abbott Laboratories.)

with a ghost image of the membranous labyrinth within. Figure 17-7 shows the outlines of the membranous labyrinth, while Figure 17-8 shows neural elements. In Figures 17-1, 17-6, 17-7 and 17-8, the right inner ear is viewed from anterior to posterior.

MACULAE AND CRISTAE. Position sense is perceived by the maculae of the *utricle* and *saccule*. Each macula (Fig. 17-9) consists of *sustentacular (supporting) epithelial cells* and two types of *sensory hair*

Fig. 17-9. A (above), Macula utriculi from guinea pig. B (below), Macula utriculi, diagram.

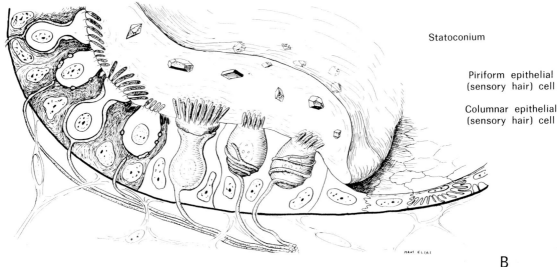

B

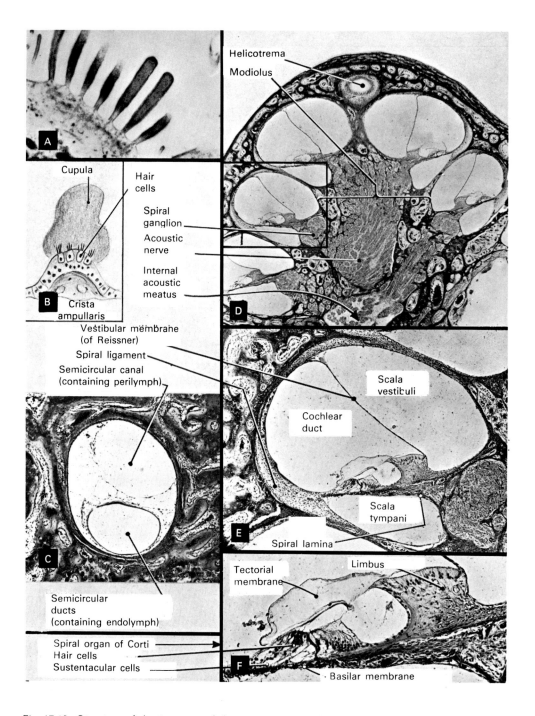

Fig. 17-10. Structure of the inner ear of the guinea pig. A, Electron microgram (by Dr. Catherine A. Smith, Ann. Otol. 65: 450-469, 1956), showing club-like microvilli (hairs) of macular hair cells. B, Crista ampullaris. C, Semicircular canal and duct. D, Cochlea. E, Cochlear duct. F, Organ of Corti.

cells *(cellular sensoria pilosa)*. The sustentacular cells together with the hair cells form a simple columnar epithelium resting on a basement membrane. The two types of epithelial hair cells are distinguished by shape and by their mode of association with the dendrites of bipolar cells from the vestibular ganglion. The Type I *pear-shaped hair cells (epitheliocytus piriformis)* sit in cup or chalice shaped terminations of the nerve endings, while the Type II *barrel shaped hair cells* (*epitheliocytus columnaris*) are contacted by a plexus of spirally wound, thread-like endings of the nerve. The free or luminal surface of both hair cell types is covered with club-shaped microvilli (the so-called hairs), embedded in the *membrana statoconiorum*. This is a membrane like gelatinous matrix containing small crystalline bodies of calcium carbonate called *statoconia*. The hairs of these cells from the utricle of the guinea pig are shown in Figure 17-10, A. When the head is inclined, the weight of the *statoconia* causes a shift in the gelatinous matrix, and thus, position in space is perceived and verified.

Communicating with the utricle are three membranous, *semicircular ducts* suspended in bony, *semicircular canals* oriented in three planes perpendicular to each other (Figs. 17-6, 17-7). Each duct terminates as an enlargement called the *ampulla*.

In each ampulla there is an epithelial elevation, the *crista ampullaris*, which is capped by a gelatinous matrix called the *cupula gelatinosa*. The microvilli of hair cells in the crista (Figs. 17-10, B and 17-11) are embedded in the cupula. When

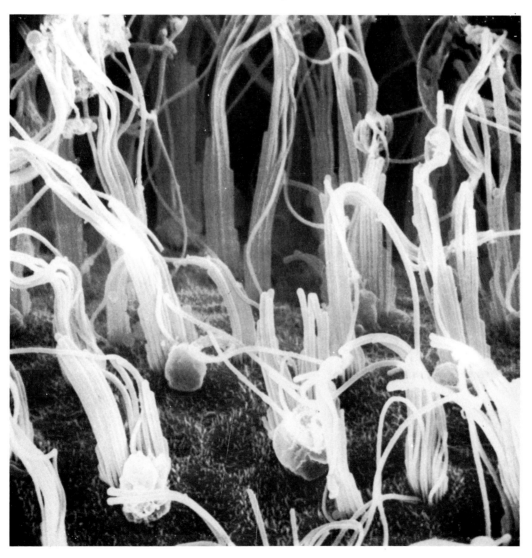

Fig. 17-11. Microvilli of hair cells of a crista ampullaris. Scanning electron microgram by Dr. Catherine A. Smith, University of Oregon.

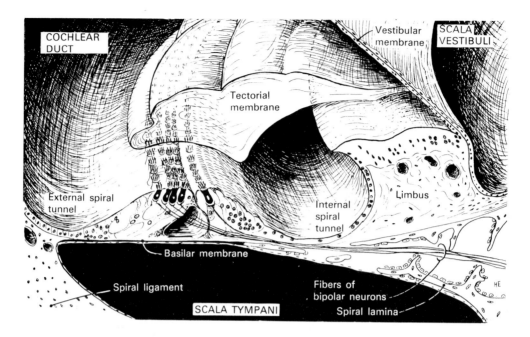

Fig. 17-12. The spiral organ (of Corti).

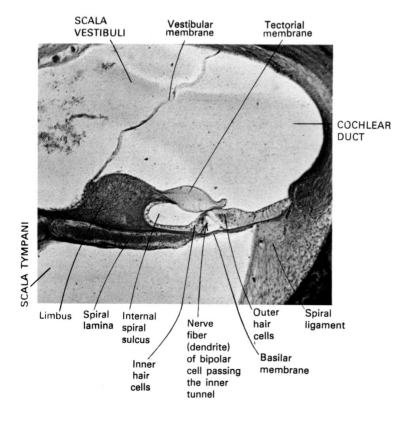

Fig. 17-12, B. Spiral organ of Corti, human. Masson stain, 185 ×.

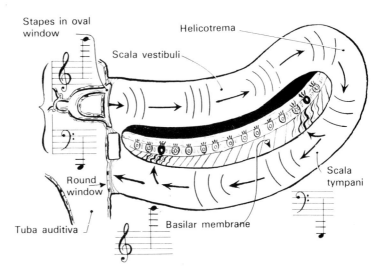

Fig. 17-13. Helmholtz's theory of hearing.

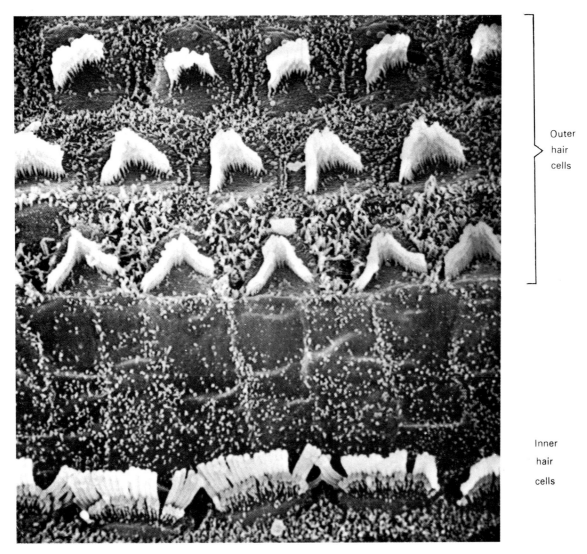

Fig. 17-14. Hairs of the hair cells of the spiral organ observed with the scanning electron microscope. Note the orderly arrangement. Courtesy of Dr. Catherine A. Smith. University of Oregon.

the head rotates, the labyrinth, including the semicircular ducts with their cristae and cupulae, follows the movement of the head; while the endolymph, due to inertia, remains almost stationary. Thus, the cupula is bent, and traction is exerted upon the microvilli of the crista hair cells; and rotation is felt, because dendrites of cells in the vestibular ganglion contact the hair cells of the cristae.

SPIRAL ORGAN (OF CORTI). Hearing is subserved by the *cochlea*, illustrated in Figures 17-10, D, E, F and 17-12. Vibrations of the air received by the tympanic membrane are transmitted by the first two auditory ossicles, the *malleus* and *incus*, to the third called the *stapes* (Fig. 17-1). The *foot plate* of the stapes rocks back and forth in the *oval window* (Fig. 17-6) at the same frequency as the sound waves. Thus the vibrations are conveyed to the perilymph in the bony canal composed of the *scala vestibuli*, the *helicotrema* and the *scala tympani* (Figs. 17-6, 17-10, D, E and 17-13). (The helicotrema is a tiny canal through which the scala vestibuli and the scale tympani communicate.) Thus the vibrations set up in the perilymph are conveyed through the bony canal to the *round window* which in turn rapidly bulges and retracts in rhythm with the pressure changes in the perilymph. These pressure changes also cause oscillations of the *basilar membrane* immediately adjacent to the scala tympani. Although large portions of this membrane vibrate with every sound, a specific pitch (frequency) will produce a greater movement in one specific area than in another (Fig. 17-13). It is known that high tones cause maximal vibration of the basilar membrane near the round window, whereas low tones cause maximal stimulation near the helicotrema. When the basilar membrane moves, the sensory hair cells attached to it move. The hairs of the hair cells (Fig. 17-14) are embedded in, or at least contact the *tectorial membrane* which is suspended in the endolymph of the cochlear duct. When the hair cells move, the hairs contacting the tectorial membrane are distorted; and a stimulus is initiated which is conveyed to the dendrites of the bipolar cells from the *spiral ganglion* (Fig. 17-8). It is assumed that the inner hair cells transmit rhythm and noise, the outer hair cells are sensitive to pitch. The spiral ganglion is located in the *modiolus*, a cone-shaped, bony pillar around which the cochlea spirals (Fig. 17-10), D). Axons from the spiral ganglion go to the brain where the vibrations originally received by the tympanic membrane are interpreted as sound.

18...
Reproductive System

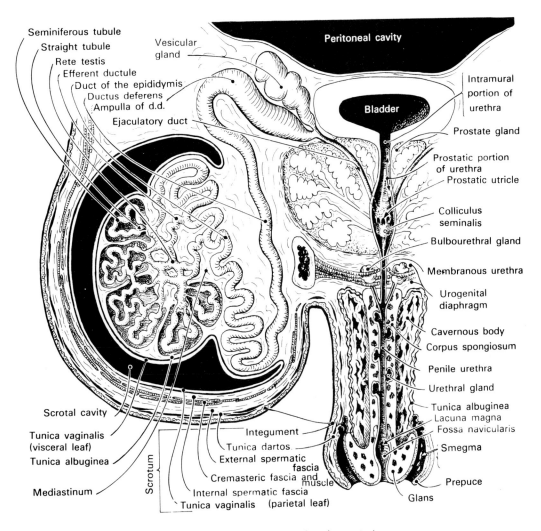

Fig. 18-1. Schematic synopsis of male genital organs.

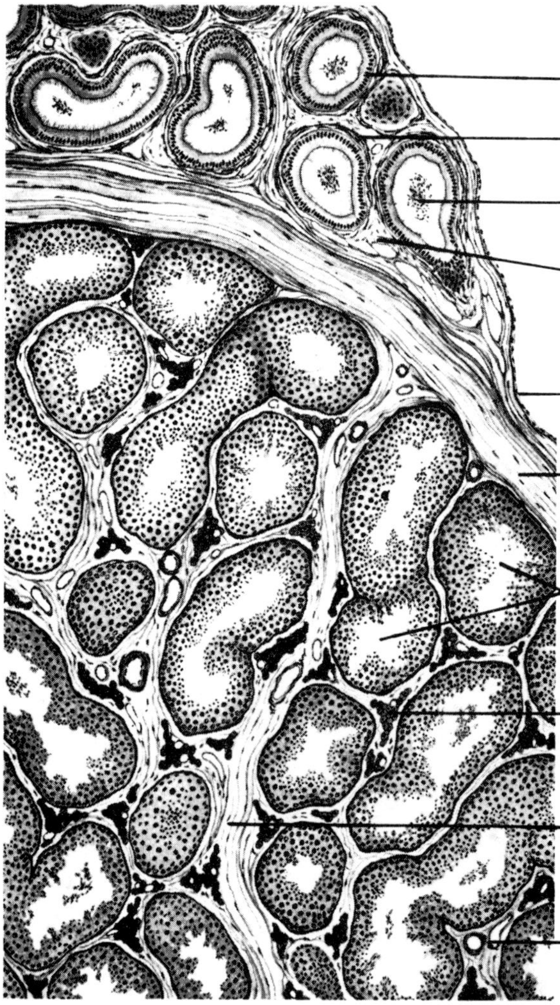

456 REPRODUCTIVE SYSTEM

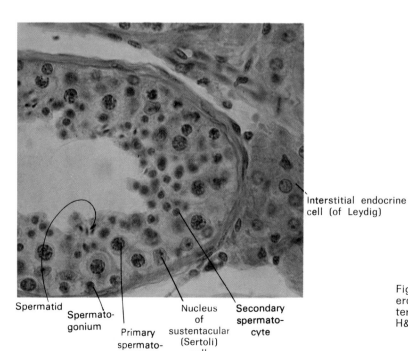

Fig. 18-2. Testis and epididymis. (Gillison, 1953) With permission of E.&S. Livingstone, Edinburgh.

- Ductus epididymidis
- Circular muscle fibers
- Spermatozoa
- Stroma of epididymis
- Tunica vaginalis
- Tunica albuginea
- Seminiferous tubules
- Interstitial endocrine cells
- Septulum
- Blood vessels

Fig. 18-3. Part of seminiferous tubule and some interstitial cells (of Leydig), H&E. 950 ×.

- Spermatid
- Spermatogonium
- Primary spermatocyte
- Nucleus of sustentacular (Sertoli) cell
- Secondary spermatocyte
- Interstitial endocrine cell (of Leydig)

Multicellular organisms age and die. To perpetuate the species, two individual cells, the *gametes*: an *ovum* from the female and a *spermatozoon* from the male, must unite to form a *zygote* from which a new unique individual develops. The gametes are produced in two cytogenic glands, the gonads: *ovary* in the female and *testis* in the male. The union of gametes and subsequent development of the zygote are facilitated through specialized sex organs.

MALE GENITAL ORGANS

Speratozoa are produced in the *seminiferous tubules* of the testes and transported through a series of passages which terminate at the distal end of the urethra (Fig. 18-1). Along this route, tubules and glands opening into them contribute various secretory products which together with the spermatozoa form the *semen* which is ejaculated into the posterior fornix of the vagina.

Testes

Each testis (Figs. 18-1 and 18-2) protruding into the cavity of the *scrotum* is invested by a serous membrane, the visceral leaf of the *tunica vaginalis*. Deep to this membrane is a thick firm capsule of collagenous and elastic fibers, the *tunica albuginea*, from which *septula* extend into the organ dividing the testis into cone-shaped lobules. In man the septula are rather rudimentary and may be absent. The base of each lobule is formed by the tunica albuginea, and the apex is directed toward the *mediastinum testis*. The lobules contain long, coiled, anastomosing *seminiferous tubules* embedded in a loose areolar connective tissue stroma in which are scattered the *interstitial endocrine cells* (Leydig), producers of testosterone (Figs. 18-2 and 18-3).

SPERMATOGENESIS. Transformation of diploid primitive *spermatogonia* into haploid *spermatozoa* is called *spermatogenesis* (Figs. 18-4, 5). Three phases characterize this process: *spermatocytogenesis, meiosis* and *spermiogenesis*. In *spermatocytogenesis,* spermatogonia are classified as TYPE A and TYPE B. Type A are stem cells which reproduce by mitosis. Some daughters produce irreversible mitotic Type B cells which differentiate into diploid *primary spermatocytes. Meiosis,* produces haploid *spermatids* via two meiotic maturation divisions of spermatocytes. *Spermiogenesis* treats spermatids to a series of cytological changes prior to releasing them from the seminiferous epithelium as definitive *spermatozoa*. The *spermatogenic epithelium* of these tubules rest on a basement membrane supported by a thin coat of dense connective tissue. The epithelium consists of (1) *sustentacular cells* (of Sertoli), extending from the basement membrane to the lumen, and (2) several layers of germ cells in various stages of development and maturation (Figs. 18-3 and 18-4).

SPERMATOGENESIS (Fig. 18-5), lasting about 74 days, commences in diploid primitive spermatogonia in contact with the basal lamina of seminiferous tubules (Fig. 18-4). Mitosis maintains a reserve of *Type A* spermatogonia stem cells plus a differentiating pool of *Type B* offspring. The *Type A* cell is distinguished by: (1) an avoid nucleus, (2) one or two nucleoli adjacent to the internal nuclear membrane, (3) a fine granular euchromatin of varying basophilic intensities that distinguish *Type A* cells as *dark* and *light* and (4) a homogeneous pale staining cytoplasm. *Type B* spermatogonia differ by a centrally placed nucleolus coated with chromatin granules and similar granules adhering to the internal nuclear membrane. Differentiation from the base to the lumen (of the seminiferous tubules) and intense mitotic activity produce larger diploid *primary spermatocytes* preparing to enter *Meiosis I* (Fig. 18-5, 18-6).

In *Meiosis I* visualization of a large number of primary spermatocytes is possible during the 22 day long prohase (consisting of *leptotene, zygotene, pachytene, diplotene* and *diakinesis*). Primary spermatocytes occupying the middle zone of the

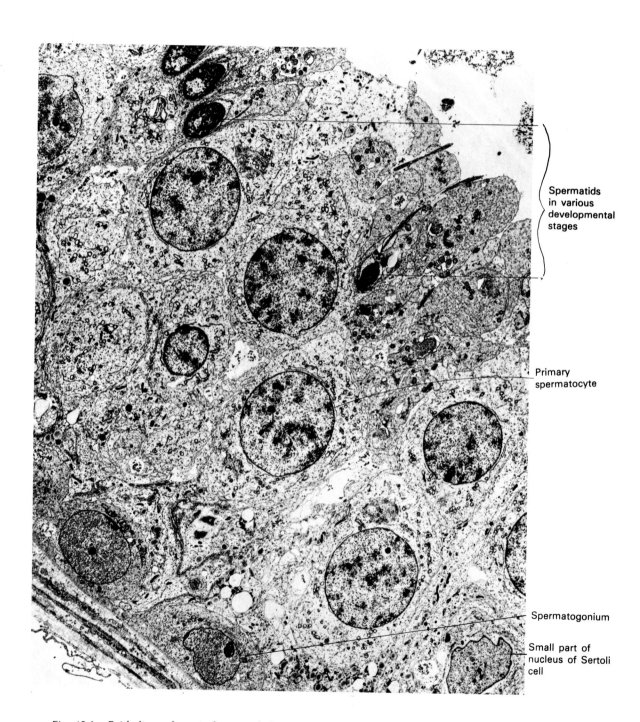

Fig. 18-4. Epithelium of seminiferous tubule. Secondary spermatocytes are not present in this segment at this time. EM by Prof. A. F. Holstein, University of Hamburg. 2025 ×.

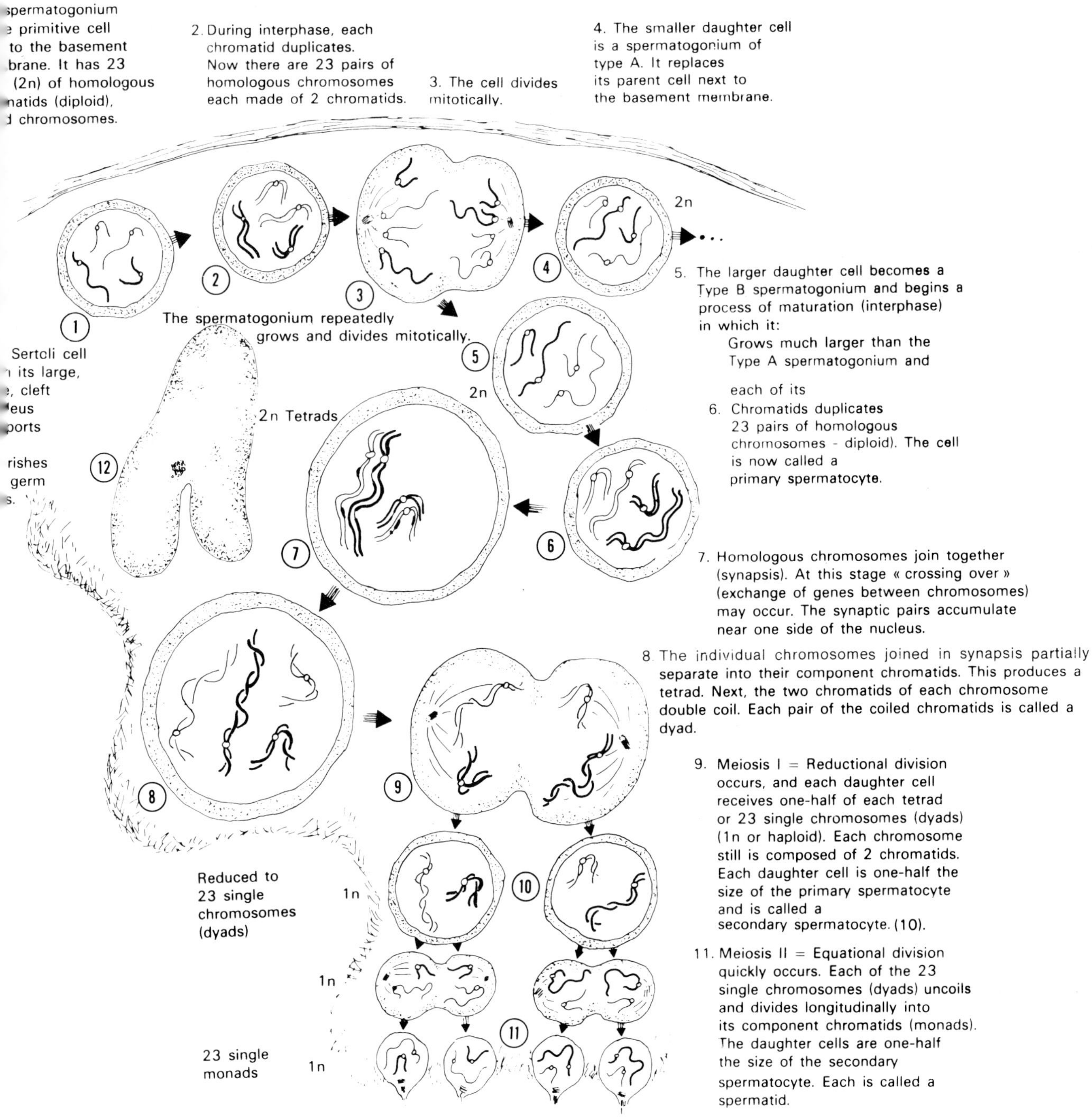

TESTES 459

spermatogonium
e primitive cell
to the basement
brane. It has 23
(2n) of homologous
natids (diploid),
d chromosomes.

2. During interphase, each chromatid duplicates. Now there are 23 pairs of homologous chromosomes each made of 2 chromatids.

3. The cell divides mitotically.

4. The smaller daughter cell is a spermatogonium of type A. It replaces its parent cell next to the basement membrane.

The spermatogonium repeatedly grows and divides mitotically.

Sertoli cell
its large,
, cleft
eus
orts

rishes
germ
s.

2n Tetrads

5. The larger daughter cell becomes a Type B spermatogonium and begins a process of maturation (interphase) in which it:
   Grows much larger than the Type A spermatogonium and
   each of its
6. Chromatids duplicates 23 pairs of homologous chromosomes - diploid). The cell is now called a primary spermatocyte.

7. Homologous chromosomes join together (synapsis). At this stage « crossing over » (exchange of genes between chromosomes) may occur. The synaptic pairs accumulate near one side of the nucleus.

8. The individual chromosomes joined in synapsis partially separate into their component chromatids. This produces a tetrad. Next, the two chromatids of each chromosome double coil. Each pair of the coiled chromatids is called a dyad.

Reduced to 23 single chromosomes (dyads)   1n

23 single monads   1n

9. Meiosis I = Reductional division occurs, and each daughter cell receives one-half of each tetrad or 23 single chromosomes (dyads) (1n or haploid). Each chromosome still is composed of 2 chromatids. Each daughter cell is one-half the size of the primary spermatocyte and is called a secondary spermatocyte. (10).

11. Meiosis II = Equational division quickly occurs. Each of the 23 single chromosomes (dyads) uncoils and divides longitudinally into its component chromatids (monads). The daughter cells are one-half the size of the secondary spermatocyte. Each is called a spermatid.

Fig. 18-5.   Diagram of spermatogenesis.

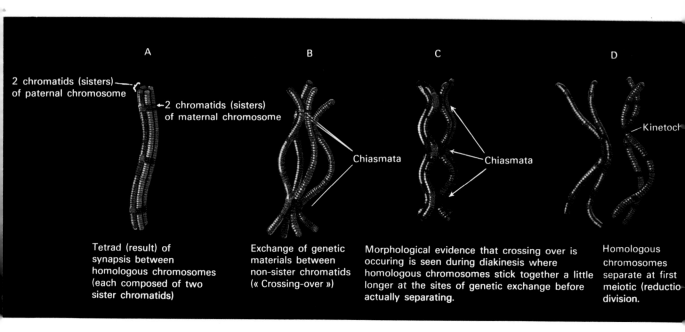

Fig. 18-6. Formation of tetrads, crossing-over and separation of chromosomes in a primary spermatocyte. Each thread shown here is a chromatid.

seminiferous epithelium, complete *Meiosis I* by exchanging genetic material between chromatid segments (*crossing over*) (Fig. 18-6). The products of *Meiosis I*, *secondary spermatocytes*, possess the haploid number of chromosomes (22 + X, or 22 + Y), but with 2N (diploid) amount of DNA. *Primary spermatocytes* entering *Meiosis I* possessed a diploid number of

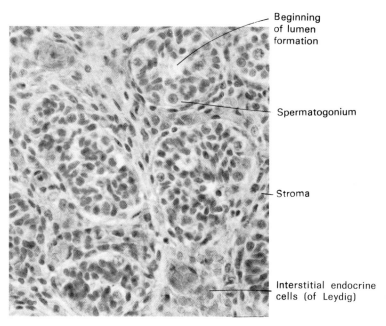

Fig. 18-7. Testicle of new born, H&E. 750 ×.

chromosomes (44 + XY) with 4N (tetraploid) the amount of DNA.

Secondary spermatocytes pass through *Meiosis II* so rapidly that they are rarely detected in histological sections of the seminiferous tubules. They give rise to *spermatids* with 23 chromosomes (22 + X, or 22 + Y) possessing N (or haploid) amounts of DNA, because DNA synthesis does not occur between *Meiosis I* and *Meiosis II*.

Slightly elongated *Spermatids*, about the size of erythrocytes, no longer divide and the chromatin becomes increasingly heterochromatic. Spermatids lie next to the lumen preparing to undergo *spermiogenesis*.

Crossing-over increases biological variations enormously over those produced solely via the sexual process. Crossing-over extends the desirability of gene recombinations down to the chromosomal level and accelerates new combination of genetic material. Consider Figure 18-6 and the many possible combinations which account for the remarkable variations in one family.

The traditional concept that karyokinesis and cytokinesis during spermatogenesis produce individual cells and spermatids is no longer espoused. The best information reveals that uniting spindle and cytoplasmic bridges persist into the late stages of spermiogenesis. Progeny from a single spermatogonium thus form communicating clusters of developing germ cells displaying remarkable synchrony.

In children, most seminiferous tubules are solid cords of sustentacular cells with interspersed spermatogonia (Fig. 18-7). A continuous lumen appears during puberty when spermatogenesis begins. In old age, the seminiferous tubules may degenerate (Figs. 18-8 and 18-9). Figure 18-8 shows the testis of an old man in which some seminiferous tubules have remained active, while others are hyalinized. Figure 18-9 demonstrates a case in which all germ cells have disappeared, while the sustentacular cells remain intact.

The last step in the production of male gametes is spermiogenesis, the metamorphosis of the haploid spermatids (Fig. 18-10 and 18-12) into spermatozoa (Fig. 18-11). The first stage in this rather compli-

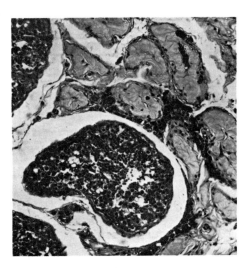

Fig. 18-8. Hyaline degeneration of seminiferous tubules (AFIP).

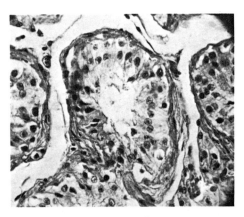

Fig. 18-9. Senile loss of spermatogenic cells with remaining sustentacular (Sertoli) cells. (From a slide prepared by Dr. J. Gant, George Washington Univ.)

cated process involves the appearance of several small *proacrosomal granules* within the golgi complex (Figs. 18-10, 18-12, 1). These coalesce into an *acrosomal granule* which becomes surrounded by a mem-

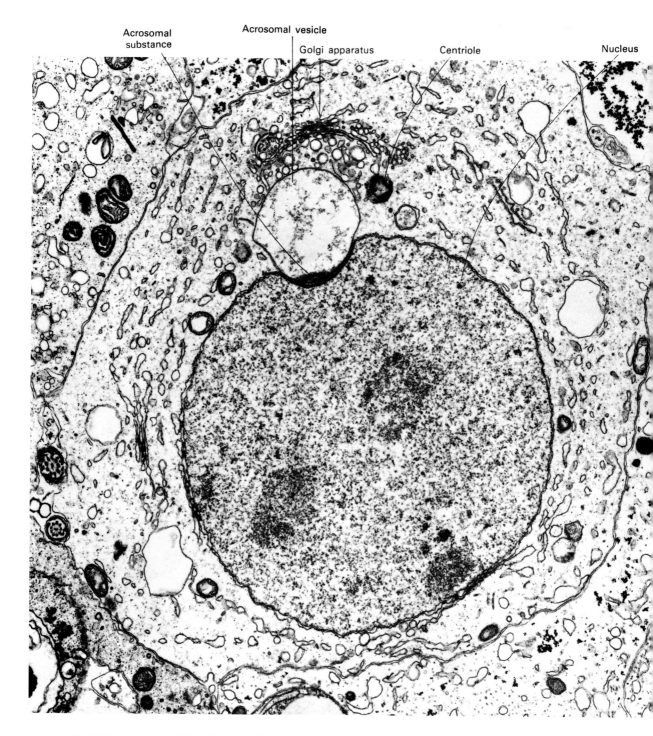

Fig. 18-10. A spermatid of 33-year-old man. 11,400 ×. Courtesy of Prof. A. F. Holstein, University of Hamburg.

brane to form the *acrosomal vesicle* (vacuole). This migrates to the opposite pole of the spermatid. The distal centriole gives rise to a long thin flagellum, while the acrosomal vesicle attaches to the nucleus and spreads to cover one-half to two-thirds of it as the *acrosomal cap* (Fig. 18-12, 2). Mitochondria migrate toward the microtubules that make up the part of the flagellum still surrounded by cytoplasm, that is, the pars intermedia between the nucleus and the cytoplasmic membrane. These mitochondria spiral end to end around the microtubules. Figure 18-12 illustrates spermiogenesis in detail. Eventually the excess cytoplasm is shed, and the mature spermatozoon (Figs. 18-11 and 18-13) consists of a head with its acrosomal cap, a middle piece where the mitochondria are concentrated, and a long slender flagel-

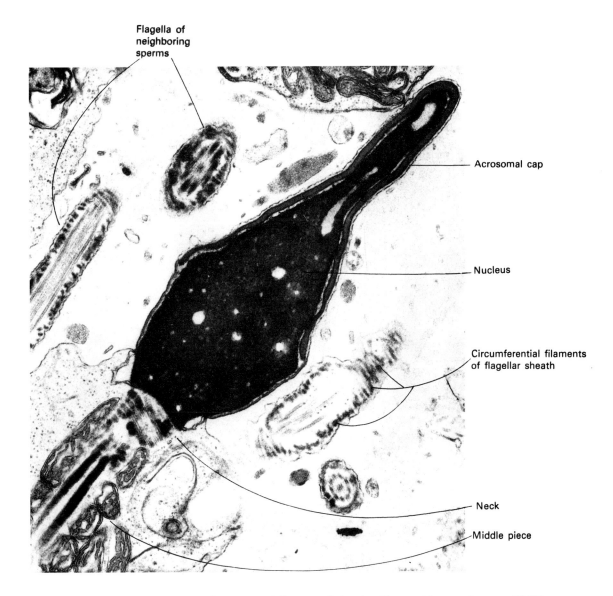

Fig. 18-11. Almost mature sperm from a seminiferous tubule of a 33-year-old man. Approx. 20,000 ×. Courtesy of Prof. A. F. Holstein, University of Hamburg.

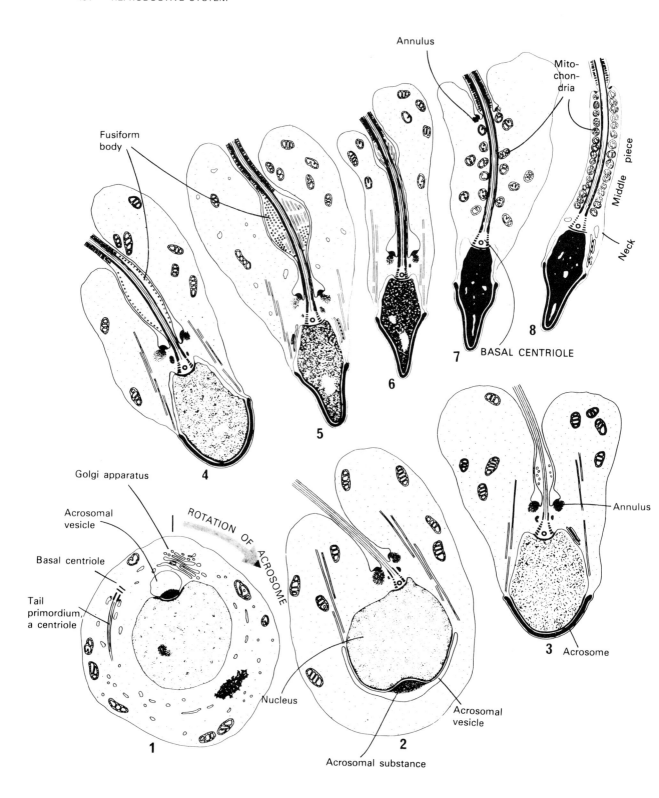

Fig. 18-12. Synopsis of spermiogenesis in man (by Prof. Dr. A. F. Holstein, University of Hamburg, and H. Wartenberg, University of Bonn), first presented at the 70th meeting of the Anatomische Gesellschäft. Düsseldorf, April. Anat. Anz. In the early spermatid (1) the Golgi complex produces the acrosomal vesicle. The acrosomal vesicle moves to the pole of the nucleus toward the epithelial basement membrane, while the centrioles move toward the tubular lumen (see also Fig. 18-4) (2). One of the centrioles becomes the flagellum, the other remains in the neck region (see also Fig. 18-11). There exists, temporarily, a fusiform body around the flagellum (3-6). A ring-shaped dense body (annulus) originates near the nucleus (2) and ends up at the base of the free flagellum (8). The mitochondria of the spermatid arrange themselves in a spiral around the middle piece of the flagellum. The nucleus undergoes condensation and its end becomes flattened.

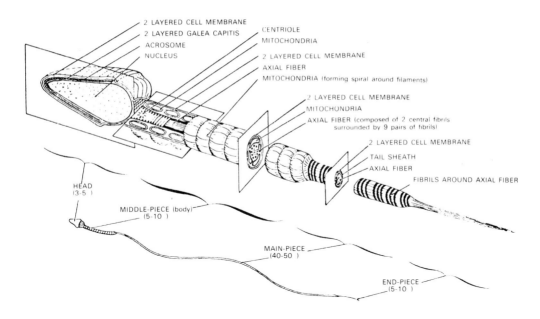

Fig. 18-13. Stereogram of human sperm.

lum[1]. The spermatozoon is released by its sustentacular cell (Fig. 18-5 [12]) to begin a long journey through the tubules of the male reproductive system.

Conducting Passages

The seminiferous tubules produce a fluid which (1) provides a medium for maturing spermatozoa and (2) is selectively reabsorbed by the *straight tubules, rete testis, efferent ductules* and by the proximal part of the *duct of the epididymis* (Ductus epididymidis) (Figs. 18-2 and 18-5, B-G). The flow of liquid produces a current which moves the spermatozoa along through the ducts. Smooth musculature in the walls of the efferent ductules, duct of the epididymis and *ductus deferens* (Fig. 18-16) also aids the movement of the spermatozoa. The histological characteristics of these passages are outlined in Table 18-1 and can be reviewed in Figures 18-15, A through L.

In addition to the secretions of the duct of the epididymis, glands whose secretions contribute to the formation of semen in-

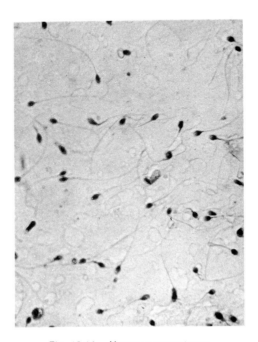

Fig. 18-14. Human spermatozoa.

---

[1] It is characteristic for the *human* spermatozoon that the front part of the head (nucleus) is flattened. In a section perpendicular to the flat front piece, it appears thin. When seen from its flat side (Fig. 18-14), one notices that it is as wide as the entire head. Due to its thinness, this part is transparent (Fig. 18-14). Every vertebrate species has spermatozoa of a specific shape.

clude the diverticula of the *ampulla of the ductus deferens* (Fig. 18-18), the seminal vesicles (Fig. 18-19), the prostate (Figs. 18-21 and 18-24), the bulbourethral glands and the tiny urethral glands (of Littre or Morgagni) (Figs. 18-26 and 18-27).

Seminal Vesicles (Vesicular Glands)

The *seminal vesicles* (*vesicular glands*) (Figs. 18-1, 18-19 and 18-20) are two small bodies posterior to the prostate and lateral to each ampulla of a ductus de-

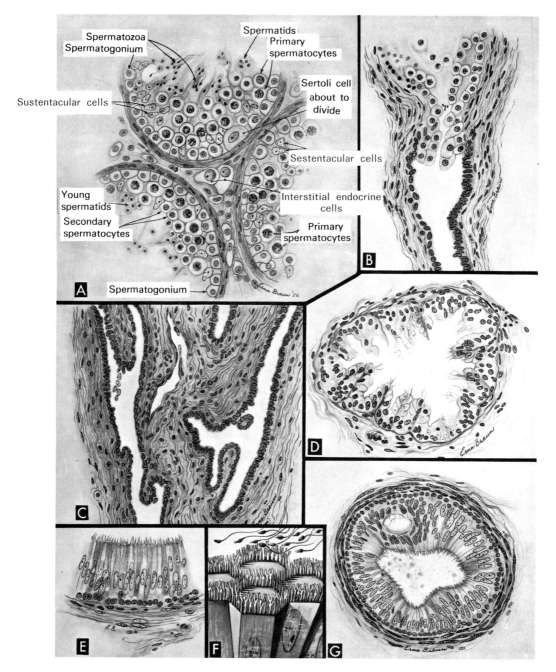

Fig. 18-15. A, Active seminiferous tubules. B, Junction of seminiferous tubule with straight tubule. C, Rete testis. D, Efferent ductule. E, F and G, Duct of epididymis. (Drawn from preparation by Dr. Roosen-Runge.)

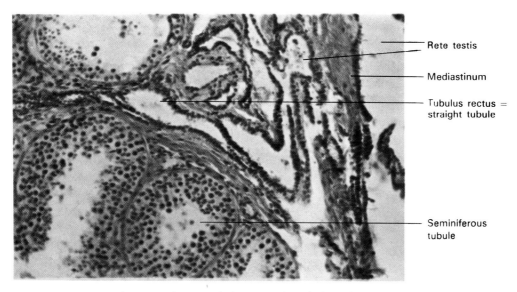

Fig. 18-15, H. Testicle, seminiferous tubules, straight tubule and rete testis in mediastinum, H&E.

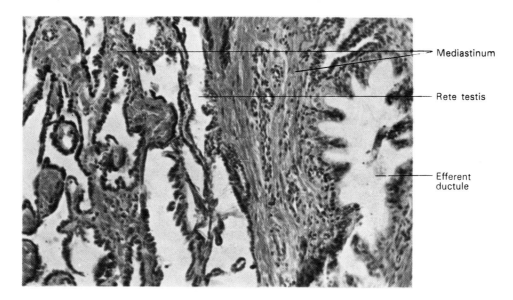

Fig. 18-15, I. Rete testis and efferent tubule, H&E.

ferens. Each gland consists of a single, highly convoluted tube which joins the terminal end of the ductus deferens to form the *ejaculatory duct*. A section cut through the seminal vescicle in several places shows a large lumen surrounded by a mucosa which forms a muralium even more extensive than that seen in the ampulla of the ductus deferens. The long, convoluted seminal vesicle is so coiled upon itself that the tunica mucosa appears fused to the tunica muscularis. There is no tela submucosa, but in a few cases, a lamina muscularis mucosae surrounds the duct dividing the bulky stroma into a lamina propria mucosae which extends into the folds of the muralium and a subjacent submucous fibromuscular layer. Figures 18-19 and 18-20 show cases in which no such subdivision occurs. The prevailing, irregularly ar-

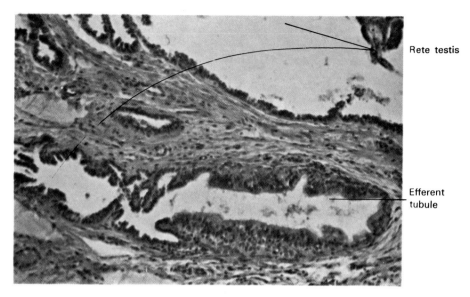

Fig. 18-15, J. Transition from rete testis to efferent tubule, H&E.

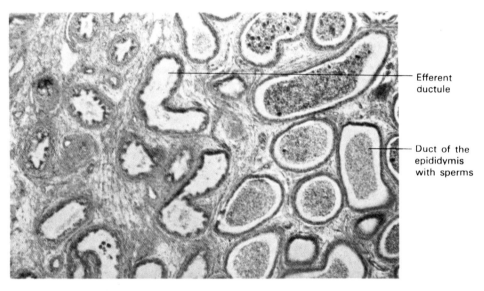

Fig. 18-15, K. Head of epididymis showing efferent tubules and sections through windings of the single duct of the epididymis.

ranged smooth musculature helps in the ejection of semen. The epithelium of the mucosa varies from simple columnar to pseudostratified. The outer tunica adventitia is an external layer of elastic connective tissue fibers.

Prostate Gland

The prostate gland (Figs. 18-21 through 18-24) situated between the urinary bladder and the urogenital diaphragm, is pierced by the urethra and both ejaculatory ducts. It is surrounded by a fibromuscular capsule that sends numerous branching septa into the body to divide the organ into 30 to 50 compound, tubulo-alveolar glands, each with its own duct opening into the prostatic urethra (Fig. 18-1). The septa are characterized by discrete bundles of smooth myocytes interweaving with connective tissue fibers. The epithelium of the alveoli varies from simple cuboidal to simple columnar to pseudostratified, according to its activity phases. While special stains can distinguish connective tissue from muscle (Figs. 18-21 to 18-23), this distinction is extremely difficult in routine slides. In men over 40, calcified bodies called *prostatic concretions* may be detected in the lumen (Fig. 18-23). These concretions are composed of calcified concentric layers resembling starch grains. During childhood, the end pieces of the prostate are solid cell cords with only a few, small lumina (Fig.18-25). The alveolar lumina appear as sexual maturity develops.

Urethra

The ducts of the prostate gland empty into the prostatic urethra, which is lined proximally by alternating patches of transitional epithelium, stratified columnar epithelium and pseudostratified epithelium. The terminal portion, and particularly the fossa navicularis, is lined by non-keratinized stratified squamous epithelium. Within the penis, the tunica mucosa shows mucous depressions and branched mucous glands, the *urethral glands* (Figs. 18-26 and 18-27).

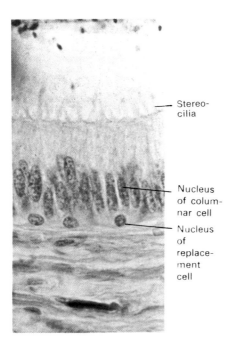

Fig. 18-15, L. Pseudostratified epithelium of ductus epididymidis with stereocilia, H&E. 1125 ×.

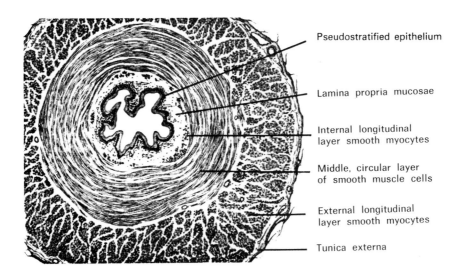

Fig. 18-16. Ductus deferens. (Gillison.)

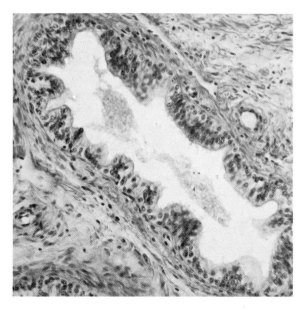

Fig. 18-17. Ductulus efferens, H&E. 380 ×.

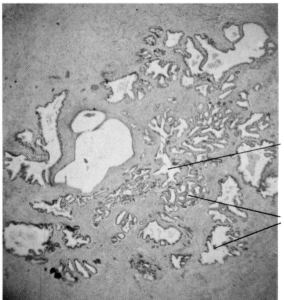

Fig. 18-18. Ampulla of ductus deferens, H&E.

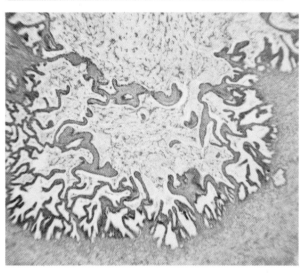

Fig. 18-19. Vesicular gland (formerly called seminal vesicle), H&E. 60 ×.

TABLE 18-1. SYNOPSIS OF MALE GENITAL DUCTS

STRUCTURE	POSITION	EPITHELIUM	MUSCLE	CONNECTIVE TISSUE
Straight tubules (Fig. 18-15, B)	In testis, connect seminiferous tubules with rete testis	Simple columnar to cuboidal	None	Mass of dense, irregular connective tissue, called "mediastinum testis", continuous with septula anteriorly and tunica albuginea posteriorly
Rete testis (Fig. 18-15, C)	In mediastinum testis, between straight tubules and efferent ductules	Simple squamous to cuboidal, some with single flagellum	None	
Efferent ductules (Figs. 18-15, D and 18-17)	Extend from rete testis in mediastinum testis to body of epididymis as cone-shaped coils which collectively form head of epididymis	Alternating groups of tall ciliated and few round basal cells. Thus lumen has irregular, wavy outline; in places pseudostratified epithelium; distinct basement membrane	Very thin layer of circular smooth cells.	Lamina propria of areolar, highly vascular connective tissue with elastic fibers
Ductus epididymidis (Fig. 18-15, E, F, and G)	Body and tail of epididymis	Pseudostratified with non-motile stereocilia arranged around tops of tall columnar cells. Lumen with smooth, regular outline. Stereocilia are very long, branched microvilli.	Thin but dense layer of circular smooth muscle cells.	Thin lamina propria; areolar connective tissue between coils of the duct
Ductus deferens (Fig. 18-16)	From tail of epididymis to prostate gland via inguinal canal	Near proximal end, like epididymis; otherwise, pseudostratified columnar around narrow, slightly folded lumen, variable distribution of stereocilia.	Three characteristic layers of smooth muscle cells: inner longitudinal, middle circular, thick outer longitudinal	Lamina propria contains many elastic fibers (causes postmortal folding of mucosa); adventitia of areolar connective tissue
Ampulla of ductus deferens (Fig. 18-18)	Local enlargement near distal end of ductus deferens posterior to prostate	Simple cuboidal or columnar to pseudostratified around dilated lumen; mucosa thrown into folds forming complicated muralium or it may form deep diverticula, having the character of alveolar glands (Fig. 18-18).	Three characteristic layers present but are thinner and less organized	Lamina propria and adventitia
Ejaculatory duct (Fig. 18-1)	From terminal end of ductus deferens through prostate gland to urethra	Simple columnar to pseudostratified; near urethra it becomes transitional.	Only at beginning	In prostate

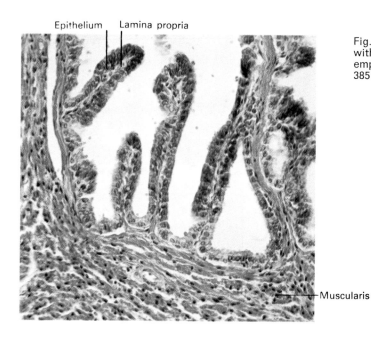

Fig. 18-20. Vesicular gland, stained with galocyanin and picric acid to emphasize the smooth musculature. 385 ×.

Bulbourethral Glands

The bulbourethral glands (Fig. 18-28, A) are compound, tubulo-alveolar, mucous glands characterized by the presence of surrounding striated myocytes of the urogenital diaphragm in which they are embedded. Irregular diameter ducts (Fig. 18-28, B), lined with tall simple columnar to stratified columnar epithelium, contain large intra epithelial patches of mucous secreting epithelium. The stroma of fibroelastic connective tissue contains both smooth and skeletal myocytes.

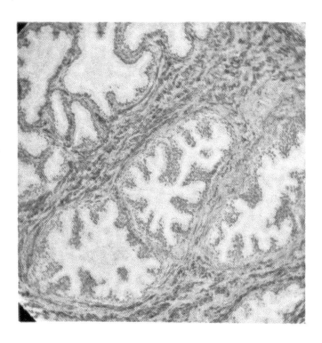

Fig. 18-21. Prostate. The stroma contains connective tissue (blue) and smooth muscle (red), Azan. 180 ×.

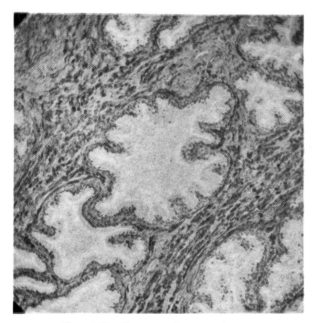

Fig. 18-22. Prostate, Azan. 190 ×.

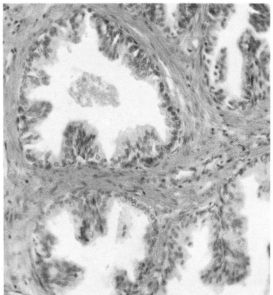

Fig. 18-24. Prostate, H&E. 395 ×.

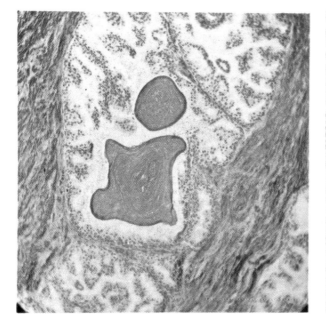

Fig. 18-23. Prostate, Azan. 180 ×. Two corpora amylacea (concretions) are seen.

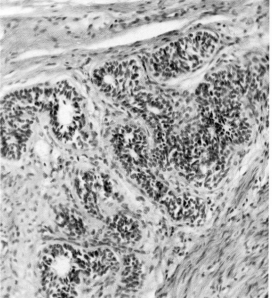

Fig. 18-25. Prostate of a boy, H&E. 290 ×.

Penis

The penis contains three cylindrical bodies of *erectile tissue*, two dorsal *corpora cavernosa* and one ventral *corpus spongiosum* (Fig. 18-26 and 18-29). The corpora cavernosa (Figs. 18-29 and 18-30) are surrounded by a dense fibrous tunica albuginea composed primarily of collagen fibers; the corpus spongiosum is surrounded by a sheath of soft fascia containing fewer collagen and more elastic fibers. The three erectile bodies together are enclosed in a sheath of areolar connective

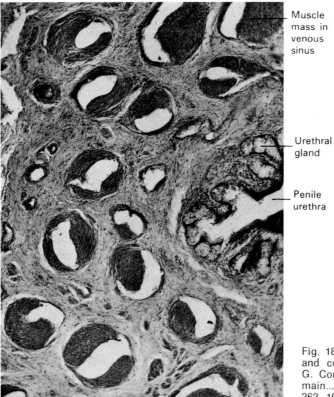

Fig. 18-26. Penile urethra (right) and corpus spongiosum. (From G. Conti, L'érection du pénis humain..., Acta Anatomica *14*: 217-262, 1952.)

tissue (fascia penis). Covering the penis is the *tunica dartos*, a kind of skin with bundles of smooth myocytes in the dermis (Fig. 18-31). The two dorsal bodies, the corpora cavernosa, are systems of anastomosing spaces separated by fibromuscular septa. Frequently, both corpora cavernosa are continuous with one another through gaps in the median tunica albuginea (Fig. 18-29).

The ventral body, the *corpus spongiosum* (Fig. 18-26), surrounding the penile

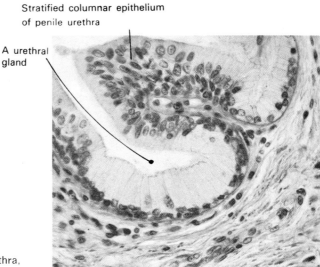

Fig. 18-27. Urethral gland of penile urethra, Azan. 635 ×.

urethra, is continuous distally with the *glans penis*. Its erectile tissue is very different from that of the corpora cavernosa but similar to that in the vagina and nasal mucosa.

Erection is achieved by the engorgement of erectile tissue with blood. The tunica albuginea is particularly strong around the corpora cavernosa of the penis (Fig. 18-30, A). However, the tunica albuginea enveloping the corpus spongiosum contains more elastic fibers, thus less re-

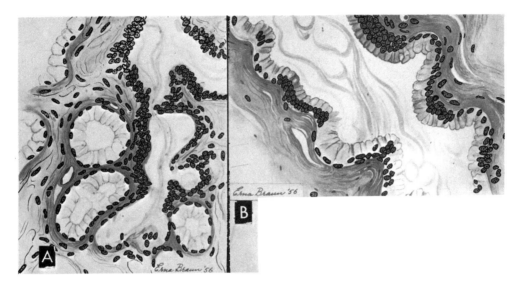

Fig. 18-28. A, Bulbourethral gland. B, Duct of bulbourethral gland.

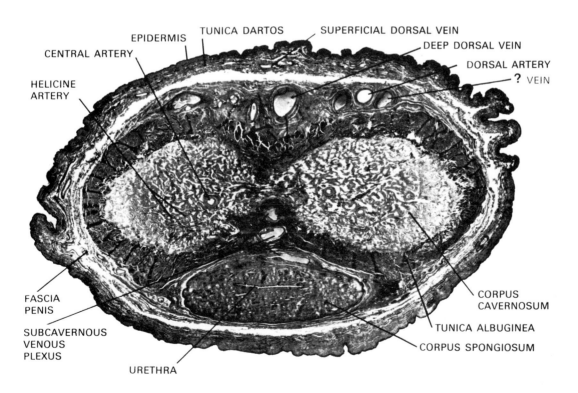

Fig. 18-29. Cross section of penis of adult.

REPRODUCTIVE SYSTEM

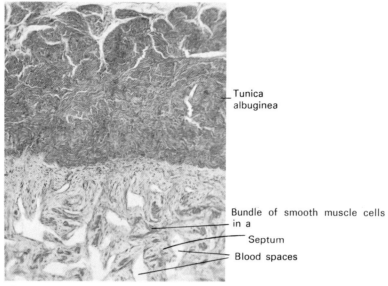

Fig. 18-30. A, Periphery of corpus cavernosum, H&E. 160 ×.

Fig. 18-30. B, Corpus cavernosum partially deflatedx (from G. Conti).

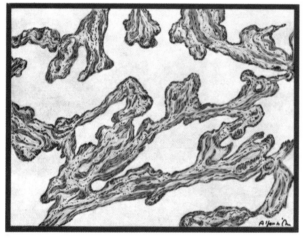

Fig. 18-30. C, Corpus cavernosum partially inflatedx (from G. Conti).

strictive, but is almost absent around the glans penis where only the dermis of the skin acts to contain the erectile tissue.

The cavernous bodies have the structure of a three-dimensional muralium composed of connective tissue and smooth muscle covered with vascular endothelium on both surfaces. The corpus spongiosum, of which the glans penis is the distal portion, is composed of large venous sinuses provided with spiral muscle sphincters. The arteries to the penis provide two kinds of blood supply: (1) one breaks up into capillary beds which provide nutrition for the organ. The blood eventually

Fig. 18-31. Tunica dartos of penis, H&E. 195 ×.

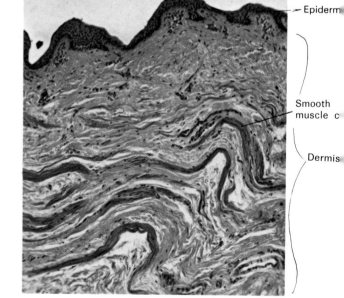

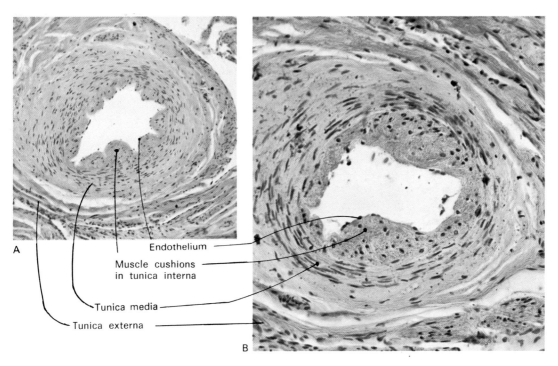

Fig. 18-32. A large (A) and a small (B) helicine artery. The magnification of A is 170 ×, that of B is 440 ×. Therefore the larger artery appears smaller in these pictures than the smaller one, H&E.

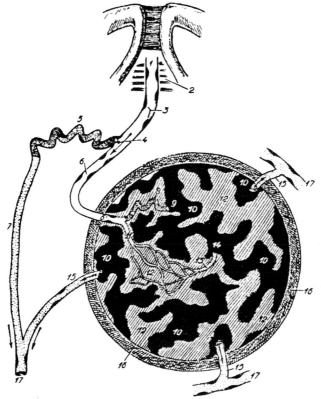

Fig. 18-33. Diagram of the vascular mechanism for the erection of a corpus cavernosum. (From G. Conti, L'érection du pénis humain et ses bases morphologico-vasculaires, Acta Anat. 14: 217-262, 1952.) 1: Internal pudendal artery and dorsal artery of penis; 2: urogenital diaphragm; 3: intimal muscle cushions in the deep artery of penis; 4: origin of AV-shunt; 5: AVA; 6: branch of deep artery of penis; 7: collecting vein; 8: helicine artery; 9 and 10: cavernous spaces; 11: nutritive artery; 12: fibromuscular septa; 13: venule; 14: its opening into a cavernous space; 15 and 17: efferent veins; 16: tunica albuginea.

flows into the spaces which anastomose with each other and flow towards the peripheral venous plexus. (2). The second kind, responsible for erection, include, as terminal branches, the *helicine arteries*, so named because they are highly coiled in the flaccid penis.

Most of the arteries which supply the penis, beginning with the internal pudendal artery down to the helicine arteries (Fig. 18-32) are provided with intimal muscle cushions in which the smooth myocytes are arranged longitudinally. These muscle cushions are normally in a state of contraction so that they bulge into the lumen and permit only a trickle of blood to reach the organ, sufficient for the nutrition of its tissues. Moreover, extra-cavernous, arterio-venous anastomoses divert blood from the bodies. Thus, the cavernous lacunae are normally almost empty, and the penis is flaccid. During sexual excitement, the intimal muscle cushions relax so that the lumina of the arteries to the penis widen to admit onrushing of large quantities of blood. At the same time, the extracavernous arteriovenous anastomoses constrict. Thus the cavernous spaces become engorged with blood, so that the corpora cavernosa become erect, due to the restriction of space by the non-extensible tunica albuginea. Enlargement of the penis causes pressure on the efferent veins so that the blood is retained in the cavernous spaces. Figure 18-33 is a synopsis of the erection mechanism.

The helicine arteries become relatively straight during erection. Essential to this process is the non-extensibility of the tunica albuginea which prevents blood from leaving the corpora cavernosa (Note its thickness and hence its strength in Fig. 18-30, A).

Erection of the corpus spongiosum (Fig.

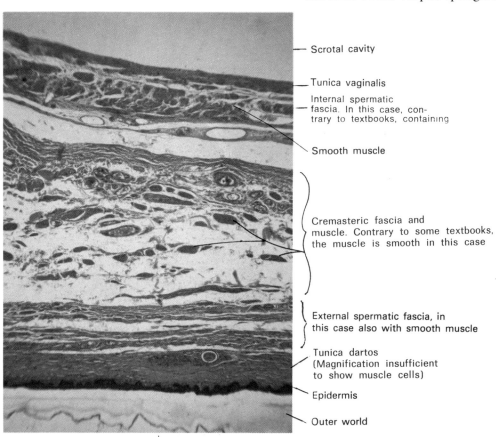

Fig. 18-34. Scrotal sac, H&E. 75 ×.

18-26 is aided by the constriction of sphincters which impede venous return. Because the tunica albuginea surrounding the corpus spongiosum is more elastic than that around the other two erectile bodies, the urethra remains open. In fact, swelling of the venous sinuses enlarges the urethra by outward traction.

The *prepuce* is a reduplication of the skin covering the glans penis (Fig. 18-1). Surgical removal of the prepuce is known as *circumcision*. The cavity between the prepuce and glans, if not cleaned frequently, accumulates *smegma* a slimy mass consisting of sebum, sweat and desquamated stratum corneum. The *tunica dartos* (Figs. 18-1 and 18-31), responsible for the corrugated appearance of the skin of the scrotum and of the flaccid penis, is a condensation of arrector pili muscles in the dermis.

Scrotal Sac

The testes lie in the cutaneous pouch called the *scrotum* (Fig. 18-34) suspended from the base of the penis. The skin of the scrotum contains bundles of smooth myocytes in the *tunica dartos*. In the midline, the dartos fascia of each side joins to form a septum which divides the scrotum into two separate compartments lined by a serous membrane, the *tunica vaginalis*. The scrotal sac consists of the external spermatic fascia, the cremasteric fascia with its bundles of skeletal myocytes, and the internal spermatic fascia. Deep to the internal spermatic fascia and almost completely surrounding the testis is a fluid filled sac called the cavity of the tunica vaginalis or scrotal cavity. The portion of this sac applied to the surface of the testis is called the *visceral layer*, continuous with the *parietal layer* of the tunica vaginalis which is covered by the internal spermatic fascia.

FEMALE GENITAL ORGANS

The human female reproductive system consists of the *ovaries*, the *uterine tubes*, the *uterus*, the *vagina*, the *external genitalia* and the *mammary glands*.

Ovaries

The female gonads, the ovaries, (Figs. 18-35 and 18-36), are cytogenic glands about the size and shape of large almonds, attached to the posterior surface of the broad ligament by an extension of peritoneum called the *mesovarium* (Fig. 18-36). As it continues from the mesovarium on to the surface of the ovary, the squamous mesothelium of the peritoneum changes to cuboidal. Previously it was called, incorrectly, *germinal epithelium*, a name given when its function was misinterpreted (Fig. 18-35, A and B).

On section the ovaries are composed of an outer cortex and an inner medulla. The *cortex* consists of: (1) dense, cellular connective tissue arranged in characteristic swirls (Fig. 18-35, A and B), (2) many *follicles* in various stages of maturity or degeneration, and (3) developing or regressing *corpora lutea* (Fig. 18-37). The more loosely arranged connective tissue of the *medulla* surrounds: (1) large convoluted bloods vessels, (2) degenerated (*atretic*) follicles (Fig. 18-28, E) and (3) degenerated corpora lutea (*corpora albicantia*) (Fig. 18-37 and 18-38, H).

The *ovogonium* (primitive ovum) (Fig. 18-35, A and B) arises in the embryo from a primordial germ cell which migrates from the yolk sac to the developing ovary. Still diploid, it grows and becomes a primary *ovocyte* (I) surrounded by follicular cells which have migrated in from the surface mesothelium. Primordial follicles form by the hundreds of thousands and in

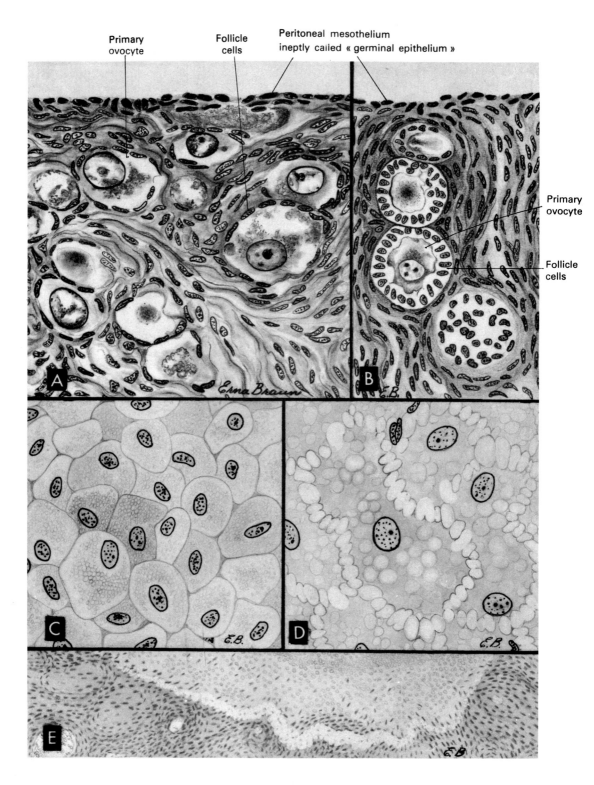

Fig. 18-35. A and B, Cortex of ovary, characterized by dense, cellular connective tissue containing young ovarian follicles. Each follicle consists of the central, primary oocyte and the peripheral follicle cells. C, The internal portion of a corpus luteum which develops from the stratum granulosum consists of finely vacuolated cells with convex and concave facets. D, During pregnancy these cells become large and foamy (nine weeks' pregnancy). E, An atretic follicle is lined by a *glassy* membrane.

the fetal ovary each primary ovocyte (I) reaches the diplotene[2] stage of meiosis and then rests. Many of these primordial follicles degenerate and disappear. Others remain in the resting stage for up to 45 years or more (until menopause when ovulation no longer occurs). Years later, follicles which have survived, start to grow again. The follicular cells multiply, the primary ovocyte (I) grows, and a thick membrane or shell, the *zona pellucida*[3], develops around it (Fig. 18-37). The stroma of the ovary surrounding the proliferating follicular cells organizes into a relatively cellular, highly vascular *theca interna* (estrogen producing cells) and a more fibrous *theca externa*. Thus the growing follicle consists of the primary ovocyte (I), the zona pellucida with its surrounding follicular cells, the cellular theca interna and the fibrous theca externa.

In the non-pregnant female between puberty and menopause (from about 12 to 50 years), a group of 15 to 20 immature follicles starts to develop just after the onset of each menstrual period under the influence of *follicule-stimulating hormone* (FSH) from adenohypaphysis (pars distalis) (Fig. 18-42). Most of the early follicles, develop only partially and then degenerate (each degenerating follicle becomes a *corpus atreticum*) (Figs. 18-35, E and 18-38, E), but one follicle continues to grow. Small spaces appearing between the ever-

---

[2] A stage of the first prophase of meiosis during which the paired bivalent chromosdmes begin to repel each other.
[3] Per, through; lux, light; hence: pellucidus, transparent, glassy.

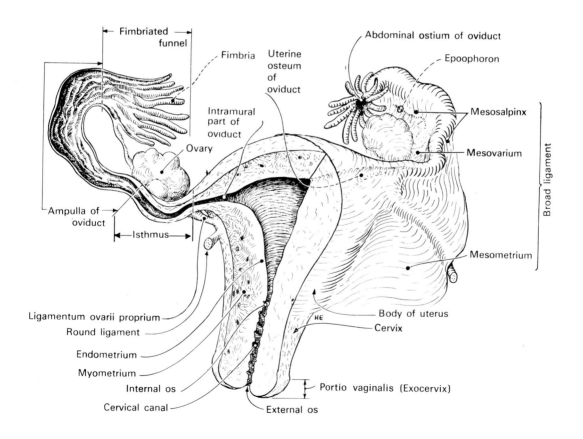

Fig. 18-36. Internal female reproductive organs in right, ventral aspect.

increasing number of follicular cells around the primary ovocyte (I) (Fig. 18-37), contain a liquid known as *liquor folliculi*. Eventually these tissue spaces coalesce to form the *antrum* (Fig. 18-37), which enlarges as the follicle continues to grow. The follicular cells lining the antrum constitute the *stratum granulosum*, a relatively loose layer of cells against a prominent basal lamina that separates them from the theca. The primary ovocyte (I) itself is surrounded by the zona pellucida around which radiate some adherent follicular cells collectively called the *corona radiata*. The ovocyte plus its corona radiata are embedded in a little mound

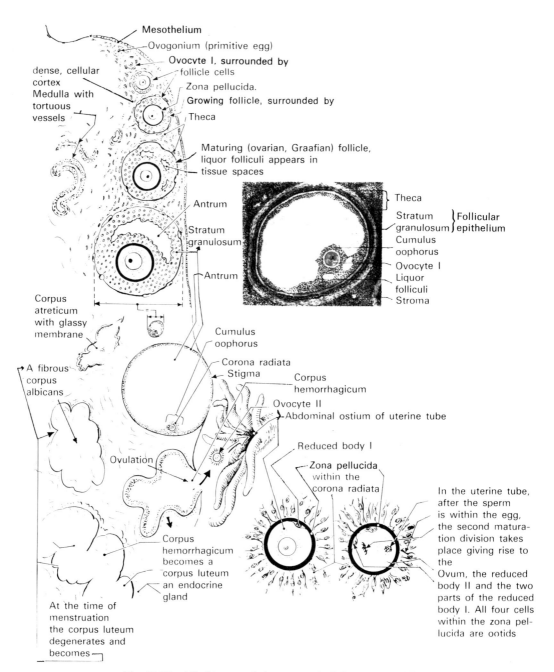

Fig. 18-37. Life history of the egg and of the ovarian follicle.

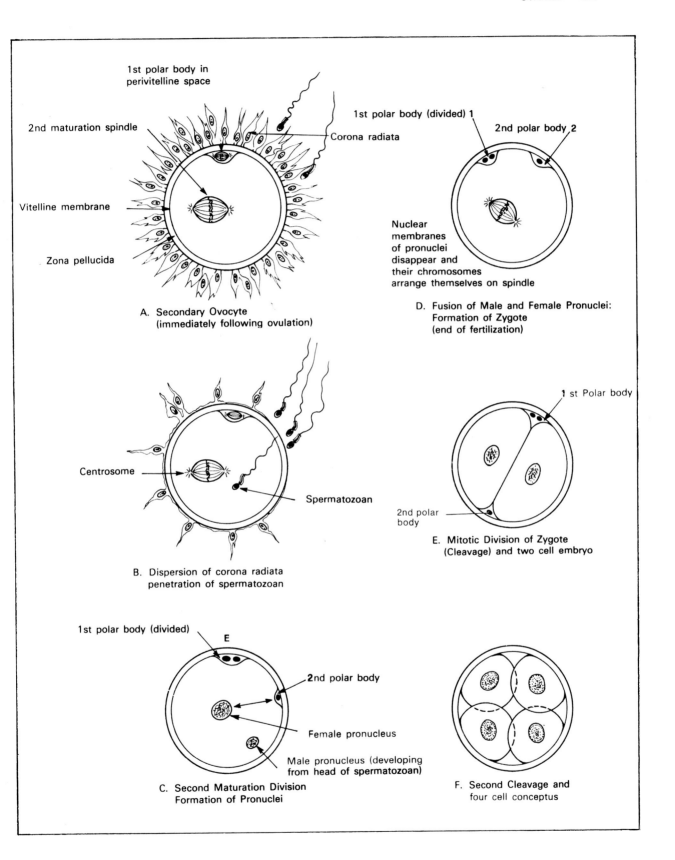

Fig. 18-37, A. Fertilization and cleavage.

of follicular cells called the *cumulus oophorus* which projects into the antrum cavity. As liquor folliculi continues to accumulate, the expanding follicle bulges on the surface of the ovary, at a point called the *stigma*, which becomes progressively more prominent. While the follicle is developing, its theca interna cells secrete increasing amounts of estrogen. Around midcycle there is a surge in the production of luteinizing hormone (LH) and FSH by the gonadotrop(h)ic endocrine cells of the adenohypophysis (Fig. 18-42). In early fetal life, *meiosis I prophases* of 3 to 5 million *primary ovocytes* (*I*) (Fig. 18-35B) progress through leptotene, zygotene and pachytene substages, but arrest in diplotene until sexual maturity. Suppression for so long is not understood. Resumption is induced by a sudden increase of LH in the first mid-menstrual cycle. A designated *primary ovocyte* (*I*) (Figs. 18-38C and D) completes meiosis I within 8 to 10 hours, just prior to *ovulation*, yielding a large *secondary ovocyte* (*II*), and a tiny *polar body I* (reduced body). The former receives almost all the cytoplasm plus a haploid set of chromosomes composed of two chromatids, thus doubling the DNA content. The cytoplasm poor *polar body I*, ultimately degenerates (Figs. 18-37 and 18-37A).

In *meiosis II*, the *secondary ovocyte* (*II*) waits in metaphase for *ovulation* and *fertilization* (Fig. 18-37 and 18-37A). The LH, required for completion of *meiosis I* and initiation of *meiosis II*, is indispensible for *ovulation* (about day 14 of a 28 day menstrual cycle). Investigations implicate collagenase and plasmin in dissolving connective tissue elements over the bulging follicle. A dehematized spot, the *stigma*, covered by disrupted cuboidal mesothelium, bulges over the thin, translucent stroma. Rupture of the *stigma* permits expulsion of the *ovocyte* (*ovulation*), along with follicular fluid and blood. Contraction of *fimbriae* and ciliary activity guides the ovum toward the infundibular *ostium* (Fig. 18-39). If it misses, the ovum plunges into the peritoneal cavity. Once safely within the distal one-third of the uterine tube, the ovocyte remains viable for 24 hours after which it succumbs to autolysis in arrested metaphase. During ovum viability, spermatozoa release enzymes which detach follicular cells protecting the ovum, thus exposing the *zona pellucida* to lysosomal activity from spermatozoon acrosomes. *Fertilization*, by definition, is the actual penetration of the *ovolemma* (cell membrane) by a spermatozoon at an entry point termed the *fertilization cone*. This stimulus to resume *meiosis II* leads to formation of a large *mature ovum* and a *polar body II* which also degenerates. Within the ovum, the sperm head expands to form a *male pronucleus* equal in size to the *female pronucleus* (ovum nucleus). Union of pronuclei reconstitutes the somatic chromosomal diploid number of a new cell, the *zygote*. Thus a new, unique human being commences life in the uterine tube (Fig. 18-40) where it undergoes repeated mitotic divisions, called *fission* (cleavage (Fig. 18-37A).

Of the three to five million ovocytes present in fetal life, only 400,000 survive at birth. The others undergo *atresia* (degeneration) at varying stages of follicular development to form minute scars, the *corpora atretica* (Figs. 18-35E and 18-39E). Atresia continues throughout life until after menopause, when follicles are difficult to find. Dead ovocytes and follicular cells are removed by phagocytic cells of the stromal connective tissue.

Following ovulation, the collapsed follicular wall forms a miniature endocrine gland, the *corpus luteum* (yellow body) as follows: (1) the basement membrane between the *stratum granulosum cells* and *theca interna cells* undergoes depolymerization, (2) capillaries and reticular fibers form delicate networks around the granulosum cells, (3) lipid droplets, mitochondria with tubular cristae and smooth

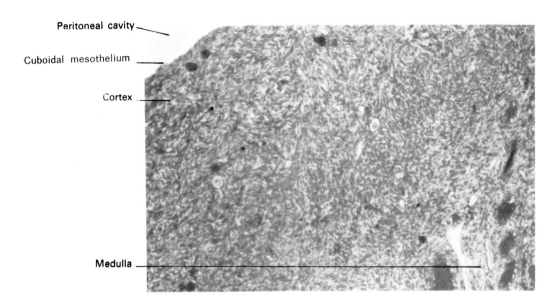

Fig. 18-38, A. Corpus luteum of pregnancy. (Detail from preceding picture).

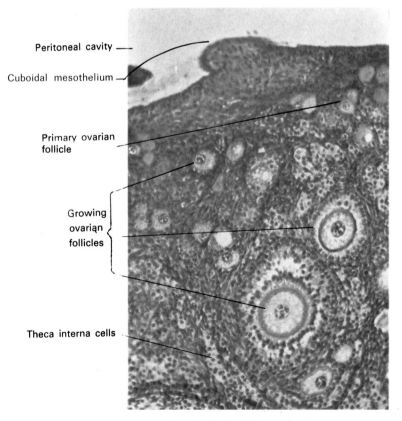

Fig. 18-38, B. Cortex of cat ovary.

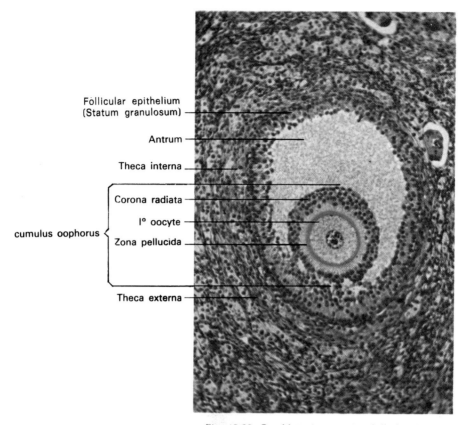

Fig. 18-38, C. Maturing ovarian follicle of cat.

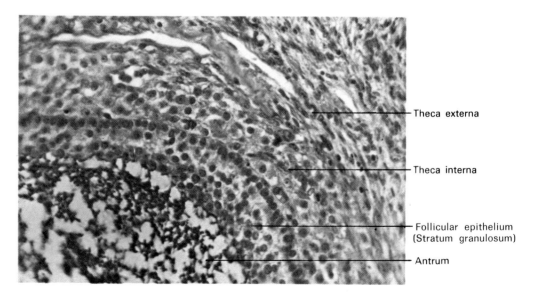

Fig. 18-38, D. Maturing follicle in *human* ovary.

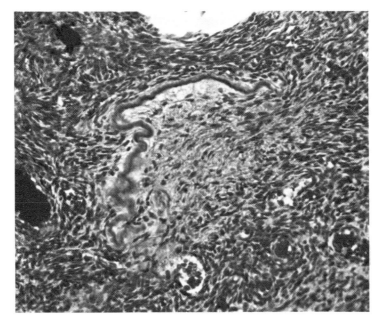

Fig. 18-38, E. *Human* atretic ovarian follicle, Azan stained.

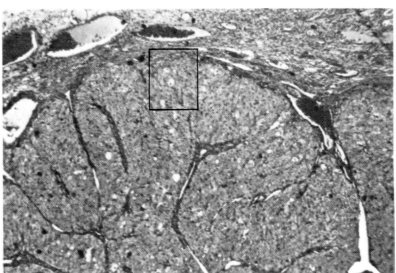

Fig. 18-38, F. *Human* corpus luteum during pregnancy.

ER accumulate in the enlarged cells (20 to 25 μm in diameter), now called *luteal cells*. Prior to formation of the corpus luteum, the collapsed follicle filled with clotted blood is known as the *corpus hemorrhagicum*.

If the ovum is not fertilized, the corpus luteum lasts 10 to 14 days as the *corpus luteum cyclicum (CLC)* (of menstruation). Its output of the hormone *progesterone* decreases after the 14th day and then the follicle involutes to form the *corpus albicans*, a white scar, which disappears after several months.

Should fertilization occur, a *corpus luteum graviditatis (CLG)* (of pregnancy) (Figs. 18-38A, F, G) increases in size. Progesterone secretion during the first few months continually decreases as the developing placenta assumes this role. The CLG is characterized by two major cell types: *granuloluteocytes* (light staining cells about 30 μm in diameter), and *thecaluteocytes* (dark cells, 15 μm) both possessing the EM morphology characteristic of steroid secreting cells. After delivery, regression in size and secretory activity transforms the CLG to a *corpus lu-*

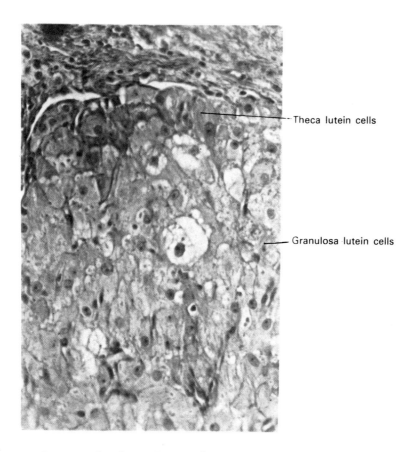

Fig. 18-38, G. Corpus luteum of pregnancy. Detail from preceding picture.

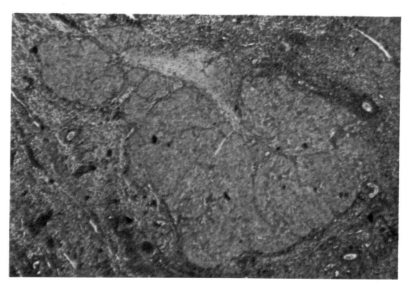

Fig. 18-38, H. Corpus albicans. All the preparations (18-38, A-F) are Azan stained.

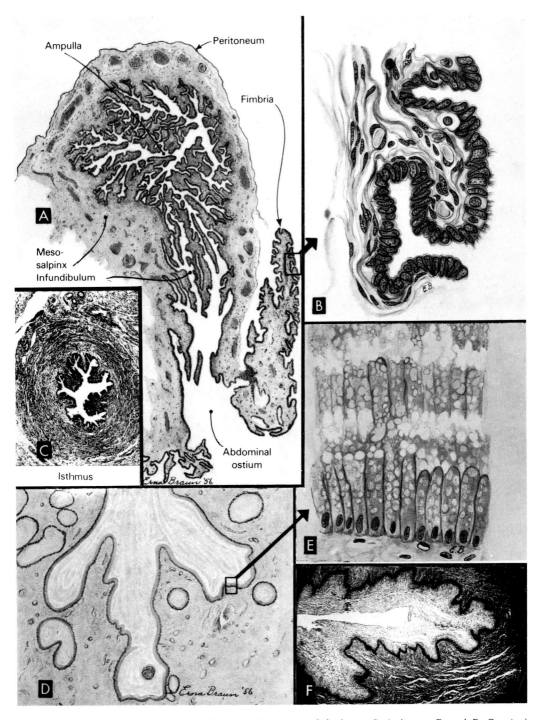

Fig. 18-39. A, Infundibulum and ampulla tubae. B, Lining of fimbria. C, Isthmus. D and E, Cervical glands at nine weeks of pregnancy showing layers of mucus which form the "mucous plug" of the cervical canal. F, Posterior vaginal fornix lined with stratified, squamous epithelium of the mucous type.

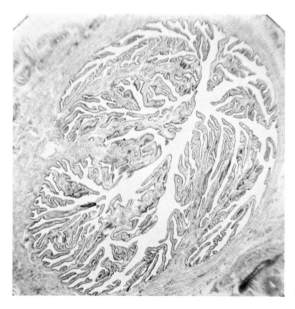

Fig. 18-40, A. Distal portion of ampulla of oviduct.

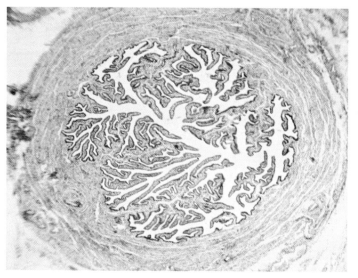

Fig. 18-40, B. Middle portion of ampulla of oviduct.

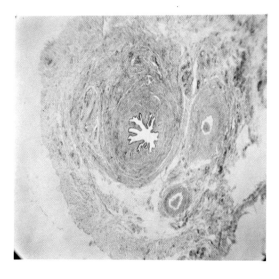

Fig. 18-40, C. Isthmus of oviduct. (All three pictures H&E. 78 ×.)

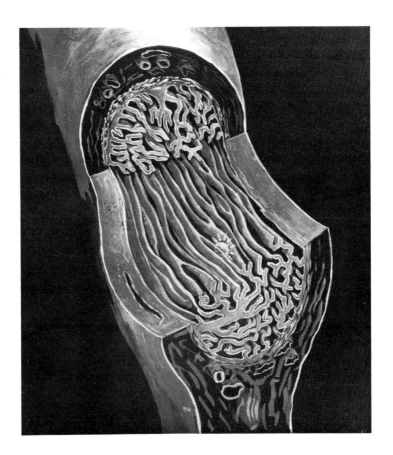

Fig. 18-41. Stereogram of uterine tube (oviduct) with zygote, zona pellucida and corona radiata.

teum regressum which involutes to form a large *corpus albicans* (Fig. 18-38H) that may persist for years.

Uterine Tube

The *uterine tube* (salpinx)[4] (Figs. 18-36, 18-39, A,B,C; 18-40 and 18-41) begins in the peritoneal cavity with the *abdominal ostium of the uterine tube* (Fig. 18-39, A). The fimbriated infundibulum of the uterine tube (opening), is the only place in the body where the outer world freely communicates with a body cavity. The uterine tube is lined with a tunica mucosa bearing a simple columnar epithelium containing ciliated epithelial cells and secretory epithelial cells with microvilli (Fig. 18-39, B).

The beat of the cilia towards the uterus protects the exposed peritoneum from infections, helps to propel the ovum or zygote toward the uterus, but at the same time created a stream against which spermatozoa must swim much as trout tend to swim upstream. The tunica mucosa is thrown into complicated folds grossly oriented in a longitudinal direction (Figs. 18-40 and 18-41). These folds are highest in the infundibulum and become progressively lower through the *ampulla tubae uterinae*, the *isthmus tubae uterinae* and the *pars uterina* (intramural). In the latter the folds (plicae tuberinae) are replaced by a few villi. The uterine tube has a cellular lamina propria mucosae, but no tela submucosa. The smooth myocytes of the muscularis, represented by an inner circular and outer longitudinal coat, is poorly developed, but nevertheless its feeble undulating movements assist transport of the ovum. The tunica serosa is

---

[4] The term salpinx is added as a synonym by the *Nomina Anatomica*, since most clinical expressions use this stem.

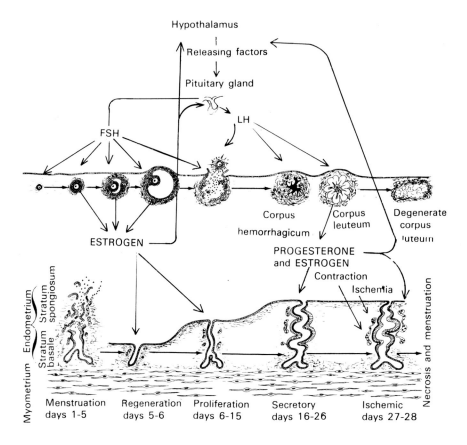

Fig. 18-42. Hypophysio-ovario-endometrial endocrine relationships.

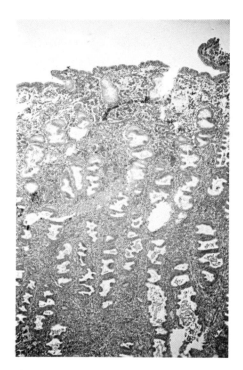

Fig. 18-43. Endometrium: First day of menstruation, H&E. 70 ×.

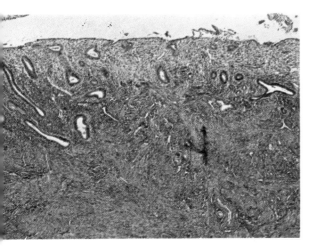

Fig. 18-44. Fifth day after menstruation: early repair, H&E. 70 ×.

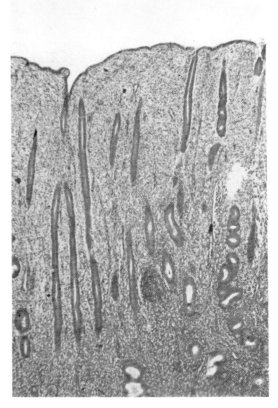

Fig. 18-46. 10th day after menstruation: late proliferative phase, H&E. 70 ×.

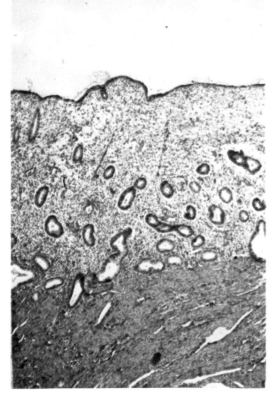

Fig. 18-45. Sixth day after menstruation: early proliferative, H&E. 70 ×.

loose and thick and blends with the mesosalpinx (Fig. 18-39, A). The uterine tube undergoes cyclic changes reflecting the phases of the menstrual cycle.

Uterus

The uterus (Fig. 18-36), a thick-walled muscular sac, is covered anteriorly, superiorly and posteriorly with a peritoneal *tunica serosa*, (*perimetrium*). A fibrous subserous coat, the *parametrium*, extends laterally between the layers of the broad ligaments. Internally the uterus is lined by a *tunica mucosa*, the *endometrium*, covered by a very thick *tunica muscularis*, the *myometrium*. The activity of the myometrium varies with the phases of the menstrual cycle (see below).

The *endometrium* of the *corpus uteri* (body) and *fundus uteri* possesses a simple columnar epithelium and helicoid, simple tubular glands. It undergoes monthly cyclic changes (Figs. 18-42, through 18-52) under the control of hormones secreted by the hypothalamus, the adenohypophysis, and the ovary. The average length of the

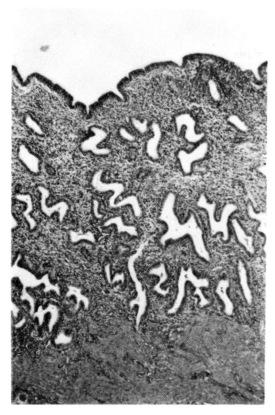

Fig. 18-47. 14th day, time of ovulation, early secretory phase, H&E. 70 ×.

Fig. 18-48. 14th day, ovulatory bleeding, another place from the same specimen as Fig. 18-47. This bleeding is spotty, H&E. 70 ×.

menstrual cycle is 28 days which by convention, is divided into five phases: *menstrual* (days 1-5), *postmenstrual (regenerative* days 5-6), *follicular (proliferative* days 6-15), *luteal secretory*, days 16-26) and *ischemic* (days 27-28).

This schedule is based on averages, and phases of endometrial growth are not exactly the same in every individual. The photomicrographs Figures 18-43 through 18-52 are accurately timed, but since they were taken from various patients, they do not conform exactly to the above schedule.

The epithelium of the tunica mucosa is simple squamous to low cuboidal during the postmenstrual and early follicular phases. Later in the cycle it becomes simple columnar with patches of ciliated cells. By convention, the lamina propria mucosae is divided into a superficial *stratum compactum endometrii*, a middle *stratum spongiosum endometrii* and a deep *stratum basale endometrii* (Fig. 18-49, B). The stratum compactum and stratum spongiosum together constitute the *stratum functionale endometrii* which is shed at the time of menstruation along with the overlying epithelium. The stratum basale, consisting of dense cellular connective tissue, penetrates the myometrium and is supplied by straight arteries. The stratum compactum and stratum spongiosum, composed of areolar connective tissue, are supplied by coiled arteries. The contraction of the latter, about the twenty-seventh day of the cycle, produces an ischemia followed by necrosis which results in a sloughing off of the stratum functionale (compactum and spongiosum). Regeneration occurs from bases of glands which remain in the stratum basale (Figs. 18-44 and 18-45). The lamina propria mucosae swells and temporarily straightens out the glands during the follicular (proliferative) phase (Fig. 18-46).

During the early luteal (secretory) phase (days 14-16, Figs. 18-47 to 18-49, B), glycogen accumulates in the basal portions of the glandular cells, displacing the nuclei distalward. Later (day 24), this secretion leaves via the cellular apex (Figs. 18-50 and 18-52), to produce an energy-rich medium in which implantation of the zygote may occur.[5]

HORMONAL RELATIONSHIPS. The functional activities, morphology and development of the female reproductive system are influenced and controlled by an intricate relationship of hormones produced by the hypophysis, hypothalamus, ovary and placenta (Fig. 18-42). The first few pubertal *menstrual cycles* and the last few

---

[5] The timing given in the text refers to averages. The timing of the figures refers to specific cases.

Fig. 18-49. A, Early secretory phase: 16th day. Glycogen-vacuoles in basal part of cells, H&E. 940 ×.

prior to *menopause* are usually anovulatory. On day one of the *menstrual phase* (Fig. 18-43), *follicle stimulating hormone* (FSH) is produced by certain basophil gonadotrophs of the hypophyseal pars distalis. FSH stimulates growth and differentiation of several ovarian follicles (arrested in meiosis I diplotene), but only one achieves maturity and prepares for ovulation. FSH production and its influence on developing follicles continues until day 14 of the menstrual cycle (Fig. 18-47, and Fig. 18-48) when the mature follicle erupts at the ovarian surface. During ovocyte development, follicular cells plus theca interna cells secrete the powerful steroid hormone, *estrogen*. Under its influence the uterus, vagina, vulva and breasts develop during puberty and continue to reflect estrogen dependancy until menopause. *Estrogen*, responsible for growth of pubic and axillary hair, controls the *proliferative phase* (Figs. 18-45, and 18-46) by stimulating growth and proliferation of the endometrial stratum functionale and influencing the hypothalamus to secrete *luteinizing hormone releasing factor* (LHrf). LHrf acts on gonadotrophic cells of the pars distalis to release *luteinizing hormone* (LH). On day 14, the titer of LH in blood is sufficient to cause ovulation and initiate formation of the *corpus luteum*. The *proliferative phase* (Fig. 18-45) is characterized by (1) growth of straight tubular glands (Fig. 18-46) from basilar remnants in the stratum basale and (2) secretion of an abundant serous fluid around day 8. On day 10 (Fig. 18-46), glycogen granules appear in the basal cytoplasm of the glandular columnar cells (Fig. 18-49A).

During the *secretory phase* (Fig. 18-49B), LH maintains the corpus leteum and stimulates it to produce the hormone *progesterone* which function in concert with *estrogen*. Progesterone prepares the uterine mucosa to receive a fertilized ovum and influences the pars distalis to cease production of FSH. Preparation of the

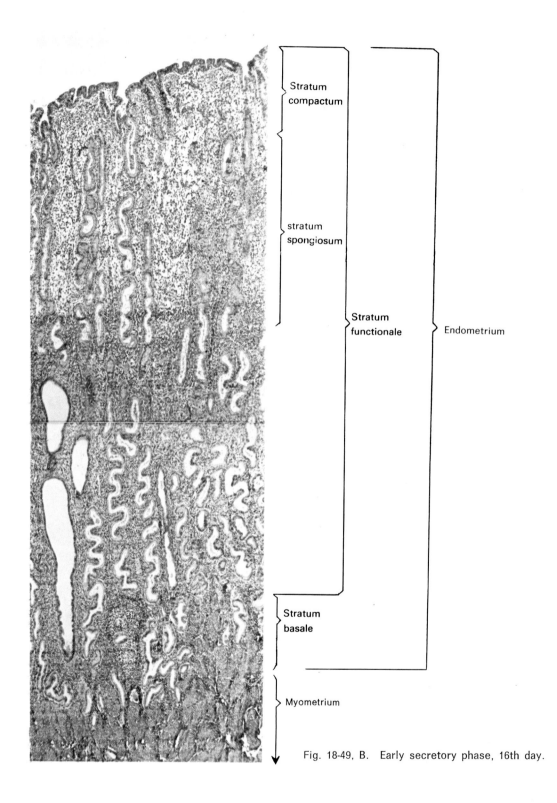

Fig. 18-49, B. Early secretory phase, 16th day.

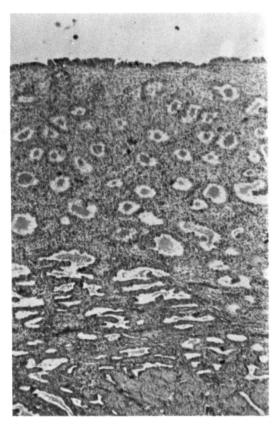 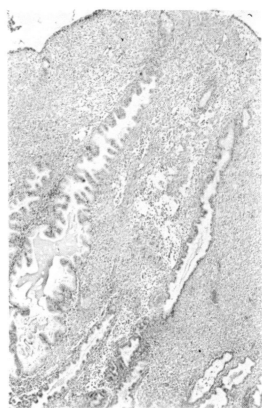

Fig. 18-50. 22nd day: late secretory phase, H&E. 70 ×.

Fig. 18-51. 28th day, premenstrual ischemia, H&E. 70 ×.

Figs. 18-43 to 18-52 are photographed from slides prepared in the laboratory of Prof. Gisela Dallenbach-Hellweg, Frauenklinik, Städtische Krankenanstalten Mannheim.

uterus involves movement of basal glycogen granules into the apical cytoplasm of the glandular epithelium (Fig. 18-52) and then into the glandular lumena already distended with fluid and mucus. The lacelike appearance of the sectioned stratum functionale (Fig. 18-50) is attributed to profiles of coiled sacculated glands and edematous connective tissue. The spongy uterine mucosa forms a most receptive implantation site for the fertilized ovum.

Following the secretory phase, a rapid series of events ushers in the *ischemic phase* (Fig. 18-51). Failure of fertilization dooms the ovum to autolysis because the corpus luteum, with a limited life span of 14 days, ceases production of progesterone. The hormonally deprived uterus reacts with rhythmic contractions of the superficial coiled arteries and constriction at their bases in the stratum basale, while white blood cells exit the vessels and pervade the connective tissue. Deprived of blood, localized ischemic necroses occur in the stratum functionale to initiate the short *ischemic phase* (Fig. 18-51). Necrosis provokes *menstruation* or hemmorhage of blood mixed with endometrial debris. The first sign of menstrual bleeding is reckoned as day 1 of the menstrual phase (Fig. 18-43).

Recapitulation: (1) *FSH* stimulates development of ovarian follicles. (2) Follicular epithelium plus theca interna cells secrete *estrogen*. (3) Estrogen indirectly stimulates secretion of *LH* (via *LHrf*). (4)

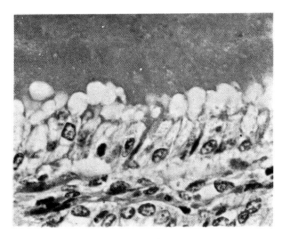

Fig. 18-52. Late secretory phase. Glycogen vacuoles at the apical end of the cells. H&E.

LH stimulates ovulation and formation of the corpus luteum. (5) Secretion of *progesterone* from the corpus luteum interrupts production of *FSH*, negating development of new ovarian follicles. (6) Without *estrogen*, *LH* secretion is inhibited, (7) thus cutting off production of *progesterone* required for maintenance and growth of the uterine stratum functionale. (8) Finally ischemia sets in and menstruation occurs with the fragmental discharge of the stratum functionale.

Fertilization, transforms the corpus luteum into a *corpus luteum of pregnancy* (Figs. 18-38F, 18-38G, 18-39), which continues production of estrogens and progesterone during the first few months of pregnancy. About a week after follicular rupture and discharge of the ovum, the *zygote* implants in the uterine wall and the *placenta* begins to form. The *syncytiotrophoblast*, the external portion of the placental trophoblast, burrows into the spongy mucosa by enzymatic activity and produces the hormone *human chorionic gonadotrophin* (HCG) which prevents the corpus luteum from regressing. Indeed the corpus luteum of pregnancy continues growth and secretion until the placenta is fully capable of independent function.

CERVIX. The cervix of the uterus (Figs. 18-36 and 18-53) is covered by tunica adventitia for most of its length, and only postero-superiorly is there a tunica serosa. The tunica mucosa exhibits *plicae palmatae*, typical ridges called palmate folds (Fig. 18-54). Near its junction with the vagina (*portio prevaginalis*). The epithelium is stratified squamous but deeper (i.e., more superiorly). The cervical canal, is lined with simple columnar epithelium of two cell types: (1) mucous exocrine cells and (2) ciliated epithelial cells. The cervix contains branched, alveolar glands lined by epithelium of simple columnar

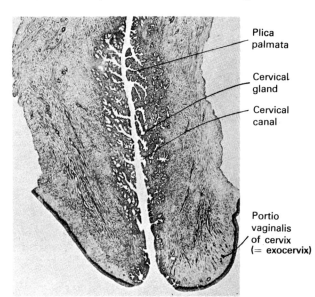

Fig. 18-53. Cervix uteri, showing mucous glands. (Stieve: Arch. Gynaek. 183: 178, 1952.)

UTERUS 499

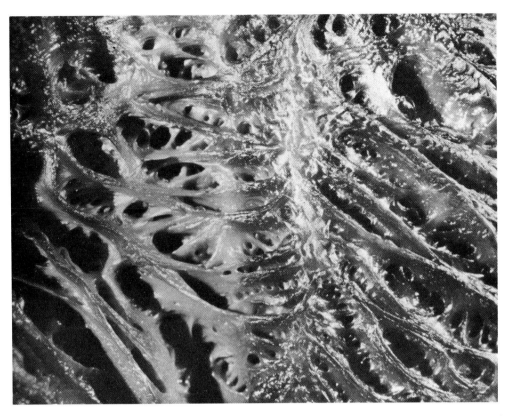

Fig. 18-54. Plicae palmatae. The depressions between these folds are exits from cervical glands. The folds prevent the mucous plug from slipping out.

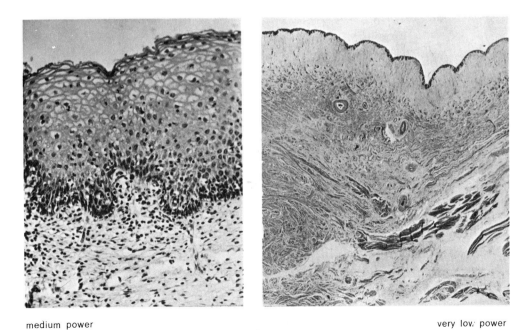

medium power   very low power

Fig. 18-55. Vagina.

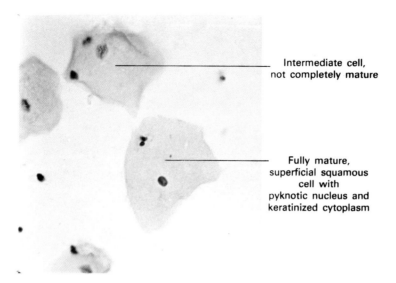

Fig. 18-56. Normal Pap smear of human vagina (courtesy of Dr. C. Uyeda, University of Arkansas for Medical Sciences). The technic of exfoliative cytology was developed by George N. Papanicolaou M.D. (1883-1962), hence term Pap smear.

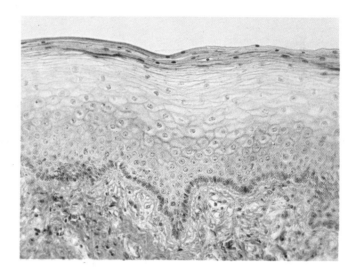

Fig. 18-56, A. Vaginal epithelium under estrogen influence.

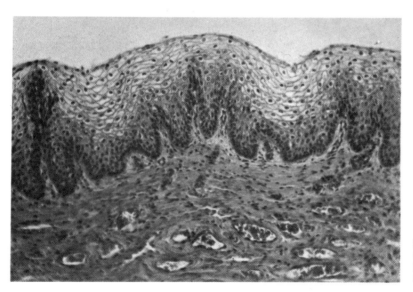

Fig. 18-56, B. Vaginal epithelium showing immature, parabasal cells the superficial layer. No hormone influence.

mucous exocrine cells (Fig. 18-39, D). The cervical glands appear quite different in structure from typical mucous glands because the epithelium is simple columnar, with spherical nuclei, just as in the urethral glands (Fig. 180-27). The depressions between the plicae palmatae (Fig. 18-54) provide receptacles for the openings of such glands. Secretion of mucus in layers occurs regularly during pregnancy (Fig. 18-39, D and E), eventually creating a "mucous plug". Cross sections of cervical glands, such as seen scattered in Figure 18-39, D, were once thought to be eggs (Naboth's ovules).

Vagina

The vagina is a musculofibrous tube lined with a stratified squamous mucous epithelium which is supported by a lamina propria mucosae of dense connective tissue (Fig. 18-55). There are no glands, per se, in the lamina propria mucosae of the vagina. However, its deeper, more loosely arranged part contains many large vessels responsible for the copious amounts of vaginal lubricant produced during sexual excitation (heretofore attributed to the cervical glands or to the greater vestibular glands). Some of the vaginal lubricant consists of desquamated watery epithelial cells. Although the smooth muscle tunica muscularis surrounding the vagina is not arranged in discrete layers, there is a predominent *stratum* (layer) *longitudinale* and a less definite *fasciculus* (bundle) *circularis*. With the exception of the posterior fornix (Fig. 18-39, F) which is covered by peritoneum, the vagina is surrounded by a tunica adventitia.

Estrogen causes the vaginal epithelium to be thick. Under its influence the cells that desquamate are fully mature, superficial squamous cells with tiny pyknotic nuclei (Figs. 18-56 and 18-56, A). Progesterone inhibits the maturation process of epithelial cells, while intermediate cells, or immature cells are shed (Fig. 18-56). In the absence of estrogen or progesterone, the cells that desquamate are immature parabasal cells (Fig. 18-56, B). Vaginal smears obtained on day 12 or 13 of the

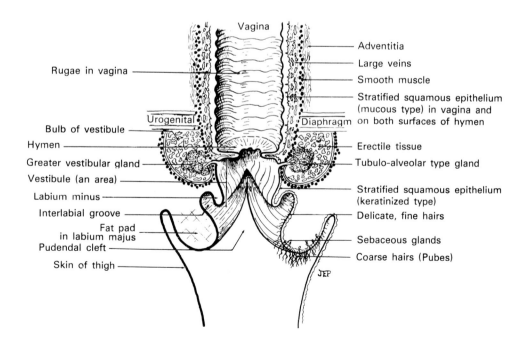

Fig. 18-57. Frontal section through female external genital organs.

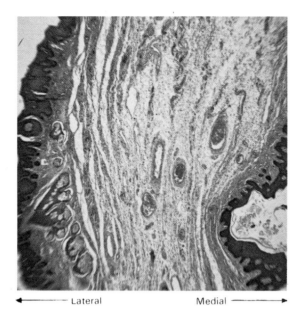

Fig. 18-58. Labium minus, H&E. 55 ×.

menstrual cycle (high estrogen) have a ratio of 0 parabasal/40 intermediate/60 superificial cells. Smears obtained on day 25 or 26 of the menstrual cycle show a maturation index of 0/70/30, a shift to mostly intermediate cells.

Female External Genital Organs

The *vulva* (*pudendum femininum*), female external genitalia (Fig. 18-57), include the vestibule of the vagina, the labia minora pudendi, the clitoris, the bulb of the vestibule, the *greater vestibular glands* (*Bartholin*), the labia majora pudendi and the mons pubis (mons veneris). The vagina communicates with the external genitalia, via the *vestibule* which is the space between the labia minora. Each *labium minus* (Fig. 18-58) is a delicate fold of connective tissue devoid of fat, covered on both surfaces with keratinized, stra-

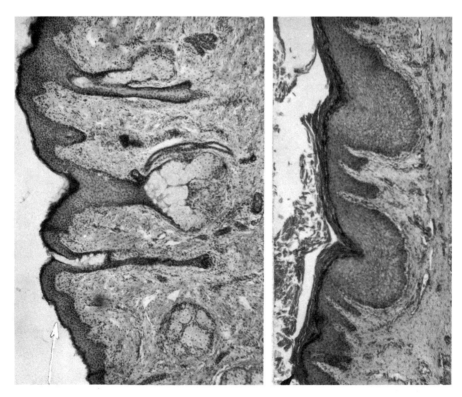

Fig. 18-59. Lateral surfaces of labium majus (left) and labium minus (right).

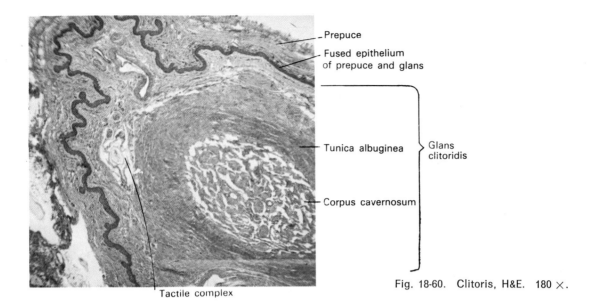

Fig. 18-60. Clitoris, H&E. 180 ×.

tified squamous epithelium. Although no hair is present, sudoriferous (sweat) and sebaceous glands occur occasionally in the lamina propria mucosae of the lateral side (Fig. 18-58, left). In most cases, however (Fig. 18-59, right), even the lateral side of the labium minus is devoid of glands. The labium majus, especially its lateral side, is covered with typical skin containing pilosebaceous complexes and sudoriferous (sweat) glands (Fig. 18-59, left). Anteriorly the labia minora come together to form a little hood (prepuce) over the clitoris and a frenulum under it. The *clitoris*, a small finger-like extension of the deeper lying bulbs of the vestibule and the more lateral crura, is the minature homologue of the penis. Although not traversed by a urethra, the clitoris, its bulb and crura have a core of erectile tissue (Fig. 18-60). The erectile tissue in the core of the clitoris, the corpora cavernosa clitoridis, has the same structure as the corpora cavernosa of the penis and similarly is surrounded by a strong tunica albuginea. Outside the tunica albuginea, there are numerous nerves and tactile corpuscles (Figs. 18-60 and 18-61). Near the base of the clitoris, the epidermis of the prepuce and of the frenulum is fused with the epidermis of the glans clitoridis so that the potential preputial cavity is obliterated (Fig. 18-61). In the nulliparous woman, the labia minora typically are hidden from view by two larger folds of skin lying on either side, the *labia majora*. After puberty their medial surfaces (next to the interlabial groove) are covered with fine, delicate hair and their outer surfaces with coarse hair. Each labium majus contains a thick pad of fat. The *mons pubis* (mons veneris) is a thick pad of fat over the symphysis pubis. After puberty its skin is covered by coarse hairs (pubes) continuous with those of the labia majora.

GREATER VESTIBULAR GLANDS. Each major (greater) vestibular gland (Bartholin's gland), tubulo-alveolar in type, is located posteriorly at the base of each labium minus. Their ducts open into the vestibule just posterolaterally to the vaginal orifice.

Breast

The breast (mamma) (Figs. 18-62 through 18-69) is a portion of integument specialized for milk production. Its growth is controlled and its function initiated by

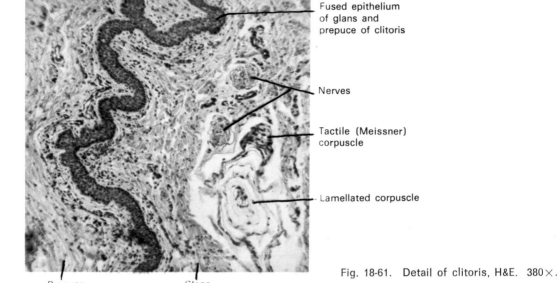

Fig. 18-61. Detail of clitoris, H&E. 380×.

hypophyseal, ovarian, placental and suprarenal hormones. The epidermis of the nipple and the surrounding areola is pigmented in the stratum basale and possesses small papillae and deep rete ridges. Each lobe of the mammary gland has an individual duct lined with an epithelium which varies from stratified squamous near the exit to simple cuboidal towards the alveoli.

At the time of puberty, rising titers of estrogen stimulate the growth and development of the fibro-fatty stroma and ducts. The areolae enlarge and become more pigmented. The breast of a nullipara (Figs. 18-64, 18-65 and 18-68) consists mainly of connective tissue. The smooth contours are due largely to the presence of fat. Ducts, without lumina, occupy minimal space. During pregnancy, alveoli sprout (Fig. 18-66), lumina appear (Fig. 18-68), and secretion of *colostrum* begins. Colostrum is a white opalescent liquid secreted for a period of one to three days at the termination of pregnancy. This fluid with rather important laxative properties, is rich in lactoproteins and antibodies, but poor in lipids. During *lactation* (Fig. 18-67), the breast is an apocrine gland. Milk production is maintained by the mechanical influence of the suckling infant. Figure 18-69, A, B and C shows mammary glands before and after lactation. Smooth myocytes are present in the nipple and around the ducts, and basket-shaped "myoepithelial" cells surround the alveoli.

Histologically, the mammary gland of the infant at the time of birth (Fig. 18-63) resembles that of its mother, because it is under the influence of maternal hormones which have crossed the placenta. The neonatal breast of either sex may even produce a secretion called "witches milk", a colostrum-like milk.

During the menstrual cycle, estrogen and progesterone cause fat deposition, stromal development, growth of lobules, ducts and alveoli. Thus the breast enlarges and becomes finely nodular. During menses and until day 10 of the menstrual cycle, the alveoli and ducts regress and the breast decreases in size.

The breast of the male is similar but smaller to that of a nullipara. Its connective tissue is mainly of the membranous type (Fig. 18-70).

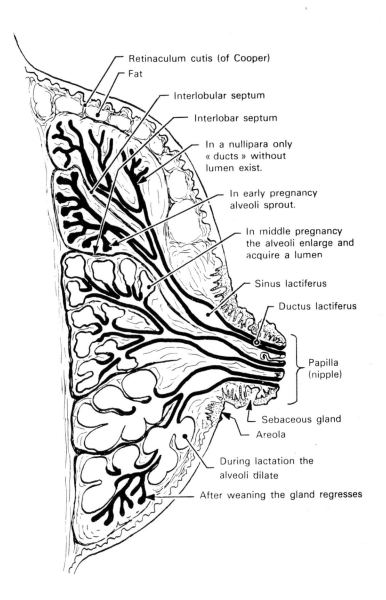

Fig. 18-62. Diagram of mamma.

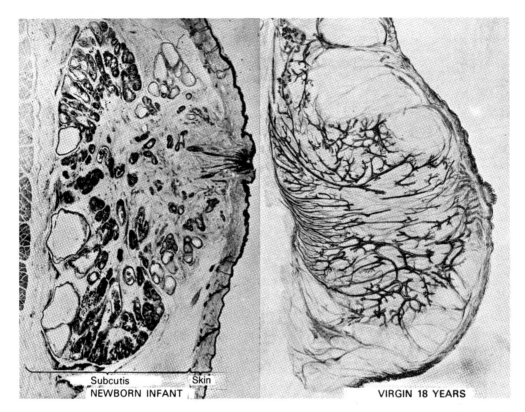

Fig. 18-63 (left). Breast of a neonata. Fig. 18-64 (right). Breast of an 18 year old nulli-gravida. (A. Dabelow: Die Milchdrüse. In Handbuch Mikr. Anat. d. Menchen, 3, 3, 1957.)

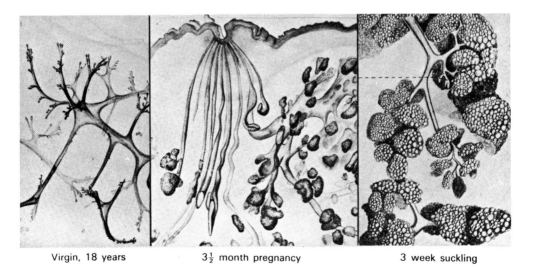

Fig. 18-65 (left). Inactive mammary gland of an 18 year old girl. Fig. 18-66 (center). Mammary gland from woman three and one half months pregnant. Fig. 18-67 (right). Mammary gland, 2 days after weaning. During lactation the alveoli are still larger, the lobules almost touch each other. (A. Dabelow: Die Milchdrüse. In Handbuch Mikr. Anat. d. Menchen, 3, 3, 1957.)

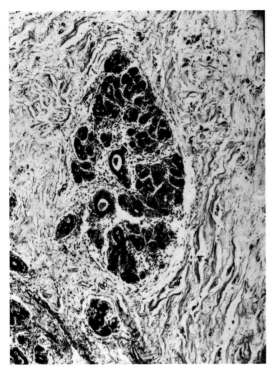

Fig. 18-68. Mammary gland of early pregnancy.

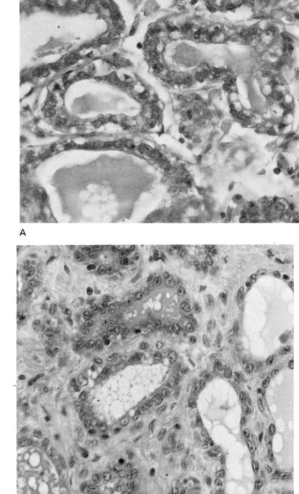

Fig. 18-69. Mammary glands before and after lactation, H&E. 750 ×. A: 6th month of gestation; B: 8th month of gestation; C: After weaning.

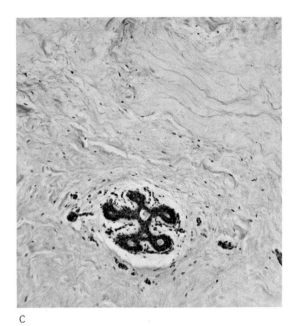

C

Fig. 18-70. Mammary gland of adult man, H&E. 750 ×.

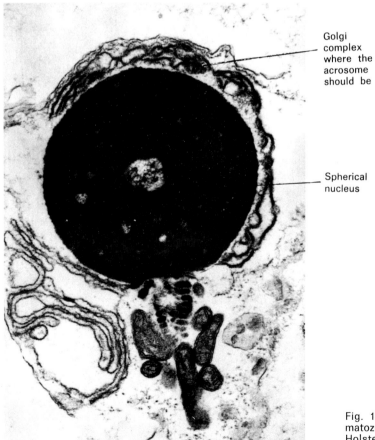

Golgi complex where the acrosome should be

Spherical nucleus

Fig. 18-71. An abnormal human spermatozoon. Courtesy of Prof. Dr. A. F. Holstein, University of Hamburg.

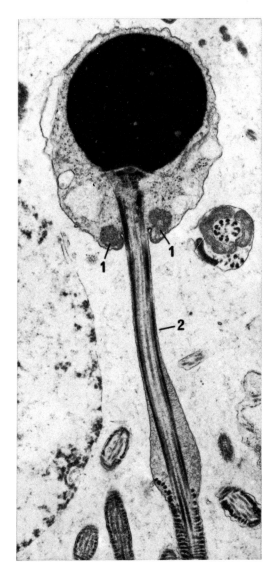

Fig. 18-72. Abnormal sperm from a 33-year-old man. 1. anulus; 2. tail without mitochondria (From A. F. Holstein, Morphologische Studien an abnormalen Spermatiden und Spermatozoen des Menschen. Virch. Arch. in press.)

Appendix I

MICROSCOPES

The objects studied in microscopic anatomy are extremely small. They cannot be seen with the naked eye. Therefore, they must be magnified by optical means.

"Magnification" is the ratio between two angles of vision. It is not the same as "enlargement" which means the ratio of the absolute diameter of an image to that of the object, such as the ratio of the width of an image on a motion picture screen to that of the frame on the film. The angle of vision can be increased by approaching the object (Fig. A-I-1). The wider the angle of vision subtended by an object is, the more detail is seen. The "simple microscope" commonly called "magnifier" permits close approach to the object.

There are basically two kinds of microscopes: light microscopes with a maximum useful magnification of 1000 x and electron microscopes with a maximum enlargement of 200,000 x. (In electron microscopy as well as in photomicrography, the term magnification, as defined above, is inappropriate, because the image is not seen with the eye, but projected on a screen or film. We are not dealing with ratios of angles of vision but with ratios of diameters.)

Light microscopes, also called optical microscopes, use the phenomenon of refraction. This means bending of light rays at the interface between 2 media of different optical densities (refractive indices). The refractive index of air or of a vacuum has the conventional value $n = 1$.

When a beam of light traveling in one medium strikes the surface of a medium of different density, forming the angle α with a line perpendicular to that surface (angle of incidence), it continues its travel into the second medium but forms the angle β with the perpendicular (angle of refraction). The ratio of the sine of the angle of incidence to that of the angle of refraction equals the ratio of light velocities in both media (Snell's law). If the first medium is air, then

$$n = \frac{\sin \alpha}{\sin \beta}$$

is the refractive index of the second medium (see Fig. A-I-2). If the body made of the second substance is a transparent slab with two parallel boundary planes, and if air is on both sides, the beam of light exits on the other side forming again the

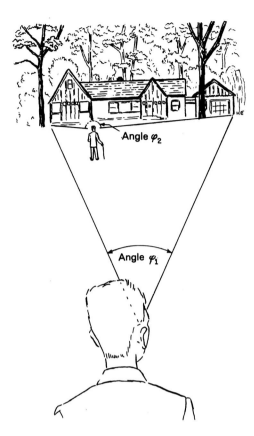

Fig. A-I-1. The sharpness of detail seen in an object depends on the angle between the lines which connect the extreme points of the object with the eye of the observer (angle of vision = apparent length). A person standing at some distance from a house sees it under an acute visual angle. He may recognize its general shape. A person viewing it from near by, i.e., under an obtuse visual angle, recognizes individual shingles, even the heads of nails driven into the siding. Magnification amounts to widening the angle of vision.

angle α. The beam thus has been shifted parallel to itself.

Simple lenses

More interesting is the case in which the second medium has the shape of a prism (Fig. A-I-3). The surface of exit is not parallel to the surface of entrance. Thus the direction of the beam is altered.

Microscopes consist of lenses. Schematically a biconvex lens can be thought of as a series of prisms (Fig. A-I-4), arranged so that parallel rays entering on one side converge on the other side in the posterior focal point F_2 of the lens*. The distance of that point from the center of the lens is called the focal distance f of the lens.

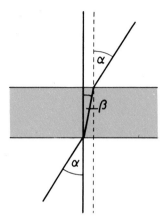

Fig. A-I-2. Refraction of a light ray when passing a plane sheet of relatively high optical density
$$n = \frac{\sin \alpha}{\sin \beta}.$$

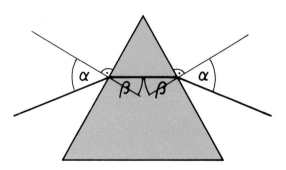

Fig. A-I-3. Refraction of a monochromatic light ray by a prism.

If the rays emerge from a point O (Fig. A-I-5), they converge in a point I on the other side of the lens. O stands for object, I for image. If we designate the distance of the object point from the optical axis

---

* The word focal point derives from focus = fire; for if you hold a combustible object such as a cigarette into the focal point of a large lens and direct sunrays on it, it will begin to burn.

as o and that of the image point as i, then the ratio $\frac{i}{o}$ is called enlargement or reduction. The closer O is to the first focal point, the greater will be the enlargement. Enlargement takes place only, if O is located at a distance $f < d_o < 2f$ from the center of the lens. The distance of the enlarged image from the center of the lens $d_i > 2f$.

Compound Miscoscope

The compound microscope (Fig. A-I-6, A), placed between an object and the eye,

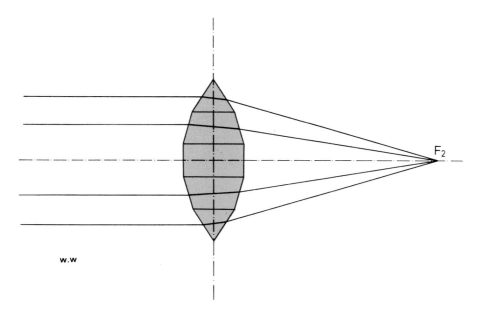

Fig. A-I-4. A lens considered as a series of prisms.

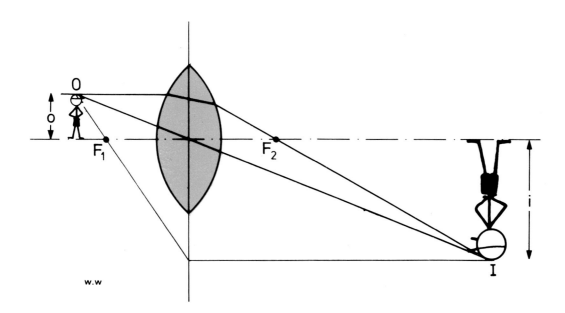

Fig. A-I-5. Production of an image by a lens.
(Enlargement or reduction)

widens even more than a single lens (magnifier) the angle under which the object appears to be seen. The objective forms an enlarged, real image of the object which cannot easily be observed with the eye. The eyepiece (ocular) refracts the rays so that they *appear* to come from a further enlarged, virtual image. By different focussing (Fig. A-I-6, B), a real image can be projected on a film or screen.

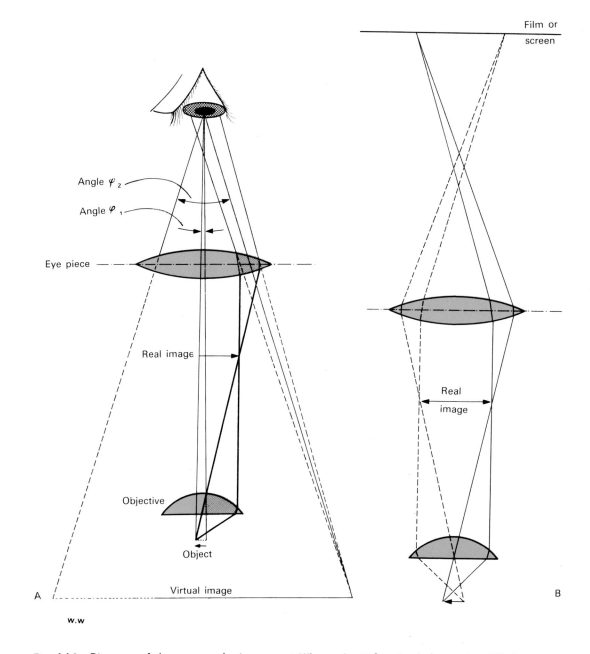

Fig. A-I-6. Diagrams of the compound microscope. When using it for visual observation (A) the microscope is focused so that the backward prolongations of light rays emitted from a point of the object cross in an imaginary point. Thus, an image appears to exist where actually none is located (virtual image). Remember that also the image seen through a plane mirror is a virtual image. For photomicrography or projection (B), the microscope is focused slightly higher, in order to produce a real image on screen or film.

In the Huygens eyepiece (Fig. A-I-7), a field glass bends the rays together to avoid loss of the marginal parts of the field. It also contributes to color correction. The real image appears above the field glass in the plane of a diaphragm. It is viewed through the eye lens which acts as a magnifier. Ocular micrometers, counting devices, and pointers, when mounted on that diaphragm, may be seen clearly, together with the image of the object. Measuring eyepieces are Huygenian eyepieces in which the eye lens is mounted in a thread so that it can be accurately focussed on the measuring or counting disk that lies on the diaphragm. These devices can not be used in most eyepieces of other types. When a Huygens eyepiece is used, the rays cross above the eye lens in a small circle called the exit pupil.

The diaphragm in the Huygens eyepiece, in addition to holding measuring and counting disks, creates a crisp outline of the field of vision. It also eliminates the peripheral part of the field.

To regain the outer image area, wide field eyepieces are being promoted. They are of usefulness in combination with plan-achromats and plan-apochromats (to be described below).

Aberrations

The main defects of simple lenses are: spherical aberration, chromatic aberration and curvature of the field. *Spherical aberration* is due to the fact that the surfaces of manufactured lenses are parts of spheres. If lenses with paraboloid surfaces could be economically ground, spherical aberration would not be a problem. In the simplified Figure A-I-4 parallel rays are drawn as if collected into one single focal point. In reality, the outer portions of a lens possess focal distances shorter than do its central

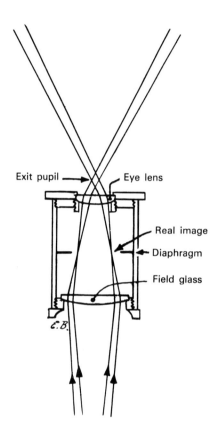

Fig. A-I-7. The Huygenian eyepiece.

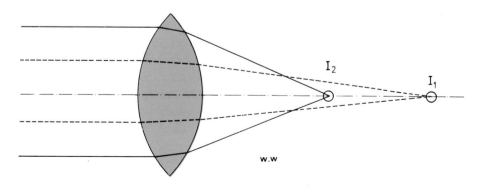

Fig. A-I-8. Spherical aberration.

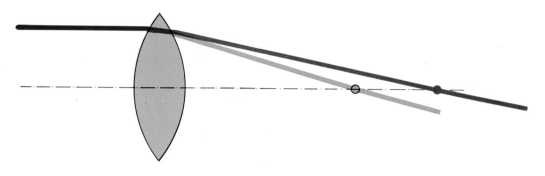

Fig. A-I-9. Chromatic aberration.

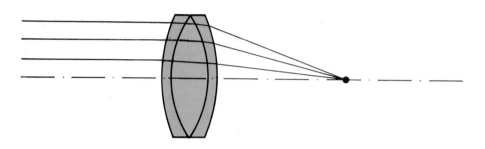

Fig. A-I-10. A lens constructed of two or three kinds of glass to eliminate spherical and chromatic aberration.

portions (Fig. A-I-8). Thus a simple lens bounded by spherical surfaces produces a series of stacked images of different sizes, resulting in haziness.

Chromatic aberration (χρῶμα = color) is due to the fact that glass has a higher refractive index to short light waves than to long (Fig. A-I-9). This results in rainbow-like color fringes around the edges of images. The optical industry eliminates much of these defects by combining lenses of crown glass with lenses of flint glass (Fig. A-I-10).

Achromatic objectives (Fig. A-I-11, A), made of several lenses of crown glass and flint glass, eliminate chromatic aberration within the spectral range of green, yellow and orange, a degree of correction sufficient for visual observation; but for black and white photography, a green filter must be used to eliminate deep red, deep blue and violet rays when using achromatic objectives.

The very expensive, apochromatic objectives (Fig. A-I-11, B) and the slightly less expensive fluorite objectives are almost perfectly corrected for the entire visible spectrum by the insertion of a fluorite lens (stippled). No filter is needed when they are used for photomicrography. Since most photographic emulsions are insensitive to infrared, and because glass is opaque to ultraviolet radiation, no further corrections to wavelengths are needed.

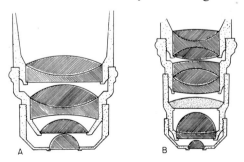

Fig. A-I-11. Longitudinal section through two oil-immersion objectives. A, Achromatic. B, Apochromatic.

Curvature of field. In spite of these corrections the image of a plane surface, such as a section mounted on a flat slide, is curved so that one must focus up and down to see clearly first the center and then the periphery of a field. In recent years, however, objectives and eyepieces have been improved so much that curvature of the field is no problem any longer.

Resolution

The *resolving power* of the microscope had been defined as the number of parallel lines individually visible in one mm. It equals

$$R = \frac{NA}{0.61 \cdot \lambda}.$$

λ is the wavelength of light. NA means numerical aperture which itself is defined as the sine of the half-angle (aperture) of the cone of light which enters into the objective from a point of the object (Fig. A-I-12), multiplied by the refractive index n of the medium between object and objective lens. Thus $NA = n \cdot \sin a$. For air $n = 1$; and for immersion oil $n = 1.52$.

More interesting than resolving *power* is tron-optical instrument*. By definition, d is the shortest recognizable distance be- is the shortest recognizable distance between two object points or pretty much the same as the diameter of the smallest particle or the width of the thinnest line just discernible.

$$d = \frac{0.61 \cdot \lambda}{NA}$$

or, to be more explicit,

$$d = \frac{0.61 \cdot \lambda}{n \cdot \sin a}.$$

From this formula it follows that for

---

* Frequently, among electron microscopists, the word resolving power is used instead of resolving limit, since they forget that the power of their instrument is greater the smaller the particles are which it resolves. The power of the instrument is usually giving thus: "The *resolution* is 3.5 Å" This designation gives immediately the desired information.

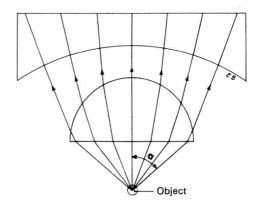

Fig. A-I-12. The aperture of a microscope objective.

visible light no particle can theoretically be visible, if its diameter is smaller than

$$d = \frac{550 \cdot 0.61}{1.52 \cdot 1} \text{ nm} = 220 \text{ nm}.$$

Actually this value is really too small, since sin a is always smaller than 1, even if the object touched the front lens of the objective. For practical purposes, then, the smallest object visible with the oil immersion measures about 1/4 of a micron.

The Condenser

If parallel rays enter into a high-power objective, only the central rays participate in image formation; the aperture is not fully utilized (Fig. A-I-13, left).

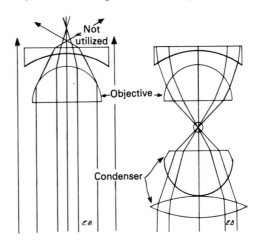

Fig. A-I-13 (left). The objective without condenser can utilize only a narrow beam of light. (Right). The condenser makes it possible that a very broad beam of parallel rays of light and the full aperture (angle a, Fig. AI-12) is utilized.

The condenser refracts parallel rays in such a manner that the entire aperture of the objective can be utilized (Fig. A-I-13, right).

Oil immersion

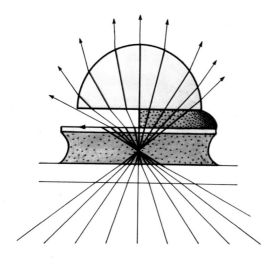

Fig. A-I-14. Comparison of a "high-dry" objective (left) with an oil immersion (right).

Figure A-I-14, right, illustrates the *oil immersion*. A drop of oil is inserted between the coverslip and the objective. Since oil, mounting medium and glass have almost the same refractive index, light rays travel on straight routes from the slide into the objective. Two refracting surfaces are eliminated. Further, the refractive index of oil is high. Thus, the value for NA is raised, and the resolving power of the objective is increased. Figure A-I-14 compares a highpower "dry" objective (left) with the oil immersion (right). To be orthodox, one should insert a drop of oil also between the condenser and the slide.

We stain histological materials differentially in order to impart different light-absorbing qualities to the structural elements of the tissues. For example, cytoplasm stained with eosin absorbs green light and transmits only red radiation, while nuclei stained with thionine are opaque to red, orange, yellow and green radiation but are transparent to blue light. It is the property of differential staining that permits microscopic observation of the finest detail. To utilize the full resolving power of the microscope, the condenser must be in the highest position, and the diaphragm must be open. Recognition of colors (important when pigments and vital stains are involved) is possible only with condenser high and diaphragm open.

When colorless objects are examined, differences in refractive index must be utilized while aperture must be sacrificed. The condenser is lowered or eliminated, or the diaphragm is narrowed.

With the condenser high and open (Fig. A-I-15, A), light passes unhindered through a transparent particle, which is therefore invisible.

Without the condenser, or with the condenser in low position or with the diaphragm narrowed, a thin bundle of almost parallel rays strikes the object. When focussing high, a particle of high refractive index, such as an oil drop, appears lighter than the background and surrounded by a sharp, dark outline (Fig. A-I-15, B); a particle of low refractive index, such as an air bubble, appears darker than the background and surrounded by a dark outline (Fig. A-I-15, C). At the interface between two media of different refractive indices, light is scattered by diffraction. The boundary appears dark (Fig. A-I-15, D).

Depth of focus

The depth of focus of a compound microscope is indirectly proportional to the product of the total magnification (magnification of objective times magnification of of eyepiece) and the numerical aperture. Depth of focus, therefore, can be increased by choice of a low-power eyepiece or by narrowing of the diaphragm. The latter process involves, obviously, loss of resolv-

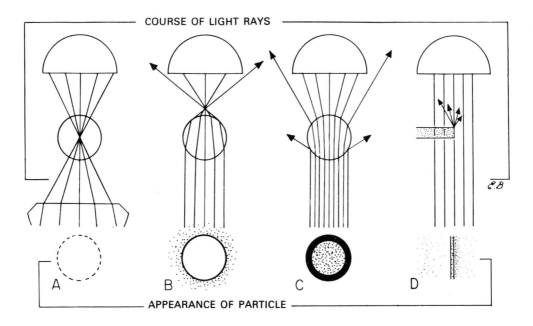

Fig. A-I-15. Diagrams to illustrate how colorless, transparent particles can be visualized by illumination with parallel rays, using differences in refractive index.

ing power. With high power objectives (when condenser is high and diaphragm open), the depth of focus is less than the usual thickness of the section. Attempts to obtain sharp photomicrographs of sections thicker than 3 micra are therefore often futile. Better results are obtained by drawing. Semithin sections of plastic embedded specimens ($\frac{1}{2} - 1\ \mu$) yield sharp images even with the strongest light-microscopic objectives. Semithin sectioning is also very useful in stereology.

Various problems in microscopy

Useless magnification. The magnification of a microscope can be increased by choice of stronger eyepieces. Since the resolving power depends on the objective only, strong eyepieces, though yielding larger images, do not show more detail than the weakest. Persons with good eyesight should use chiefly a Huygens eyepiece which magnifies five times. To relieve eye strain, a 10 × eyepiece is often useful, although it does not reveal more detail. Higher ocular magnification is of value only to persons with retinal deficiencies, beginning cataract, or astigmatism.

Microscope design. Even the largest and most impressive-looking microscope of modern design is not better than the simplest old-fashioned student model, if provided with modern lenses. For the optical quality of a microscope depends on the condenser, objective and eyepiece only. Inclined eyepieces are frequently detrimental to posture and cause back strain. Most built-in light sources deprive the viewer of the gentle, uniform light of the sky and of strong light needed for microprojection. Many modern devices impart rigidity to the microscope and detract from its portability.

The old-fashioned, inclinable, monocular microscope with a straight tube, however, is versatile, optically efficient and portable. It can be used for visual observation, projection, photography and drawing.

Figure A-I-16 shows a microscope built about the year 1910. This microscope, equipped with filter holder, oblique illumination device, rotating and mechanical stage, and a condenser and stage that can be centered can be carried about with ease. It is equally well adapted for the most critical visual observation as for photomicrography.

In Figure A-I-16 it is shown horizontally inclined for microprojection. The fact that it can be used upright or inclined at any desired angle makes it comfortable to use in many different situations. Unfortunately microscopes of this versatility are rarely manufactured today.

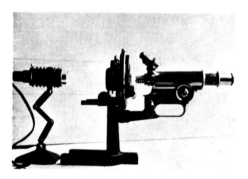

Fig. A-I-16. Old-fashioned microscope of high efficiency and versatility. In this illustration it is used as a microprojector.

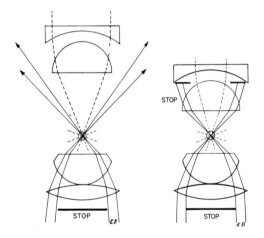

Fig. A-I-17 (left): Dark-field illumination with a low-power objective. (Right): Dark-field illumination with a high-power objective. This must be provided with an upper stop to block out peripheral light rays.

Fig. A-I-18. Shape of a metal disk which, when inserted into the substage, coverts an ordinary microscope into an ultramicroscope.

Ultramicroscope

An ultramicroscope or dark-field microscope is an ordinary, compound microscope equipped with a dark-field condenser (Fig. A-I-17). Direct light is prevented from entering into the microscope by an opaque disk blocking the central rays. Only side rays can strike the particle which scatters some of the light so that the particle is seen luminous on a dark background. Even a particle too small to be resolved can thus signal its presence without revealing, however, its size, color, and shape. The dark-field microscope is useful to study living protozoa and blood cells. These relatively large objects behave, in a dark field, as composites of millions of ultramicroscopic particles. High-power objectives, to be usable for ultramicroscopy, must be provided with a ringshaped diaphragm (upper stop) to prevent marginal rays from entering the objective (Fig. A-I-17, right). A ring such as shown in Figure A-I-18, bearing an opaque disk in the center, when inserted into the filter holder of the substage, will convert a conventional microscope into an ultramicroscope.

Polarization

Regularity of arrangement of submicroscopic particles can be revealed by polarization. A conventional microscope can be converted into a polarizing microscope by inserting a polarizer below the object and an analyzer above it. Figure A-I-19 shows the principle of polarization. Light swings in all directions perpendicular to the direction of the light ray (A). The polarizer (B) may be a Nicol prism or a polaroid film. It contains molecules in parallel arrangement. They act in such a way that only that component of light which swings parallel to the axes of these molecules can pass (C).

A second Nicol prism or polaroid film placed above the object is known as the analyzer (E). If the analyzer is placed in such a position that its molecules are parallel to those of the polarizer, we say, in laboratory language, "the Nicols are parallel". In this position, all the polarized light passes through the analyzer. If the molecules of the analyzer (E) run perpendicularly to those in the polarizer, i.e., when "the Nicols are crossed," all light is blocked. The field appears dark.

If a transparent, isotropic object is placed between "crossed Nicols", it remains invisible; but if an anisotropic or birefringent object (D), i.e., one containing molecules in parallel arrangement, is placed obliquely between crossed Nicols, it appears luminous on a dark background. Submicroscopic structure can be revealed by the polarizing microscope.

Figure A-I-20 shows a tangential section of facial skin of a man (labium superius) photographed under polarized light. It reveals regular arrangement of keratin molecules in the inner root sheaths of mustache hairs and parallel arrangement of molecules in collagen fibers.

Fig. A-I-20. Tangential section of facial skin from labium superius (mustache) photographed in polarized light.

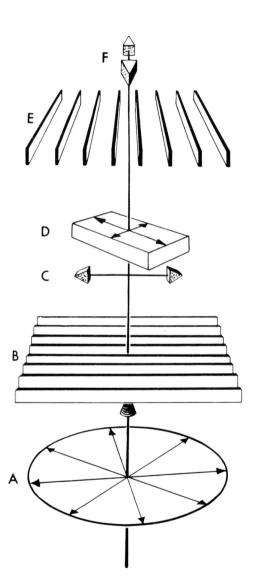

Fig. A-I-19. Diagram to illustrate the principle of polarization.

Phase Microscopy

According to Snell's law, the wave length λ of light is indirectly proportional to the refractive index n of the medium through which it passes (Fig. A-I-21). By proper choice of the thickness of a film of known refractive index, one can retard light by any desired fraction of its wavelength. Utilizing this phenomenon, a transparent ring of a highly refractive material of given thickness is placed onto the substage. and another ring of the same optical properties, but of different dimensions, can be placed onto the objective in appropriate position. Differences in the refractive indices in the object can thus be emphasized. Phase microscopy is used extensively for the study of living cells.

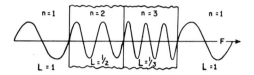

Fig. A-I-21. The velocity of light is inversely proportional to the density (refractive index) of the medium through which it travels. Thus, the wavelength is shortened in a denser medium. Phase microscopy is based on this phenomenon. (From Schmidt, W. J.: Handbuch Mikrosk.-Technik, Bd. 1, Teil 1, Frankfurt, 1956.)

A narrowed diaphragm (Figs. A-I-15 3-3 and 6-27) can acchieve an equally good effect. Oblique illumination which can be produced by displacing the condenser diaphragm sideways gives even better, three dimensional images (Fig. 9-36) than phase contrast.

Fluorescence Microscopy

Fluorescence is conversion of an eletromagnetic radiation into radiation of longer wave lengths. The invisible ultraviolet radiation, only usable with a quartz condenser and quartz slides, can be converted into visible light of specific colors by specific filters. Dark blue light is more frequently used, since it passes through glass. This light, only faintly visible, can be converted into bright green, yellow or red light by certain substances, emphasized by special filters. Fluorescence microscopy can reveal many important physiological and cytological processes not recognizable by ordinary illumination.

Camera Lucida

The camera lucida permits simultaneous viewing of a microscopic image and a piece of paper and thus makes it possible to trace accurately the outlines of that image. It consists of a "beamsplitter" mounted on the eyepiece and a mirror. Figure A-I-22 shows a homemade camera lucida. A coverslip (preferably round for safety) is pasted on top of the eyepiece with plastilin or chewing gum at a 45-degree angle. At a distance of 10 to 12 cm (4 to 5 inches) a mirror is mounted, on a ring stand, also 45-degrees inclined. Light from a piece of paper showing a pencil is reflected by the mirror onto the coverslip; and while most of it passes through the coverglass, about 10 per cent of the light is reflected into the eye of the observer. The coverslip permits passage of all the light from the microscope into the eye. Thus, the observer sees the preparation and the drawing superimposed on one another. Whenever dimensional precision is required in drawing, as, for example, for reconstructions from serial sections, the camera lucida is of essence. Counting and measuring grids and scales can be drawn on paper to be observed, together with the slide, through the camera lucida. In more sophisticated camerae lucidae the "drawing tubes" are focusable and provided with zoom devices.

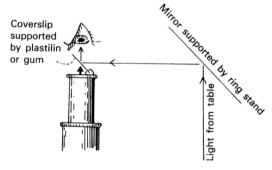

Fig. A-I-22. A home-made camera lucida.

Figure A-I-23 illustrates a few principles of microscopy.

Electron microscopy

Electron microscopy, because of the extremely short wavelength of electron beams, reveals detail more than 200 times finer than that visible in the light microscope.

Since 1952 electron microscopy has become indispensable to the microanatomist.

Its resolving power is far superior to that of the light microscope, because the wavelength of electron beams is only 0.04 nm. Basically, the instrument is built according to the same optical principles as the light microscope, but glass lenses are substituted by electromagnetic fields. The electron beams (wavelength $\lambda = 0.04$ nm $= 0.4$ Å, invisible to the eye, fall on a fluorescent screen so that the image becomes faintly visible.

Fortunately photographic emulsions are sensitive to electron beams, whereby detail is revealed on the negatives which the eye can not discern on the fluorescent screen.

Objects to be observed must be embedded in hard plastics and cut with glass or diamond knives. "Staining" is accomplished by impregnation with heavy metals to obtain contrast. The so-called ultra-thin sections, ideally 50 nm thick, are, compared with the enormous enlargements, thick slices. They are about three times thicker than the diameter of a synaptic vesicle or of a ribosome. The depth of focus of the electron microscope is so great that the entire slice called "ultrathin section" is in focus simultaneously. Estimation of quantitative data of small particles (10-50 nm in diameter) becomes very difficult because

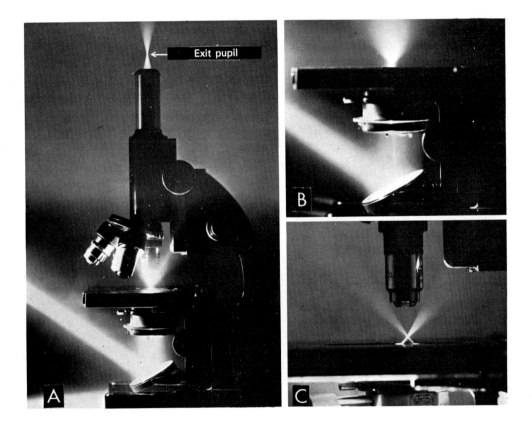

Fig. A-I-23. A, The illuminating system and the exit pupil of a microscope visualized by blowing smoke into the path of the light rays. B, Parallel light entering from the left is coverted by the condenser into a broad cone of light (above) the apex of which is located at the approximate level of the object, i.e., slightly above the stage to account for the thickness of the slide. C, An ordinary condenser has been converted into a dark-field condenser by pasting a piece of black paper below the center of the bottom lens of the condenser. No direct rays enter into the objective, but a particle located at the point where the rays intersect could signal its presence to the observer by scattering some of the light (diffraction). It would thus appear bright on a dark background.

of this depth of focus which is very pleasing from an esthetic point of view, but often leads to misinterpretations. At high primary enlargements of electron microscopy (higher than 10,000×), the rules of section stereology (see Appendix II) no longer apply, for the images of small particles resolvable at high magnifications are projections rather than sections. However, up to 10,000 × of primary enlargement, the thickness of the slices is still negligible; and section stereology still applies.

The very great depth of focus of the electron microscope is, however, of considerable advantage in the observation of surfaces, natural or fractured, and of opaque or pseudo-opaque specimens, such as in scanning electron microscopy and in freeze-fracture-etching.

Appendix II

QUANTITATIVE MICROSCOPY
(Stereology and Morphometry)

The most frequently used method of microanatomy is the observation of "sections" which are really very thin slices cut through a three-dimensional organ (Fig. A-II-1, A, C). They are produced with a microtome after the organ has been hardened. Mounted on glass slides (Fig. A-II-1, D), they are usually stained to emphasize contrast. A true section would be a plane without thickness (Fig. A-II-1, B). True sections can be used in the study of metals, because these substances are opaque; but the parts of living organisms (as well as rocks) are transparent enough so that the thickness of the slices must often be considered in histology as well as in geology. In histology, then, we deal with slices (Fig. A-II-1, C and D), but we call them "sections". Since the advent of electron microscopy, even metallurgists must struggle with the thickness of slices, since metal foils are transparent to electron beams.

An organ is composed of various parts (Fig. A-II-1, E). A section through an organ (Fig. A-II-1, F) produces a section of each part which the knife encounters (Fig. A-II-1, G). The word "section", therefore has more than one meaning. To make things simpler, modern microscopists and electron microscopists often reserve the word "section" for the entire slice; while the section through a specific component is called a "profile" of that feature. This useful term will be frequently employed below.

Until 1961 histological descriptions were based on intuitive interpretations of "sections". Histology was a descriptive science. Since the foundation of stereology in 1961, however, histology gradually approaches the state of an exact science.

For a physiologic and pathologic evaluation of histologic preparations, mere verbal description does not suffice. Such words as "enlarged", "shrunken" and "more numerous than normal" are no longer acceptable, except in preliminary reports. The pathologist and the histophysiologist must attempt to obtain information about size, shape and orientation of the parts within an organ. Also volume ratios and the number of parts per unit volume are often of interest.

In the following brief outline of quantitative stereology, procedures and formulae, but not their derivations, are given.

Basic Definitions

The words morphometry and stereometry are often used synonymously with stereology. In reality, they mean three different areas of mathematics. *Morphometry* is measurement of structures. A tailor who measures you for a suit practices morphometry. The focussing devices of many photographic cameras which measure distance are morphometric instruments. However, morphometry is often the aim of stereology. In morphometry we ask the questions "how large?" and "how many?". Often, stereology is the only method by which the answers to these questions can be obtained. *Stereometry* is the same as solid geometry. It is that part of mathematics which calculates volumes and surface areas of exactly defined solids, such as cubes, spheres, cones, etc.

Stereology is the three-dimensional interpretation of flat images (sections and projections) by the criteria of geometric probability. It can also be defined as extrapola-

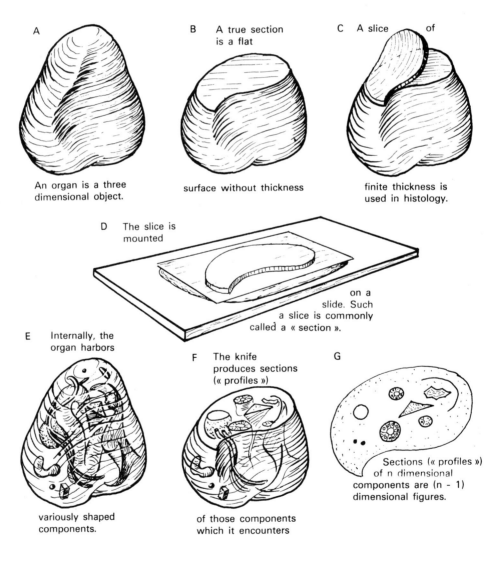

Fig. A-II-1. Basic facts about making sections.

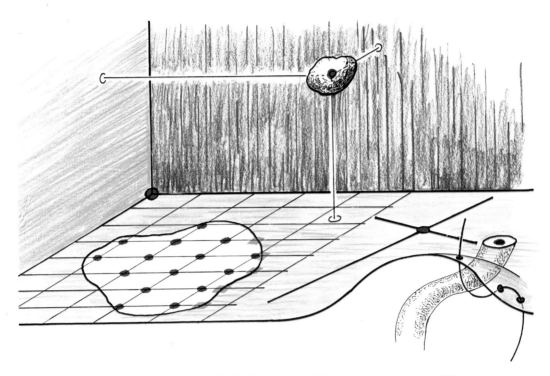

Fig. A-II-2. Various methods of creating, defining and using *points* (P).

tion from two- to three-dimensional space. It is a branch of applied mathematics; but it is a very easy science to learn. Its rules and formulae are of utter simplicity. They are not approximations, but mathematically precise. Most of them have been derived independently by several workers who were confronted with similar problems.

Stereology deals with points of zero dimensions (P), lines and their one-dimensional lengths (L), areas of two dimensions (A or S) and volumes of three dimensions (V). A point P may exist as a dimensionless spot in space, such as the center of a renal glomerulus, of a fat cell or of a lysosome. It may be a test point used for measurements. It can be generated by the intersection of two lines on a surface or by the intersection of a line with a surface in space or by the intersection of three surfaces in space. It can also be the center of a profile, (Fig. A-II-2). A line of length L may be straight or curved in space, such as the axis of a glandular tubule, of a skeletal muscle fiber, or the total length of all capillaries in a large organ. It may be a test line or the trace of a membrane or of a sheet, i.e. its intersection with the cutting plane, (Fig. A-II-3).

The letter A is used in stereology for a test area within which features are counted, such as a square or a circle of known size (Fig. A-II-4). S signifies the area of a curved surface in space, such as the basement membrane which surrounds a thyroid follicle (Fig. A-II-5).

The number of features counted in a test area is given by the letter n or P while the number of tissue components, such as pancreatic islands, nuclei, microbodies etc. per unit volume is N.

The volume V may be a unit volume, such as a cubic millimeter; it may be the total volume of the entire kidney or the volume fraction occupied by a component of the organ such as the glomeruli, or of

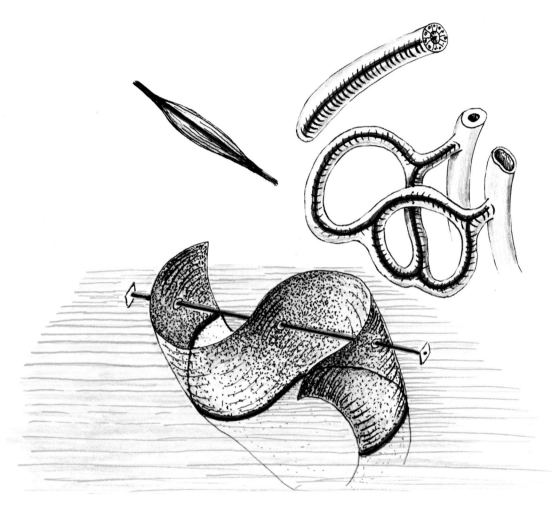

Fig. A-II-3. Various meanings of the terms *line* and *legth*, L or *l*.

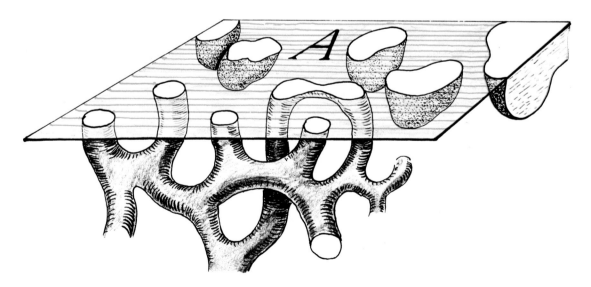

Fig. A-II-4. The use of a test area A to count profiles.

the medulla compared with cortex. It also may be the average volume of nuclei, mitochondria or lysosomes, etc. Since in stereology we extrapolate from two- to three-dimensional space and take all our measurements on plane surfaces, test volumes do not exist as operational devices.

Dimensional reduction. A true *section*, without thickness, *through an n-dimensional object produces an (n — 1) dimensional profile.* Thus, a point having no dimensions will have practically no chance to be cut at all by a section. But a granule may be caught between the two cutting surfaces of a slice. A line represented, perhaps, by a fiber yields a dot (point of zero dimensions) when cut by a plane. A surface such as an extremely thin membrane yields a line on section (one dimension). A solid, such as a sympathetic ganglion or a cell, for example, has an area as its profile.

Conversely: *An n-dimensional figure in a section* is, in general, *a profile of an (n+1)-dimensional structure.* (These relationships are illustrated in Figure A-II-6).

Terminology

Section: a plane without any thickness at all generated by one cut through a solid (Fig. A-II-7, A).

Slice: a histological or ultrathin "section" of finite thickness t generated by two cuts (Fig. A-II-7, B).

Slab: a portion of an object between two specified cutting planes. A slab may be opaque or translucent; for macroscopic anatomy it can be 1 cm thick, as for example a slice of brain as we use in neuroanatomy courses. A slab may consist of discarded, thin slices between two cutting planes. The distance between these two planes is, likewise, denoted by the letter t (Fig. A-II-7, C).

Profile: section or slice through an individual component of a solid (Fig. A-II-7, D).

Trace: section of a two-dimensional surface, such as a membrane, or an interface (boundary between two components of a space filling aggregate). A trace appears as a line (Fig. A-II-7, E).

Intercept: length of that segment of a test line, that falls on the area of a profile (Fig. A-II-7, F).

Intersection: point of crossing of a test line with a trace (Fig. A-II-7, G, also green dots in C).

Axial ratio: quotient of length over width of a profile (Fig. A-II-7, H).

The aim of stereology is to determine from sections and slices: three-dimensional shapes, orientation in space, volumes and volume ratios, curved surface areas, cur-

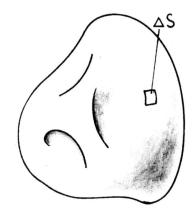

Fig. A-II-5. A curved surface of area S.

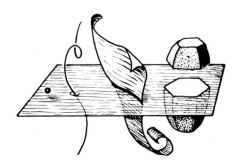

Fig. A-II-6. The principle of dimensional reduction: A plane, when cutting an n-dimensional structure, produces an (n — 1)-dimensional figure. The presence in the section of an n-dimensional figure indicates the existence in space of an (n+1)-dimensional structure.

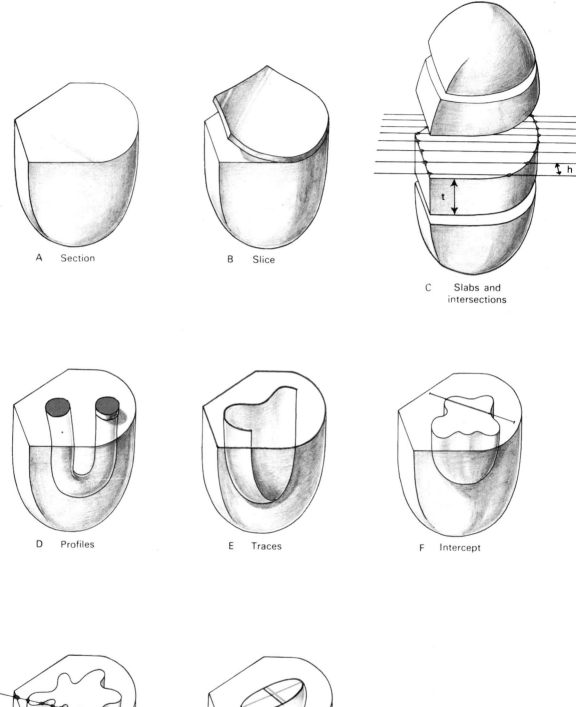

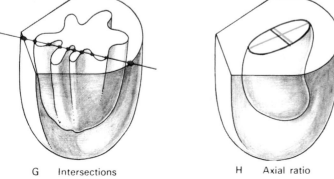

Fig. A-II-7. Stereographic explanations of basic terminology.

vature of surfaces, lengths of curved linear elements, size distributions of corpuscles or organelles, and number of particles per unit volume. We shall proceed in the order of difficulty of understanding, beginning with the easiest (volumes) and ending with the most difficult of problems (number per volume or volume density).

It is assumed in most phases of stereology that the tissue components are *randomly* arranged in space. If not, random sections must be employed. The following rules apply only to random sampling or to isotropic construction of an organ. Many organs, however, are anisotropic in structure which means that they show oriented arrangement of parts or gradients of density. Only determination of volumes and volume ratios are independent of arrangement.

Stereology is practiced by superimposing grids of lines or squares and circles over an image. This can be done in many different ways: A glass disk with an engraved pattern can be inserted into the eyepiece, a pattern can be reflected into the path of vision by a camera lucida or the slide can be projected on a paper on which the desired pattern is drawn. Also a pattern engraved on a sheet of plexiglass can be superimposed on a photographic print of the object. The latter method is frequently used in electron microscopy.

In addition to these simple methods there are others, such as automatic or semiautomatic scanning devices and so forth.

The choice of the specific pattern is a matter of taste and also may depend on the kind of material examined.

Figure A-II-8 shows a few such patterns.

Test patterns with many points, lines and fields, some times a few hundred, are often used by electron microscopists because of the minuteness of detail in their pictures.

While test patterns a — e employ squares as their basic shape, there are others which emphasize the equilateral triangle (f, g, h).

It has been shown experimentally that greater efficiency, that is, equal accuracy in less time, is achieved by a simple pattern, frequently and systematically displaced to new positions. The cause for this effect is that a simple pattern can be scanned faster by a human observer than a complicated one. Better yet is the use of a simple pat-

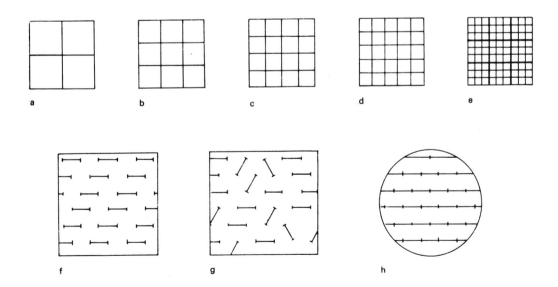

Fig. A-II-8. Some frequently used patterns for stereological point counts.

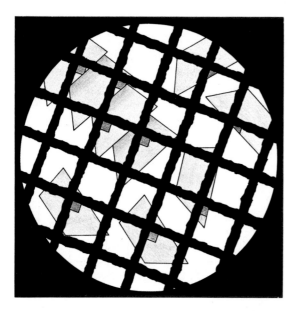

Fig. A-II-9. Unbiased sampling of observational areas in electron microscopy.

tern with systematic displacement of the specimen. Systematic displacement of the specimen is achieved in light microscopy by a mechanical stage with click-stops (sampling stage) which makes it impossible for the observer to focus on a point of interest. Independence from human bias is very important. In the absence of a sampling stage one can guard against bias by moving the specimen equal distances without looking into the microscope.

The writer of this appendix prefers pattern c, since it is coarse enough for easy observation and since its 25 points facilitate expressing the result in percentages.

Unbiased sampling in electron microscopy may be achieved by mincing fixed tissue into tiny blocks making it almost impossible that they be oriented in a desired direction. After cutting, the observer should not use only areas of special interest; instead he should decide to take a picture of every area which, for example, is located in the upper left hand corner of every square formed by grid bars (if a section is located there), (Fig. A-II-9).

Volumes and Volume Ratios

Direct volume determination. The easiest method of determining the volume of an organ is the immersion method (Fig. A-II-10). Put a container with water on a scale and weigh it. Then immerse the organ suspended with a very thin thread so that it is fully covered by water and does not touch the bottom of the container. The new weight, minus the weight of the container plus water, expressed in grams equals the volume of the organ expressed in cubic centimeters. Then, drop the organ on the bottom of the container and weigh again. The new weight minus water and container, divided by the volume is its specific gravity. After fixation, dehydration and infiltration with the embedding medium, the process should be repeated to determine shrinkage or swelling. The surrounding embedding medium must be blotted off before weighing. This will provide a correction factor for future stereological results.

Determination of volume by serial planimetry. After sectioning an organ serially,

Fig. A-II-10. Direct volume determination by the water displacement and weighing method.

one can measure planimetrically a *complete* series of slabs of equal thickness. If the area of slab i equals A_i, then the sum of all these areas multiplied by their common thickness equals the volume of the solid:

$$\sum_{i=1}^{n} A_i = V \qquad (1)$$

Direct planimetry is most easily performed by superimposing a grid of test-points in square arrangement over the slab. Any specific test-point hits or does not hit the structure to be measured (Fig. A-II-11). Each test-point corresponds to the area of the little square formed by four test-points. Hits (red dots in Fig. A-II-11) for each position of the grid are counted transposing the grid frequently. By repeating this procedure several times, the point-count will give the area to any desired degree of precision. Doubtful points, i.e. those which fall on the periphery (blue square) are counted as one half.

Volume ratios. The determination of the fraction which one tissue component occupies in the total volume of an organ is the easiest of all stereological operations. It is based on the Delesse principle which states that the areas of profiles of several tissue components are related as the volumes occupied in space by these components. Using the point-count method of planimetry, we superimpose a grid of lines over the sections and count the crossing points of the grid lines which "*hit*" the component of interest. The equation.

$$P_P = V_V \qquad (2)$$

means: The number of points which hit profiles of the component divided by the total number of test-points equals the volume fraction of that component in the entire organ (number of black round dots in Fig. A-II-12 divided by 25, i.e. the total

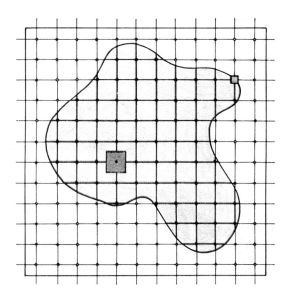

Fig. A-II-11. Planimetry by counting points.

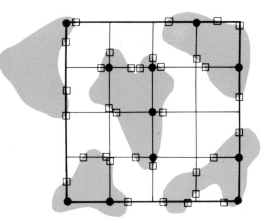

Fig. A-II-12. "Hits" (black dots) versus "non hits" give the volume fraction of the blue feature in the entire specimen. "Intersections" (squares) of lines of known length with the periphery of the blue profiles give the interphase surface area per volume.

number of test points of the grid in this specific position). Figure A-II-13 shows, as an example, a "section" of spleen. Grid points hitting profiles of white pulp, red pulp and the connective tissue framework give information on the volume ratios of these components in the organ as a three-dimensional solid. The position of the grid or the position of the slide must be changed frequently.

Volume ratios can be measured also by

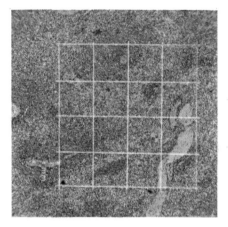

Fig. A-II-13. Point grid superimposed on a "section" of spleen to determine the volume ratios of white pulp: red pulp: frame-work.

of all the intercepts divided by the total length of all the test lines used is

$$L_L = V_V \qquad (3)$$

This intercept method is used conveniently for very small details, such as the space within cristae mitochondriales or the cisterns of the endoplasmic reticulum. More efficient than a test-line thrown randomly over the image is the use of an array of parallel lines of known length (Fig. A-II-15).

When volume ratios are to be determined by automatic scanners, the intercept method is most convenient. If the specimen is composed of two phases of strikingly different color (for example green for connective tissue and red for parenchyma, such as in a Goldner stain), a beam of light coupled with a spectrometer meanders over the specimen, while the times of illumination of phase 1 and phase 2 are recorded.

The word intercept and intersection (see below) should not be confused.

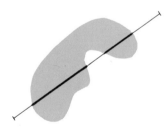

Fig. A-II-14. The intercept.

the *intercept* method. Basically this means that a line of known length is thrown randomly over the image (Fig. A-II-14). That portion of the line which falls upon a profile of the feature to be evaluated is called an intercept. The combined length

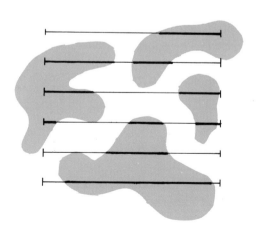

Fig. A-II-15. The intercept method for the determination of volume fractions.

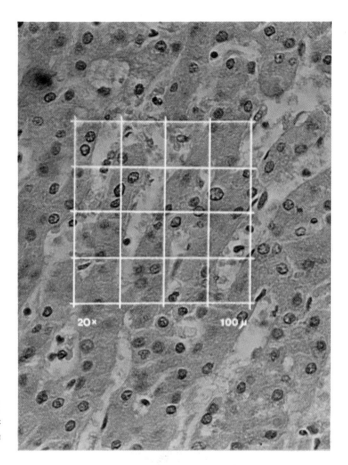

Fig. A-II-16. Grid of 10 lines, each 100 μ long superimposed on a "section" of liver to determine the surface area of the sinusoidal walls by intersection count.

Surface Area per Volume

An organ may contain a great number of parts of the same kind (such as follicles, alveoli or capillaries), and it may be necessary to estimate the total surface area through which diffusion can take place. Let us take the sinusoids in the liver as an example. The total volume of the liver can be measured. If we knew the total surface of sinusoids in a very small fraction of the organ, say in a volume of 1 mm³, we could compute the total sinusoidal surface in the entire liver, since the liver possesses an isotropic structure.

Again we can use the same reticule which we have taken before, a square divided into 16 little squares. By means of a stage micrometer we determine the length of a side of the large square. In the case of Figure A-II-16, it measures 100 μ. All the lines together measure 1000 μ or 1 mm. We count the number of *intersections* of the horizontal lines and then those of the vertical lines with traces of the sinusoidal walls. (In Fig. A-II-12 intersections are shown as empty squares.) In the position shown in Figure A-II-16, we count 37 intersection points of horizontal and vertical lines with traces of sinusoidal walls along a stretch of 1 mm. In stereology this finding is expressed thus:

$$S/V = 2P/L \text{ or } S_V = 2P_L \quad (4)$$

If S_V stands for surface per volume, the formula

$$S/V = 74/mm^3$$

should give us, in our case, a sinusoidal surface area of 74 mm² for one cubic millimeter of liver. If our intersection count were accurate, we could compute the total surface area of all the sinusoids in the liver knowing the measured volume of the liver. Of course, to achieve accuracy, we must repeat the intersection count very often. Since the number of intersections is proportional to the sine of the angle between the test-line and the traces, it is customary to rotate the test-grid twice by 60°.

Although less efficient and usually less accurate, the following procedure is mathematically just as valid for surface area per unit volume determination: It is the measurement of the length L of a trace per unit area by means of a curvimeter (see Fig. A-II-12, circumference of blue spots within the large square). If this procedure is applied,

$$S_V = \frac{4\,L}{\pi \cdot A} \qquad (5)$$

Figure A-II-12 combines the procedures for volume and surface area determination. Solid dots designate hits for volume determination. Empty squares which are intersections, or the measured length of the traces serve for surface area determination.

The *absolute* surface area of a solid or of an individual particle can be determined, if a complete series of sections or slabs exists, and if the average thickness t of the slices or the slabs is known (Figs. A-II-7, C and A-II-17). Absolute surface determination is accomplished by superimposing an array of parallel, equidistant lines over the sections. It is not necessary, in that case, to know the length of the test lines; but the distance of their separation h in Figure A-II-7, C and A-II-17 must be known. In such cases,

$$S = 2 \cdot P \cdot h \cdot t. \qquad (6)$$

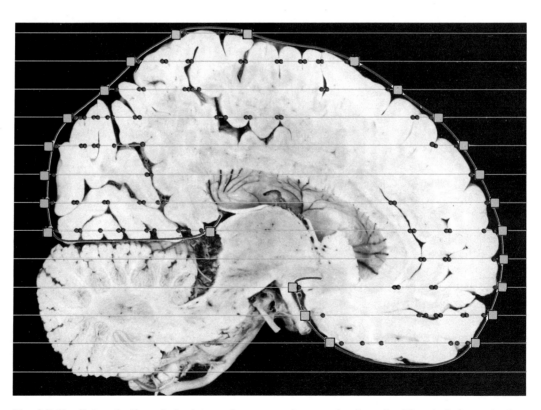

Fig. A-II-17. Determination of absolute surface area of cortex by formula (6). And determination of index of folding: number of red dots divided by number of blue squares.

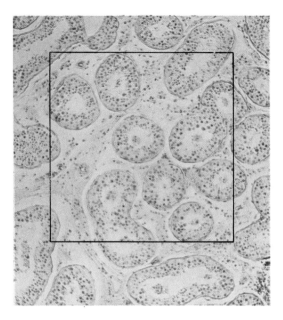

Fig. A-II-18. Determination of length of seminiferous tubules per unit volume of testicle. In this case, the test area A = 0.25 mm².

ing them out on a glass plate and measuring them with a yardstick. This method presents technical obstacles that are often insurmountable. However, using the technique of Alexander the Great, the histologist, instead of untying the Gordian knot, cuts it. Counting the number of intersected elements per unit area, he very easily obtains reliable results. Let us take as an example the length of the seminiferous tubules of the testicle (Fig. A-II-18). We superimpose a test square of known area over the image and count the number of profiles of tubules within the area. Profiles which project over the left and upper side of the square are considered "in"; those which are intersected by the right and lower sides are considered "out". The number of profiles equals P, and the total test area used equals A. The length

P is the number of intersection points (green in Fig. A-II-7, C) of these lines (blue in Fig. A-II-7, C) with the traces of the surface of the object. An example for the determination of the absolute surface area of the cerebral cortex is given in Figure A-II-17. The red dots are intersection points P with the green test-lines of constant distance h. The thickness of the slabs is t. *The series must be complete.* Formula (6) applies.

Length per Unit Volume

It is often interesting to learn how long a tube or a fiber is of which only profiles can be seen in slides. For example, the total average length of the nephron, the combined length of the seminiferous tubules or the combined length of capillaries in a bulky organ may present a problem of physiological interest. In former times, such problems were studied by maceration and teasing out of the long objects, stretch-

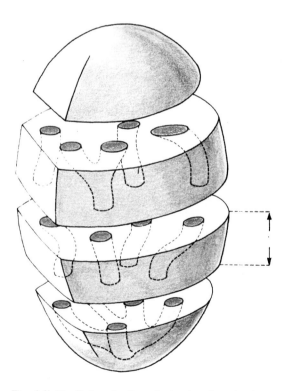

Fig. A-II-19. Determination of absolute length of a system of curved, branched and anastomotic cylinders by counting of profiles on every slab of constant thickness in a complete series of slabs, using formula (8).

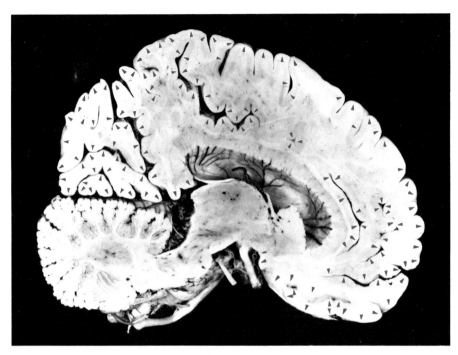

Fig. A-II-20. Determination of absolute total length of gyri of cerebral cortex, superficial plus deep.

of seminiferous tubules per unit volume is

$$L_V = 2 P_A \qquad (7)$$

The *absolute* length of a system of linear structures can be determined by counting all the profiles on every slab of a complete series of slabs. This length is

$$L = 2 P \cdot t \qquad (8)$$

(Fig. A-II-19). A practical example of the application of this method is shown in Figure A-II-20. This picture illustrates how the total length of the gyri of the cerebral cortex, superficial and hidden, can be determined by counting all the cortical convexities (red darts) on a complete series of slabs of average thickness t. One counts these convexities P on one side, and always on the same side of each slab. Again, the total length of all the gyri is $L = 2 P t$.

Size Distribution of Particles

Let us take the simplest solid, a sphere, as an example. Assume that spheres of equal size are distributed in an organ (as, for instance, fat cells in adipose tissue). Any section through a sphere is a circle whose radius r depends on the radius R of the sphere itself and on the distance h of the cutting plane from the center of the sphere (Fig. A-II-21) so that $r = \sqrt{R^2 - h^2}$. Thus, a slice ("section") through a mass of spheres will show large and small circles. At first, we measure the diameter of the largest circle which can be assumed to pass approximately through the center of a sphere. Let this diameter equal D. If all

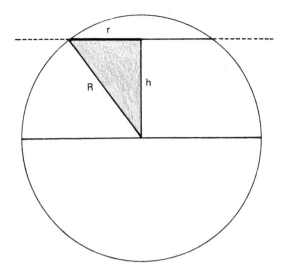

Fig. A-II-21. Sectioning a sphere.

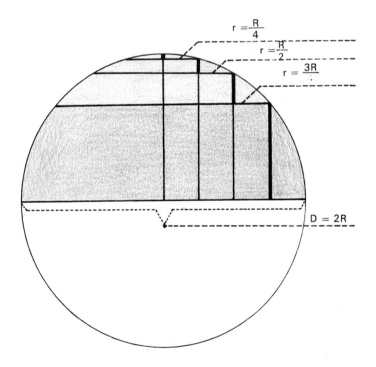

Fig. A-II-22. Size distribution of sectional circles for spheres of equal size.

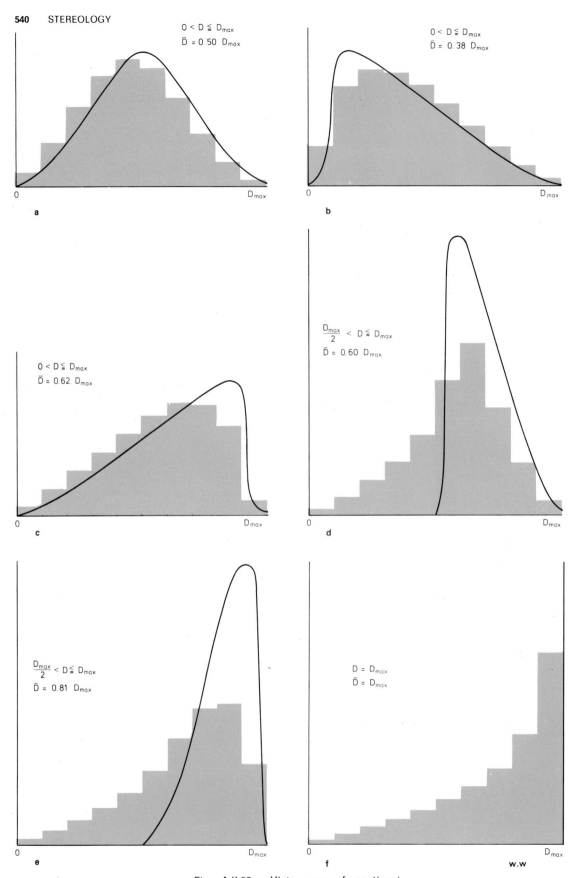

Fig. A-II-23. Histograms of sectional diameters and curves to determine size distribution of particles, (black lines).

spheres in the organ are of equal size, 13.4% or less of all circles in the section have diameters d smaller than $\frac{D}{2}$. Or if we divide the circles into 4 size classes,

for 3.2% of circles,
$$0 < d \leq 0.25 \, D$$
for 10.2% of circles,
$$0.25 < d \leq 0.5 \, D$$
for 20.5% of circles,
$$0.5 < d \leq 0.75 \, D \text{ and}$$
for 66.1% of circles,
$$0.75 < d \leq D$$

This distribution is illustrated in Figure A-II-22 (dark, vertical lines).

However, in most cases the particles are not of equal size, but show a size distribution so that every particle is neither too small nor too large for proper physiological function. Such particles are renal glomeruli, thyroid follicles, pancreatic islands, specific cell types, various organelles and other structures.

If more than 13.4% of sectional profiles have diameters d smaller than $\frac{D}{2}$, we must conclude from the previous observation that the particles in the tissue are not all equal in size; and we are confronted with a size distribution of particles, hitherto the most difficult problem of stereology.

Algebraic and analytical methods, to obtain a size distribution of particles in space from the size distribution of their sectional profiles, have been worked out but require considerable arithmetic or computer effort for their solution. The standard curves devised by Hennig and Elias* permit an almost instantaneous solution of the problem. The procedure is the following: Diameters of every sectional profile of the particles are classified into 10 size classes with the upper limits of the classes being $0.1 \, D_{max}$, $0.2 \, D_{max}$, $0.3 \, D_{max}$, ... $0.9 \, D_{max}$ and D_{max}. After a sufficient number (say 1000) of profiles have been classified, one constructs a histogram (blue in Figs. A-II-23, a, b, c, d, e, f). To each histogram of sectional diameters belongs a curve (black) for the size distribution of the particles with an indication of the *average* diameter $\bar{D}$ of all the spheres. This latter value is indispensible for the calculation of the number of particles per unit volume. Six such combinations are shown in Figure A-II-23. None of these will ever be perfectly achieved in an actual situation. To arrive at the true size dis-

---

* Hennig, A. and Elias, H., A rapid method for the visual determination of size distribution of spheres from the size distribution of their sections. J. Microsc. 93: 101-107, 1970.

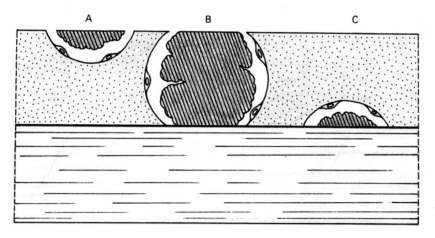

Fig. A-II-24. The phenomenon of lost polar caps.

tribution in a specific situation one can approximate the correct distribution by interpolation. For more sophisticated work, the original publication can be consulted.

The diameters of sections (profiles) of spheroid particles approach the value 0 toward the poles regardless of the lower size limit of the particles. Therefore, from a mathematical point of view, a curve for sectional distribution always passes through the origin. In practical cases, this is almost never accomplished, because the smallest profiles are not recognizable or fall out of the slices. Thus, size classes 0.1 and 0.2 D_{max} are usually empty. An explanation of this effect, employing renal glomeruli is given in Figure A-II-24. Particle A may fall out of the slice while particle C is too small or too indistinct for identification. The histogram which the observer constructs from sections is, in addition, affected by statistical irregularities. Our visual method is superior to algebraic and analytical methods, but it requires the observational histogram to be smoothed out to the left so that the curve passes through the origin. We regain, by this procedure, the lost polar sections.

The easiest method of classifying sectional sizes is to superimpose a light circle on the image by the optico-electrical particle size classifier described by Schwartz & Elias, J. Microsc. *91*: 57-59, 1970.

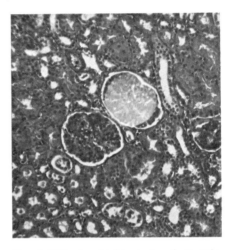

Fig. A-II-26, A. Light circle of particle size classifier superimposed on profile of a glomerulus.

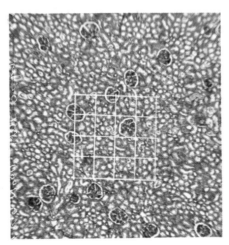

Fig. A-II-26, B. Count of profiles of particles in known area A.

Fig. A-II-25. Particle size classifier.

An illuminated iris diaphragm is reflected into the microscope by means of a camera lucida (Fig. A-II-25). By depression of the lever in a specific position that corresponds to a size class, an electrical counter is activated, one for the size class, the other for the total number of profiles measured. Histograms such as seen in Figure A-II-23 are constructed using the numbers which these counters have registered. The light circle superimposed on the profile of a glomerulus is seen in Figure A-II-26, A.

Number of Particles per Unit Volume

The number of particles per unit volume N_V can be determined by counting profiles per area n_A. This quotient appears in the equation

$$N_V = \frac{n}{A(\bar{D} + t - 2h)} \qquad (9)$$

The formula is useful only, if $\bar{D}$ is known. If the particles are spheroid, $\bar{D}$ must be determined first using the method described in the preceding sub-chapter. $\bar{D}$ must stand in the denominator because a large particle may appear in several consecutive sections. t can be neglected in semithin sections (500 nm). h in this formula indicates the altitude of the "lost" polar sections. While we were able to compensate for their loss during size determination by smoothing the profile-curve to the left, in a count of profiles the calculated N_V will appear very slightly smaller than N_V of reality.

The number of profiles per test area n/A can be determined by superimposing a square of known area on the image of the slide (Fig. A-II-26, B). Again, the profiles which intersect the left and upper margin are counted; those that intersect the right and lower margin are ignored.

In all these formulae n or P are accumulated throughout the observation, and the letter A in the denominator designates the *total* test area used throughout the experiment.

For anisodiametric particles whose shape varies greatly from that of a sphere, D is the average height, the so called mean caliper diameter of that particle considering any possible position of it. Imagine a flat object, such as a brick. Its height is low when it lies flat on the ground; standing up, it is about 3 1/2 times as high. Balanced on one corner with the most distant corner vertically above it, it is 4 times higher. There is one mean height for any kind of particle. This is $\bar{D}$ in the above formula. This $\bar{D}$ can be found for many shapes in J. E. Hilliard's paper in Proc. 2nd Int. Cong. for Stereol., New York, Springer Verlag, 1967: 211-218 and in E. E. Underwood "Quantitative Stereology" Addison-Wesley Publ., Reading, Mass., 1970 (tables 4-1 and 4-2). Among biological particles, endothelial nuclei, and nuclei of muscle fibers are highly anisodiametric; but most of them are ellipsoids in light microscopy. We shall see below how such shapes can be determined from their profiles in section.

Parallel cylinders

The number of parallel, straight, circular cylinders of average diameter D per cross sectional area $A = l^2$ (Fig. A-II-28) can be calculated from a count of their longitudinal sections. Σn is the total number of counted profiles along a cumulative stretch of length $L = \Sigma l$.

$$\frac{N}{A} = \frac{\Sigma n}{\bar{D} \cdot L} \qquad (9')*$$

---

* This number is out of order because the formula has been developed during the last phase of production of this book.
J. Minosc. *107*: 199-202, 1976.

Identification of Shape

Points, lines and surfaces. By virtue of the principle of dimensional reduction, a point (granule) would not be cut at all by a mathematical plane, except in extremely rare cases. However, since in histology we deal with slices rather than with true sections, granules will be found within a slice, between its two cutting planes. A granule will be in focus at oil immersion with the condenser up and with the diaphragm open for a depth of 1 μ or at the most 2 μ only. It can be identified as a granule, because it can be put out of focus within the slice.

A line, when intercepted by a plane, yields a point. Fibers, fibrils and filaments have the quality of lines. Since we deal with slices of finite thickness, a slice contains a short segment of the fiber. When cut perpendicularly, this fiber segment will appear in the microscope as a point. This point will remain in focus throughout the thickness of the "section". A fiber segment

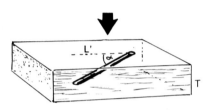

Fig. A-II-27. Dependence of apparent length (L') on inclination and section thickness.

inclined toward the cutting plane will appear as a rod whose apparent lengths varies with the angle of inclination α. Its projection lenght L' as observable in the microscope (with diaphragm narrow) is

$$L' = t \cot \alpha \qquad (10)$$

(Fig. A-II-27).
With the condenser high and the diaphragm open, the inclined fiber segment seems to shift its position when focussing up and down. The entire fiber segment within the slice will appear as a short rod with the condenser low or the diaphragm narrow. Only then, can L' be measured. Applying rules of probability it can be shown that, if a mass of fibers randomly running through space is cut with a microtome at thickness t, approximately 50 per cent of fiber segments included in the slice ("section") will appear shorter than t; about 50 per cent will appear longer. Very long lines will be very rare, i.e., about 1 per cent of all fiber segments will appear longer than 7 t and only 0.5 per cent longer than 10 t. Upon superficial inspection, the long lines appear more numerous than they are because each occupies more area than several dots together. Therefore actual counts must be made within a specified area to identify fibers correctly. In high power electron microscopy, the linear elements with which we deal are microtubules, protofilaments of collagen, filaments of actin and myosin, etc. Because of the considerable thickness of an "ultrathin section" compared to the width of the field and because of the great depth of focus, the rules just spelled out must be weighed more carefully and more rigorously than in light microscopy.

Since fibers are often curved, they may wind within the slice, thus raising the number of longer segments above 50 per cent. When the number of lines longer than t exceeds 60 per cent, it is probable in light microscopy that among the suspected fibers there are in reality some bands.

A two-dimensional surface, when intercepted by a plane, yields a line, which is called a "trace" of the surface. Membranes have the geometric quality of surfaces. Hence, if one finds in a slide numerous lines but few dots and commas, it is probable that these lines are traces of membranes.

Solids of various shapes. Fortunately, most organs are constructed of components of similar, three-dimensional shape. Homogeneous construction is a physiologic advantage, because only similar parts can act in unison. Because of their uniform architecture they yield themselves to geometrico-statistical analysis. (Anisotropic organs will be discussed below).

An exception to this general rule are the mitochondria. They are most dissimilar in shape, i.e. pleomorphic. It appears that a determination of their shape is not possible by stereology, but requires either serial ultrathin sectioning with reconstruction or semithin sectioning under high-voltage electron microscopy and a stereoscopic treatment, i.e., using parallax. The pleomorphism of mitochondria makes determination of N_V impossible for the time being.

Most other components of living systems can be approximated to spheres, ellipsoids, disks, cylinders, prisms or sheets. And these basic shapes yield themselves easily to stereological identification, measurement and quantification.

The shape of a secton ("profile") of a solid depends on the three-dimensional shape of this solid and on the angle of cutting. A solid has three dimensions: length, width and height. A section through it has only two dimensions: length and width. We call the quotient $\frac{\text{length}}{\text{width}}$ its axial ratio Q. If many three-dimensional objects of equal shape, randomly distributed in space, are cut by a plane, the axial ratios of their sections ("profiles") will show a characteristic distribution.

Spheres. One hundred per cent of sections through spheres will have an axial ratio of $Q = 1$; i.e., they are all circles. Thus, if one finds only circles in a section, these may have resulted from cutting many spheres.

Circular cylinders. Many circles in sections could also be "profiles" of parallel, circular cylinders cut transversely, such as we see them when looking at the open end of a pack of cigarettes. In histology a transverse section through a peripheral nerve exhibits this aspect; so does a transverse section through a medullary ray in the kidney. If the plane of cutting is oblique to the longitudinal direction of these parallel circular cylinders, all their "profiles" will be ellipses of equal shape. Their common axial ratio will be equal to the cosecant of the angle which the cutting plane forms with the longitudinal direction of the cylinders

$$(Q = \operatorname{cosec} \alpha), \qquad (11)$$

as for example in an oblique section through renal medulla.

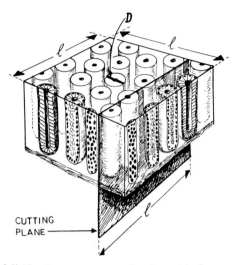

Fig. A-II-28. Tunica mucosa of colon with the surface epithelium removed to demonstrate the principle of number determination of parallel cylinders from a count of their longitudinal sections.

When circular cylinders are randomly distributed and oriented in space (as, for example, the seminiferous tubules, Fig. A-II-18), 75 per cent of their sections will be "short" ellipses, i.e., their axial ratios Q will lie between 1 and 2 ($1 \leq Q \leq 2$); 25 per cent will be more oblong. In histologic "sections", however, the long ellipses will appear just slightly more numerous than they would in cutting planes of thickness zero. In fact, their absolute length will be augmented by $t \cot \alpha$ (Fig. A-II-29). The distribution of axial ratios of sections through randomly cut circular cylinders is shown in Fig. A-II-a. Cylindrical tissue components (such as tubules, arteries, the gut as a whole) are often confined to a restricted space. Therefore, they may be twisted, convoluted and curved. For this reason, the number of very long ellipses will be even lower than the value of 2 per cent predicted for straight cylinders. Actually, if in the human organism a cylinder is straight, as, for example, certain

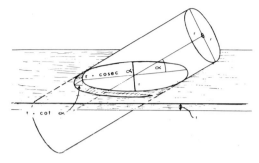

Fig. A-II-29. Sectioning a cylinder.

large arteries and ducts, it will be alone (i.e., unaccompanied by structures of its own kind), except for nerves which are bundles of cylindrical fascicles. Randomly arranged cylinders in man are always crooked.

Rotatory ellipsoids. Shapes intermediate between the sphere and the circular cylinder are those of prolate (egg-shaped) rotatory ellipsoids, as are many nuclei in columnar epithelium, or oblate (lens-shaped), as are nuclei of squamous epithelium. When sectioned in any direction, they yield ellipses. There is a distribution of the axial ratios of their sections characteristic for each shape of rotatory ellipsoid, as shown in Figure A-II-30, b-d; b for prolate; c and d for oblate ellipsoids. We assume that the ellipsoids are equal in shape and randomly oriented; or that several sectional planes in random directions are used.

Tri-axial ellipsoids. Tri-axial ellipsoids are oblong, flattened ovoids. Many endothelial nuclei exhibit this shape. The sections of ellipsoids of specific degrees of flattening and elongation should show characteristic distributions of axial ratios. However to this date, no satisfactory stereological solution for these shapes has been found. To determine such shapes it remains necessary to produce or search for exact longitudinal and cross sections.

Elliptic cylinders. In contrast to circular cylinders whose exact cross sections are circles (Fig. A-II-30, a) those of elliptic cylinders are ellipses, so that they approach the shape of bands. The taeniae coli have this shape, so do many veins and venous sinuses, as well as certain Platyhelminthes. Distribution of axial ratios for sections of elliptic cylinders of various flatnesses are shown in Figure A-II-30, e and f.

Another shape parameter is the degree of wrinkling or folding. Figure A-II-31 shows a section through a nucleus of a goblet cell. In the light microscope this nucleus would appear as an oval (red enveloping line). The degree of wrinkling can be determined by superimposing an array of lines over the image. One counts the number of intersections P of these lines with the trace of the nuclear envelope (green circles) and then the number of intersections p with the enveloping oval (red squares). The quotient $I = \dfrac{P}{p}$ is the *index of folding*. This index assumes great significance in the comparative anatomy of the cerebral cortex, where it is determined on macroscopic slabs of brains (Fig. A-II-17, number of red dots/number of blue squares).

Still another shape parameter is the mean surface curvature to be discussed below.

Networks or plexuses and branched cylinders. Heretofore we have discussed cylinders which are, in essence, isolated; but in the living organism there exist many cylindrical structures which branch. Exam-

Fig. A-II-30. Histograms (yellow) of the distribution of axial ratios ($1 \leqslant Q \leqslant Q_{max}$) of sections (profiles) through ellipsoids of specific shapes (b-d) and through cylinders of specific shapes (a, e and f). By measuring and counting length and width of profiles, one can draw a conclusion as to the common shape of these bodies, if they are all similar and randomly distributed and orientated in space.

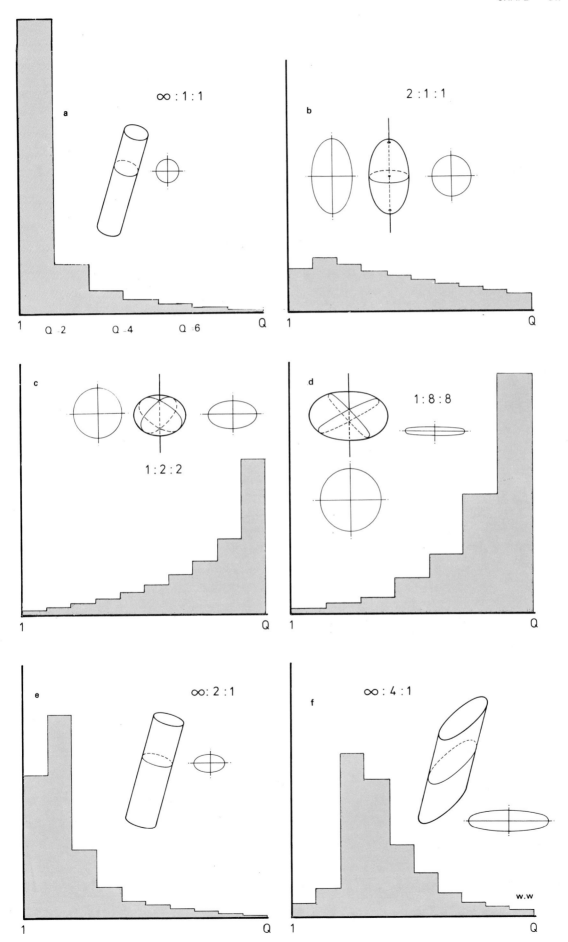

ples of branched cylinders are ducts of glands and end arteries. In order to identify branching it will be necessary that one finds, occasionally, a Y-shaped or a T-shaped figure in a "section" (see, for example, the bronchi of the lung, Figs. 13—18, A and 13—18, C). The number N of points of branching per unit volume is proportional to the number n of Y-shaped or T-shaped figures per test area and indirectly proportional to the average internodal length I, the thickness of the slice (if translucent) and to the thickness T of the branches. The exact relationship has not yet been worked out.

In the case of networks or plexuses H-shaped figures and loops are occasionally found.

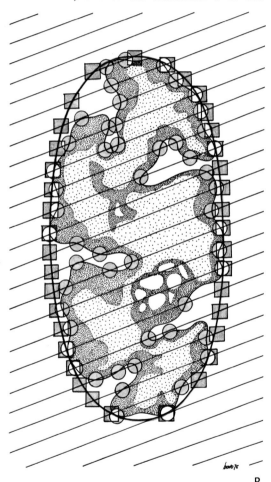

Fig. A-II-31. Determination of index of folding for the nucleus of a goblet cell.

Laminae. The greatest amount of flatness is possessed by a sheet, plate or lamina of infinite extension and of finite thickness. An epithelium would have these properties. While a basement membrane appears in the light microscope often as a surface without thickness, in electron microscopy it acquires the geometric properties of a lamina of measurable thickness. If such a flat object is said to be of infinite expansion, we mean that it reaches beyond the boundaries of the microscopic or electron-microscopic field. All sections of it are stripes of "infinite" length. The width of such a stripe varies as the cosecant of the angle of sectioning. As an example of such a lamina we present, in Figure A-II-32 the renal glomerular basement membrane and in Figure A-II-33 the section of a sheet of noodle dough (lasagna), embedded in black gelatin. Most of its windings extend beyond the border. Since such plates in the living organism are usually buckled, their sections exhibit various widths from place to place. Let us visualize a lamina which is of equal thickness throughout and which is irregularly folded, and let us assume that its real, constant thickness equals T. If we measure the width of its sectional stripe at regular intervals, we find 86.6 per cent of the places to have a width W, so that $T < W < 2T$; while, at 13.4 per cent of measured locations, the stripe is wider, so that $W > 2T$. The renal glomerular basement membrane (Fig. A-II-32) is a lamina of finite thickness. Its sections exhibit various widths dependent on the angle of cutting. Since at this time no statistical measurements have been made of the widths of the sections, we do not yet know whether this particular membrane is of even or variable thickness.

Muralia. A muralium which consists of interconnected walls appears in sections as a maze of interconnected stripes without end in any direction. A classical case of muralium is the liver (see Chapters 8 and 12, also Fig. A-II-16). In fact the study of liver structure provided the impetus for the

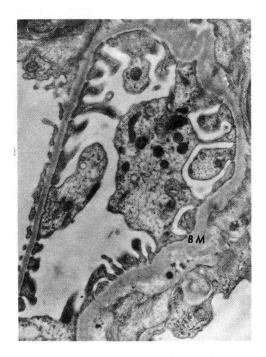

Fig. A-II-32. Glomerulus of rat showing dependence on the angle of inclination of the width of the section of the basement membrane (BM). (Electron microgram by Dr. Arnold Schecter, Harvard University.)

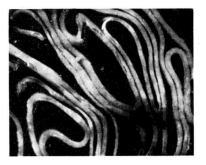

Fig. A-II-33. Section through a sheet of noodle dough (lasagna).

foundation of stereology. Numerical rules for the identification and topological classification from random sections have not yet been worked out, but muralia are easily recognized by the presence of numerous loops in sections.

Branched sheets. Intermediate between a muralium and a very flat, oblate lens is a branched lamina of finite extension, such as the vascular lamina of the renal glomerulus. Its sections are very long, branched stripes with an axial ratio distribution intermediate between that for a very flat, oblate lens and for a sheet of finite extension (see Chapter 14).

This account exhausts the list of shapes known to exist in human organs which had been considered stereologically up to 1976. As new objects are observed, hitherto unknown shapes are likely to be discovered requiring further analysis.

The methods of mathematical stereology are applicable only where a great number of similar objects are present in the tissue or when a statistical randomness of sectioning planes is employed.

Anisotropy

Isotropic organs are those of homogeneous structure, so that any random section through any region of the organ presents the same kind of image. Among such organs are the liver, the lung, the spleen, the large salivary glands and the prostate. It is impossible when viewing a section through an isotropic organ to determine the region or the direction of cutting. Such organs are ideal for stereological examination.

Anisotropic organs are characterized by gradients of density of certain components, regional variations of structure and/or preferred orientation of components.

The extreme of preferred orientation is parallel arrangement of parts, such as the fibers in muscles and nerves, or the cells in the stratum corneum.

Other organs less strictly organized show a preferred orientation such as the convoluted tubules of the renal cortex. Although their course is irregular, these tubules "prefer" a direction perpendicular to the capsule, due to their restriction in space between medullary rays (Fig. 14-8, B).

Gradients of density exist in organs divisible into cortex and medulla.

Gradients of density may be combined with preferred orientation. A classical example is the cerebral cortex where the density of neurons is different in every layer and where the pyramidal cells are oriented perpendicular to the surface of the brain. In the adrenal gland, the medulla is of isotropic construction. Not so the cortex: neighboring cell groups and cords of the zona fasciculata run parallely to each other and perpendicularly to the capsule. In the kidney, the medulla consists of "straight" tubules and of vasa recta which run parallel each to its neighbors. Yet, very low power views, such as Figure 14-4 show that, on a large scale, the "straight" tubules and the vasa "recta" are curved. In the kidney, there exists a characteristic gradient of glomeruli. They are absent from the subcapsular cortex down to 1mm depth, numerous in the main part of the cortex and absent from the medulla. There is also a gradient of size of glomeruli, those nearer the capsule being relativelly small; while the juxtamedullary glomeruli are much larger.

There are various methods of sampling to deal with anisotropy. When examining organs of small animals, it is recommended that one analyze quantitatively entire series of sections through the organ, as well as study complete series of sections perpendicular to each other, for example, a series of sagittal sections through the left and a series of transverse sections through the right kidney should be taken. Or, when studying the cerebral cortex, *a few* sections *perpendicular* to the surface *plus a complete* series of sections parallel to the surface can be evaluated for each cortical area.

If the organ is large, this method is not practicable. A human kidney is too large for serial sectioning. Here is a recommended method to master quantitative kidney histology: The entire kidney, after hardening, is divided into 1000 to 1500 little cubes. These are all thrown into a dish; and by a lottery, 50 cubes are extracted from the mixture. These 50 blocks are embedded, and one section through each is used for quantitative stereological counts and measurements. Each of these 50 sections is analyzed throughout using a "sampling stage", and the results from all 50 sections are treated collectively. By this method, and after previous determination of the volume of the organ as a whole, one can determine the size distribution of glomeruli, the number per unit volume of kidney, their total number in the entire kidney, the area of glomerular filtration surface for the entire kidney, the average length of the nephron, etc. Anisotropy has been artificially abolished.

Surface areas of cerebral cortices have been measured, stereologically, in many mammals. Brains differ in shape from the very broad and relatively short brains of whales through the long and narrow brains of foxes. The external shape of the brain determines a preferred orientation of gyri and sulci. In every such case the brain must be divided into two hemispheres; one must be cut transversely, the other horizontally. Figure A-II-34 shows as an example a frontal and a horizontal section through the cerebral hemispheres of a coyote on which an array of parallel testlines of equal distance has been superimposed, in 3 directions of space perpendicular to each other. In A and B the same section is shown with test-lines perpendicular to each other. Calculating the absolute value of $S = 2P \cdot h \cdot t$ one obtains the highest values of S for transverse testlines (A), a slightly lower value for vertical test-lines (B) and the lowest value for longitudinal test-lines (C). The arithmetic mean of the three values equals the absolute surface area of one hemisphere.

While all our formulae are based on the theoretical assumption of accuracy, in any actual experiment both irregularities of structure, the accident of placing the testgrid over an image as well as deviations from the intended cutting direction intro-

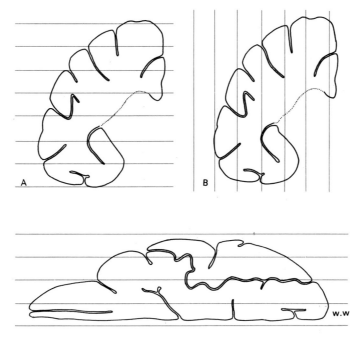

Fig. A-II-34. Transverse and horizontal sections of cerebral hemispheres of coyote (Canis latrans). Test lines for surface determination run in three directions of space perpendicular to each other.

Curvature

The mean surface curvature of a tissue element is a characteristic of shape. It can be determined stereologically. It assumes importance in problems of coherence of a tissue, for example when the interdigitations of epithelial cells are considered, and in problems of surface increase. To determine the mean surface curvature K one sweeps a line remaing parallel to itself over the section and counts the number of tangent points T of the trace with the moving line (Fig. A-II-35). One must also make an intersection count P/L. Then

$$K = \frac{T L \pi}{2 P A} \quad (11)$$

gives the mean surface curvature.

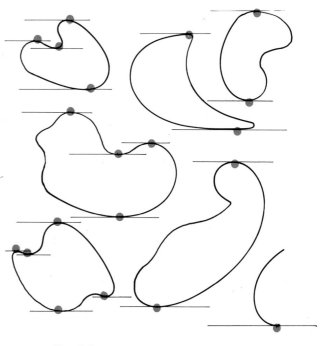

Fig. A-II-35. Tangent count for curvature determination.

The Holmes effect

Wrong estimation of size due to overlapping of opaque particles is called the Holmes effect. It provides great difficulties

duce statistical mistakes which must be considered for each individual experiment.

in high resolution electron microscopy, when electron-opaque particles such as glycogen granules, ribosomes or melanosomes overlap.

Direct morphometric measurements and the application of stereometry (= solid geometry)

There are cases in nature, and particularly in pathological material, which exhibit great structural variations from place to place. In such cases, quantitative studies must be made separately, within very small areas, so that sampling methods will not produce misleading results. An example is a typical, ring-shaped adenocarcinoma of the colon (Fig. A-II-36). In such cases, the specimen must be carefully oriented, so that the length of the glands and of the cancer follicles, the number of cells per gland and follicle and the sizes of the cells can be determined from place to place. Glands must be cut longitudinally to measure their length l (section perpendicular to the mucosal surface). For the determination of the glandular diameter d, tangential sections through the mucosa must be used. The same cutting direction may be used to determine the number of glands per unit area of mucosa. As Figure A-II-36 shows, length and density vary from spot to spot. A micrometer is used to determine l and d. The surface area of a gland is, then, $S = l \cdot d \cdot \pi$. The number of cells per gland is found by superimposing a square of known area A on tangential sections of glands and counting the number of nuclei in that area n/A. Then, the number of cells per gland is

$$N = l \cdot d \cdot \pi \frac{n}{A} \qquad (12)$$

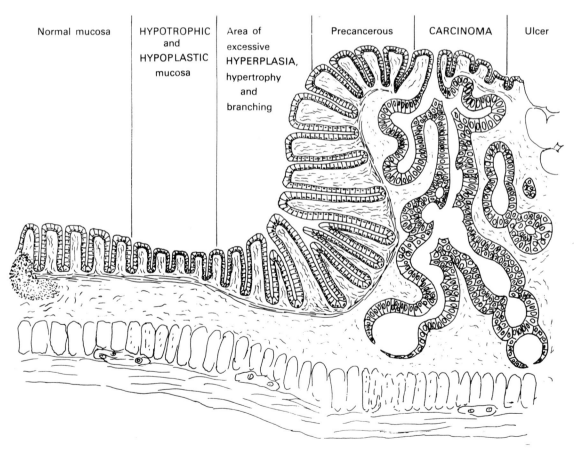

Fig. A-II-36. Adenocarcinoma of the colon, an example of great structural variation from region to region.

Stereology has not been used for this determination, but direct morphometry (measurement of length and diameter) and stereometry (calculation of the surface area by a well known formula of solid geometry) has been used. Knowing the sizes of glands and cancer follicles as well as the number of cells per gland or follicle, one can determine the pathogenetic processes which have occurred: hypotrophy (decrease of cell size), hypoplasia (decrease of cell number), hypertrophy (increase of cell size), hyperplasia (increase of the number of cells).

In each individual case one must select the specific method to solve a specific problem in a specific kind of organ or lesion.

When morphometric data can be obtained by direct measurement, this is to be preferred to the indirect way of stereology. For example, when the size distribution of liver cell nuclei is to be determined, one can use very thick slices (20-50 μ) stained with hematoxylin only and measure all those nuclei directly whose equator is included within the slice. In electron microscopy, this procedure can be applied, for example, to synaptic vesicles.

Infinite versus finite objects

An object is of infinite extension for stereological purposes if it reaches on all sides beyond the field of vision, as do sections through most organs when viewed through the microscope. When stereological counts are undertaken on an infinite object, they must be referred to a limited or finite test-system.

Limited objects, such as a brain viewed with the naked eye can be tested by systems of points and lines which are of infinite or indeterminate extension. In that case, however, a complete series of sections or of slabs is needed; and the test-points and test-lines must be equidistant.

Reconstruction from serial sections

If a structure is present in the singular only and if it is of complicated shape, geometrico-statistical methods are of little use. For the determination of the shape of such objects, reconstruction from serial sections is the only, though very time-consuming, method available at this time. Reconstruction depends on the production of large, uninterrupted series of well-stretched slices of uniform thickness. These can be obtained only under optimal conditions by highly skilled technicians.

A tracing is made of the projected image of the structure to be examined from every section. If projected on wax plates, the images must be cut out, and successive cut out tracings are pasted together. Thus a model of the object is obtained.

Another method of reconstruction from serial sections is serial section cinematography where a motion picture camera is focused on the surface of the paraffin block in which the object is embedded, mounted in a microtome. This surface, after appropriate staining, is photographed on one frame. One obtains, in this manner, a photographic series of sections parallel to the knife and discarded, but the newly exposed surface is stained and photographed on the next motion picture frame. This photographic series of sections are in perfect register with each other free from any distortions. Reconstructions can be made by projecting one motion picture frame after the other on wax plates. Many other methods exist for serial section reconstruction which we can not describe here.

Mathematical stereology in general is to be preferred when a population of similar objects are to be characterized by average parameters. It provides efficient and easy methods which yield results of relatively high precision.

The following is a list of the most frequently used stereological formulae.

Basic formulae of stereology

Volume in volume by point counts:
$$V_V = P_P \qquad (1)$$

Volume in volume by intercept measurement (linear scanning):
$$V_V = L_L \qquad (2)$$

Surface area in volume by intersection counts:
$$S_V = 2 P_L \qquad (3)$$

Absolute surface area of limited objects by intersection counts: (h = constant distance between test lines, t = thickness of slabs), only applicable to *complete* series of slabs.
$$S = 2 P h t \qquad (4)$$

Length in volume by profile count:
$$L_V = 2 P_A \qquad (5)$$

Absolute length (three-dimensional curve confined to limited portion of space):
$$L = 2 P t \qquad (6)$$

Number of particles in volume (n = number of profiles, h' = height of lost caps):
$$N_V = \frac{n}{A (\overline{D} + t - 2h')} \qquad (7)$$

Mean absolute surface curvature obtained from tangent and intersection counts:
$$K = \frac{\pi T L}{2 P A} \qquad (8)$$

Number of parallel cylinders of diameter D per cross sectional area $A = l^2$ from a count of their longitudinal sections along the test-stretch $L = \Sigma l$
$$\frac{N}{A} = \frac{\Sigma n}{D \cdot L} \qquad (9)$$

Stereology is a very young and active field. New developments are to be expected.

Index

Abdominal ostium, uterine tube 167
aberrations 515
accommodation, eye 434, 435
acetylcholine 145
acetylcholinesterase 145
acidophil endocrine cells 411
acidophilic granulocyte 106
acini 178, 267
acrosome 463
actin 60, 125
action potential, nerve 145
adaptation, eye 439
adenine 36
adenohypophysis 410, 411
adipocytes 77
adipose tissue 79
adrenal glands 418
adrenergic synapse 145
adventitia 185
afferent arteriole glomerulus 377
aggregated lymph nodules 308
agranular endoplasmic reticulum 25
agranulocytes 105
aldosterone 380, 397, 421
alpha endocrine cells 344
alveolar ducts 364
alveolar macrophages 368
alveolar sacs 364
alveoli, lung 365
alveolus 178
amebism 49
ameboid movement 105
ameboidism 49
ameloblasts 279
amitosis 29
ampulla of Vater 335
ampulla, ductus deferens 465
ampulla, semicircular duct 451
ampulla, uterine tube 491
anagen 261
anal sphincters 315
analyser 522
anastomosis 182
androgen 423
angiotensin II 380
angstrom 2
anisotropic 548
anisotropic (A) disk 123
annulus [anulus] fibrosus 90, 208
anoxia, 323
ansa nephroni 375, 395
anterior epithelium, cornea 433

anterior limiting lamina (membrane) 433
anthracosis 369
antidiuretic hormone 397, 411, 412
antigonadotro(h)ic effect 428
antrum 482
apocrine secretion 179
apocrine sweat glands 254
aponeuroses 80
apparato reticolare interno 28
appendages, skin 249
appositional growth 87
aqueous humor eye 433
arachnoid granulations 160
arachnoid mater 157, 159
arcuate arteries 378
area cribrosa, kidney 401
area cribosa, retina 432
area densa, smooth muscle 119
areola, breast 504
areolar connective tissue 78
argentaffin cells 7, 309
argyrophil cells 7
argyrophil fibers 70
arrector pili muscle 259, 264
arteries 185, 193, 194, 195
arteriole 194
arteriovenous anastomosis 205
articular cartilage 101
ascending limb, nephron 375, 395
aster 183
astral rays 29
astrocytes 149
atretic follicles 479
atrioventricular node, bundle 212
atrioventricular valves 212
atrium, lung 364
atropine 434
auditory ossicles 447
auditory tube 447
auricle 447
autophagocytic vacuoles 21
autophagosomes 21
autosomes 30
axial ratio 528
axial section 5
axis cylinders 162
axolemma 140
axon 137
axon hillock 137
axoplasm 140
azan 8
azurophilic granules 105

Baldness 260
bands, skeletal muscle 124
basal corpuscles 59
basal metabolism 416
basement membrane 52, 239
basila membrane 447
basket cells 147
basophil endocrine cell 412
basophilic granulocyte 106
basophilic stains 6
beta endocrine cells 344
biconvex lens 512
bile ducts 332
bilirubinemia 239
biogenic amines 301
biopsy, liver 320
bipolar cell 439
bipolar nerve cells 140
blackhead 256
bladder, gall 336
bladder, urinary 401
blastema, suprarenal 421
blind spot 441
blood 103
blood vessels, straight (vasa recta) 399
body, Herring's 411
bone 81, 90
bone growth 100
bone lining cells 91
bone marrow 93
bone marrow vasculature 111
bone modeling 100
bone, compact 92
bone, spongy 92
bone, trabecular 92
bony (osseous) muralium 182
boutons terminaux 145
brain sand 427
breast 503
bronchi 357
bronchial passages 357
bronchiolar exocrine cells 357
bronchioles 357
brown fat 77
brush border 60, 394
bulb of the vestibule 502
bulbar conjunctiva 442
bulbourethral glands 472

Calcium metabolism 418
calculus, dental 279
calyces; major, minor 373, 374
camera lucida 523
canal, Haversian 92
canal, Schlemm 434
canal, Volkman 92
canal, central 92
canal, perforating 92
canaliculi, bile 320, 332
canaliculi, intercellular 293
canaliculi, osseous 91
canals of Herring 332
capillaries 185
capillaries, fenestrated 188, 190
capillaries, nonfenestrated 188
capsule 169
capsule, liver, Glisson's 323
capsule, renal 373
carbaminohemoglobin 104
cardiac conduction system 211
cardiac glands 290

cardiac muscle 128, 131, 133
cardiac skeleton 208
cardiovascular system 185
carotene 239
carotid artery 206
cartilage 81, 85, 87
cartilage formation 86
cartilage growth 87
cartilage, secondary 97
cartilagenous joints 100
catagen 261
caveolae 119, 188
cecum 315
cell division 29
cell membrane 15, 117
cell movements 49
cell shapes 41
cell sizes 41
cell, APUD 311
cell, Paneth 305
cell, Purkinje 297
cell, Rouget 188
cell, Schwann 162
cell, endothelial 185
cellula sensoria pilosa 450
cement line 92
cementum 275
central nervous system 149
central vein, liver 323
centrioles 29
centroacinar cells 342
centromere 36
centrosome 28
cerebral cortex 157
cerumen 255, 447
ceruminous glands 255, 447
cervix, uterus 498
chambers of the eye 431
choanae 349
cholangioles 332
cholecystokinin 314
cholinergic synapsis 145
chondroblasts 85, 87
chondrocyte aggregates 86
chondrocytes 85
chondrogenic layer 86
chondroitin sulfates 66, 86
choroid plexus 149, 150
choroid, eye 432
chromaffin substances 7
chromaffin system 425
chromatids 30, 457
chromatin 15
chromatin granules 31
chromatophilic substance 137
chromoidial substance 25
chromophilic 7
chromophobe endocrine cells 411
chromophobic 7
chromophobic islet cells 344
chromosomes 30
chyme 308
cilia 60
ciliary muscle 435
ciliary processes 434
ciliary sweat glands, Moll 442
circadian rhythm 29
circular folds of Kerckring 307
circumcision 479
cisterna nucleolemmae 20
clara cells 357, 363

classification of epithelia 56
clear endocrine cells 423
cleavage 484
clefts, Schmidt-Lantermann 165
clitoris 503
clotting 104
coelomic cavity 168
collagen 67
collagen TYPE I 67, 90
collagen TYPE II 69, 86
collagen TYPE III 69
collagen TYPE IV 54, 69
collagen TYPE V 69
collagenous glycoproteins 54
collateral nerve fibers 138
collecting tubules, kidney 397
colliculus axonis 137
colloid, thyroid 414
colon 315
colony stimulating factor (CSF) 108
colostrum 504
columnar epithelium 56
comedo 256
common bile duct 332, 336
compact bone 92
complexus basalis 433
conception 30
conceptus 30
conchae 349
conducting passages, lung 347
condenser 517
conducting arteries 193, 195
cones retinal 439
conjuctival sac 442
connective tissue 65
connective tissue cells 72
connective tissue, areolar 78
connective tissue, dense irregular 79
connective tissue, loose 78
connective tissue, mucous 78
connective tissue, reticular 78
conus ligament 208
convoluted tubules, nephron 374, 375
cordae tendineae 212
cornea 431, 433
corona radiata 482
coronary sinus 212
corpora albicantia 479, 487
corpora arenacea 428
corpora cavernosa 473
corpus atreticum 481, 484
corpus hemorrhagicum 484
corpus luteum 487
corpus neuroni 137
corpus spongiosum 473, 474
corpus uteri 493
cortex, ovary 479
cortex, renal 373
cortex, suprarenal 418
cortical labyrinth 374
corticotrop(h)ic endocrine cells 411
crescents 269
cretinism 416
crista ampularis 451
cristae 22, 449
crossing over 461
crypts of Lieberkuhn 305
cuboidal epithelium 56
cumulus oophorus 484
cupula gelatinosa 451
curvature 550
curvature of field 517

cushions, muscle, intimal 199
cuticle 264
cylinders, branched 545
cylinders, circular 544
cylinders, elliptic 545
cylinders, parallel 542
cystic duct 332
cytocentrum 28
cytokinesis 29
cytolysosomes 21
cytoplasm 20, 117, 140
cytoplasmic reticulum 25
cytopodia 385
cytopodial slits 390
cytoskeleton 29, 183

Definitive islet cells 344
dehydration 8
delta endocrine cells 344
dendrites 137
dense cellular connective tissue 80
dense endocrine cell 423
dense irregular connective tissue 79
dense regular connective tissue 80
density, gradients of 549
dentinal tubules 275
dentine 275
deoxyribonucleic acid 7, 8
deoxyribose 35
depolarization, nerve 145
dermal papillae 246
dermatoglyphics 249
dermis 239, 246
descending limb, nephron 375, 395
desmosomes 28, 43, 328
dextrin 269
diabetes insipidus 399
diabetogenic effect 424
diad 133
diapedesis 105
diaphragms, capillaries 190
diastole 198
differentiation 50, 51
digestive system, apparatus 265, 267
dihydroxyphenylalanine 77
dilator pupillae 437
dimensional reduction 528
dioptric apparatus 433
diploid 30
diplosome 29
diplotene 481
direct cell division 29
disjunction 461
disks, skeletal muscle 124
distal convoluted tubule 395
distributing arteries 192
dodecaploid 29
drumstick, sex chromosome 31
ducts, liver 331
ductus choledochus 332
ductus deferens 465
ductus epididymidis 465
duodenal glands, submucosal, Brunner 306
duodenal papillae 336
dura mater 157, 159
dura mater encephali 159, 432
dura mater spinalis 159
dust cells 368
dyad 461

Ear 445
ear wax, cerumen 255

eccrine (merocrine) secretion 179
edema 193
effect, Holmes 550
effectors 145
efferent arteriole, glomerulus 375, 378
efferent ductules, testis 465
efferent nerve endings 145
ejaculatory duct 467
elastic cartilage 89
elastic fibers 69
elastin 69
ellipsoid, retinal rod 439
ellipsoids, rotary 545
ellipsoids, triaxial 545
enamel prisms 279
end arteries 207
endocardium 208, 213
endochondral ossification 97
endocrine glands 180, 409
endocrine system 409
endolymph 449
endometrium 493
endomysium 127
endoneurium 161
endoplasmic reticulum 7, 25
endorphin 311
endosteum 92
endotendinium 128
endothelium 55, 185
enlargement 513
entactin 54
enteric plexuses 282
enterochromaffin cells 309
enteroendocrine cells 311, 309
enteroglucagon 311
eosin 6, 8
eosinophil 106
ependymal cells 150
epicardium 208, 209
epidermal ridges 248
epidermis 237, 239
epididymis, duct of 465
epidural space 159
epigenome 31
epiglottis 281
epimysium 127
epinephrine 437
epinephrine cell 423
epineurium 161
epiphyseal cartilage 99
epitendinium 128
epithelia, simple 56
epithelia, stratified 56
epithelia, surface specializations 61
epitheliocytus bacilifer, (rod) 438
epitheliocytus columnaris, Type II (ear) 451
epitheliocytus conifer, (cone) 439
epitheliocytus piriformis, Type I (ear) 450
epithelioid cells 206
epithelium 53, 56
epithelium, columnar 56
epithelium, cuboidal 56
epithelium, pseudostratified 56, 57
epithelium, squamous 56
eponychium 264
equatorial division 461
equatorial plane 38
erectile tissue 473
erection 205, 475
erythrocyte aggregates 104
erythrocytes 104
erythrocytopoesis 108

erythropoetin 108
esophageal cardiac glands 282
esophageal glands 282
esophagus 282
estrogen 423, 487, 495
estrogen dominated smooth muscle 498
euchromatin 15, 31, 73
euploid 39
evolution, lung 370
exfoliative cytology, PAP smear 500
exocrine cells with acidophilic granules 305
exocrine glands 177
exocytosis 50, 188
external auditory meatus 447
external elastic membrane 185
external genital organs, female 502
external limiting membrane 149
exteroceptors 142
extrahepatic bile ducts 332
extramedullary hemocytopoesis 101
eye 431
eyelashes, cilium 442
eyelids 441
eyepiece, Huygens 515

Falx cerebri 160
fascia adherens 131
fascia, Camper's and Scarpa's 249
fat cells 77
fat stain 8
fat tissue 79
fenestrated endothelium 388
fertilization 30, 484
fertilized ovum 497
fiber formation 70
fibers, Muller's 438
fibers, Sharpey's 92
fibers, argyrophil 70
fibers, collagen 67
fibers, connective tissue 67
fibers, elastic 69
fibers, perforating 92
fibers, reticular 70
fibrillogenesis 70
fibrils 29
fibrin 104
fibrinogen 103
fibroblast 70, 73
fibrocartilage 90
fibrocytes 73
fibroma 33, 444
fibrous astrocytes 149
fibrous joints 100
filaments 29
filiform papillae 274
filtration barrier 390
finger prints 249
fixative, Bouin's 7
fixative, Carnoy's 7
fixatives 7
flagellum 61
flight or fight 425
focus, depth of 518
folding, index of 545
foliate papillae 274, 278
follicle 182
follicle stimulating hormone (FSH) 412, 481, 484, 495
follicular endocrine cells 418
foreign body giant cell 47, 80
formalin 7
fovea centralis 439
foveolae gastricae 289

free ribosomes 25
fundic glands 290
fungiform papillae 274
fuzzy coat (glycocalyx) 60

Gall bladder 336
gametes 30, 457
gamma cells 411
ganglia 160
gap junction 47
gastric endocrine cells 295
gastric glands 290
gastric pits 290
gastric rugae 297
gastric sulci 297
gastrin 295, 311
gastroinhibitory polypeptide 312
gemmules 138
genes 31
genome 31
giant cell, Langhan's 80
giant cell, foreign body 47, 80
giant cells (Betz) 158
glands, Bartholin's 503
glands, classification 177
glands, cytogenic 180
glands, general 174
glands, structural forms 181
glands clitoridis 502
glans penis 475
glassy membrane of Bruch 433
glassy membrane, ovary 484
glaucoma 433, 444
glial cells 149
gliocytes, central 410
glomerular capsule 375, 383
glomerulus 375, 383
glomus 206
glottis 351
glucagon 311, 344
glucocorticoids 423
glycocalyx 47
glycogen 29
glycosaminoglycans 86
glycosyl 28
glysine 67
goblet cells 305
golgi complex 20, 28
golgi type II cells 158
gomphosis 100
goose bymps 260
granular endoplasmic reticulum 25
granulo-luteocytes 487
granulocyte 105
granulocytopoesis 108
granulomeres 106
gray matter 151
great alveolar cell 367
great epithelial cells 367
great stellate neurons 158
greater vestibular glands 503
ground substance 66, 86
growth hormone (STH) 412
guanine 36, 77
guanophores 77, 435, 437
gum, gingiva 279

Hair 256
hair cells, Type I (pear-shaped) 450
hair cells, barrel shaped, Type II 451
haploid 30, 457
haustra 315

heart 208
helicine arteries 478
helicotrema 454
hematocrit 103
hematoxylin 6, 8
hemes 239
hemidesmosome 46
hemocytoblast 108
hemocytopoesis 107
hemocytopoetic tissue 93
hemostatis 106
heparan sulfate proteoglycan 54, 59
heparin 66, 75, 106
hepatic duct 332
hepatic lacunae 320
hepatic lobules 329
hepatic veins 327
hepato-pancreatic ampulla 335
hepatocytes 320
heterochromatin 15, 31, 73
hilus, kidney 373
histamine 75, 106, 311
histiocytes 73
histology 1
holocrine secretion 179
homologous pairs 30
horizontal neuronal cells 439
hormones 409
hormones, digestive tract 308
human chorionic gonadotrop(h)in 497
hyaline cartilage 87
hyalomeres 106
hyaluronic acid 66, 86
hyaluronidase 66
hydrochloric acid 293
hydrocortisone 423
hydrostatic pressure 193
hydroxyapatite 90
hypercalcemic factor 418
hyperglycemic factor 344
hyperplasia 416
hypodermis 249
hypoglycemic factor 344
hyponychium 264
hypophyseal portal circulation 413
hypophysis cerebri 410
hypothalamo-hypophyseal tract 410
hypothalamus 410, 498

Implantation 484
incisures of myelin 165
inclusions 20
index of folding 545
indirect cell division 29
inferior meatus 349
infiltration 8
infinite vs finite 552
infundibular stalk 410
inner ear 447
insertion corpuscle 119
insulin 344
integument 237
interalveolar pores 365
interalveolar septa 365
intercalated discs 131
intercalated ducts 271, 342
intercellular bridges 240
intercellular canaliculi 293, 342
intercellular substance 54, 66
intercept 528
interlobar arteries 378
interlobular arteries 378

intermediate filaments 29
intermediate junction 46, 320
intermedin (MSH) 413
intermitotic period 34
internal elastic membrane 185
internal limiting layer of glial cells 438
internuncial neurons 145
interoceptors 142
interphase 29
interprismatic substance 279
intersection 528
interstitial cell stimulating hormone (ICSH) 412
interstitial endocrine cells (Leydig) 457
interstitial growth 87
interterritorial matrix 89
interventricular septum 212
intestinal glands 305
intracellular canaliculi 342
intraepithelial glands 177
intrapapillary capillary loops 246
intrinsic factor 295
involution, parathyroid 418
involution, thymus 228
iridial portion, retina 418
iridocorneal angular spaces 433
iris 435
islets of Langerhans 342
isotropic 548
isotropic (I) disk 123
isthmus, tubae uterinae 491

Joints 100
junction, esophageal-cardiac 289
junctional complex 46, 320
juxtaglomerular complex 380, 383
juxtamedullary nephrons 377

Karyotype 35, 37
keratan sulfate 86
keratin 64, 244
keratinization 64
keratofilaments 183
keratosulfate 66
kidney 373
kinetochore (centromere) 27

Labia majora pudendi 502
labia minora pudendi 502
lacrimal glands 443
lactation 504
lacteal 307
lacunae 86, 91, 264
lacunae, Howship's 91
lacunae, erosion 91
lamellae, inner circumferential 93
lamellae, interstitial 93
lamellae, outer circumferential 93
lamina basalis (densa)
lamina chroidocapillaris 433
lamina dentinalis 279
lamina fibroreticularis 54
lamina fusca 432
lamina lucida (rara) 54
lamina muscularis mucosae 169, 267
lamina propria intima 185
lamina propria mucosa 173, 267
lamina propria serosa 169
lamina propria subserosa 267
lamina suprachoroidae 432
lamina vitrea 433
laminae 547
laminae hepatis 320

laminin 54
lanugo hairs 259
large intestine 314
large veins 202
laryngeal cartilages 351
larynx 351
law, Snell's 511
lecithin 365
length per unit volume 536
length, absolute 537
lens 431, 434
lens fibers 434
leucocytes 77
leuteinizing hormone (LH) 412, 484, 495
ligaments 80
limen, nose 349
limiting membranes 167
limiting plate, liver 324
line, Z (telephragma) 124
lines, Langer's 248
lipochrome 77, 437
lipofuchsin pigment 140
lipophores 77, 437
lips 268, 279
liquor folliculi 482
liver 318
liver acini 329
liver sinusoids 321
lobes 183
lobes, kidney 373
lobule, lung 365
lobules 183
lobules, renal 376
lobules, thyroid 414
loop, nephron 375, 395
loose connective tissue 78
lumen 55
lumen, glomerular capsule 375
lymphocyte 105
lymphocytopoesis 114
lysine 67
lysosomes 20

Macrophages (fixed, free) 73
macrophagocytus nomadicus 73
macrophagocytus stabilis 73
macula adherens 43, 46, 131
macula communicans 47
macula densa 375, 383
macula lutea 439
maculae 447
maculae adherentes 240, 320
magenstrasse 297
magnification 511
magnification, useless 519
malleus 454
maltose 269
mamma (breast) 503
mammotrop(h)in 412
mast cells (tissue basophils) 75
master gland 410
mastoid air cells 447
matrix 86
meatus inferior 349
meatus, middle 349
meatus, superior 349
medistinum testis 457
medium sized arteries 194, 195
medulla, kidney 373
medulla, ovary 479
medulla, surarenal 423
medullary cavity 374

medullary rays 374
megacaryocytes 107
megacaryocytopoesis 112
meiosis 30, 457, 460
melanin 29, 77, 239
melanocyte stimulating hormone (MSH) 413
melanocytes 77, 239, 244, 431
melanosomes 77, 246
melatonin 428
membrana statoconiorum 451
membrane, Bowman's 433
membrane, Descemet's 433
membranous connective tissue 71
membranous ossification 94
menstrual cycle 473
merocrine secretion 50, 179
merocrine sweat glands 251
meromyosin 125
mesangium 384
mesenchymal cells 71
mesenchyme 78
mesosalpinx 493
mesothelium 55, 169, 267
mesovarium 479
messenger RNA (mRNA) 40
metachromatic 75
metaphase 38
meter 2
microbodies 21
microfibrils 183
microfilaments 29, 183
microglia 149
micrometer 1
micropinocytosis 49
micropinocytotic vesicles 119
microscope design 519
microscope, compound 513
microscope, darkfield 520
microscope, light 511
microscope, optical 511
microscopy, electron 523
microscopy, fluorescence 523
microscopy, phase 522
microscopy, quantitative 524
microtome 3
microtubules 29, 182, 183
microvilli 60, 300
mineralocorticoids 421
mitochondria 22, 117
mitosis 29, 38
mitotic apparatus 38
mitotic cycle 38
modiolus 454
monad 461
monocytes 73, 105
monocytopoesis 114
monosomic 39
mons pubis 503
mons veneris 503
morphometry 524, 525
mosaic 31
motilin 312
motor end-plate 127, 145
motor endings 145
mucigen 289
mucopolysaccharides 66
mucoserous glands 251
mucous 269
mucous acini 269
mucous connective tissue 78
mucous neck cells 290

mucous plug 501
mucus 269
multipolar nerve cells 140
multivesicular bodies 21
muralia 547
muralium 181
muscle 117
muscle contractility 117, 125, 127
muscle fascia 127
muscle spindle 144
muscularis mucosa 169
myelin 156
myelinated fiber 162
myeloblast 108
myeloid metaplasia 107
myenteric plexus (Auerbach's) 282
myocardium 128, 208, 213
myocyte, smooth 118
myoepithelial cells 118, 504
myofibers 183, 209, 211
myofibril 124
myofilament 125, 127, 183
myomere, (sarcomere) 126
myosatellite cells 128
myosin 60, 125
myxedema 416
myxomatous tissue 78

Nail fold, proximal 264
nail plate 264
nails 264
nanometer 1
nasal cavity 349
nasal pyramid, external 349
necropsy, liver 320
nephron 374
nerve 161
nerve cell body 137
nerve cells, bipolar 140
nerve cells, multipolar 140
nerve cells, primary sensory 143
nerve cells, secondary sensory 143
nerve cells, sensory 144
nerve cells, unipolar 142
nerve fascicle 161
nerve fiber 137
nervous plexuses 307
nervous tissue 135
networks 545
neuroendocrine cells, lung 371
neurofiber node (Ranvier) 162
neurofibrils 137, 183
neurofilaments 183
neuroglial cells 149
neurohypophysis 410
neurolemmal sheath 162
neuromuscular junction 145
neuromuscular synapse 127, 145
neuron 135
neuropil 147
neurosecretory substance 405, 411
neurotrasmitter 145
neutrophilic granulocyte 106
nexus 47
nissl substance 137
nitrogenous base 35
node, neurofiber (Ranvier) 162
non-striated muscle tissue 118
non-myelinated nerve fibers 162
norepinephrine cell 423
nose 349

nostrils 349
nucleokinesis 29
nucleolar fusion 20
nucleolar organizer regions 38
nucleolonema 20
nucleolus 15
nucleotide 36
nucleus 15
nucleus solitarius 274
nullisomic 40

Objective, achromatic 516
objective, apochromatic 516
oblique section 6
obstructive jaundice 323
ocular 514
odontoblasts 275
oil immersion 518
olfactory cilia 351
olfactory glands 349
olfactory mucosa 349
olfactory neurosensory cells 349
oligodendroglia 149
optical density 511
optic disc 441
optic papilla 441
ora serrata 441
oral cavity 268
oral mucous membrane 268
organelles 20
organs, composition 181
os membranaceum lamellosum 97
os membranaceum reticulofibrosum 95
osmium tetroxide 9
osmotic pressure 193
osseous labyrinth 454
osseous tissue 90
osseous trabeculae 100
ossification 94
ossification center, primary 99
ossification center, secondary 100
osteoblasts 90
osteoclasts 91
osteocytes 91
osteogenesis 94
osteoid 90
osteoprogenitor cells 90
ostium, uterine tube 491
outer ear 419
oval window 447
ovary 457, 479
ovocyte, primary I 479
ovocyte, secondary II 484
ovogenesis 484
ovogonium 479
ovotids 484
ovulation 484, 495
ovules, Naboth's 501
ovum 457, 484
oxyhemoglobin 104
oxyphil endocrine cells 418
oxytocin 411

Palate 279
palpebrae 448
palpebral conjunctiva 442
pancreas 342
pancreatic islets 342
pancreatitis 352
pancreozym 313, 342
paniculus adiposus 239, 249
papillary ducts, kidney 397

papillary layer, dermis 246
papillary muscle 212
parafollicular endocrine cells 416
paraganglia 425
parallel section 5
parametrium 493
paranasal sinuses 349
paraportal ductules 332
parathyroid gland 417
parathyroid hormone 412
paraventricular nuclei 411
paraxial section 5
parenchyma 79, 181
parietal exocrine cells 290, 293
parotid duct 271
parotid gland 271
pars cutanea, lips 268
pars distalis, hypophysis 410
pars filamentosa 20
pars granulosa 20
pars intermedia, hypophysis 410, 413
pars intermedia, lips 268
pars interna, tubae uterina 491
pars mucosa, lips 268
particles, number per unit volume 542
particles, size distribution of 537
patches, Peyer's 308
pectinate ligaments 434
penis 473
pepsinogen 290
pericardial cavity 208
pericardial fluid 209
pericarditis 209
pericardium 209
perichondrium 86
pericyte 187
peridontal membrane 275
perilymph 447
perimetrium, uterus 503
perimysium 127
perineurium 161
perinuclear cisterna 20
periosteum 92
peripheral nerve fiber 162
perisinusoidal space (Disse) 321
peritendinium 128
peritoneum, visceral 171
perivascular mesangial cells 383, 384
peroxisomes 21
perpendicular section 6
phagocytosis 20, 49
phagosome 20
pharynx 281
pia mater 157, 159
pigment cells 77
pigments 29
pilosebaceous organ 259
pineal body 427
pinocytosis 20, 49, 188
pinocytotic vesicle 20, 60
piriform cells 158
pitch fibers 351
pituicytes 410
pituitary gland 410
placenta 497
plasma cells 75, 289
plasma, blood 103
plasmalemma 15, 117, 140
plasmodium 47
platelet formation 114
platelets, blood 106
pleura, parietal 171

pleura, visceral 171
plexus 182, 207
plexuses 545
plicae circulares 300, 307
plicae palmatae 498
plicae semilunares 315
plicae tuberinae 491
pluripotential stem cell 108
pneumocyte, Type I 365, 367
pneumocyte, Type II 367
podocytes 385
podocytes membrane 385
polar bouies 484
polar cushion cells (Polkissen) 383
polarization 520
polarizer 521
polynucleotide 35
polypeptide hormones 311
polyploid 39
polyribosomes 25
portal canal 326
portal lobules 329
portal triad 326
posterior epithelium, cornea 433
posterior limiting lamina (membrane) 433
posterior lobe, hypophysis 410
postmitotic interval 34
postsynaptic membrane 145
potio prevaginalis, cervix 498
precartilage 87
premitotic interval 36
prepuce 502
presbyopia 435
presynaptic membrane 145
primary lysosomes 20
principal endocrine cells 417
principal exocrine cells 290
prism, Nicol 521
proerythroblast 108
profile 528
progesterone 423, 497, 501
prolactin (LTH) 412
proline 67
pronucei; male, female 484
prophase 38
proprioceptors 143
prostate gland 468
protein synthesis 40
proteoglycans 86
protoplasmic astrocytes 149
proximal convoluted tubules 394
pseudostratified epithelium 55, 56
ptyalin 269
ptyocytosis 188
pubes 503
pudendum feminium 502
pupil, eye 435, 437
purines 35
pyloric glands 295
pyloric sphincter 297
pyloris 297
pyrimidines 36

Radial glial cells 438
receptors, nerve 144
reconstruction from serial sections 552
rectum 315
reduction 513
reduction division 460
refraction 511
refractory period, nerve 145
regeneration, liver 331

releasing factors 413, 498
renal columns 373
renal corpuscle 374
renal lobes 373
renal papilla 374, 399
renal pelvis 373
renal pyramids 373
renal sinus 373
reproductive system 455
residual body 21
resolution 517
resolving power 517
respiratory apparatus 347
respiratory bronchioles 364
respiratory epithelial cells 365
respiratory epithelium 349
respiratory mechanics 369
respiratory mucosa 349
respiratory spaces 347
respiratory system 349
rete ridges, pegs 246
rete testis 465
reticular connective tissue 78
reticular fibers 70
retina 431, 437
retina, ciliary portion 438
retinacula cutis 79, 239, 249
retinal layers 438
retroperitoneal 169
rhodopsin 439
ribonucleic acid 8
ribosomal RNA (rRNA) 40
ribosomes 25
rods, retinal 438
rough endoplasmic reticulum 25
rouleaux 104
round window 454

Saccule 449
salivary glands 268, 271
salpinx 491
sarcolemma 117
sarcomere, (myomere) 126
sarcoplasm 117
sarcosome 117
satellite cells 160
scala vestibuli 454
sclera 431
scleral venous sinus 434
scrotum 457, 479
sebaceous glands 255
sebum 255
secondary lysosome 20, 21
secretin 342
section 524, 528
sections, tissue 3, 4
segregation, random 461
sella turcica 410
semen 457
semicircular canals 451
semiluna serosa 269, 273
seminal vesicles 445
seminiferous tubules 457
sensory ganglia 160
sensory hair cells 449
septa 181
septal cell 367
septum membranaceum 208
seromucous glands 269
serotonin 75, 311
serous acini 269
serous demilunes 273

serous glands (von Ebner) 274
serous membrane 167
serum, blood 104
sex chromosome (Barr body) 30, 106
sex hormones, suprarenal 423
shape, identification of 542
sheath, Neumann's 279
sheets, branched 548
shunt, AVA 205
simple columnar epithelium 57
simple cuboidal epithelium 57
simple epithelia 56
simple squamous epithelium 57
sinuatrial node 211, 212
sinuses 188
sinusoidal capillaires 421
sinusoids 188, 192
skeletal muscle tissue 123
skeletal myocytes, branched 274
skeletal tissue 81
skin (cutis) 239
slab 528
slice 528
small intestine 299
small sized arteries 194
smegma 479
smooth endoplasmic reticulum 25
smooth muscle cells 118
solids of various shapes 543
somatic cells 30
somatostatin 311
somatotrop(h)in (STH) 412
space of Disse 321
spaces of Fontana 434
spermatid 461
spermatocytogenesis 457
spermatogenesis 457
spermatogonia 457
spermatozoa 457
spermiogenesis 457
sphenoid bone 410
sphincter of Oddi 335
sphincter of the ampulla 335
sphincter pupillae, muscle 437
spinal ganglia 160
spiral organ (Corti) 454
splenic sinuses 192
spongiosa, primary 95
spongiosa, secondary 100
spongy bone 92
squamous cell carcinoma 188
squamous epithelium 56
stain, Goldner's 8
stain, Mallory's 8
staining qualities 6
stains 8
stapes 454
starch 269
statoacoustic organ 447
statoconia 451
stellate macrophage cells (Kupffer) 320
stem cell, blood 108
stereocilia 62
stereology 524, 525
stereology formulas 553
stereometry 525
stereoscopic vision 441
stigma, ovary 484
stomach 289
stomadeum 410
straight tubules, testis 465
stratified columnar epithelium 57

stratified cuboidal epithelium 57
stratified epithelia 56
stratified squamous epithelium 57, 63
stratum basale endometrii 494
stratum basale, skin 239
stratum compactum endometrii 494
stratum corneum, skin 64, 239, 243
stratum disjunctum, skin 244
stratum fibrosum, skin 249
stratum functionale endometrii 494
stratum granulosum, ovarian follicle 482
stratum granulosum, skin 243
stratum lucidum, skin 243
stratum papillare, skin 246
stratum pigmentosum, eye 439
stratum spongiosum endometrii 494
stratum subendotheliale 213
striated border 60, 300
striated salivary ducts 271
striated skeletal myocytes 123
stroma 79, 181
subarachnoid space 159
subdural space 159
sublimate 7
sublingual glands 271, 273
sublobular veins, liver 327
submandibular glands 271, 273
submucous plexus (Meissner's) 282
subpodocytic lacunae 388
substantia propria, choroid 432
substantia propria, sclera 431
sudoriferous glands 251
sulcus sclerae 433
superficial fascia 249
supraoptic nuclei 411
suprarenal capillaries 421
suprarenal glands 418
suprarenal stroma 423
surface area, absolute 535
surface area, per volume 534
surface increase 172
surfactant 365
suspensory ligament, lens 434
sustentacular cells (Sertoli) 457
sustentacular cells, olfactory 349
sustentacular epithelial cells, ear 449
sutures 100
sweat glands 251
symphysis 100
symplasm 47
synapse 145
synapse en passage 147
synapsis 38, 458
synaptic clefts 145, 147
synaptic vesicles 140, 145, 147
synchondroses 100
syncytiotrophoblast 494
syncytium 47
syndesmoses 100
syndrome, Down's 39
syngamy 484
synovial joints 100
synovial membrane 101
synthetic phase (S phase) 34
systole (198)

Tangential section 5
target organs 410
tarsal glands, Meibom 442
tarsal plate 442

tartar (dental calculus) 279
taste buds 273
tectorial membrane 454
teeth 275
tela chroidea 150
tela subcutanea 237, 239, 249
tela subendocardialis 212
tela subepicardiaca 209
tela submucosa 267
tela subserosa 169, 267
telephragma (Z line) 124
telodendria 140
telogen 264
telophase 38
tendons 80
tenia coli 315
tentorium cerebelli 160
terminal bars 62, 259
terminal bronchioles 357
terminal cisternae 125
terminal web 60
territorial matrix 89
testis 457
tetrad 460, 461
tetraidothyronine 414, 416
tetrasomic 39
theca interna, externa 481
theory of hearing, Helmholtz 453
theory, dualistic 108
theory, monophyletic 108
theory, polyphyletic 108
theory; Neura, Haekel's 445
thin limb, nephron 375, 376
thrombocytes 106
thromboplastin 106
thymic corpuscles, Hassall's 227
thymine 36
thyrocalcitonin 414
thyroglobulin 415
thyroid follicle 414, 418
thyroid gland 414
thyroid stimulating hormone (TSH) 412
thyrotrop(h)ic endocrine cells 412
thyroxine. T4 414, 416
tight junctions 320
tissue basophils (mast cells) 75
tongue 274
tonofibrils 240
tonofilaments 29, 240
tonsils 275, 281
trabeculae 92, 181
trabeculae carnae 212
trabecular bone 92
trace 528
trachea 354
trachealis muscle 354
transendothelial transport 188
transfer RNA (tRNA) 40
transformation of cells 50
transitional epithelium 57
transudate 49
transverse tubules 125
triad 125
trigona fibrosa 208
triiodothyronine, T3 414
trisomy 21 39
triatiated thymidine 36
tropocollagen 67
tropomyosin 125
troponin 125
tubular pole, glomerulus 375
tubulin 29

tunica adventitia 169
tunica albuginea 182, 457
tunica dartos 479
tunica externa 185
tunica fibrosa bulbi 431
tunica interna (sensoria) bulbi 431
tunica media 185
tunica mucosa 169, 267
tunica muscularis 267
tunica serosa 267
tunica vaginalis 457, 479
tunica vasculosa bulbi 437
tympanic cavity 419
tympanic membrane 447
tyrosinase 77

Ultramicroscope 520
ultrathin 3
undifferentiated islet cells 345
unipolar nerve cells 142
unit membrane 15
urethra 406, 469
urethral glands 469
urinary bladder 401
urinary organs 373
uterine tube 491
utricle 449, 451
uvea 431
uvula 281

Vagina 501
vallate papillae 274
valves of Heister 342
vas sinusoideum 320
vasa afferentia 378
vasa efferentia 378
vasa recta 399
vasa varosum 203
vascular pole, glomerulus 375
vasculature, bone marrow 111
vasoactive intestinal peptide 311, 312
vasopressin 411
veins 185, 205
vena comitans 200
venous sinuses 160
venous sinuses, nose 349
ventricular folds 351
venules 201
vermiform appendix 314
vesicles, micropinocytotic 119
vesicles, superficial 119, 188
vestibule, vagina 502
vibrissae 349
villi 307, 300
visceral leaf, glomerulus 385
visceral peritoneum 305
visual purple 439
vitamin B12 295
vitreous body 435
vocal cords, false 351
vocal cords, true 351
vocal ligament 351
vocalis muscle 351
volume determination, direct 531
volume determination, serial planimitry 531
volume determination, serial planimitry 531
volume ratios 532
vulva 502

Wax, ear 447
white matter 151
witch's milk 504

Xanthophores 437

Zona externa, kidney 374, 379
zona fasciculata 422
zona glomerulosa 421
zona interna, kidney 374, 379
zona pellucida 481

zones, hepatic lobule 329
zonula adherens 46
zonula occludens 190
zygote 30, 451, 484
zymogenic cells 290
zymogenic granules 269, 342